Theoretical and Mathematical Physics

The series founded in 1975 and formerly (until 2005) entitled *Texts and Monographs in Physics* (TMP) publishes high-level monographs in theoretical and mathematical physics. The change of title to *Theoretical and Mathematical Physics* (TMP) signals that the series is a suitable publication platform for both the mathematical and the theoretical physicist. The wider scope of the series is reflected by the composition of the editorial board, comprising both physicists and mathematicians.

The books, written in a didactic style and containing a certain amount of elementary background material, bridge the gap between advanced textbooks and research monographs. They can thus serve as basis for advanced studies, not only for lectures and seminars at graduate level, but also for scientists entering a field of research.

Malte Henkel • Haye Hinrichsen • Sven Lübeck

Non-Equilibrium Phase Transitions

Volume I: Absorbing Phase Transitions

Springer

Malte Henkel
Laboratoire de Physique des Matériaux (LPM)
Nancy Université
CNRS
Nancy
France

Haye Hinrichsen
Fakultät für Physik und Astronomie
Universität Würzburg
Germany

Sven Lübeck
Theoretische Physik
Universität Duisburg-Essen
Germany

ISSN 1864-5879
ISBN 978-94-007-9660-7
ISBN 978-1-4020-8765-3 (eBook)

Published by Springer Science+Business Media B.V.
P.O. Box 17, 3300 AA Dordrecht, The Netherlands
In association with
Canopus Academic Publishing Limited,
27 Queen Square, Bristol BS1 4ND, UK

www.springer.com and www.canopusbooks.com

Malte Henkel, Haye Hinrichsen, Sven Lübeck

Non-equilibrium phase transitions

Volume 1 – Absorbing phase transitions

September 16, 2008

Springer

Preface

"The career structure and funding of the universities [...] currently strongly discourages academics and faculties from putting any investment into teaching – there are no career or financial rewards in it. This is a great pity, because [...] it is the need to engage in dialogue, and to make things logical and clear, that is the primary defence against obscurantism and abstraction."

B. Ward-Perkins, *The fall of Rome*, Oxford (2005)

This is the first volume of a planned two-volume treatise on non-equilibrium phase transitions. While such a topic might sound rather special and academic, non-equilibrium critical phenomena occur in much wider contexts than their equilibrium counterparts, and without having to fine-tune thermodynamic variables to their 'critical' values in each case. As a matter of fact, most systems in Nature are out of equilibrium. Given that the theme of non-equilibrium phase transitions of second order is wide enough to amount essentially to a treatment of almost all theoretical aspects of non-equilibrium many-body physics, a selection of topics is required to keep such a project within a manageable length. Therefore, Vol. 1 discusses a particular kind of non-equilibrium phase transitions, namely those between an active, fluctuating state and absorbing states. Volume 2 (to be written by one of us (MH) with M. Pleimling) will be devoted to ageing phenomena. The book is intended for readers who are familiar with general principles of statistical mechanics, at the level of a standard university course, and who have had some previous exposure to equilibrium critical phenomena and the renormalisation group. We aim at a presentation as self-contained and as accessible for the non-expert as possible.

This work was conceived and begun by two of us (MH & SL) in the summer of 2005. A little later, during the participation of two of us (MH & HH) in the workshop *Principles of Dynamics of Non-Equilibrium Systems* at the Newton Institute in Cambridge in spring 2006, the formal decision to write was made and a large part of the work was produced. We thank the Newton Institute and the organisers M. Evans, C. Godrèche, S. Franz and D.

Mukamel for the stimulating atmosphere which provided substantial encouragement with this project. Later, MH enjoyed the warm hospitality of and thanks cordially the Dipartimento di Fisica of the Università di Firenze and INFN - Sezione di Firenze, of the Centro de Física Teórica e Computacional (CFTC) at the Complexo Interdisciplinar of the Universidade de Lisboa, of the Instituut voor Theoretische Fysica at the Katholieke Universiteit Leuven and the Department of Theoretical Physics at the University of Saarbrücken for their support, which permitted him to make progress.

Our views on non-equilibrium physics have been formed by many friends and colleagues, sometimes through joint authorship on a paper, sometimes through intensive discussions and sometimes by some piece of advice. We are grateful to all of them, whether or not their contributions can be gleaned from the pages of this volume. It is a pleasure to gratefully thank P. Alnot, F. Baumann, B. Berche, G. Bonhomme, A. Capelli, E. Carlon, C. Chatelain, S.R. Dahmen, D. Dhar, E. Domany, S.B. Dutta, T. Enß, G. Foltin, P. Fulde, A. Gambassi, F. Ginelli, C. Godrèche, P. Grassberger, F. Hucht, W. Janke, H.-K. Janssen, D. Karevski, W. Kinzel, J.K. Krüger, R. Livi, J.-M. Luck, M. Lücke, C. Maes, S.S. Manna, J.F.F. Mendes, R.J. Meyer, G. Ódor, H. Park, M. Paeßens, I. Peschel, A. Picone, I.R. Pimentel, M. Pleimling, A. Politi, V.B. Priezzhev, J. Ramasco, J. Richert, V. Rittenberg, R. Sanctuary, M.A.P. Santos, W. Selke, C.A. da Silva Santos, U. Schollwöck, R. Schott, G.M. Schütz, S. Stoimenov, U.C. Täuber, L. Turban, E. Vincent, D.E. Wolf, K.D. Usadel, J. Unterberger, C. Wagner, R.D. Willmann and J.-B. Zuber.

As everyone knows who is trying to compile a list of works from the literature, the exploding quantity of publications makes the production of a complete bibliography a task beyond human capabilities. The references we included are those which we needed in writing this volume and we sincerely apologise to any authors whose important contribution we might not have taken into account or might have covered inadequately.

We thank F. Hucht and the Institute of Theoretical Physics of the University of Duisburg for friendly support. The project has been overseen with diligence and patience by T. Spicer and we thank him and C. Caron for their help in bringing the first part of this work to completion.

Nancy, Würzburg, Gerlingen,
April 2008

Malte Henkel
Haye Hinrichsen
Sven Lübeck

Contents

Chapter 1
Introduction

Curiously, and interestingly, there is an almost dialectical relationship between equilibrium and non-equilibrium systems. Almost all systems occurring in Nature are open systems coupled to external reservoirs such that the exchange of energy, particles, or other conserved quantities between the system and the reservoirs leads to currents through the system and which may drive the temporal evolution of the system. Such currents may be realised differently, for example as an electric current, or as a flow of particles, or heat conduction, or even as spin transport, to name just a few. On the microscopic level, such non-equilibrium effects manifest themselves as a breaking of **detailed balance**, which means that between pairs of microstates there is in general a non-vanishing flow of probability. Then one may also say that an intrinsic microscopic irreversibility will produce a macroscopic non-equilibrium behaviour.

In *equilibrium* systems, one considers situations where such currents have decayed away to zero such that internal probability currents are absent and the thermodynamic state variables of the system are fixed by the corresponding properties of the reservoirs. For the theoretical description of such systems, it has been understood for more than a century, mainly through the efforts of Boltzmann and Gibbs, how to formulate a general statistical description of equilibrium systems in terms of the probabilities of the micro-states and how to derive from this the thermodynamic behaviour at thermal equilibrium. Unfortunately, such a canonical and generally valid formalism does not yet exist for general non-equilibrium systems, in spite of recent efforts. In practise, for each non-equilibrium system afresh, first one has to find the probability distribution of the micro-states or some equivalent information. A common way of doing this is to write down either a Fokker-Planck equation or, more generally, a master equation for the probability distribution or alternatively a stochastic Langevin equation for some averages of physical observables. Such equations are in general no longer fully microscopic, but should rather be seen as some kind of coarse-grained description on a mesoscopic level, large enough with respect to the truly microscopic level which involves the details of the individual motion of atoms and molecules and yet still small compared to the scales of macroscopic observation. Hence the discussion of non-equilibrium behaviour will be almost always formulated in terms of *phenomenological models*. Of course it must then be understood to what extent such an approach is justified and useful and this will be one of the main themes of this book.

Since in contrast to equilibrium systems, for non-equilibrium systems *time* is an essential degree of freedom, one of the first questions is under what conditions a relaxation towards an equilibrium steady-state may occur. It turns out that, again, the condition of detailed balance in combination with some ergodicity requirements is sufficient. On the other hand, for choices of the dynamics which no longer respect detailed balance, a much richer behaviour, both at the steady-states as well as for the time-dependent relaxation processes, is to be expected. Of particular interest is the frequently found situation where a complex system is made up of many strongly interacting parts and where a *collective behaviour* over large scales results [422, 576]. At equilibrium, this occurs if the system undergoes a second-order phase transition which requires the fine-tuning of its thermodynamic variables (e.g. temperature fixed to the 'critical value' T_c) in order to be realised. In many non-equilibrium situations, however, this kind of constraint can be considerably relaxed.

At equilibrium, second-order phase transitions are the consequence of dynamically created long-range correlations, even if the original microscopic interactions are short-ranged.[1] Probably the best-known example of an equilibrium second-order phase transition is related to the **Ising model** which is used to describe liquid-gas transitions or the order-disorder transition in uniaxial magnets. In simple systems with a conventional phase transition, such as those described by the Ising model, the long-range correlations close to the transition are known to be **universal**, that is independent of microscopic 'details' of the model. This universality was originally introduced experimentally from observations that the critical exponents (and certain amplitude ratios) which describe the divergence of observables, such as the magnetic susceptibility, in the immediate vicinity of the critical point, turned out to have essentially the same values for a large range of *a priori* very different systems. By now, conceptually understood from the renormalisation-group description of critical phenomena, the concept of universality has become one

[1] The first physical example of a second-order equilibrium phase transition (in three dimensions) seems to have been discovered by Cagniard de la Tour in 1822 who observed **critical opalescence** in mixtures of alcohol and water, which become opaque close to the critical point when the size of the spatial correlations becomes comparable to the wavelength of visible light. The terminology of a second-order critical point as it is used today goes back to Andrews (1869). On the other hand, the first equilibrium phase-transition in *two* dimensions was found only in 1970 by Thomy and Duval for Xenon adsorbed on graphite.

of the pillars of statistical physics. In particular, second-order phase transitions are now cast into distinct **universality classes** such that the critical behaviour of different systems (experimental or theoretical), if they only have the same symmetry properties, should have the same critical exponents and the same scaling functions. For instance, generically the values of the critical exponents are expected to depend only on the spatial dimensionality of the system, the number of components of the order-parameter and the global symmetry of the interactions (any other property of the system being considered as an 'irrelevant detail' in the sense of the renormalisation group). One may therefore use the critical exponents to label the distinct universality classes. One of the remaining open questions is whether there exists a classification of possible universality classes. A breakthrough in this direction was the discovery of **conformal invariance** and the relevance of the Virasoro algebra (previously discovered in a string-theory context) for equilibrium critical phenomena in two dimensions, which not only complements many of the more qualitative insights of the renormalisation group by precise quantitative information on the values of critical exponents, any correlation function at criticality or on the scaling functions but also provides a (partial) classification of possible kinds of second-order transitions in two dimensions through a list of the modular-invariant partition functions of the so-called minimal models.

Much of what is known about equilibrium phase-transitions can be extended to the non-equilibrium case as well. Even if new types of continuous phase transitions, which can only be observed under non-equilibrium conditions, are found, the central concept of *universality*, which proved to be so useful at equilibrium, can as well be applied to non-equilibrium systems. The present two-volume work is intended as an introduction to and as a survey of the field of non-equilibrium phase transitions. Specifically, in volume 1 we focus on situations where one may consider the microscopic dynamics to be fundamentally irreversible such that detailed balance is broken and the steady-states cannot be equilibrium states. In this situation, one has second-order phase transitions from fluctuating (ordered) states into so-called **absorbing states**. By definition, a system can never leave an absorbing state once it has reached it. Analogously to the equilibrium phase-transitions, such **absorbing phase transitions** exhibit universal features, determined by symmetry properties and conservation laws, which allows one to cast these into universality classes. So far, only a few classes of non-equilibrium phase transitions into absorbing states are known. Therefore, an ambitious task would be to specify all possible universality classes of second-order absorbing non-equilibrium phase transitions.

The most important class of absorbing phase transitions is **directed percolation** (DP). Directed percolation was originally introduced, as a model for directed random connectivity, in 1957 by Broadbent and Hammersley [87]. It exhibits a phase transition that is characterised by a well-defined set of universal critical exponents. The same type of transition occurs in models for the spreading of an infectious disease, certain catalytic reactions, percolation in porous media, intermittent turbulence and even in the context of certain hadronic interactions. Since all these models share the same symmetry properties and follow essentially the same reaction-diffusion scheme, they all belong generically to a single universality class, called the *directed percolation universality class*, irrespective of the microscopic details of their dynamical rules. In view of its robustness, the directed percolation universality class is widely considered as the analogue of the Ising model for the field of absorbing phase transitions. In contrast to equilibrium critical phenomena, where a lot of information on critical behaviour has come from the *integrability* of many two-dimensional classical models (or their one-dimensional quantum analogues) [50], only relatively few non-equilibrium models have turned out to be integrable and directed percolation does not appear to belong to them. Although directed percolation is very easily defined, its critical behaviour is highly non-trivial. This is probably one of the reasons why directed percolation continues to fascinate.

The attentive reader will have noticed that we have been referring to 'models' only. Indeed, it has been (and still continues to be) one of the most challenging questions in the field why the apparently so robust universality class of directed percolation has been so difficult to encounter reliably in experimental studies on non-equilibrium phase transitions. This might be about to change with important recent progress, coming from studies on intermittent turbulence, which we shall describe in Chap. 3.

The standard reference for directed percolation was, for more than two decades, a review by Kinzel [390]. Meanwhile, enormous progress in our understanding of non-equilibrium phase transitions was made, new numerical and analytical methods have been developed and new universality classes of absorbing phase transitions have been identified. Still, many open problems remain, see for example the list compiled by Grassberger [243] (Sect. 5.13). Similarly, the book by Marro and Dickman [462] devoted several chapters to non-equilibrium phase transitions into absorbing states. Three relatively recent reviews [287, 496, 433] described absorbing phase transitions from different points of view, two of them having been written by two of us (HH and SL). In particular with studies of systems in reduced dimensionality, as is currently becoming increasingly fashionable in connexion with expected nano-scale applications, the consideration of *integrable* non-equilibrium systems provides different kinds of insight, see [13, 565, 271]. Field-theory methods will be covered in a forthcoming book by Täuber, see [593, 592] for shorter reviews. For phase-transitions in systems with external currents, see either [462] or the review [557]. Similarly, phase transitions related to Langevin equations with so-called multiplicative noise will not be considered in this book because such systems do not have an absorbing state in a strict sense and we refer to the review [232]. A recent workshop at the Newton Instute at Cambridge discussed many advanced topics on the dynamics of non-equilibrium critical phenomena.[2] The exciting new developments on the foundations of non-equilibrium statistical mechanics *near to* equilibrium cannot yet be applied to the topics discussed here, see [453].

In Vol. 2, written by one of us (MH) together with M. Pleimling, a different kind of non-equilibrium criticality will be taken up, which arises when a system is rapidly brought out of an equilibrium initial state by suddenly changing the value of a thermodynamic control parameter to 'quench' the system into the ordered phase (with at least two equivalent stationary states) or onto a critical point of the stationary state. The resulting **ageing** phenomena have been empirically studied since prehistoric times (for example in metallurgical materials processing) but a systematic study of its scaling and dynamical symmetry properties was only started in 1978 with the book by Struik [585].

The purpose of this volume is to give an overview of the field of absorbing phase transitions and to present the present knowledge coherently and in a self-contained way. By emphasising phenomenological scaling descriptions, we hope that this volume will be useful as an accessible and comprehensive introduction for beginners, as well as a reference for researchers. In Chap. 2 we recall basic notions, especially about scaling, the renormalisation group and universality, from the field of equilibrium critical phenomena. Besides establishing a consistent notation, another purpose of this chapter is to prepare the study of non-equilibrium phase-transitions by providing the background for establishing later analogies with the already well-understood equilibrium case. Then, in Chap. 3 we introduce the model of directed percolation, arguably representing the most important universality class of non-equilibrium absorbing phase transitions. A variety of lattice realisations of the DP universality class is presented and we also discuss in detail various attempts to find experimental realisations of this universality class. The phenomenological scaling theory for absorbing phase transitions is discussed in detail in Chap. 4, taking for the first time the metric factors systematically into consideration. Using the model of directed percolation as a scaffold, we discuss a large variety of generally applicable methods for the analysis of theoretical models. Having thus developed a general scaling picture, we then turn to absorbing critical phenomena different from directed percolation in Chap. 5 and present a survey of recent results.

A lot of background and detailed numerical information, usually being considered folklore knowledge but rather widely scattered in the literature, is collected in a large set of tables gathered in this volume (see the list of tables to find them rapidly) and in the appendices for, hopefully, easy reference. Problem sections with exercises of quite variable degrees of difficulty are provided at the end of each chapter; their solutions are outlined in a special

[2] See `http://www.iop.org/EJ/journal/-page=extra.focus4/jstat` for the proceedings of the workshop, in the form of a series of reviews.

section at the end of this volume. These exercises are essentially analytical, since we consider it natural that readers might wish to write computer programmes simulating one or the other of the models discussed by themselves, even without our explicit request.

Progress in physics depends crucially on a close interaction of theory and experiment. However, so far absorbing phase transitions were primarily the focus of theoretical physicists while the subject started to attract the attention of experimentalists only recently. This book – written by theoretical physicists – reflects this situation as it discusses experimental applications in a rather limited way. Still, we hope that the phenomenology presented in detail will motivate readers to go on to more specialised techniques for theoretical analysis or else to receive some inspiration on how to conduct experiments in order to put the concepts developed here to a test. Physics is a natural science, after all. And why should mathematicians not find some inspiration from the model-specific results presented here? At times, it might be useful to recall a sentence of Lichtenberg (1779):

> *"Keine Untersuchung muß für zu schwer gehalten werden und keine Sache für zu sehr ausgemacht."*
> G.C. Lichtenberg, *Aphorismen*, Insel (Frankfurt, 1976)

One might add the more contemporary voice of Bottéro (1982) [74]:

> *"Yes, the university of sciences is useless; for profit, yes, philosophy is useless, ... history, oriental studies and assyriology are useless, entirely useless. That is why we hold them in such high esteem!"*
> J. Bottéro, *Apologie d'une science inutile*, Akkadica **30**, 12 (1982)

Problems

1. A simple case study of an absorbing phase transition may be formulated in terms of a population of $n(t)$ individuals with a reproduction rate λ and a death rate 1.

Consider the balance between the term increasing and reducing the probability $P_n(t)$ of having exactly n individuals at time t and write down the **master equation** for $P_n(t)$. Use the solution to analyse, in dependence of λ, the indefinite survival of the population. Study the influence of the initial conditions by comparing the cases (i) of a single individual and (ii) an initial Poisson distribution $P_n(0) = \frac{\mu^n}{n!}e^{-\mu}$.

2. Derive for the population model of the previous exercise 1 the mean particle-number $\langle n\rangle(t)$ and its variance $\sigma^2(t) = \langle n^2\rangle(t) - \langle n\rangle^2(t)$. How do you interpret your results ?

Chapter 2
Survey of Equilibrium Critical Phenomena

Equilibrium and non-equilibrium critical phenomena are similar in many respects. Therefore, this book starts with a brief survey of some basic concepts of equilibrium critical phenomena, providing this background in a self-consistent way and establishing basic notations. This chapter may be used to recall the main features of modern theories of phase transitions for beginners who are not entirely familiar with the concepts of scaling and universality.

Although this introduction is kept as general as possible, for the sake of concreteness we shall use the language of ferromagnetic phase transitions. Following the historical perspective, we discuss the concepts of scaling and universality. In particular, critical exponents, generalised homogeneous functions, scaling forms and universal amplitude combinations are introduced. The basis for a deeper understanding of scaling and universality is provided by Wilson's renormalisation group theory [629, 630], which is a topic on its own and not presented in this book. Instead, we focus on its implications on universal scaling and illustrate the main results, e.g. by sketching how renormalisation-group theory allows one to identify the relevant system parameters which determine the universality class. The renormalisation group also provides tools for computing critical exponents as well as universal scaling functions and explains the existence of an upper critical dimension. As a reference and preparation for the second volume of this book we also comment on the extension of scale-invariance in equilibrium systems to conformal invariance and recall the fluctuation-dissipation theorem in the context of relaxation phenomena. For a rigorous substantiation of scaling, universality and conformal invariance the interested reader is referred to established textbooks (e.g. [520, 525, 640, 122, 168, 270]) and review articles [632, 219, 621, 120, 112]. The fluctuation-dissipation theorem is discussed in various textbooks, see e.g. [144, 122].

2.1 Phase Transitions in Equilibrium Systems

2.1.1 Notations

Studying continuous phase transitions, one is often interested in the asymptotic behaviour of certain quantities. In this book, we shall follow the standard notation used by most authors, i.e., the symbols $\propto$, O, $\sim$, and $\simeq$ denote that two functions $f(x)$ and $g(x)$ are[1]

$$f(x) \propto g(x) \quad \Longleftrightarrow \frac{f(x)}{g(x)} = c\,, \quad \forall\, x \qquad \textbf{(proportional)}\,,$$

$$f(x) = g(x) + \mathrm{O}(x^n) \Longleftrightarrow \left| \frac{f(x) - g(x)}{x^n} \right| < c\,, \forall\, x > x_0 \qquad \textbf{(order of)},$$

$$f(x) \sim g(x) \quad \Longleftrightarrow \lim_{x \to x_c} \frac{f(x)}{g(x)} = c \quad \textbf{(asymptotically proportional)},$$

$$f(x) \simeq g(x) \quad \Longleftrightarrow \lim_{x \to x_c} \frac{f(x)}{g(x)} = 1 \qquad \textbf{(asymptotically equal)}\,.$$

In the following, the mathematical limit $x \to x_c$ corresponds to the physical situation that a phase transition is approached (usually $x \to 0$ or $x \to \infty$).

For the specification of numerical estimates and their error bars we use the standard bracket notation $x(y)$, where y denotes the expected statistical error in the last digit. For example, the estimate $0.2765(3) = 0.2765 \pm 0.0003$ means that the true value is expected to be between 0.2762 and 0.2768.

For easy reference, frequently used symbols and abbreviations are listed on pages 343 and 344.

2.1.2 Phase Transitions in Ferromagnetic Systems

Phase transitions in equilibrium systems are characterised by singularities in the free energy and its derivatives [503, 639, 417]. This singularity causes a discontinuous behaviour of various physical quantities when the transition point is approached. Phenomenologically, the phase transition is described by an **order parameter**, having a non-zero value in the ordered phase whereas it vanishes in the disordered phase [407]. Prototype systems for equilibrium phase transitions are simple ferromagnets, superconductors, liquid-gas systems, ferroelectrics, as well as systems exhibiting superfluidity.

---

[1] This notation differs from the traditional **Landau notation** [406, 2], which uses the symbol $\sim$ for 'asymptotically equal' instead of 'asymptotically proportional'.

The paradigmatic example is the transition between the ferromagnetic and paramagnetic phases in the **Ising model.**[2] The standard version of the Ising model is defined on a d-dimensional lattice with $\mathcal{N}$ sites and associated spins $\sigma_i = \pm 1$ at site i. A given configuration $\{\sigma\}$ of spins has the energy

$$E(\{\sigma\}) = -\frac{J}{2} \sum_{(i,j)} \sigma_i \sigma_j - h \sum_i \sigma_i \,, \tag{2.1}$$

where $J > 0$ is a coupling constant and h denotes an external homogeneous magnetic field. The first sum runs over all pairs of nearest-neighbour lattice sites (i, j). Since the energy functional is invariant under reversal of all spins and the external field, the Ising model has the finite group $\mathbb{Z}_2$ as a **global symmetry**. The group $\mathbb{Z}_2$ is the cyclic group with two elements which is used to describe reflection symmetries. In the present case the group elements are the identity and the simultaneous reversal of all spins and the external field.

According to the postulates of (classical) equilibrium statistical mechanics for a system in contact with a thermal reservoir at temperature T each configuration of spins $\{\sigma\}$ occurs with a probability proportional to $e^{-E(\{\sigma\})/k_{\mathrm{B}}T}$. All physical quantities of interest can be derived from the **partition function**

$$Z_{\mathcal{N}} = \sum_{\{\sigma\}} \exp(-E(\{\sigma\})/k_{\mathrm{B}}T) \,, \tag{2.2}$$

in which the sum runs over all possible configurations $\{\sigma\}$ and $\mathcal{N}$ denotes the total number of sites. At thermal equilibrium the system is then in a state that minimises the **Gibbs potential**[3] $G = -k_{\mathrm{B}}T \ln Z_{\mathcal{N}}$. In the Gibbs functional $G = G(T, h)$, temperature T and the external field h are the independent thermodynamic parameters, while the **free energy** $F = F(T, M) = G + Mh$ depends on T and the magnetisation

$$M = \frac{1}{\mathcal{N}} \sum_i \sigma_i . \tag{2.3}$$

Only in the limit $\mathcal{N} \to \infty$, thermodynamic observables, such as specific heats or susceptibilities, derived from either G or F, converge to each other (see exercise 3).

The Ising model without external field exhibits a phase transition at a certain critical temperature T_c. The transition separates the **ferromagnetic phase** with non-zero magnetisation from the **paramagnetic phase** with zero magnetisation. This phase transition is caused by the competition between the ordering due to the energy minimisation (surface tension) on the

---

[2] Definitions and some equilibrium critical properties of several common spin systems are collected in appendix A.

[3] We shall usually work with units such that the Boltzmann constant $k_{\mathrm{B}} = 1$.

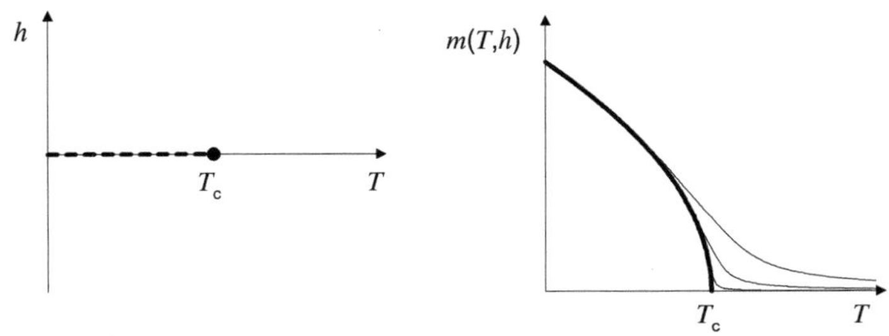

Fig. 2.1 Sketch of the phase diagram of a ferromagnet (left) and the behaviour of the corresponding order parameter (right). The phase diagram comprises a line of first-order transitions at zero field (dashed line) that ends in a critical point at temperature T_c. The temperature dependence of the magnetisation is sketched for $h = 0$ (bold line) and three different values of the the external field $h > 0$. Reprinted with permission from [433]. Copyright (2004) World Scientific Publishing Company.

one hand and the entropy production by thermal fluctuations on the other hand.

Following Landau, the average magnetisation per site

$$m(T, h) = \frac{1}{\mathcal{N}} \sum_i \langle \sigma_i \rangle = -\frac{1}{\mathcal{N}} \frac{\partial G}{\partial h} \tag{2.4}$$

is the **order parameter** of the ferromagnetic phase transition [407]. Furthermore the temperature T is the control parameter of the phase transition and the external field h is **conjugate** to the order parameter.

The phase diagram of the Ising model is sketched in Fig. 2.1. Due to the $\mathbb{Z}_2$-symmetry of ferromagnets the transition occurs at zero external field h. The phase diagram exhibits a transition line along the temperature axis, terminating at the **critical point** T_c. Crossing the boundary for $T < T_c$ by varying h the magnetisation changes discontinuously, i.e. the system undergoes a **first-order phase transition**. The magnitude of the discontinuity decreases if one approaches the critical temperature. At $T = T_c$ the magnetisation varies continuously but its derivatives are discontinuous. Here the system is said to undergo a **continuous** or **second-order phase transition**. For $T > T_c$ no singularities of the free energy occur and the systems changes continuously from a state of positive to a state of negative magnetisation.

Crossing the critical point in the horizontal direction by varying the temperature T, the transition can be characterised as follows. Without an external field ($h = 0$) the high-temperature or paramagnetic phase is characterised by a vanishing magnetisation. Decreasing the temperature, a phase transition takes place at the critical temperature and for $T < T_c$ one observes an ordered phase which is spontaneously magnetised (see Fig. 2.1). Figure 2.2

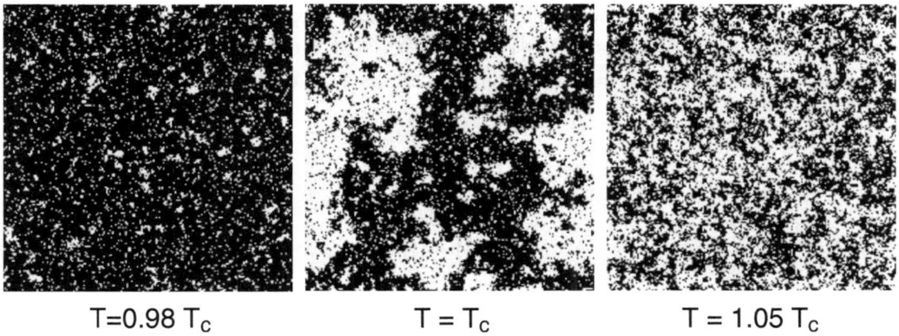

Fig. 2.2 Typical configurations of the two-dimensional Ising model at thermal equilibrium below, at, and above criticality.

shows snapshots of typical equilibrium configurations below, at and above criticality.

2.1.3 Power-law Scaling

Systems with continuous phase transitions sufficiently close to the critical point are habitually characterised by power laws. For example, in a ferromagnetic system the order parameter $m = m(T, h)$ varies as (for $\tau > 0$)

$$m(T, 0) \sim \tau^{\beta} \tag{2.5}$$

with the **reduced temperature**

$$\tau = (T_c - T)/T_c, \tag{2.6}$$

where β is a so-called **critical exponent**.[4] As we shall see below, critical exponents play a key role as they characterise specific types of critical behaviours at continuous phase transitions.

For a non-zero external field h, the magnetisation increases smoothly with decreasing temperature. At the **critical isotherm** ($\tau = 0$) for $h \to 0$ the magnetisation obeys a power law

$$m(T_c, h) \sim |h|^{1/\delta} \tag{2.7}$$

[4] Note that in order to have a common notation with the absorbing phase transitions in the rest of this volume, τ is defined here with the opposite sign as usually found in the literature.

with another critical exponent δ. Further critical exponents are introduced to describe the singularities of the order parameter **susceptibility** $\chi = \chi(T,h)$, the **specific heat** $C = C(T,h)$, as well as the **correlation length** $\xi = \xi(T,h)$:

$$\chi(T,0) \sim |\tau|^{-\gamma}\,, \tag{2.8}$$

$$C(T,0) \sim |\tau|^{-\alpha}\,, \tag{2.9}$$

$$\xi(T,0) \sim |\tau|^{-\nu}\,. \tag{2.10}$$

Another important quantity is the **correlation function**

$$G(\boldsymbol{r}_1,\boldsymbol{r}_2;T,h) \;=\; \langle\sigma_1\sigma_2\rangle \;-\; \langle\sigma_1\rangle\langle\sigma_2\rangle\,, \tag{2.11}$$

where $\boldsymbol{r}_1$ and $\boldsymbol{r}_2$ denote the vectors pointing to the sites 1 and 2. For translationally and rotationally invariant systems, the correlation function depends only the distance $r = |\boldsymbol{r}_1 - \boldsymbol{r}_2|$ but not on the direction.

At criticality, the correlation function decays as

$$G(r;T_c,0) \;\sim\; r^{-d+2-\eta}\,, \tag{2.12}$$

defining yet another critical exponent η. Here, d denotes the dimensionality of the system. Away from criticality, the correlation function is characterised by an exponential decay of the form

$$G(r;T,0) \;\sim\; r^{-\vartheta}\,\mathrm{e}^{-r/\xi}\,, \tag{2.13}$$

where ξ is the correlation length. Unlike the previously introduced exponents, ϑ may take different values below and above the transition. For example, in the two-dimensional Ising model, one has $\vartheta = 2$ above and $\vartheta = 1/2$ below the transition point [467].

The exponents α, β, γ, δ, η, and ν are called **critical exponents**. Notice in (2.8-2.10) the equality of the critical exponents below and above the critical point, which is an assumption of the scaling theory.

Sometimes it may happen that an exponent is zero. Such a vanishing exponent corresponds either to a discontinuity of the corresponding quantity at the critical point or to a **logarithmic singularity** since

$$\lim_{s\to 0}\frac{|\tau|^{-s}-1}{s} \;=\; -\ln|\tau|\,. \tag{2.14}$$

In many cases it is notoriously difficult, given either experimental or numerical data, to distinguish between a logarithmic singularity and a small absolute value of the exponent.

The phenomenological scaling theory for ferromagnetic systems was developed by several authors in the 1960s (e.g. [210, 189, 627, 368, 512, 250]) and has been confirmed by experiments as well as simulations. In particular, the scaling theory predicts that the six critical exponents introduced above are connected by four scaling laws

$$\alpha + 2\beta + \gamma = 2 \qquad \text{(Rushbrooke)}, \tag{2.15}$$
$$\gamma = \beta\,(\delta - 1) \qquad \text{(Widom)}, \tag{2.16}$$
$$\gamma = (2 - \eta)\,\nu \qquad \text{(Fisher)}, \tag{2.17}$$
$$2 - \alpha = \nu\, d \qquad \text{(Josephson)}\,. \tag{2.18}$$

In this way the critical behaviour of an equilibrium system is determined by only *two* independent critical exponents. In the following we shall recall how these scaling laws can be derived.[5]

The Josephson law is special in so far as it includes the spatial dimension d of the system. Such scaling relations are termed **hyperscaling relations** and are valid only below the so-called **upper critical dimension** d_c, as will be explained in Sect. 2.3.3.

2.2 Scale-Invariance and Universality

2.2.1 Scale-Invariance

It is generally recognised that thermodynamic potentials, such as the density of the Gibbs potential, can be decomposed, via $g(\tau, h) = g_{\text{reg}}(\tau, h) + g_{\text{sing}}(\tau, h)$, into a *regular part* $g_{\text{reg}}(\tau, h)$ without any critical behaviour and a *singular part* $g_{\text{sing}}(\tau, h)$ one is interested in.[6] The phenomenological scaling theory for equilibrium phase transitions rests on the assumption that close to the critical point the singular part of a thermodynamic potential is asymptotically given by a generalised **homogeneous function** (see e.g. [266]). Mathematically, a function $f(x_1, x_2, \ldots)$ is called a generalised homogeneous function if it satisfies the relation

$$\lambda\, f(x_1, x_2 \ldots) = f(x_1 \lambda^{s_1}, x_2 \lambda^{s_2}, \ldots) \tag{2.19}$$

for all positive values of $\lambda \in \mathbb{R}_+$. The exponents $s_1, s_2, \ldots$ are usually termed **scaling powers** or **scaling dimensions** and the variables $x_1, x_2, \ldots$ are called **scaling fields**. In physics, the singular part of the **Gibbs potential**

[5] From the convexity of the equilibrium free energy, one may derive the following exponent inequalities: $\alpha + 2\beta + \gamma \geq 2$, $2 - \alpha \leq d\nu$, $\gamma \geq (2 - \eta)\nu$ and $\alpha + \beta(\delta + 1) \geq 2$.

[6] The analysis of the scaling behaviour always refers to the *singular* parts of the thermodynamics potentials. In the following, we shall always assume that any regular parts have been subtracted off.

per spin $g_{\text{sing}}(\tau, h)$ is assumed to scale asymptotically as

$$g_{\text{sing}}(\tau, h) \simeq \lambda^{\alpha-2} \; \tilde{g}(\tau\,\lambda,\; h\,\lambda^{\beta\delta}), \tag{2.20}$$

where $\tilde{g}(x, y)$ is the corresponding scaling function. The scaling power of the conjugate field is often denoted as the **gap exponent** $\Delta = \beta\,\delta = \beta + \gamma$. It should be emphasised that (2.20) is only asymptotically valid, i.e., only when τ and h tend to zero. Away from this limit (as e.g. in any numerical simulation or experiment) one expects corrections to occur.[7]

It can be shown that Legendre transforms and partial derivatives of generalised homogeneous functions are again generalised homogeneous functions. Thus with the Gibbs potential, all thermodynamic functions that are expressible as derivatives of thermodynamic potentials (such as the magnetisation, specific heat, etc.) are generalised homogeneous functions as well.

Consider, for example, the magnetisation and the corresponding susceptibility. Differentiating (2.20) with respect to the conjugate field h one has

$$m(\tau, h) = -\left(\frac{\partial\, g_{\text{sing}}}{\partial\, h}\right)_T \simeq \lambda^{\alpha-2+\beta\delta} \; \tilde{m}(\tau\,\lambda, h\,\lambda^{\beta\delta}), \tag{2.21}$$

$$\chi(\tau, h) = -\left(\frac{\partial^2\, g_{\text{sing}}}{\partial\, h^2}\right)_T \simeq \lambda^{\alpha-2+2\beta\delta} \; \tilde{\chi}(\tau\,\lambda, h\,\lambda^{\beta\delta}), \tag{2.22}$$

where the scaling functions are given by

$$\tilde{m}(x, y) = -\left(\frac{\partial\, \tilde{g}(x, y)}{\partial\, y}\right)_x, \qquad \tilde{\chi}(x, y) = -\left(\frac{\partial^2\, \tilde{g}(x, y)}{\partial\, y^2}\right)_x. \tag{2.23}$$

As usual, the indices at the brackets indicate which of the quantities are kept fixed.

2.2.2 Scaling Functions and Data Collapses

Since the relation (2.19) for generalised homogeneous functions is valid for all positive values of λ, one can make a particular choice in order to derive a certain **scaling form**. For example, we may choose $\lambda = |x_1|^{-1/s_1}$ to obtain an expression in which the first argument of the r.h.s. is constant. In this manner the function f, which depends on n arguments $x_1, \ldots, x_n$, is reduced to

---

[7] In view of these *co*-variant transformations of physical observables under scale-transformation, the reader might wonder why this situation is commonly referred to as scale-*in*-variance. Indeed, at $T = T_c$ the microscopic configurations (see middle panel of Fig. 2.2) are *on average* scale-invariant, which on a more formal level is reflected by the scale-invariance of the partition function.

$$f(x_1, \ldots, x_n) = |x_1|^{1/s_1} \tilde{f}(1, |x_1|^{-1/s_1} x_2, \ldots |x_1|^{-1/s_1} x_n) \quad (2.24)$$

with a scaling function $\tilde{f}$ which depends effectively on $n - 1$ arguments. Plotting $|x_1|^{-1/s_1} f(x_1, \ldots, x_n)$ as a function of these $n-1$ arguments various data sets for different x_1 should collapse onto a single manifold given by $\tilde{f}$. In particular, for $n = 2$ different data sets *collapse* onto a single curve described by $\tilde{f}$.

Data collapses are not unique because there are several possibilities to choose λ, leading to different but mathematically equivalent scaling forms. For example, for $n = 2$ the scaling forms

$$|x_1|^{-1/s_1} \, f(x_1, x_2) = \tilde{f}(\pm 1, x_2 \, |x_1|^{-s_2/s_1}) \,, \quad (2.25)$$

$$|x_2|^{-1/s_2} \, f(x_1, x_2) = \tilde{\tilde{f}}(x_1 \, |x_2|^{-s_1/s_2}, \pm 1) \quad (2.26)$$

are both equally legitimate [266]. As an example, let us consider (2.21-2.22). Choosing $\lambda = 1/|\tau|$ one obtains for a vanishing external field $h = 0$

$$m(\tau, 0) \simeq \tau^{-(\alpha - 2 + \beta\delta)} \; \tilde{m}(-1, 0) \,, \quad (2.27)$$

$$\chi(\tau, 0) \simeq |\tau|^{-(\alpha - 2 + 2\beta\delta)} \; \tilde{\chi}(\pm 1, 0) \,, \quad (2.28)$$

where the magnetisation is defined for $\tau < 0$ only. Notice that one generally expects $\tilde{\chi}(+1, 0) \neq \tilde{\chi}(-1, 0)$, i.e., the amplitudes of the susceptibility are different below ($T < T_c$) and above ($T > T_c$) the transition. Comparing these equations with (2.5) and (2.8), one has

$$\beta = -\alpha + 2 - \beta\,\delta \,, \qquad \gamma = \alpha - 2 + 2\,\beta\,\delta \,, \quad (2.29)$$

reproducing the Rushbrooke (2.15) and Widom (2.16) scaling laws. The Fisher scaling law can be obtained in a similar way from the scaling form of the correlation function while the Josephson law requires the combination of both thermodynamic scaling forms and correlation scaling forms and is left as an exercise (see e.g. [640] or Sect. 2.5).

Scaling theory implies even more. Consider for instance the M-h-τ equation of state (2.21). Choosing $h\lambda^{\beta\delta} = 1$ in (2.21) one finds

$$m(\tau, h) \simeq \lambda^{-\beta} \; \tilde{m}(\tau\,\lambda, h\,\lambda^{\beta\delta})\big|_{\lambda = h^{-1\beta\delta}} = h^{1/\delta} \; \tilde{m}(\tau\,h^{-1/\beta\delta}, 1) \,. \quad (2.30)$$

At the critical isotherm ($\tau = 0$) we recover the relation $m(T_c, h) \sim |h|^{1/\delta}$ in (2.7). Furthermore, the equation of state may be written in the rescaled form

$$m_h \simeq \tilde{m}(\tau_h, 1) \qquad \text{with} \qquad m_h = m\,h^{-1/\delta} \,, \quad \tau_h = \tau\,h^{-1/\beta\delta} \,. \quad (2.31)$$

In this way, the equation of state is described by the single curve $\tilde{m}(x, 1)$ and all M-h-τ data points will collapse onto the single curve $\tilde{m}(x, 1)$ if one plots the rescaled order parameter m_h as a function of the rescaled control parameter τ_h. Such a data collapse is shown in the left panel of Fig. 2.3 using

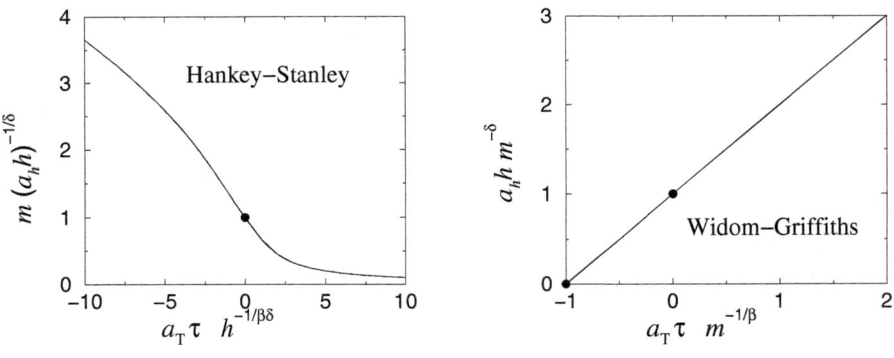

Fig. 2.3 Data collapse of the M-h-τ equation of state derived within mean-field theory for the same three values of h as in Fig. 2.1. Left: $\tilde{m}(x,1)$ according to the Hankey-Stanley scaling form (2.31). Right: $\tilde{h}(x,1)$ as described by the Widom-Griffiths scaling form (2.31). In both cases all points collapse exactly onto a single curve. The circles mark the normalisation conditions (2.37) and (2.41). Reprinted with permission from [433]. Copyright (2004) World Scientific Publishing Company.

the data of Fig. 2.1 (see Sect. 2.3.1 for the derivation). The function $\tilde{m}(x,1)$ is usually referred to as the **Hankey-Stanley** scaling function [266].

A different scaling form is obtained if one considers instead of the Gibbs potential the Helmholtz[8] potential or free energy $f = g + mh$ via a Legendre transformation. Since Legendre transformations of generalised homogeneous functions are again generalised homogeneous, the singular part of the Helmholtz potential obeys the scaling form

$$f_{\text{sing}}(\tau, m) \simeq \lambda^{\alpha-2}\ \tilde{f}(\tau\,\lambda, m\,\lambda^{\beta}). \tag{2.32}$$

This equation leads to the scaling form of the conjugate field

$$h(\tau, m) = \left(\frac{\partial f_{\text{sing}}}{\partial m}\right)_T \simeq \lambda^{-\beta\delta}\ \tilde{h}(\tau\,\lambda, m\,\lambda^{\beta}). \tag{2.33}$$

Choosing $m\,\lambda^{\beta} = 1$, one finds

$$h_m \simeq \tilde{h}(\tau_m, 1) \qquad \text{with} \qquad h_m = h\,m^{-\delta}, \quad \tau_m = \tau\,m^{-1/\beta}. \tag{2.34}$$

The function $\tilde{h}(x,1)$ is often called the **Widom-Griffiths** scaling function [627, 250], as shown in the right panel of Fig. 2.3.

Both scaling functions, $\tilde{h}(x,1)$ and $\tilde{m}(x,1)$ are analytic in the neighbourhood of $x = 0$, i.e., at the critical temperature. The Hankey-Stanley scaling

[8] While considering the Gibbs potential with h fixed is natural for ferromagnets, the study of the Helmholtz potential is more useful for fluids or alloys, where m is proportional to the concentration/density difference of the pure states.

form is just the order parameter curve as a function of the control parameter in a fixed conjugate field and thus it is more natural and perhaps more elegant to study. But often the mathematical formulae of the Hankey-Stanley functions are rather complicated, whereas the Widom-Griffiths scaling forms are analytically tractable. Therefore $\tilde{h}(x,1)$ is often calculated within certain approximation schemes, e.g. ϵ- or $1/n$-expansions within a renormalisation-group approach.

2.2.3 Universality Classes

According to the four scaling laws (2.15-2.18), equilibrium phase transitions are characterised by two independent critical exponents. In the 1950s and 1960s, it was experimentally recognised that quantities like T_c depend sensitively on the details of the interactions whereas the critical exponents are universal, i.e., they do not depend on microscopic details but only on a small number of general features such as the dimension or the symmetries of the system. This led to the celebrated **universality hypothesis** which was first clearly formulated by Kadanoff [369], but based on earlier works including e.g. [217, 345, 618, 251, 63]. The universality hypothesis reduces the large variety of critical phenomena to a small number of equivalence classes, so-called **universality class**es, which depend only on a few fundamental parameters. All systems belonging to a given universality class have the same **critical exponents** and the corresponding **scaling functions** become identical near the critical point.

For short-range interacting equilibrium systems, the fundamental parameters determining the universality class are the symmetry of the order parameter and the dimensionality of space [369, 251].[9] The specific nature of the transition, i.e. the details of the interactions such as the lattice structure and the range of interactions (as long as it remains finite), do not affect the scaling behaviour. For instance, ferromagnetic systems with a preferred direction of magnetisation are characterised by a one-component order parameter ($n = 1$) and belong to the universality class of the Ising ferromagnet [148, 134]. The **Ising universality class** includes also liquid-gas transitions at the end points of the coexistence line of first-order transitions separating the liquid from the gaseous state [314, 86, 302, 568, 570, 569, 41], binary mixtures of liquids [34], as well as systems exhibiting an order-disorder transition in alloys such as beta-brass which are described by a scalar order parameter (see e.g. [18]). Even certain phase transitions occurring in high-energy physics are expected to belong to the Ising universality class. For example, in the electroweak theory the early universe exhibits a transition from a symmetric high-temperature phase to a spontaneously broken Higgs phase [377]. The

[9] Implicitly, one assumes here the absence of so-called *marginal* scaling fields.

predicted line of first-order phase transitions terminates at a second-order point which is argued to belong to the Ising universality class.

Ferromagnetic systems, which can be magnetised in two different directions, are characterised by a two-component order parameter ($n = 2$). In this case ferromagnetic ordering breaks the rotational symmetry below the critical temperature, giving rise to the so-called **XY universality class**. Representatives of this universality class are the magnetic **XY model** [148, 134] (see appendix A for the definition),[10] superconductors, as well as liquid crystals that undergo a phase transition from a smectic-A to a nematic phase [263, 262, 127]. The most impressive prototype, however, is the superfluid transition of ^{4}He along the λ-line. Due to its characteristic features like the exceptional purity of the samples, the weakness of the singularity in the compressibility of the fluid, as well as the reasonably short thermal relaxation times, superfluid ^{4}He is more suitable for high-precision experiments than any other system [7, 23]. For example, heat-capacity measurements under micro-gravity conditions of liquid helium within 2 nK of the λ-transition provide the extremely precise estimate $\alpha = -0.0127 \pm 0.0003$ [424]. Thus, the λ-transition of ^{4}He offers an exceptional opportunity to test theoretical predictions, obtained from renormalisation-group theory or simulations.[11] An excellent recent review on micro-gravity experiments for measuring critical exponents and amplitudes, their interpretation and comparison with theoretical results for the O(n) model with $n = 1, 2$ is [41].

The **Heisenberg universality class** describes isotropic ferromagnetic systems that are characterised by a three-component order parameter ($n = 3$) [148, 134]. Besides the Ising, XY, and Heisenberg universality classes, various other classes with O(n) global symmetry and $n \geq 4$ have been discussed in the literature (where O(n) stands for the Lie group of orthogonal transformations in n dimensions). For instance, the $n = 5$ universality class is expected to be relevant for the description of high-T_c superconductors [643, 155] whereas $n = 18$ is reported in the context of superfluid ^{3}He [364]. Furthermore, the limit $n \to 0$ corresponds to the critical behaviour of polymers and self-avoiding random walks [521]. The other limiting case $n \to \infty$ corresponds to the exactly solvable **spherical model** [61, 577, 366]. In appendix A, we recall

---

[10] In very thin films of magnetic materials, interactions with the substrate may break the full rotation-symmetry of the spins. If the spins are oriented perpendicular to the substrate, the phase transition is in the Ising universality class while if the spin are parallel to the surface, one has XY-behaviour.

[11] For example, re-summed field-theoretic six- and seven-loop expansions give $\alpha = -0.0129(6)$ [392] or $\alpha = -0.0112(21)$ [344] and extensive Monte Carlo simulations combined with re-summed high-temperature series lead to $\alpha = -0.0151(3)$ [101]. While the overall agreement between different theoretical methods and experiments is satisfying, it is remarkable that exponent differences of the order $\Delta\alpha \approx 1 - 3 \cdot 10^{-3}$ are by now considered significant. Substantial further effort will be needed to resolve the remaining small discrepancies.

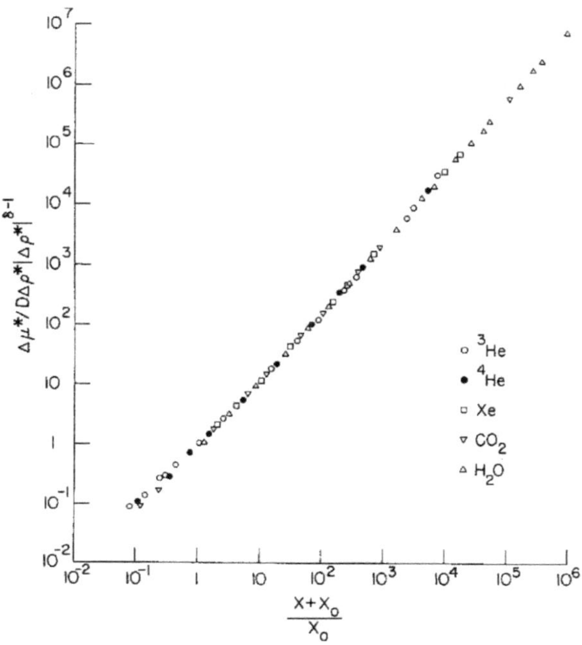

Fig. 2.4 Experimental evidence for universal scaling. The figure displays the universal Widom-Griffiths scaling function $\tilde{H}(x, 1)$ for five different fluids, undergoing a liquid-gas phase transition. The scaling variable is defined as $x = \Delta T |\Delta\varrho|^{-1/\beta}$ and x_0 is related to the amplitude B of the power-law for the coexistence curve $\Delta\varrho = B\,\Delta T^{\beta}$. The figure is taken from [570].

the definitions of these models and summarise some of their main equilibrium properties.

2.2.4 Experimental Evidence of Universality

An impressive example of universal scaling plots, perhaps the most striking experimental evidence for the concept of universality at all, is shown in Fig. 2.4. It presents data for five different fluids undergoing a fluid-gas transition. All fluids are characterised by different inter-atomic interactions. The data are plotted according to the Widom-Griffiths scaling form. As can be seen, the rescaled data of the chemical potential of five different gases collapse onto a single universal curve, providing evidence that all these transitions should belong to the Ising universality class [570].

2.2.4.1 Metric Factors

In the scaling theory presented above, the exponents are universal but the scaling functions are still written in a form that contains non-universal elements. These non-universal features can be absorbed by introducing three non-universal parameters, called **metric factors**. For example, we may relate the scaling function $\tilde{m}$ to a standardised universal scaling function $\tilde{M}$

$$c_m\, m(\tau,h) \;\simeq\; \lambda^{-\beta}\;\tilde{M}(-c_T\tau\,\lambda, c_h h\,\lambda^{\beta\delta}) \tag{2.35}$$

which we shall denote by the corresponding capital letter. Physically, the metric factors encode the units which may differ from model to model. In this way, any scaling function m can be reduced to a universal scaling function $\tilde{M}$ which is the same for all models belonging to the same universality class, while the non-universal system-dependent features are contained in metric factors like c_m, c_T, and c_h [535].

Since the above scaling form is valid for all positive values of λ, the number of metric factors can be reduced by a scale transformation. For the sake of convenience it is customary to choose $c_m^{1/\beta}\lambda \mapsto \lambda$, giving

$$m(\tau,h) \;\simeq\; \lambda^{-\beta}\;\tilde{M}(-a_T\tau\,\lambda, a_h h\,\lambda^{\beta\delta}) \tag{2.36}$$

with $a_T = c_T c_m^{-1/\beta}$ and $a_h = c_h c_m^{-\delta}$, respectively. Moreover, it is convenient to normalise the universal scaling function $\tilde{M}$ by the conditions

$$\tilde{M}(-1,0) \;=\; 1\,, \qquad\qquad \tilde{M}(0,1) \;=\; 1\,, \tag{2.37}$$

so that the non-universal metric factors can be determined by the amplitudes of

$$m(\tau,0) \simeq \lambda^{-\beta}\;\tilde{M}(-a_T\tau\,\lambda,0)\Big|_{a_T\tau\,\lambda=-1} \;=\; (a_T\tau)^{\beta}\,, \tag{2.38}$$

$$m(0,h) \simeq \lambda^{-\beta}\;\tilde{M}(0,a_h h\,\lambda^{\beta\delta})\Big|_{a_h h\,\lambda^{\beta\delta}=1} \;=\; (a_h h)^{1/\delta}\,. \tag{2.39}$$

Similarly, any other scaling function can be standardised by introducing suitable metric factors. For example, the universal Widom-Griffiths scaling form is given by

$$a_h\, h(\tau,m) \;\simeq\; \lambda^{-\beta\delta}\;\tilde{H}(-a_T\tau\,\lambda, m\,\lambda^{\beta})\,. \tag{2.40}$$

Here the universal scaling function $\tilde{H}$ is usually normalised by the conditions

$$\tilde{H}(-1,1) \;=\; 0\,, \qquad\qquad \tilde{H}(0,1) \;=\; 1\,, \tag{2.41}$$

which correspond to (2.37), i.e., the metric factors are again determined by the amplitudes of the power-laws $m(\tau,0) \sim (a_T\tau)^{\beta}$ and $m(0,h) \sim (a_h h)^{1/\delta}$, respectively.

2.3 Mean-Field and Renormalisation Group Methods

2.3.1 Mean-Field Theory of Ferromagnetic Systems

To illustrate the handling of universal scaling functions and metric factors, let us consider a mean-field theory of a simple ferromagnet (with a scalar order parameter). Following Landau, the free energy density is given by [407]

$$f(\tau, m) - f_0 = -\frac{1}{2} b_2 \tau m^2 + \frac{1}{4} b_4 m^4, \tag{2.42}$$

where the factors $b_2 > 0$ and $b_4 > 0$ are system-dependent non-universal parameters. Variation of the free energy with respect to the magnetisation yields the equation of state

$$h = -b_2 \tau m + b_4 m^3. \tag{2.43}$$

At zero-field, one has

$$m = 0 \qquad \text{or} \qquad m = \sqrt{\frac{b_2}{b_4}\tau} \implies \beta = \frac{1}{2}, \ a_T = \frac{b_2}{b_4}, \tag{2.44}$$

whereas at the critical isotherm

$$m = \sqrt[3]{\frac{1}{b_4} h} \implies \delta = 3, \ a_h = \frac{1}{b_4}. \tag{2.45}$$

From (2.43) one derives the scaling function

$$\tilde{H}(x, y) = x y + y^3, \tag{2.46}$$

leading to the universal **Widom-Griffiths** scaling form

$$a_h h m^{-\delta} = \tilde{H}(-a_T \tau m^{-1/\beta}, 1) \qquad \text{with} \qquad \tilde{H}(x, 1) = 1 + x. \tag{2.47}$$

Since the magnetisation is a cube-root function ($\delta = 3$) the universal **Hankey-Stanley scaling form** for $y > 0$ is given by

$$\tilde{M}(x, y) = \frac{\left(27y + \sqrt{108x^3 + 729y^2}\right)^{1/3}}{3\, 2^{1/3}} - \frac{2^{1/3}\, x}{\left(27y + \sqrt{108x^3 + 729y^2}\right)^{1/3}}. \tag{2.48}$$

This example demonstrates that the Hankey-Stanley scaling form is usually more complicated than the corresponding Widom-Griffiths form.

Using these mean-field results, we generated the curves shown in Fig. 2.1 and Fig. 2.3. In addition the normalisations (2.37) and (2.41) are marked in Fig. 2.3 by a black dot.

2.3.2 Universal Amplitude Ratios

Universal properties are ascribed not only to critical exponents and scaling functions but also to certain **amplitude combinations** (see [536] for an excellent review and [41] for an update on experimental results). These quantities are very useful in identifying the universality class of a phase transition since amplitude combinations are often more sensitive to numerical differences between the classes than critical exponents. Furthermore, the measurement of amplitude combinations in experiments or numerical simulations serves as a reliable test of theoretical predictions and approximations.

As an example, let us consider the singular behaviour of the susceptibility $\chi(\tau, h)$ when approaching the transition point from above and below

$$\chi(\tau,0) \simeq a_{\chi,+}\,(-\tau)^{-\gamma} \quad \text{if } T > T_c \text{ or } \tau < 0, \tag{2.49}$$

$$\chi(\tau,0) \simeq a_{\chi,-}\,\tau^{-\gamma} \quad \text{if } T < T_c \text{ or } \tau > 0. \tag{2.50}$$

The amplitudes $a_{\chi,+}$ and $a_{\chi,-}$ involve non-universal metric factors as well as special values of universal scaling functions. To see this, consider the universal scaling form of the susceptibility derived from (2.36)

$$\chi(\tau,h) = \left(\frac{\partial\, m(\tau,h)}{\partial\, h}\right)_T \simeq a_h\,\lambda^{\gamma}\;\tilde{\mathrm{X}}(-a_T\tau\,\lambda, a_h h\,\lambda^{\beta\delta}) \tag{2.51}$$

with $\tilde{\mathrm{X}}(x,y) = \partial_y \tilde{M}(x,y)$. Setting $a_T|\tau|\lambda = 1$, one obtains for the amplitudes

$$a_{\chi,+} = a_h\, a_T^{-\gamma}\,\tilde{\mathrm{X}}(+1,0) \tag{2.52}$$

$$a_{\chi,-} = a_h\, a_T^{-\gamma}\,\tilde{\mathrm{X}}(-1,0)\,. \tag{2.53}$$

Clearly, the amplitudes $a_{\chi,+}$ and $a_{\chi,-}$ are non-universal but the *ratio*

$$\frac{a_{\chi,+}}{a_{\chi,-}} = \frac{\tilde{\mathrm{X}}(+1,0)}{\tilde{\mathrm{X}}(-1,0)} \tag{2.54}$$

is a universal quantity. For example, the mean-field behaviour of the susceptibility takes the form

$$\chi(\tau,0) = \frac{1}{-b_2\,\tau + 3\,b_4\,m^2}\,, \tag{2.55}$$

leading to [cf. (2.44)]

$$a_{\chi,+} = \frac{1}{b_2}, \qquad a_{\chi,-} = \frac{1}{2\,b_2}, \qquad \Longrightarrow \qquad \frac{a_{\chi,+}}{a_{\chi,-}} = 2\,. \tag{2.56}$$

Similarly, one can define a large variety of other amplitude ratios. Well-known and experimentally significant is the quantity [536]

$$R_\chi = \Gamma\, D_c\, B^{\delta-1}\,. \tag{2.57}$$

Here, Γ, D_c, B are the traditional, but unfortunately unsystematic, notations for the amplitudes of

$$\chi \sim \Gamma\,|\tau|^{-\gamma}\big|_{T>T_c}\,, \qquad h \sim D_c\, m^\delta\big|_{T=T_c}\,, \qquad m \sim B\,\tau^\beta\big|_{T<T_c}\,. \tag{2.58}$$

These amplitudes correspond to the values [see (2.36)]

$$\Gamma = a_{\chi,+} = a_h\, a_T^{-\gamma}\, \tilde{\mathrm{X}}(1,0)\,, \qquad D_c = a_h^{-1}\, \tilde{M}(0,1)^{-\delta}\,, \qquad B = a_T^\beta\, \tilde{M}(-1,0)\,. \tag{2.59}$$

Using the normalisations $\tilde{M}(-1,0) = \tilde{M}(0,1) = 1$ one finds for the amplitude combination

$$R_\chi = \tilde{\mathrm{X}}(1,0) \tag{2.60}$$

which is obviously a universal quantity, reducing to $R_\chi = 1$ within mean-field theory. These two examples demonstrate how the universality of amplitude combinations emerges naturally from the universality of the scaling functions, i.e., universal amplitude combinations are nothing but special values of universal scaling functions.

The phenomenological concepts of scaling and universality have been tested in a large variety of systems with remarkable success. Nevertheless they have certain shortcomings. For example, in many cases there is no way of determining the critical exponents and scaling functions explicitly. To remedy these shortcomings different methods such as Wilson renormalisation group theory are needed, as will be sketched in the following.

2.3.3 Remarks on Renormalisation-Group Theory

A deeper understanding of scaling and universality is provided by Wilson's **renormalisation-group** (RG) theory [629, 630]. In equilibrium systems the RG theory relates the critical point to a fixed point of a certain transformation of the system's Hamiltonian (for introductions to this topic see e.g. [219, 520, 640, 525]). In case of the instructive real-space RG [368] the transformation rescales a microscopic length scale, e.g. the lattice spacing a, by a factor b ($a \mapsto b\,a$) combined with the elimination of those degrees of freedom that correspond to the range between a and $b\,a$. This rescaling will change the system's properties away from the critical point, where the system exhibits only finite characteristic length scales. However, at criticality there are no finite correlation lengths and thus the properties of the system remain unaffected by the rescaling procedure. In this way, criticality corresponds to a fixed point of the renormalisation transformation. Exercise 4 illustrates the procedure in the context of the one-dimensional Ising model.

Denoting the system's Hamiltonian by $\mathcal{H}$ and the rescaled Hamiltonian by $\mathcal{H}'$ the renormalisation transformation is described by an appropriate operator $\mathcal{R}$ that maps $\mathcal{H}$ to $\mathcal{H}'$:

$$\mathcal{H}' = \mathcal{R} \circ \mathcal{H} . \tag{2.61}$$

Fixed point Hamiltonians $\mathcal{H}^*$ satisfy the equation

$$\mathcal{H}^* = \mathcal{R} \circ \mathcal{H}^* . \tag{2.62}$$

It turns out that different fixed point Hamiltonians correspond to different universality classes [219]. For the sake of concreteness consider the reduced Ising Hamiltonian (cf (2.1))

$$\tilde{\mathcal{H}} = -\frac{\mathcal{H}}{k_{\mathrm{B}}T} = \frac{J}{2\,k_{\mathrm{B}}T}\sum_{\langle i,j\rangle}\sigma_i\sigma_j + \frac{h}{k_{\mathrm{B}}T}\sum_i \sigma_i \tag{2.63}$$

with the spin variables $\sigma_i = \pm 1$, the nearest neighbour interaction coupling J, and the homogeneous external field h, taking the first sum over all pairs of neighbouring spins on a given d-dimensional lattice. The partition function for a system of $\mathcal{N}$ spins is

$$Z_{\mathcal{N}}(K_1, K_2) = \sum_{\{\sigma\}} \mathrm{e}^{\tilde{\mathcal{H}}(K_1,K_2)} , \tag{2.64}$$

where the sum runs over all possible spin configurations, introducing the couplings

$$K_1 = \frac{J}{2\,k_{\mathrm{B}}T} , \qquad K_2 = \frac{h}{k_{\mathrm{B}}T} . \tag{2.65}$$

More generally, a given Hamiltonian may be written as a sum over the couplings

$$\tilde{\mathcal{H}}(\underline{K}) = \sum_i K_i\,\mathcal{O}_i , \tag{2.66}$$

where $\mathcal{O}_i$ are the operators appearing in the Hamiltonian.

A **renormalisation-group transformation** reduces the number of lattice sites

$$\mathcal{N} \mapsto \mathcal{N}' = \mathcal{N}/b^d \tag{2.67}$$

and leads to a rescaled Hamiltonian characterised by the couplings

$$K_1' = K_1'(K_1, K_2), \qquad K_2' = K_2'(K_1, K_2), \qquad K_3' = K_3'(K_1, K_2), \dots . \tag{2.68}$$

Here the $K'_{i>2}$ account for additional coupling terms of the renormalised Hamiltonian which may appear as a result of the renormalisation transformation even if they are not present in the initial Hamiltonian. These so-called RG **recursion relations** generate trajectories in the space spanned by the couplings, i.e., the couplings $\underline{K}$ flow under successive renormalisation

$$\underline{K} \underset{\mathcal{R}}{\longrightarrow} \underline{K}^{(1)} \underset{\mathcal{R}}{\longrightarrow} \underline{K}^{(2)} \underset{\mathcal{R}}{\longrightarrow} \dots \underset{\mathcal{R}}{\longrightarrow} \underline{K}^{(n)} \underset{\mathcal{R}}{\longrightarrow} \underline{K}^{(n+1)} \underset{\mathcal{R}}{\longrightarrow} \dots \tag{2.69}$$

along the RG trajectories towards a certain fixed point $\underline{K}^*$. If the system is not initially at criticality the couplings will flow towards a trivial fixed point, e.g., a fixed point that corresponds to zero or infinite temperature.

Linearising the problem close to a fixed point yields

$$\begin{aligned}
\mathcal{R} \circ \tilde{\mathcal{H}}(\underline{K}^* + \delta\underline{K}) &= \tilde{\mathcal{H}}^* + \mathcal{R} \circ \sum_m \delta K_m \, \mathcal{O}_m \\
&\approx \tilde{\mathcal{H}}^* + \sum_m \delta K_m \sum_n \mathcal{L}_{m,n} \, \mathcal{O}_n \\
&= \tilde{\mathcal{H}}^* + \sum_n \left(\sum_m \delta K_m \, \mathcal{L}_{m,n} \right) \mathcal{O}_n \\
&= \tilde{\mathcal{H}}^* + \sum_n \delta K_n' \, \mathcal{O}_n ,
\end{aligned} \tag{2.70}$$

where $\mathcal{L}$ can be thought of as a first-order approximation of $\mathcal{R}$. Assuming that the diagonalised operator $\mathcal{L}$ has the eigenoperators Ω_i and eigenvalues Λ_i such that $\mathcal{L}\Omega_i = \Lambda_i \Omega_i$ we find that the couplings transform in the diagonal representation ($\delta\kappa$) according to $\delta\kappa_i' = \Lambda_i \delta\kappa_i$, thus

$$\mathcal{R} \circ \tilde{\mathcal{H}}(\underline{K}^* + \delta\underline{K}) = \tilde{\mathcal{H}}^* + \sum_i \Lambda_i \, \delta\kappa_i \, \Omega_i . \tag{2.71}$$

The couplings $\delta\kappa_i$ are called **scaling fields** and their recursion relation can be expressed by the rescaling factor b as

$$\delta\kappa_i' = b^{y_i} \, \delta\kappa_i \qquad \text{with} \qquad \Lambda_i = b^{y_i} , \tag{2.72}$$

where the y_i are the **scaling exponents**. Successive renormalisation steps correspond to

$$\delta\kappa_i \underset{\mathcal{R}}{\longrightarrow} b^{y_i} \, \delta\kappa_i \underset{\mathcal{R}}{\longrightarrow} b^{2y_i} \, \delta\kappa_i \underset{\mathcal{R}}{\longrightarrow} b^{3y_i} \, \delta\kappa_i \underset{\mathcal{R}}{\longrightarrow} \dots . \tag{2.73}$$

Thus, the renormalisation flow in the vicinity of a given fixed point $\tilde{\mathcal{H}}^*$ depends on the exponents y_i. For $y_i > 0$ ($\Lambda_i > 1$) the corresponding scaling field $\delta\kappa_i$ is called **relevant** since successive renormalisation transformations drive the system away from $\tilde{\mathcal{H}}^*$. In case of a negative exponent $y_i < 0$ ($\Lambda_i < 1$) the system approaches the fixed point under repeated transformations and the scaling field is said to be **irrelevant**. **Marginal** scaling fields correspond to $y_i = 0$ ($\Lambda_i = 1$) and require higher than linear order in the expansion, and may lead to critical exponents depending continuously on the corresponding interaction parameters. In this way each fixed point is characterised by its associated scaling fields and by a domain of attraction which corresponds to the set of points flowing eventually to the fixed point. This set forms a hypersurface in the space of couplings and is termed the **critical surface**.

In summary, a fixed point $\tilde{\mathcal{H}}^*$ is approached if all associated relevant scaling fields are zero, otherwise the system flows away from $\tilde{\mathcal{H}}^*$. Examples of relevant scaling fields in ferromagnetism are the reduced temperature τ and the conjugate field h. Criticality is only achieved for $\tau \to 0$ and $h \to 0$, therefore we may identify

$$\delta\kappa_1 = \delta\kappa_T = a_\tau \tau\,, \qquad y_1 = y_\tau > 0\,, \tag{2.74}$$

$$\delta\kappa_2 = \delta\kappa_h = a_h h\,, \qquad y_2 = y_h > 0\,, \tag{2.75}$$

and $y_{i>2} < 0$. Moreover, all Hamiltonians that differ from the fixed point $\tilde{\mathcal{H}}^*$ only by irrelevant scaling fields flow towards $\tilde{\mathcal{H}}^*$. For example the five magnetic materials presented in Fig. 2.4 differ only by irrelevant scaling fields. It is this irrelevance of higher-order scaling fields that causes universality.

Although the linear recursion relations (2.72) describe the RG trajectories only in the vicinity of fixed points they provide some insight into the topology of the entire RG flow (see Fig. 2.5). These RG flow diagrams are useful to illustrate the RG transformations schematically and provide a classification scheme in terms of fixed point stability. The stability of a given fixed point is determined by the number of relevant and irrelevant scaling fields. Unstable fixed points are characterised by at least one relevant scaling field since Hamiltonians arbitrarily close to the fixed point will flow away under successive RG iterations. *Ordinary critical points* correspond to singly unstable fixed points, i.e., unstable with respect to the control parameter (e.g. temperature) of the phase transition. **Tricritical points** are characterised by a second instability. An applied external field conjugate to the order parameter gives rise to an additional instability of the fixed point.

Furthermore, the stability of fixed points depends on the spatial dimensionality d of a system. It turns out that above a certain dimension d_c, called the **upper critical dimension**, the scaling behaviour is usually determined by a trivial fixed point with classical mean-field exponents, whereas a different fixed point with non-classical exponents determines the scaling behaviour below d_c. This sudden change of the scaling behaviour is caused by an exchange of the stability of the corresponding fixed points below and above d_c [520, 621, 83]. At the upper critical dimension d_c both fixed points are identical and marginally stable and the corresponding scaling be-

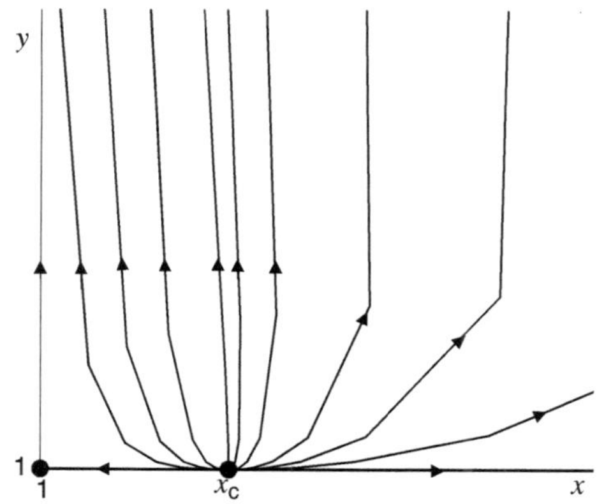

Fig. 2.5 Sketch of the renormalisation-group flow of an Ising ferromagnet (see [(2.63)]) on a hierarchical lattice. Hierarchical lattices are iteratively constructed lattices [60, 379] where the so-called Migdal-Kadanoff scheme of renormalisation is exact [476, 370]. The flowlines show the motion of the coupling constants $x = \exp(2K_1)$ (temperature-like variable) and $y = \exp(K_2)$ (field-like variable) under successive iterations of the recursion relations $x' = (x^2+y^2)(x^{-2}+y^2)(1+y^2)^{-2}$ and $y' = y^2(1+x^2y^2)(x^2+y^2)^{-1}$. The flowlines are attracted and repelled by the fixed points $(1,1)$, $(1,\infty)$, $(\infty,1)$, and $(x_c,1)$ with $x_c = 3.38298\ldots$. The phase transition corresponds to the non-trivial fixed point $(x_c,1)$ with the eigenvalues $\Lambda_1 \approx 1.6785$ and $\Lambda_2 \approx 3.6785$, leading to $\beta \approx 0.16173$ and $\delta \approx 15.549$. Reprinted with permission from [433]. Copyright (2004) World Scientific Publishing Company.

haviour is characterised by mean-field exponents modified by logarithmic corrections [622, 632]. We shall discuss the scaling behaviour of certain non-equilibrium phase transitions at the upper critical dimension in detail in the following chapters.

2.3.4 Scaling Laws Induced by Renormalisation-Group Theory

Let us now demonstrate how scaling emerges from the renormalisation transformation. To this end it is essential to note that the partition function is invariant under the renormalisation operation $\mathcal{R}$ [219]

$$Z_{\mathcal{N}}(\delta\kappa_T, \delta\kappa_h, \delta\kappa_3, \ldots) = Z_{\mathcal{N}'}(\delta\kappa'_T, \delta\kappa'_h, \delta\kappa'_3, \ldots) . \tag{2.76}$$

Therefore, the free energy[12] per degree of freedom transforms according to

$$f(\delta\kappa_T, \delta\kappa_h, \delta\kappa_3, \ldots) = b^{-d} f(\delta\kappa'_T, \delta\kappa'_h, \delta\kappa'_3, \ldots) . \tag{2.77}$$

Combining this equation with (2.72) and using the identities (2.74) and (2.75) yields the scaling form

[12] From now on, the term 'free energy' is meant to include 'Gibbs potential', 'Helmholtz potential' and so on.

$$f(a_T\tau, a_h h, \delta\kappa_3, \ldots) \simeq b^{-d} f(b^{y_\tau} a_T\tau, b^{y_h} a_h h, b^{y_3} \delta\kappa_3, \ldots) . \qquad (2.78)$$

Introducing $\lambda = b^{y_\tau}$, one obtains the scaling form of the free energy [see (2.20)]

$$f(a_T\tau, a_h h, \delta\kappa_3, \ldots) \simeq \lambda^{\alpha-2} f(a_T\tau\,\lambda, a_h h\,\lambda^{\beta\delta}, \delta\kappa_3\,\lambda^{\phi_3}, \ldots) , \qquad (2.79)$$

where one has identified the exponents

$$y_\tau = \frac{1}{\nu}, \quad y_h = \frac{\beta\,\delta}{\nu}, \quad y_{i>2} = \frac{\phi_{i>2}}{\nu} . \qquad (2.80)$$

The possible additional scaling fields $\delta\kappa_{i>2}$ deserve some comments. It turns out that irrelevant scaling fields ($\phi_{i>2} < 0$) may cause corrections to the asymptotic scaling behaviour [620]. For example, choosing $\lambda = 1/|a_T\tau|$ one obtains at zero field

$$\begin{aligned} f(a_T\tau, 0, \delta\kappa_3) &\simeq |a_T\tau|^{2-\alpha} f(\pm 1, 0, \delta\kappa_3\,|a_T\tau|^{|\phi_3|}) \\ &\approx |a_T\tau|^{2-\alpha} f(\pm 1, 0, 0) \\ &\quad + |a_T\tau|^{2-\alpha+|\phi_3|}\,\delta\kappa_3\,\partial_x f(\pm 1, 0, x)\big|_{x=0} + \ldots . \end{aligned} \qquad (2.81)$$

The non-universal corrections to the leading-order term $\sim |\tau|^{2-\alpha}$ are called **confluent singularities** and they determine the size of the critical region. In many cases, confluent singularities have to be taken into account, in order to obtain reliable estimates of the universal critical parameters. Impressive examples of confluent singularity effects of superfluid Helium are reviewed in [7]. The above expansion of $f(\pm 1, 0, x)$ implies that the free energy is an analytic function in $\delta\kappa_3$. If the free energy is non-analytic, the scaling field $\delta\kappa_3$ is termed a **dangerous irrelevant variable** [218, 534]. In that case, the free energy exhibits e.g. a power-law divergence

$$f(x, y, z) = z^{-\mu} \hat{f}(x, y) \qquad (2.82)$$

characterised by the exponent $\mu > 0$. Singularities of this type occur for example in the mean-field regime of the well-known Landau-Ginzburg-Wilson Hamiltonian for short-range interacting ferromagnets (see e.g. [80]). There, the dangerous irrelevant variable corresponds to the coupling constant of the ϕ^4 interactions. The non-analytic behaviour leads to the modified scaling form of the free energy

$$\begin{aligned} f(a_T\tau, a_h h, \delta\kappa_3) &\simeq \lambda^{-\nu d} f(a_T\tau\,\lambda, a_h h\,\lambda^{\beta\delta}, \delta\kappa_3\,\lambda^{\phi_3}) \\ &= \lambda^{-\nu d-\mu\phi_3}\,\delta\kappa_3^{-\mu}\,\hat{f}(a_T\tau\,\lambda, a_h h\,\lambda^{\beta\delta})\Big|_{h=0} \\ &= |a_T\tau|^{\nu d+\mu\phi_3}\,\delta\kappa_3^{-\mu}\,\hat{f}(\pm 1, 0) . \end{aligned} \qquad (2.83)$$

Compared to the standard behaviour $f \sim |\tau|^{2-\alpha}$, the above result reflects the breakdown of the hyperscaling law $2-\alpha = \nu d$. Moreover, dangerous irrelevant variables may also cause the breakdown of common finite-size scaling within the mean-field regime $d > d_c$ [82, 534, 201]. This is well-established in equilibrium, and can be seen analytically in the $n \to \infty$ limit of the $O(n)$ model [82, 447, 201]. We shall come back to this point in detail in Sect. 5.10, where we consider non-equilibrium phase transitions in high-dimensional systems.

The situation is different when the scaling field $\delta\kappa_3$ is *relevant*, i.e., $\phi_3 > 0$. In that case the free energy at zero field is given by

$$f(a_T\tau, 0, \delta\kappa_3) \;\simeq\; |a_T\tau|^{2-\alpha}\, f(\pm 1, 0, \delta\kappa_3\, |a_T\tau|^{-\phi_3})\,. \tag{2.84}$$

For sufficient small arguments ($|\delta\kappa_3|\tau|^{-\phi_3}| \ll 1$) the relevant scaling field leads again to corrections to the asymptotic scaling behaviour. However, approaching the transition point ($\tau \to 0$) the scaling argument diverges and gives rise to a different critical behaviour, i.e., the system crosses over to a different universality class. Finally, a *marginal* scaling field may generate logarithmic corrections via

$$\begin{aligned} \delta\kappa_3\, |a_T\tau|^{-\phi_3} \;&=\; \delta\kappa_3\, \exp\left(-\phi_3\, \ln|a_T\tau|\right) \\ &\underset{\phi_3\to 0}{\longrightarrow}\; \delta\kappa_3\, (\,1 - \phi_3\, \ln|a_T\tau| \,+\, \ldots)\,. \end{aligned} \tag{2.85}$$

Often, these logarithmic contributions mask the power law singularities and can make the analysis of experimental or numerical data notoriously difficult.

Analogous to the free energy, the renormalisation group determines the scaling form of the correlation length ξ. Performing a renormalisation transformation, the correlation length ξ, like all length scales, is decreased by the factor b,

$$\xi' \;=\; b^{-1}\,\xi\,. \tag{2.86}$$

It is essential for the understanding of phase transitions that fixed points are characterised by an infinite (or trivial zero) correlation length since ξ satisfies at a fixed point

$$\xi' = \xi\,. \tag{2.87}$$

In this way, a singular correlation length characterizes a critical point. In other words, *scale-invariance is the hallmark of criticality.*

The scaling form of the correlation length is obtained from (2.86)

$$\xi(\delta\kappa_T, \delta\kappa_h, \delta\kappa_3, \ldots) \;=\; b\,\xi(\delta\kappa_T', \delta\kappa_h', \delta\kappa_3', \ldots)\,, \tag{2.88}$$

yielding with $\lambda = b^{y_\tau}$ and (2.72)

$$\xi(a_T\tau, a_h h, \delta\kappa_3, \ldots) \;\simeq\; \lambda^{\nu}\;\; \xi(a_T\tau\,\lambda, a_h h\,\lambda^{\beta\delta}, \delta\kappa_3\,\lambda^{\phi_3}, \ldots)\,. \tag{2.89}$$

2.3.5 Field-Theory and ϵ-Expansion

Instead of the real-space renormalisation considered so far, it is often more convenient to work in momentum space. This can be achieved by reformulating the above derivations in terms of Fourier transforms. We refer the interested reader to the reviews in [632, 219, 520]. The momentum-space formulation allows one to study a perturbative RG theory, leading to a field-theoretical formulation in terms of Feynman graphs. The appropriate small parameter for the perturbation expansion is the dimensionality difference to the upper critical dimension $\epsilon = d_c - d$ [631]. This so-called **ϵ-expansion** gives systematic corrections to mean-field theory in powers of ϵ.

The ϵ-expansion provides a powerful tool for calculating the critical exponents and the scaling functions. For example, the exponents ν and γ for n-component magnetic systems with short range interactions and global $O(n)$-symmetry can be found from a two-loop calculation to second order in $\epsilon = 4 - d$ (see e.g. [83, 8, 648])[13]

$$\begin{aligned}\nu &= \frac{1}{2} + \epsilon \frac{n+2}{4(n+8)} + \epsilon^2 \frac{n+2}{8(n+8)^3}(n^2+23n+60) + O(\epsilon^3) \\ \gamma &= 1 + \epsilon \frac{n+2}{2(n+8)} + \epsilon^2 \frac{(n+2)}{4(n+8)^3}(n^2+22n+52) + O(\epsilon^3)\,. \end{aligned} \tag{2.90}$$

Furthermore, the Widom-Griffiths scaling function can be written as a power series in ϵ

$$\tilde{H}(x,1) = 1 + x + \epsilon \tilde{H}_1(x,1) + \epsilon^2 \tilde{H}_2(x,1) + O(\epsilon^3)\,. \tag{2.91}$$

As expected, the mean-field scaling behaviour (2.47) is retrieved for $\epsilon = 0$.

These scaling functions become more complicated with increasing order. For example, the first-order term is given by [84]

$$\begin{aligned}\tilde{H}_1(x,1) = \frac{1}{2(n+8)} \Big[& 3(x+3)\ln(x+3) + (n-1)(x+1)\ln(x+1) \\ & + 6x\ln 2 - 9(x+1)\ln 3 \Big] \end{aligned} \tag{2.92}$$

while for the second-order term $\tilde{H}_2(x,1)$ we refer to the reviews [83, 8]. Thus the ϵ-expansion provides estimates of almost all quantities of interest as an asymptotic expansion in powers of ϵ around the mean-field values. Unfortunately it is impossible to estimate the corresponding error bars within this approximation scheme since the extrapolation to larger values of ϵ is uncontrolled.

---

[13] Presently, these expansions have been carried up to seven-loop order. The numerical evaluation of the resulting ϵ-series (which *a priori* are divergent) requires the use of non-trivial resummation methods.

A detailed analysis reveals that the critical exponents are more accurately estimated than the scaling functions and the corresponding amplitude ratios. For example, the ϵ^2-approximation (2.90) for the susceptibility exponent of the two-dimensional Ising model ($n = 1$, $\epsilon = 2$) yields $\gamma \approx 1.642$. This value differs by 6% from the exact value $\gamma = 7/4$ [43]. On the other hand, the amplitude ratio of the susceptibility (2.54) can be expanded as [486]

$$\frac{\tilde{X}(+1,0)}{\tilde{X}(-1,0)} = 2^{\gamma(\epsilon)-1}\,\frac{\gamma(\epsilon)}{\beta(\epsilon)} \tag{2.93}$$

suggesting the estimate 81.14... for $\epsilon = 2$. This result differs significantly (115%) from the exact value 37.69... [43, 154]. The different accuracy reflects a conceptual difference between the universality of critical exponents and the universality of scaling functions. As pointed out in [536], the universality of exponents arises from the linearised RG flow in the vicinity of the fixed point, whereas the scaling functions are obtained from the entire, i.e., non-linear, RG flow. More precisely, the relevant trajectories from the fixed point of interest to other fixed points determine the actual functional form of universal scaling functions. This also explains why exponents between different universality classes may change only slightly while the scaling functions and therefore the amplitude combinations vary much more significantly. Thus, the identification of a system's universality class by considering the scaling functions and amplitude combinations instead of critical exponents is expected to be more sensitive and provides a useful diagnostic tool. Hence, one of the most convincing demonstrations of universality is by means of a universal data-collapse of various systems, as illustrated for experimental data in Fig. 2.4.

2.3.6 Surface Critical Phenomena

In the vicinity of surfaces and boundaries, new and important aspects of critical behaviour arise as will be summarised briefly in the following. Throughout this book, many of the quantities (especially exponents) related to surfaces and boundaries will be distinguished by an index '1' from their bulk counterparts.

Restricting ourselves to the particularly simple case of a flat surface with a free boundary, one expects that for the density of the Gibbs potential

$$g = g_b(T, h, V, \mathcal{N}) + \frac{2}{L} g_1(T, h, h_1, V, \mathcal{N}) + \ldots, \tag{2.94}$$

where g_b is the density of the *bulk Gibbs potential* studied so far and g_1 is the **surface Gibbs potential**, often loosely referred to as the **surface free**

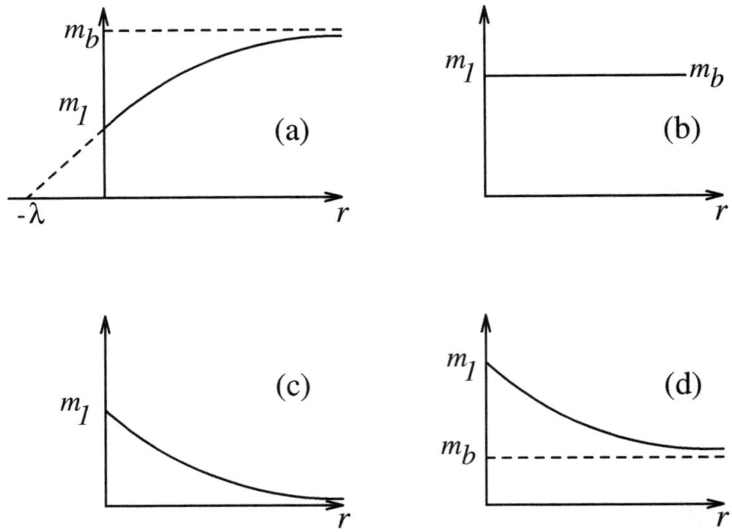

Fig. 2.6 Schematic local order parameter profiles $m(r)$ in dependence of the temperature and the extrapolation length λ. After [68], reproduced from [270] with kind permission of Springer Science and Business Media.

energy density. Besides the bulk magnetic field h a surface magnetic field h_1 located at the surface itself has to be included as well. The factor 2 takes the presence of two free interfaces into account. Using mean-field results as an example, Fig. 2.6 shows schematic profiles of the order parameter as a function of the distance r from the boundary. Denoting by $T_{c,b}$ the usual *bulk critical temperature*, the following different kinds of behaviour are seen [68, 183, 523].

1. For $T < T_{c,b}$ and a positive **extrapolation length** $\lambda > 0$, as shown in Fig. 2.6a, the transition from the disordered state to this state is called the **ordinary transition** O. Here $m_b = m(\infty)$ is the magnetisation deep inside the bulk and $m_1 = m(0)$ is the magnetisation at the surface.
2. For $T = T_{c,b}$, the extrapolation length is infinite. In this case the transition from the disordered state to this state is called the **special transition** SP. It is a surface tricritical point, see Fig. 2.6b.
3. For $T_{c,b} < T < T_{c,s}$, the extrapolation length is negative and $T_{c,s}$ is some surface critical temperature. The transition from the disordered state into this state is called the **surface transition** S, see Fig. 2.6c.
4. For $T < T_{c,b}$ and a negative extrapolation length, the transition from the state of Fig. 2.6c to this state is called the **extraordinary transition** E, see Fig. 2.6d.

Note, however, that in two dimensions in the absence of the surface field and for short-ranged interactions only the ordinary transition can exist.

We illustrate the distinct thermodynamic and critical behaviour near to a surface in the case of the local order parameter $m = m(r)$. One must distinguish the *bulk* magnetisation m_b, the **surface magnetisation** m_1 and the **excess magnetisation** m_s. These quantities scale with three distinct critical exponents

$$\begin{aligned} m_b &= m(\infty) = -\frac{\partial g_b}{\partial h} \sim \tau^{\beta} \\ m_1 &= m(0) = -\frac{\partial g_1}{\partial h_1} \sim \tau^{\beta_1} \\ m_s &= \int_0^{\infty} \mathrm{d}r\, [m(r) - m_b] = -\frac{\partial g_1}{\partial h} \sim \tau^{\beta_s} . \end{aligned} \tag{2.95}$$

For an ordinary transition, there is only a relevant surface magnetic field h_1. Hence one has the scaling form for the singular part of the Gibbs potential density

$$g_1^{\rm sing} \simeq |\tau|^{2-\alpha_s} W_1^{\pm}\left(h|\tau|^{-\beta-\gamma}, h_1|\tau|^{-\Delta_1}\right) , \tag{2.96}$$

where $\Delta_1 = (\nu/2)(d - \eta_{\|})$ is the **surface gap exponent**. Analogous to the hyperscaling condition $g_b^{\rm sing} \sim \xi_b^{-d}$, which is valid below the upper critical dimension $d < d_c$, one expects for a free surface $g_1^{\rm sing} \sim \xi_b^{-(d-1)}$ where ξ_b is the bulk correlation length. Hence $\alpha_s = \alpha + \nu$. It is now obvious that the exponent $\beta_s = \beta - \nu$ of the excess magnetisation is related to bulk exponents, too. On the one hand, the surface magnetisation exponent β_1 *cannot* be expressed in terms of bulk critical exponents. However, it is related to the surface exponents of the two-point correlation functions via the scaling relation $\beta_1 = (\nu/2)(d - 2 + \eta_{\|})$. This means that the ordinary transition is characterised by a single independent surface exponent.

The surface exponents of the correlators are defined as follows. Using the notation $\boldsymbol{r}_s = (\boldsymbol{r}_{\|}, 0)$ and $\boldsymbol{r} = (\boldsymbol{r}_{\|}, r_{\perp})$, one has correlations parallel and perpendicular to the free surface at $r_{\perp} = 0$

$$G_{\|}(\boldsymbol{r}_{\|} - \boldsymbol{r}'_{\|}) := G(\boldsymbol{r}_s, \boldsymbol{r}'_s) \;\;,\;\; G_{\perp}(r_{\perp}, r'_{\perp}) := G(\boldsymbol{r}, \boldsymbol{r}')\big|_{r_{\|} = r'_{\|}} . \tag{2.97}$$

The surface exponents are then defined exactly *at* the critical point $\tau = h = h_1 = 0$

$$G_{\|}(\boldsymbol{r}_{\|}) \sim |\boldsymbol{r}_{\|}|^{-d+2-\eta_{\|}} \;\;,\;\; G_{\perp}(r_{\perp}, r'_{\perp}) \sim |r_{\perp} - r'_{\perp}|^{-d+2-\eta_{\perp}} . \tag{2.98}$$

These definitions apply to the ordinary, special and extraordinary transitions.[14] Moreover, the exponents are related by the scaling relation $2\eta_{\perp} = \eta + \eta_{\|}$. In the O($n$)-model, a two-loop calculation gives at the ordinary transition to second order in $\epsilon = 4 - d$

[14] Exponents belonging to the surface transition can be defined by replacing τ by $\tau_s := (T_{c,s} - T)/T_{c,s}$.

$$\eta_{\|} = 2 - \epsilon\frac{n+2}{n+8} - \epsilon^2\frac{n+2}{2(n+8)^3}(17n+76) + \mathrm{O}(\epsilon^3) \tag{2.99}$$

We refer to the excellent reviews [68, 183, 184, 523] for more systematic expositions. In appendix A, values of some surface exponents are listed for several spin systems.

2.3.7 Finite-Size Scaling

Having looked in the previous section into the local scaling behaviour near to a flat surface, we now recall the main features of a system confined to a *finite* geometry [220, 37, 201, 536]. The Gibbs functional/free energy $G(T, V, \mathcal{N})$ of the system can be written as (V=volume, A=surface area)

$$G(T, V, \mathcal{N}) = V g_{\mathrm{b}}(T, \rho) + A g_1(T, \rho) + \dots \tag{2.100}$$

where T is the temperature, $\mathcal{N}$ the number of particles and we have suppressed other thermodynamic variables. Writing this, a limit $V \to \infty$, $\mathcal{N} \to \infty$ such that $\rho = \mathcal{N}/V$ is kept fixed is implied, that is, the **bulk Gibbs potential** density $g_{\mathrm{b}}(T, \rho)$ is defined by

$$g_{\mathrm{b}}(T, \rho) = \lim_{\mathcal{N}\to\infty, V\to\infty} G(T, V, \mathcal{N})/V \quad \text{with } \rho = \mathcal{N}/V \text{ fixed.} \tag{2.101}$$

A similar limit is implied for the definition of the density of the **surface Gibbs potential** $g_1(T, \rho)$.

How many of such terms, describing the contribution of surfaces, edges and so on, should one expect? At least for systems sufficiently far away from a critical point, it has been proposed that an answer might be formulated in terms of **Minkowski functionals** [396].[15] According to **Hadwiger's theorem**, any continuous motion-invariant and additive functional can be decomposed into Minkowski functionals, viz. $f(\Omega) = \sum_{n=0}^{d} f_n M_n(\Omega)$. Furthermore, the $d+1$ Minkowski functionals $M_n(\Omega)$, $n = 1, \dots, d$, are complete in d dimensions [261, 469].[16] If the mathematical theory of Minkowski functionals should apply to statistical systems at equilibrium, one would expect that the expansion (2.100) should contain $d+1$ terms depending algebraically on L, up to exponentially small corrections (often, the finite size is $L \sim \mathcal{N}^{1/d}$).

---

[15] By definition, a Minkowski functional is a continuous map $M : \Omega \to \mathbb{R}$ of a geometric object $\Omega \subset \mathbb{R}^d$ embedded into an d-dimensional space which is (i) *motion-invariant*, that is $M(\Omega) = M(g\Omega)$ if g is a translation or rotation and (ii) satisfies the *additivity property* $M(\Omega_1 \cup \Omega_2) = M(\Omega_1) + M(\Omega_2) - M(\Omega_1 \cap \Omega_2)$.

[16] In $3D$, the four independent Minkowski functionals are: (i) volume, (ii) area, (iii) integral mean curvature and (iv) Euler characteristic.

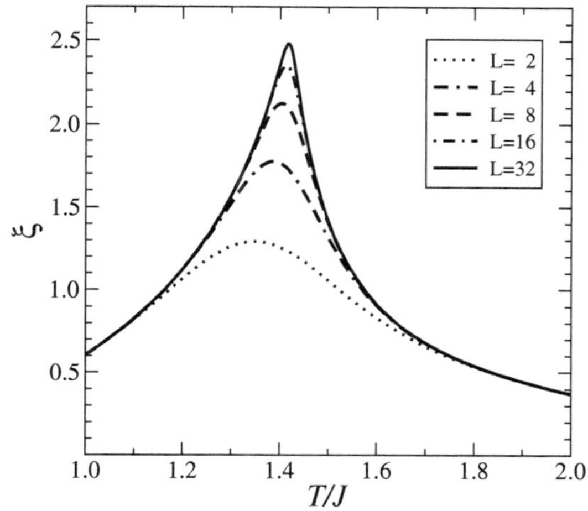

Fig. 2.7 Finite-size behaviour of the correlation length ξ_L in the $2D$ Ising model in infinitely long slabs of finite width $L = [2, 4, 8, 16, 32]$ from bottom to top.

As an illustration of what may happen to a finite system, which on an infinite lattice undergoes a second-order phase transition, we show in Fig. 2.7 the correlation length ξ_L of the $2D$ Ising model [97] as a function of the temperature and for a vanishing external field $h = 0$, considered on an infinitely long slab of *finite* width L with periodic boundary conditions.

There are two basic observations to be made:

1. Rather than a divergence of the form $\xi_\infty \sim |\tau|^{-\nu}$, the finite-size data ξ_L show a maximum at some pseudo-critical temperature $T_c(L)$. For sufficiently large system sizes, one expects
$$\left(T_c(L) - T_c(\infty)\right)/T_c(\infty) \sim L^{-\lambda} \text{ if } L \to \infty. \tag{2.102}$$
This defines the **shift exponent** λ.
2. The curves for finite L are broadened with respect to the bulk behaviour. One may describe this in terms of a **rounding temperature** $T^*(L)$, defined such that if $|T - T_c|/T_c \geq |T^* - T_c|/T_c$, then the finite-size correlation length is close to the bulk value: $\xi_L(T) \simeq \xi_\infty(T)$. For sufficiently large systems, one may define the **rounding exponent** θ via
$$\left(T^*(L) - T_c(\infty)\right)/T_c(\infty) \sim L^{-\theta} \text{ if } L \to \infty. \tag{2.103}$$
Asymptotically, θ characterises the size of the finite-size scaling region.

The basic hypothesis of finite-size scaling, as first formulated by Fisher and Barber [220], asserts that *there should be only one relevant length describing the rounding and shifting of the thermodynamic singularities.* Hence $\xi(T^*(L)) \sim L$, which implies
$$\lambda = \theta = 1/\nu \tag{2.104}$$

		bulk	finite-size
specific heat	$C_{\rm sing}$	$\lvert\tau\rvert^{-\alpha}$	$L^{\alpha/\nu}$
magnetic susceptibility	χ	$\lvert\tau\rvert^{-\gamma}$	$L^{\gamma/\nu}$
correlation length	ξ	$\lvert\tau\rvert^{-\nu}$	L
free-energy density	$g_{\rm sing}$	$\lvert\tau\rvert^{2-\alpha}$	L^{-d}
order parameter$^{(a)}$	M	τ^{β}	$L^{-\beta/\nu}$
latent heat$^{(a)}$	$\ell_{\rm h}$	$\tau^{1-\alpha}$	$L^{(\alpha-1)/\nu}$

Table 2.1 Bulk scaling near to $\tau = 0$ and finite-size scaling close to T_c and for L large. $^{(a)}$ Special care is needed to define non-vanishing lattice expressions for M and ℓ_h in the ordered phase where $\tau > 0$, see [270].

which means that the shift and the rounding are of the same order. For a list of experimental tests of finite-size scaling in thin magnetic films, see [270].

Close to a critical point, finite-size effects will be important in the so-called **finite-size scaling limit** where one takes simultaneously $L \to \infty$ and $T \to T_c$ such that the **finite-size scaling variable** $\mathfrak{z} := L/\xi_\infty(T)$ remains finite. Changing $\mathfrak{z}$ from large values to values $\mathfrak{z} \ll 1$, one goes over from the bulk critical behaviour to the finite-size region where shift and rounding effects will be very important. For example, the bulk scaling behaviour *close to* the critical point $\tau = 0$ can be traded for a singular finite-size behaviour which holds for a finite value of $\mathfrak{z}$. Table 2.1 gives some examples and indicates how the systematic study of finite-size effects may be turned into a tool for the determination of critical points and exponents. For further details on this, see appendix F.

For a systematic analysis of finite-size scaling, it appears natural to generalise (2.77) and to write the singular part of the Gibbs functional g as[17]

$$g(\tau, h; L) \sim A_1 |\tau|^{2-\alpha} \tilde{W}^{\pm}(A_2 h |\tau|^{-\beta-\gamma}; L/\xi_\infty), \tag{2.105}$$

where the metric factors $A_{1,2}$ are related to the metric factors $a_{\tau,h}$ used above, $\tilde{W}^{\pm}$ are universal scaling functions and $\xi_\infty \sim \xi_0 \tau^{-\nu}$ (as $\tau \to 0$) is the bulk correlation length. Note that there is no extra metric factor associated with L/ξ_∞ [220, 535].[18] Similarly, one expects for the (connected) correlation function

$$\begin{aligned} G(\boldsymbol{r}; \tau, h) &:= \langle s_{\boldsymbol{0}} s_{\boldsymbol{r}} \rangle - \langle s_{\boldsymbol{0}} \rangle \langle s_{\boldsymbol{r}} \rangle \\ &\sim D_1 r^{2-d-\eta} X^{\pm}(\boldsymbol{r}/\xi_\infty, D_2 h |\tau|^{-\beta-\gamma}), \end{aligned} \tag{2.106}$$

[17] We implicitly assume here either a slab geometry with only one finite direction or else a fully finite hypercube geometry. Otherwise, the shape-dependence of the scaling functions must be included in the discussion which may involve further non-universal metric factors. See the literature for further details [130, 185, 595].

[18] This holds true if $d < d_c$, but is invalid for $d \geq d_c$, due to the presence of dangerous irrelevant variables, see [82, 447] for explicit spherical model calculations.

where $X^{\pm}$ are universal functions and D_1, D_2 are metric factors. The magnetic susceptibility is obtained from the static fluctuation-dissipation theorem (see Sect. 2.4)

$$\chi_\infty(\tau, h) = \int \mathrm{d}\boldsymbol{r} \;\; G(\boldsymbol{r}; \tau, h) \sim D_1 \xi_\infty^{2-\eta} \tilde{X}^{\pm}(D_2 h |\tau|^{-\beta-\gamma}). \tag{2.107}$$

The same scaling behaviour is also expected for the total correlation function

$$\Gamma(\boldsymbol{r}; \tau, h) := \langle s_{\boldsymbol{0}} s_{\boldsymbol{r}} \rangle \sim D_1 r^{2-d-\eta} Z^{\pm}(\boldsymbol{r}/\xi_\infty, D_2 h |\tau|^{-\beta-\gamma}) \tag{2.108}$$

with a new universal scaling function $Z^{\pm}$. In the limit $r \to \infty$, $\Gamma(\boldsymbol{r}) \to m_\infty^2$, where m_∞ is the magnetisation per spin. Thus

$$m_\infty^2(\tau, h) \sim D_1 \xi_\infty^{2-d-\eta} \tilde{Z}^{\pm}(D_2 h |\tau|^{-\beta-\gamma}). \tag{2.109}$$

From the bulk scaling behaviour one has, taking the required derivatives

$$\begin{aligned} m_\infty(\tau, h) &\sim A_1 A_2 |\tau|^{\beta} W_1^{\pm}(A_2 h |\tau|^{-\beta-\gamma}) \\ \chi_\infty(\tau, h) &\sim A_1 A_2^2 |\tau|^{-\gamma} W_2^{\pm}(A_2 h |\tau|^{-\beta-\gamma}), \end{aligned} \tag{2.110}$$

where $W_1^{\pm}(x) = \partial_x W^{\pm}(x)$ and $W_2^{\pm}(x) = \partial_x W_1^{\pm}(x)$. Comparing coefficients, one recovers the bulk scaling relations and obtains relationships between the universal scaling functions $W_{1,2}^{\pm}$ and $X^{\pm}$ and $Z^{\pm}$. As the result, the metric factors become related and one can identify the universal combinations [535]

$$\begin{aligned} A_1 \xi_0^d &=: Q_1 \\ A_2 / D_2 &=: Q_2 \\ D_1 A_1^{-\psi} A_2^{-2} &=: Q_3 \quad \text{if} \quad \psi = 1 + \gamma/(d\nu) \end{aligned} \tag{2.111}$$

whose universality follows from the universality of the scaling functions. This universality is a little stronger than the universality of, say, the ratio of the specific heat amplitudes above and below T_c, since one has now a so-called **hyperuniversality relation** [535]

$$\lim_{\tau \to \pm 0} g^{\mathrm{sing}}(\tau) \xi^d(\tau) = u^{\pm} = \text{ universal.} \tag{2.112}$$

This is known as *two-scale factor universality* or the **Privman-Fisher hypothesis**. One often writes this statement in the following form

$$\begin{aligned} g^{\mathrm{sing}}(\tau, h) &= L^{-d} Y(C_1 \tau L^{1/\nu},\, C_2 h L^{(\beta+\gamma)/\nu}) \\ \xi^{-1}(\tau, h) &= L^{-1} S(C_1 \tau L^{1/\nu},\, C_2 h L^{(\beta+\gamma)/\nu}) \end{aligned} \tag{2.113}$$

with the *same* metric factors C_1, C_2 in both observables and universal finite-size scaling functions $Y(x_1, x_2)$ and $S(x_1, x_2)$. In particular, one expects the following amplitudes to be universal [535]

$$Y(0,0), \quad S(0,0), \quad \lim_{\mathfrak{z}\to\pm\infty} Y(\mathfrak{z},0)S^{-d}(\mathfrak{z},0), \quad \lim_{\mu\to\pm\infty} Y(0,\mu)S^{-d}(0,\mu) \quad (2.114)$$

Since universal amplitudes often vary much more between distinct universality classes than critical exponents [536], the consideration of those amplitudes in a specific system may be of diagnostic value. In $2D$ their values can be calculated from conformal invariance, see Sect. 2.5.

2.4 Fluctuation-Dissipation Theorem

In the second volume of this book we shall study time-dependent phenomena, in particular *ageing* which occurs far from the stationary states. In these studies, the precise relationship between correlators and responses will play an important role. At equilibrium, this relation is given by the celebrated **fluctuation-dissipation theorem.** This theorem relates the response of an equilibrium system to an applied external field with the *internal* fluctuations of the same equilibrium system. Such fluctuations are usually described in terms of time-dependent equilibrium correlation functions

$$C_{AB}(t) := \langle A(t)B(0)\rangle_0 \tag{2.115}$$

of two time-dependent observables A and B, where we assumed time-translational invariance. When considering a perturbation of an equilibrium system through an external field, it is useful to distinguish three basic situations:

(i) **response**, which considers the temporal evolution of a system under the influence of a time-independent force (Fig. 2.8a)

(ii) **relaxation**, where an external force is turned off and the free decay of a system is studied (Fig. 2.8b) and

(iii) **alternation**, where the effects of an oscillatory force (Fig. 2.8c) are investigated.

The discussion below follows closely [144] and concentrates first on the response and later on its relation to the equilibrium correlations of the system. The fluctuation-dissipation theorem is first presented abstractly, followed by concrete physical illustrations from several distinct relaxation phenomena.

Initially, one considers a quantum system in thermal equilibrium at temperature T which is described by a **density matrix**

$$\rho_0 := Z_0^{-1} \exp\left(-\frac{1}{T}H_0\right) , \tag{2.116}$$

where H_0 is the Hamiltonian and $Z_0 = \mathrm{tr}[\exp(-H_0/T)]$ the partition function. Then, starting at time $t = 0$, one perturbs the system with a steady

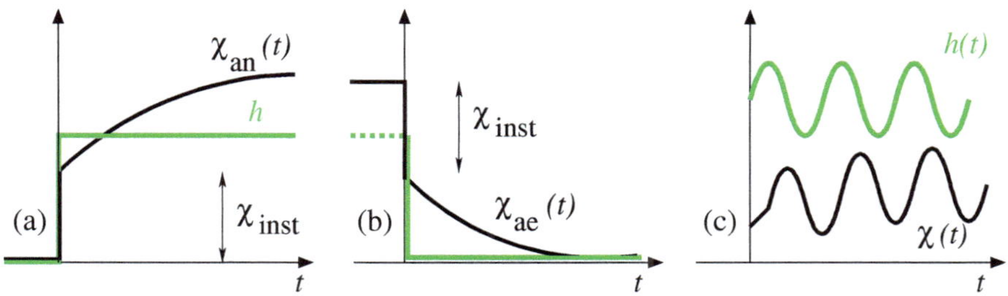

Fig. 2.8 Schematic equilibrium responses χ to (a) a constant external field h, (b) the turning-off of a constant external field and (c) an oscillatory external field $h(t)$. The instantaneous response χ_{inst} and the time-dependent 'an-elastic' and 'aftereffect' responses $\chi_{\text{an,ae}}(t)$ are indicated.

external force F which couples to an operator A. For times $t > 0$ the Hamiltonian becomes

$$H = H_0 - AF \tag{2.117}$$

and the density matrix $\rho(t)$ is determined by the **Liouville equation**

$$\frac{\partial \rho(t)}{\partial t} = -\mathrm{i}\mathcal{L}\rho(t) \tag{2.118}$$

with the initial condition $\rho(0) = \rho_0$, where the **Liouville operator** is defined by[19]

$$\mathcal{L}. = \frac{1}{\hbar}[H,.] \, . \tag{2.119}$$

Now one considers how the time-dependent average $\langle B\rangle(t)$ of some operator B is perturbed by the force F, quantified through the **response function**

$$\mathcal{R}_{BA}(t) := \left.\frac{\delta\langle B\rangle(t)}{\delta F}\right|_{F=0} . \tag{2.120}$$

In the context of linear-response theory it can be shown that for perturbations around equilibrium, for classical systems (such that $\hbar \to 0$), this function is given by

$$\begin{aligned} \mathcal{R}_{BA}(t) &= \frac{\mathrm{i}}{\hbar}\int_0^t \mathrm{d}\tau \, \langle [A(0), B(\tau)]\rangle_0 \\ &= T^{-1}\langle B(0)A(0)\rangle_0 - T^{-1}\langle B(t)A(0)\rangle_0 \, , \end{aligned} \tag{2.121}$$

where the average $\langle . \rangle_0$ is taken for the unperturbed system at temperature T. This relationship is the celebrated **fluctuation-dissipation theorem**.

---

[19] For classical systems, replace the commutator by the Poisson bracket.

Proof: Here we outline how the fluctuation-dissipation theorem can be derived, following [144]. Under a small perturbation (2.117), the Liouville operator can be decomposed as $\mathcal{L} = \mathcal{L}_0 + \mathcal{L}_A$. Writing the density matrix as $\rho(t) = \rho_0 + \delta\rho(t)$ the Liouville equation yields an equation for $\delta\rho(t)$. The basic assumption of linear response theory is that terms of the order $\mathcal{L}_A\delta\rho(t)$ are sufficiently small to be neglected. It is then straightforward to show from (2.118) that

$$\delta\rho(t) = \int_0^t \mathrm{d}t'\, e^{-\mathrm{i}(t-t')\mathcal{L}_0}\left(-\mathrm{i}\mathcal{L}_A\rho_0\right) = \frac{\mathrm{i}}{\hbar}F\int_0^t \mathrm{d}t'\, e^{-\mathrm{i}(t-t')\mathcal{L}_0}\left[A(0),\rho_0\right].$$

Similarly, one shows that the average of an observable B is, to leading order

$$\langle B\rangle(t) = \mathrm{tr}\,\left(\rho(t)B(0)\right) = \langle B\rangle_0 + \frac{\mathrm{i}}{\hbar}F\int_0^t \mathrm{d}t'\,\mathrm{tr}\,\left(e^{-\mathrm{i}(t-t')\mathcal{L}_0}\left[A(0),\rho_0\right]B(0)\right)$$

Next, one resorts to the **Heisenberg picture** and introduces time-dependent operators through (see exercise 9)

$$A(t) := e^{\mathrm{i}t\mathcal{L}_0}A(0) = e^{\mathrm{i}H_0 t/\hbar}A(0)e^{-\mathrm{i}H_0 t/\hbar}. \tag{2.122}$$

Furthermore, the operators A and B can be assumed to be Hermitean since they represent physical observables. The proof of the Hermiticity of the Liouville operator (2.119) is left as an exercise. Then, using the cyclicality of the trace

$$\begin{aligned}
\langle\delta B\rangle(t) &= \langle B\rangle(t) - \langle B\rangle_0 \\
&= \frac{\mathrm{i}F}{\hbar}\int_0^t \mathrm{d}\tau\,\mathrm{tr}\,\left(\rho_0\left[A(0),B(\tau)\right]\right) \\
&= -\frac{\mathrm{i}F}{\hbar}\int_0^t \mathrm{d}\tau\,\mathrm{tr}\,\left(\left[\rho_0,A(0)\right]B(\tau)\right) \\
&= -\frac{\mathrm{i}F}{\hbar}\int_0^t \mathrm{d}\tau\,\mathrm{tr}\,\left(\left(\rho_0 A(0) - e^{H_0/T}A(0)e^{-H_0/T}\rho_0\right)B(\tau)\right) \\
&= \frac{\mathrm{i}F}{\hbar}\int_0^t \mathrm{d}\tau\,\mathrm{tr}\,\left(\rho_0\int_0^{1/T}\mathrm{d}\beta\,\frac{\mathrm{d}}{\mathrm{d}\beta}\left(e^{\beta H_0}A(0)e^{-\beta H_0}\right)B(\tau)\right) \\
&= \frac{\mathrm{i}F}{\hbar}\int_0^t \mathrm{d}\tau\,\mathrm{tr}\,\left(\rho_0\int_0^{1/T}\mathrm{d}\beta\,\left(e^{\beta H_0}\left[H_0,A(0)\right]e^{-\beta H_0}\right)B(\tau)\right).
\end{aligned}$$

From the second line we get the first part of (2.121). We stress that in these transformations is it essential that there exists an equilibrium Hamiltonian H_0 which gives the density matrix ρ_0 through (2.116). If one introduces

$$\frac{\mathrm{d}A}{\mathrm{d}t}(0) := \frac{\mathrm{i}}{\hbar}\left[H_0, A(0)\right]$$

the response function $\mathcal{R}_{BA}(t)$ becomes

$$\mathcal{R}_{BA}(t) = \int_0^t \mathrm{d}\tau \int_0^{1/T} \mathrm{d}\beta \left\langle e^{\beta H_0} \frac{\mathrm{d}A}{\mathrm{d}t}(0) e^{-\beta H_0} B(\tau) \right\rangle_0 .$$

Let us now assume that the unperturbed state is *stationary* which means that two-time averages should satisfy time-translation invariance, viz.

$$\langle A(t)B(t+\tau)\rangle_0 = \langle A(0)B(\tau)\rangle_0$$

such that these correlators do not depend on t. This further implies

$$\left\langle \left(\frac{\mathrm{d}}{\mathrm{d}t}A(t)\right) B(t+\tau) \right\rangle_0 = -\left\langle A(t) \left(\frac{\mathrm{d}}{\mathrm{d}t}B(t+\tau)\right) \right\rangle_0$$

and the response function finally becomes

$$\begin{aligned} \mathcal{R}_{BA}(t) &= -\int_0^t \mathrm{d}\tau \int_0^{1/T} \mathrm{d}\beta \left\langle e^{\beta H_0} A(0) e^{-\beta H_0} \frac{\mathrm{d}B}{\mathrm{d}t}(\tau) \right\rangle_0 \\ &= \int_0^{1/T} \mathrm{d}\beta \left\langle e^{\beta H_0} A(0) e^{-\beta H_0} \left(B(0) - B(t)\right) \right\rangle_0 \\ &= \int_0^{1/T} \mathrm{d}\beta \left\langle A(-\mathrm{i}\hbar\beta) \left(B(0) - B(t)\right) \right\rangle_0 \end{aligned} \qquad (2.123)$$

which is the quantum version of the fluctuation-dissipation theorem. Since $\beta \le 1/T$, the argument of A tends to zero in the classical limit $\hbar \to 0$ and one recovers (2.121). This completes the proof. □

We need to point out here that in the definition (2.120) of the response function, the external force F was assumed to be applied statically at *all* times τ, $0 \le \tau \le t$. For later considerations of non-equilibrium systems, it is more useful to consider the response to a short impulse $\delta F(s)$ at time s. The **response function** to such an impulse is defined by

$$R_{BA}(t,s) = \left. \frac{\delta \langle B \rangle(t)}{\delta F(s)} \right|_{F=0} . \qquad (2.124)$$

At equilibrium, the response function $R_{BA}(t,s) = R_{BA}(t-s)$ only depends on the time-difference $\tau = t - s$. The fluctuation-dissipation theorem (2.121) may then be written in the form

$$R_{BA}(\tau) = \frac{\mathrm{d}\mathcal{R}_{BA}(\tau)}{\mathrm{d}\tau} = -\frac{1}{T}\frac{\partial}{\partial \tau}\langle B(\tau)A(0)\rangle_0 = -\frac{1}{T}\frac{\partial C_{BA}(\tau)}{\partial \tau} \qquad (2.125)$$

to which we shall refer when coming to the study of ageing phenomena in the second volume of this book.

Secondly, we consider relaxation effects. In this situation, the system is initially in an equilibrium state characterised by the density matrix $\rho(0) = Z^{-1}\exp(-H/T)$ and the time-dependent density matrix is given by the formal solution $\rho(t) = \exp(-\mathrm{i}\mathcal{L}_0 t)\rho_0$ of the Liouville equation, where the Liouville operator $\mathcal{L}_0$ is determined from the unperturbed Hamiltonian H_0. One considers the relaxation of the average $\langle B\rangle(t)$ from its initial value $\langle B\rangle_0 = \mathrm{tr}[\rho(0)B]$ towards its stationary value $\langle B\rangle(\infty) = \mathrm{tr}[\rho_0 B]$. This is described by the **relaxation function**

$$
\begin{aligned}
\Phi_{BA}(t) &:= \mathcal{R}_{BA}(0) - \mathcal{R}_{BA}(t) \\
\langle B\rangle(t) &= \langle B\rangle_0 + F\Phi_{BA}(t)\,.
\end{aligned}
\tag{2.126}
$$

Thirdly, we may also consider oscillating forces of the form $F\cos\omega t$. A straightforward calculation shows that to first order in F

$$
\langle \delta B\rangle(t) = \frac{F}{\hbar}\int_0^t \mathrm{d}\tau\; \cos(\omega(t-\tau))\; \langle [A(0), B(\tau)]\rangle_0\,. \tag{2.127}
$$

In many situations, and in keeping with the above treatment, one is mainly interested in non-transient effects. A simple way to obtain them is to wait 'sufficiently long' such that the response can be described faithfully through a single Fourier component with angular frequency ω and one may mimic this by extending the upper limit from t to ∞ in the above equation. This then leads to

$$
\begin{aligned}
\langle \delta B\rangle_{\mathrm{nt}}(t) &= \frac{F}{\mathrm{i}\hbar}\int_0^\infty \mathrm{d}\tau\; \cos(\omega(t-\tau))\; \langle [A(0), B(\tau)]\rangle_0 \\
&= \Re\left(\chi_{BA}(\omega) F e^{-\mathrm{i}\omega t}\right)\,,
\end{aligned}
\tag{2.128}
$$

where we defined the *complex* **alternating susceptibility**

$$
\chi_{BA}(\omega) = \chi'_{BA}(\omega) - \mathrm{i}\chi''_{BA}(\omega) = \frac{1}{\mathrm{i}\hbar}\int_0^\infty \mathrm{d}\tau\; \exp(\mathrm{i}\omega\tau)\; \langle [A(0), B(\tau)]\rangle_0 \tag{2.129}
$$

whose real part $\chi'_{BA}(\omega)$ describes the frequency-dependent response in phase with the external force while its imaginary part $\chi''_{BA}(\omega)$ describes the out-of-phase response. In Fig. 2.8c, we schematically indicated that the linear response to an external oscillating field $h(t)$ is in general out of phase with respect to $h(t)$ and besides the oscillating part, it may also contain contributions which vary monotonously as a function of time (in experiments, one usually averages over at least one period of the external field $h(t)$).

If the reference system is at equilibrium, the three functions $\mathcal{R}_{BA}(t)$, $\Phi_{BA}(t)$ and $\chi_{BA}(\omega)$ are related. The first two are related by the fluctuation-dissipation theorem (2.121) while the remaining ones are related as follows. Denoting by

$$\overline{f}(s) = \int_0^\infty \mathrm{d}t\, e^{-st} f(t) \tag{2.130}$$

the **Laplace transform** $\overline{f}(s)$ of a function f and comparing the first identity of (2.121) with the definition (2.129), one easily finds that

$$\chi_{BA}(\omega) = \lim_{\varepsilon\to 0}(-\mathrm{i}\omega+\varepsilon)\overline{\mathcal{R}}_{BA}(-\mathrm{i}\omega+\varepsilon) = \Phi_{BA}(0) - \lim_{\varepsilon\to 0}(-\mathrm{i}\omega+\varepsilon)\overline{\Phi}_{BA}(-\mathrm{i}\omega+\varepsilon). \tag{2.131}$$

These relations are known as the **response-relaxation relations**. It is instructive to further illustrate the content of the fluctuation-dissipation theorem. For an oscillating external force, the power absorbed or the average rate of work done on the system is

$$Q(t) = -\frac{\mathrm{d}}{\mathrm{d}t}\langle H_0 - AF\cos\omega t\rangle = \langle A(t)\rangle \omega F \sin\omega t\,. \tag{2.132}$$

Since experiments are often carried out in the non-transient domain, one recalls the definition

$$\langle A(t)\rangle_{\mathrm{nt}} = \langle A\rangle_0 + F\chi'_{AA}(\omega)\cos\omega t + F\chi''_{AA}(\omega)\sin\omega t\,. \tag{2.133}$$

Experiments actually measure the averaged power $\bar{Q}$ absorbed over at least one cycle, of period $2\pi/\omega$, and one has

$$\bar{Q} = \frac{\omega}{2\pi}\int_0^{2\pi/\omega} \mathrm{d}t\, Q_{nt}(t) = \frac{1}{2}\omega F^2 \chi''_{AA}(\omega) \tag{2.134}$$

which makes the relationship of the imaginary part $\chi''(\omega)$ with dissipation explicit.

In the vicinity of an equilibrium phase transition, the extension of the equilibrium scaling arguments towards **dynamical scaling** (see exercise 13) predicts the scaling behaviour of the alternating susceptibility, in the low-frequency limit $\omega \to 0$

$$\chi'(\omega,\tau) \sim |\tau|^{-\gamma} \;\;,\;\; \chi''(\omega,\tau) \sim \omega|\tau|^{-\gamma-\nu z}. \tag{2.135}$$

It is convenient to take magnets as physical systems of reference (as we shall do from now on), although many studies on relaxation phenomena are performed in different physical contexts, for example through dielectric and/or elastic measurements. In Table 2.2 we illustrate the above abstract discussion by giving concrete physical interpretations of the quantities discussed. At the same time, this table provides a short glossary between magnetic, dielectric and an-elastic relaxations.

A few more comments are required concerning an-elastic relaxation [144]. It is a well-known fact that a mechanical system subjected to a stress shows first an elastic response for small stress, where Hooke's law is valid (the

symbol	magnetic	dielectric	an-elastic
force	magnetic field $\boldsymbol{H}_z$	electric field $\boldsymbol{E}_z$	stress tensor $V\sigma_{zz}$
conjugate operator	magn. moment $\boldsymbol{\mu}_z$	el. dipole moment $\boldsymbol{d}_z$	strain ε_{zz}
interaction Hamiltonian	$-\boldsymbol{H}_z \cdot \boldsymbol{\mu}_z \cos\omega t$	$-\boldsymbol{E}_z \cdot \boldsymbol{d}_z \cos\omega t$	$-V\sigma_{zz}\varepsilon_{zz}\cos\omega t$
macroscopic observable	magnetisation M	polarisation P	an-elastic strain $\varepsilon_{\rm an}$
response function	magnetic response	polarisation response	creep function
relaxation function	magnetic aftereffect		elastic aftereffect
alternating susceptibility	magn. susc. $\chi(\omega)$	permittivity $D(\omega)$	compliance $J(\omega)$
power absorbed	$\chi''(\omega)$	$D''(\omega)$	internal friction $F(\omega)$

Table 2.2 Glossary for magnetic, dielectric and an-elastic relaxation, listing quantities corresponding to the force F, its canonically conjugate operator A and the associated macroscopic observable $\langle A\rangle$, the response function $\mathcal{R}_{AA}(t)$, the relaxation function $\Phi_{AA}(t)$, and the alternating susceptibility $\chi_{AA}(\omega)$.

induced strain is proportional to the applied external stress), and a plastic regime with a non-linear response which will not be considered here. Also, we have ignored completely the important tensorial aspects[20] of stress and restrict our analysis to uniaxial stresses and strains.

An-elastic relaxation as discussed above in terms of the response functions is measured in so-called **creep experiments**. In response to a uniaxial, homogeneous and constant stress σ_0 applied from time $t = 0$ onward the corresponding strain is given by $\varepsilon(t) = \varepsilon_{\rm inst} + \varepsilon_{\rm an}(t)$, see Fig. 2.8a. Here the component $\varepsilon_{\rm inst}$ builds up almost instantaneously and obeys Hooke's law, whereas the an-elastic strain $\varepsilon_{\rm an}(t)$ evolves slowly in time and shows a saturation (much as the magnetisation or polarisation in magnetic or dielectric systems). The measured response function is called the **creep function** $\mathcal{R}_{\rm an}(t) = \lim_{\sigma_0\to 0} \varepsilon_{\rm an}(t)/\sigma_0$. Similarly, in relaxation experiments one observes an instantaneous jump in the strain, followed by a slow decay of the an-elastic strain, called the **an-elastic aftereffect**, see Fig. 2.8b. Therefore, an an-elastic material completely recovers upon removal of an external stress, which is very different from what is seen e.g. in plastic materials. Physically, this is plausible since an-elastic relaxation comes from the stress-modulated motion of defects in materials, which on a microscopic level should be essentially random.

2.5 From Scale-Invariance to Conformal Invariance

In the second volume, we shall study the possibility of extending a scale-symmetry towards larger symmetry groups. The basic idea is perhaps most easily gleaned from the extension of scale-invariance at an equilibrium critical point towards conformal invariance.

[20] See e.g. [252] for an introduction.

In order to do so, we first reconsider scale-invariance. While we had previously started from the assumption that the thermodynamic potentials are generalised homogeneous functions which may be justified through renormalisation group arguments, we wish to take here the scaling of (connected) correlation functions as a starting point. For a simple ferromagnet at a critical point, consider the **two-point functions**

$$\begin{aligned} G_\sigma(\boldsymbol{r}_1-\boldsymbol{r}_2) &= \langle\sigma(\boldsymbol{r}_1)\sigma(\boldsymbol{r}_2)\rangle - \langle\sigma(\boldsymbol{r}_1)\rangle\langle\sigma(\boldsymbol{r}_2)\rangle \\ G_\epsilon(\boldsymbol{r}_1-\boldsymbol{r}_2) &= \langle\epsilon(\boldsymbol{r}_1)\epsilon(\boldsymbol{r}_2)\rangle - \langle\epsilon(\boldsymbol{r}_1)\rangle\langle\epsilon(\boldsymbol{r}_2)\rangle \end{aligned} \tag{2.136}$$

where $\sigma(\boldsymbol{r})$ and $\epsilon(\boldsymbol{r})$ stand for the densities of the order parameter and the energy at site $\boldsymbol{r}$, respectively. In terms of the reduced variables $\tau=(T_c-T)/T$ and h, dynamical scaling asserts that under a length rescaling with a constant rescaling factor b (suppressing metric factors) G_σ and G_ϵ are generalised homogeneous functions

$$\begin{aligned} G_\sigma(\boldsymbol{r};\tau,h) &= b^{-2x_\sigma}G_\sigma(\boldsymbol{r}/b;\tau b^{y_\tau},hb^{y_h}) \\ G_\epsilon(\boldsymbol{r};\tau,h) &= b^{-2x_\epsilon}G_\epsilon(\boldsymbol{r}/b;\tau b^{y_\tau},hb^{y_h})\,. \end{aligned} \tag{2.137}$$

Here x_σ and x_ϵ are the scaling dimensions of σ and ϵ while y_τ, y_h are, by abuse of language, called **renormalisation-group eigenvalues**. Rather than using the RG to derive (2.137), one may also adopt an axiomatic approach to scale-invariance and then use (2.137) as the *definition* of scaling. In other words, one *assumes* G_σ and G_ϵ to be generalised homogeneous functions. The densities $\sigma(\boldsymbol{r})$ and $\epsilon(\boldsymbol{r})$, which are conjugate to the scaling fields h and τ, respectively, are called **scaling operators** [122].

Let us recall how to recover the scaling behaviour of the thermodynamic potentials from (2.137). This is obtained from the static fluctuation-dissipation theorem (FDT) discussed in the previous section (exercise 11)

$$\begin{aligned} \chi &= \frac{1}{T}\sum_{\boldsymbol{r}} G_\sigma(\boldsymbol{r}) \simeq \frac{1}{T}\int \mathrm{d}^d\boldsymbol{r}\, G_\sigma(\boldsymbol{r})\,, \\ C &= \frac{1}{T^2}\sum_{\boldsymbol{r}} G_\varepsilon(\boldsymbol{r}) \simeq \frac{1}{T^2}\int \mathrm{d}^d\boldsymbol{r}\, G_\varepsilon(\boldsymbol{r})\,. \end{aligned} \tag{2.138}$$

Integrating (2.137), one obtains for the scaling of the susceptibility per site

$$\chi(\tau,h) = b^{d-2x_\sigma}\chi(\tau b^{y_\tau},hb^{y_h}). \tag{2.139}$$

Recall that $\chi = -\partial^2 g/\partial h^2$, where $g = G/\mathcal{N}$ is the density of the Gibbs potential. Therefore, integrating twice with respect to h, one arrives at

$$g(\tau,h) = b^{d-2x_\sigma-2y_h}g(\tau b^{y_\tau},hb^{y_h}). \tag{2.140}$$

Similarly, one has for the scaling of the specific heat

$$C(\tau, h) = b^{d-2x_\varepsilon} C(\tau b^{y_\tau}, h b^{y_h}) \tag{2.141}$$

and consequently, since $C = -\partial^2 g/\partial t^2$, up to factors which are constant and non-zero close to criticality, one obtains

$$g(\tau, h) = b^{d-2x_\varepsilon - 2y_\tau} g(\tau b^{y_\tau}, h b^{y_h}). \tag{2.142}$$

Comparing the two forms for the scaling of g, one arrives at $x_\varepsilon + y_\tau = x_\sigma + y_h$. In fact, a similar relationship in the scaling of the Gibbs potential (free energy) could be derived for any pair of scaling fields coupled to their conjugates.

Since the above constant is independent of the scaling operators, it should have simple value. A plausible choice in view of the scaling of the density of the Gibbs potential $g = G/\mathcal{N}$ is (equivalent to hyperscaling)

$$x_\varepsilon + y_\tau = x_\sigma + y_h = d \tag{2.143}$$

and finally the scaling law for the Gibbs potential density becomes

$$g(\tau, h) = b^{-d} g(\tau b^{y_\tau}, h b^{y_h}) \,. \tag{2.144}$$

The relationship with the conventional equilibrium critical exponents can be read from

$$x_\varepsilon = d - y_\tau = \frac{1-\alpha}{\nu} \quad , \quad x_\sigma = d - y_h = \frac{\beta}{\nu}. \tag{2.145}$$

From these considerations, it appears natural to assume that the co-variance of certain two-point correlation functions under dilatation really comes from the co-variance of the local operators $\sigma(\boldsymbol{r})$ or $\epsilon(\boldsymbol{r})$ from which they are built. If we let $\phi(\boldsymbol{r})$ stand for any local scaling operator, the covariance conditions under dilatations (2.137) can be reduced to $\phi(\boldsymbol{r}) = b^{-x}\phi(\boldsymbol{r}/b)$, where $x = x_\phi$ is the scaling dimension of the field ϕ.

While so far we have only studied *global* scale-transformation with a constant dilatation factor b, one may enquire whether there is a generalisation of global scale-invariance to a **local scale-invariance**, such that $b = b(\boldsymbol{r})$ becomes space-dependent, quite analogous to the generalisation of global to local gauge symmetry in particle physics. Using again the notation of local fields $\phi(\boldsymbol{r})$, the natural generalisation of global scale-invariance (2.137) is

$$\phi(\boldsymbol{r}) \mapsto \phi'(\boldsymbol{r}) = J(\boldsymbol{r})^{x/d} \phi(\boldsymbol{r}/b(\boldsymbol{r})) \,, \tag{2.146}$$

where $J(\boldsymbol{r})$ is the Jacobian of the transformation $\boldsymbol{r} \mapsto \boldsymbol{r}' = \boldsymbol{r}/b(\boldsymbol{r})$. From this, the transformation of correlators built from $\phi(\boldsymbol{r})$ is easily derived. Indeed, such extended space-time symmetries occur in many known situations, for example in Maxwell's equations in a vacuum, which are not only scale- but also conformally invariant. It is therefore tempting to ask whether scale-invariance in critical equilibrium systems may be extended to **conformal**

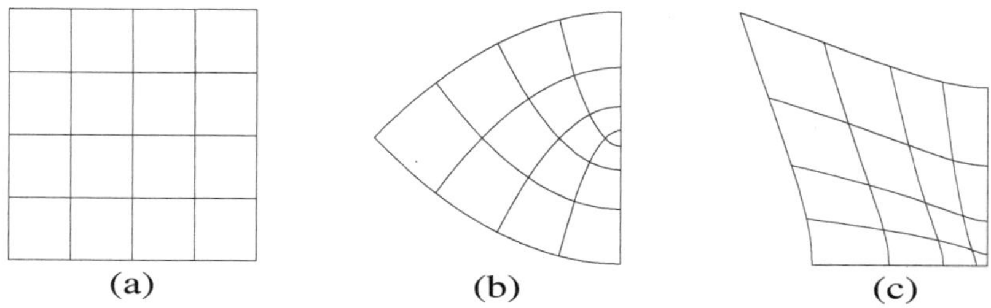

Fig. 2.9 Coordinate transformations. The transformation from the square lattice in **(a)** onto the lattice in **(b)** is conformal while the transformation onto the lattice in **(c)** is not. Reproduced from [270] with kind permission of Springer Science and Business Media.

invariance. Indeed, this extension appears to be true quite generally and appears to depend essentially on the locality of the underlying field-theory.

The discovery of conformal invariance in equilibrium phase-transitions in the 1970's [527] and the tremendous development in $2D$ in the 1980's, starting with the work of Belavin, Polyakov and Zamolodchikov [51] and of Cardy [117, 118] has been one of the most fruitful recent developments in equilibrium statistical physics, especially in two spatial dimensions. While in $d > 2$ dimensions, conformal invariance determines the form of the two- and three-point correlators, in $2D$ the form of any n-point correlator is determined in terms of the scaling dimensions. Furthermore, there exists a class of so-called **conformal minimal models** which contain a *finite* number of independent primary scaling operators whose scaling dimensions are known rational numbers. Hence all exponents, amplitudes and scaling functions can be found, the cross-over between different critical points can be analysed and much more.

In the following, we give a short summary of those properties of conformal invariance which will be useful for possible extensions to non-equilibrium phase-transitions in Vol. 2 of this book. For more detailed information, see e.g. [120, 122, 168, 270, 112].

1. By definition, a **conformal transformation** $\boldsymbol{r} \mapsto \boldsymbol{r}'$ maps the entire space $\mathbb{R}^d$ into itself such that *angles are preserved* while length scales may vary. Locally, a conformal transformation is a combination of rotations, dilatations and translations. For example, in Fig. 2.9 the mapping from the square lattice in panel (a) to panel (b) is conformal since all angles are locally conserved, while for the non-conformal mapping from (a) to (c) they are not. Geometrically, a non-conformal mapping can be recognised by its shear.

The predictive power of conformal invariance depends on the dimensionality of the system. In $d > 2$ spatial dimensions, the conformal group is finite-dimensional and is generated by translations $\boldsymbol{r} \mapsto \boldsymbol{r} + \boldsymbol{a}$, rotations $\boldsymbol{r} \mapsto \mathcal{D}\boldsymbol{r}$, where $\mathcal{D}$ is a rotation matrix in d dimensions, dilatations (scale-transformation) $\boldsymbol{r} \mapsto b\boldsymbol{r}$, and the so-called **special conformal transforma-**

tions

$$\boldsymbol{r} \mapsto \boldsymbol{r}' = \frac{\boldsymbol{r} + \boldsymbol{a}r^2}{1 + 2\,\boldsymbol{a}\cdot\boldsymbol{r} + a^2r^2}. \tag{2.147}$$

The geometric meaning of the special transformation (2.147) becomes clear when rewriting it as

$$\frac{\boldsymbol{r}'}{r'^2} = \frac{\boldsymbol{r}}{r^2} + \boldsymbol{a}, \tag{2.148}$$

which is the combination of an inversion $\boldsymbol{r} \mapsto \boldsymbol{r}/r^2$ followed by a translation $\boldsymbol{r} \mapsto \boldsymbol{r}+\boldsymbol{a}$ and again an inversion. In $d > 2$ dimensions, there are $\frac{1}{2}(d+1)(d+2)$ independent types of conformal transformations.

2. Because of the small number of conformal transformations in $d > 2$ dimensions, the imposed constraints should not be too strong. In *two* dimensions, however, the conformal group is generated by *infinitely many* generators. Here it is useful to work with complex coordinates

$$z = r_1 + \mathrm{i}r_2 \ , \ \ \bar{z} = r_1 - \mathrm{i}r_2 \tag{2.149}$$

instead with two-dimensional vectors $\boldsymbol{r} = (r_1, r_2)$. In this formulation, *any* transformation $z \mapsto w(z)$ or $\bar{z} \mapsto \bar{w}(\bar{z})$ where w and $\bar{w}$ are, respectively, complex analytic and anti-analytic functions, is a conformal transformation. For illustration, consider again Fig. 2.9, where the conformal transformation $w_{a\to b}(z) = z^2$ is clearly analytic while the non-conformal transformation $w_{a\to c}(z) = z|z|$ is not.[21]

Writing $w(z) = z + \epsilon(z), \bar{w}(\bar{z}) = \bar{z} + \bar{\epsilon}(\bar{z})$, the infinitesimal conformal transformations are formally given by

$$\epsilon_n = -\epsilon z^{n+1} \ , \ \ \bar{\epsilon}_n = -\bar{\epsilon}\bar{z}^{n+1}. \tag{2.150}$$

When these transformations are applied to a scaling operator $\phi = \phi(z, \bar{z})$, they generate the following infinitesimal changes $\delta\phi = \epsilon\ell_n\phi$ and $\bar{\delta}\phi = \bar{\epsilon}\bar{\ell}_n\phi$ with the explicit Lie algebra generators

$$\ell_n = -z^{n+1}\partial_z - \Delta(n+1)z^n \ , \ \ \bar{\ell}_n = -\bar{z}^{n+1}\partial_{\bar{z}} - \overline{\Delta}(n+1)\bar{z}^n\,. \tag{2.151}$$

Here the non-derivative terms express the transformation of the local field $\phi(\boldsymbol{r}) = \phi(z, \bar{z})$ itself and the terms containing derivatives describe the changes in the coordinates z and $\bar{z}$. The real numbers Δ and $\overline{\Delta}$ are (misleadingly) referred to as *complex* scaling dimensions or **conformal weights** of the scaling operator ϕ. Indeed, its scaling dimension x and its spin s are given by

$$x = \Delta + \overline{\Delta} \ , \ \ s = \Delta - \overline{\Delta}\,. \tag{2.152}$$

---

[21] In complex coordinates, only the conformal **Möbius transformations** $z \mapsto (\alpha z + \beta)/(\gamma z + \delta)$, $\alpha\delta - \beta\gamma = 1$ (also called **projective transformations**) have an analogue in $d > 2$. Moreover, they are the only conformal transformations which map the entire complex plane $\mathbb{C} \cong \mathbb{R}^2$ surjectively *onto* itself.

The generators (2.151) satisfy the following commutation relations

$$
\begin{aligned}
[\ell_n, \ell_m] &= (n-m)\,\ell_{m+n} \\
[\bar{\ell}_n, \bar{\ell}_m] &= (n-m)\,\bar{\ell}_{m+n} \\
[\ell_n, \bar{\ell}_m] &= 0.
\end{aligned}
\tag{2.153}
$$

This algebra is also called the **loop algebra** and decomposes into the direct sum of two commuting Lie algebras, one generated by the set $\langle \ell_n \rangle_{n\in\mathbb{Z}}$ and the other by $\langle \bar{\ell}_n \rangle_{n\in\mathbb{Z}}$. Because of this simple structure, it is often enough to consider merely the z-dependence of correlators.

3. Next, we remind the reader of the conformal invariance of Laplace's equation $\Delta_{\rm L}\phi = (\partial_{r_1}^2 + \partial_{r_2}^2)\phi = 0$ in two dimensions. In complex coordinates, the Laplace operator becomes $\Delta_{\rm L} = 4\partial_z\partial_{\bar{z}}$. The conformal invariance of Laplace's equation can be simply expressed through the commutator (and similarly for $\bar{\ell}_n$)

$$
[\Delta_{\rm L}, \ell_n] = -4(n+1)z^n \frac{\partial^2}{\partial z \partial \bar{z}} - 4\Delta(n+1)nz^{n-1}\frac{\partial}{\partial \bar{z}}
\tag{2.154}
$$

which means the following. *If* the conformal weight of the function $\phi(z,\bar{z})$ vanishes, viz. $\Delta = 0$, then $[\Delta_{\rm L}, \ell_n]\phi = -(n+1)z^n\Delta_{\rm L}\phi$. Therefore, solutions of the Laplace equation $\Delta_{\rm L}\phi = 0$ with a vanishing conformal weight and hence of vanishing scale dimension $x = \Delta + \overline{\Delta} = 0$ are mapped under any $2D$ conformal transformation into other solutions $\phi \mapsto (1+\epsilon\ell_n)\phi$ of the Laplace equation, for any integer $n \in \mathbb{Z}$.

4. Invariance under the finite-dimensional subgroup of the conformal group fixes some correlation functions built from a certain class of scaling operators, which are referred to as **quasiprimary** [51]. By definition, a quasiprimary scaling operator transforms covariantly according to (2.151) under projective conformal transformations. Covariance under translations, dilatations and the special transformation generated by ℓ_{-1}, ℓ_0 and ℓ_1 implies the so-called **projective Ward identity** for n-point correlation functions built from quasiprimary scaling operators $\phi_i = \phi_i(z_i, \bar{z}_i)$

$$
\sum_{i=1}^{n} \frac{\partial}{\partial z_i}\langle \phi_1 \ldots \phi_n \rangle = 0 \;, \quad \sum_{i=1}^{n}\left(z_i \frac{\partial}{\partial z_i} + \Delta_i \right)\langle \phi_1 \ldots \phi_n \rangle = 0
$$
$$
\sum_{i=1}^{n}\left(z_i^2 \frac{\partial}{\partial z_i} + 2\Delta_i z_i \right)\langle \phi_1 \ldots \phi_n \rangle = 0 \quad . \tag{2.155}
$$

A similar set of equations holds for the dependence on the variables $\bar{z}_i$.

We illustrate how the conditions (2.155) determine the form of the conformally covariant two-point function $\Phi(z_1, z_2; \bar{z}_1, \bar{z}_2) = \langle \phi_1(z_1,\bar{z}_1)\phi_2(z_2,\bar{z}_2)\rangle$. To this end it is enough to study the dependence on z_1 and z_2 explicitly. First, from translation invariance it is clear that $\Phi = \Phi(z)$ with $z = z_1 - z_2$.

Then dilatation-invariance implies

$$\ell_0 \Phi(z) = (-z\partial_z - \Delta_1 - \Delta_2)\, \Phi(z) = 0 \tag{2.156}$$

with the obvious solution $\Phi(z) = \Phi_0 z^{-\Delta_1-\Delta_2}$, whereas invariance under the special transformation gives

$$\begin{aligned} \ell_1 \Phi(z) &= \left(-\left(z_1^2 - z_2^2\right)\partial_z - 2\Delta_1 z_1 - 2\Delta_2 z_2\right) \Phi(z) \\ &= \left(-z^2\partial_z - 2\Delta_1 z\right)\Phi(z) + 2z_2\left(-z\partial_z - \Delta_1 - \Delta_2\right)\Phi(z) = 0\,, \end{aligned} \tag{2.157}$$

where we used the decomposition $z_1^2 - z_2^2 = (z_1 - z_2)^2 + 2z_2(z_1 - z_2)$. The last term in the second line of (2.157) vanishes because of dilatation-invariance (2.156). Next, multiply (2.156) by z and subtract it from (2.157). This leads to the result

$$(\Delta_1 - \Delta_2)\, z\Phi(z) = 0 \tag{2.158}$$

which means that the conformal weights of the two scaling operators have to be equal. Combining these results and restoring the conjugate part as well, the two-point function of quasi-primary scaling operators must be [527]

$$\langle \phi_1\, \phi_2 \rangle = \delta_{\Delta_1,\Delta_2}\, \delta_{\overline{\Delta}_1,\overline{\Delta}_2} \phi_0\, (z_1 - z_2)^{-2\Delta_1}\, (\bar{z}_1 - \bar{z}_2)^{-2\overline{\Delta}_1}\,, \tag{2.159}$$

where ϕ_0 is an arbitrary normalisation constant. For *scalars* $\phi_{1,2}$, one has $\Delta_i = \overline{\Delta}_i = x_i/2$ and the covariant two-point function becomes, reverting to real coordinates

$$\langle \phi_1(\boldsymbol{r})\, \phi_2(\boldsymbol{0}) \rangle = \phi_0\, \delta_{x_1,x_2} \left(r_1^2 + r_2^2\right)^{-x_1} = \phi_0\, \delta_{x_1,x_2}\, |\boldsymbol{r}|^{-2x_1}\,. \tag{2.160}$$

Similarly, one can determine the conformal three-point function [527]

$$\begin{aligned} \langle \phi_1 \phi_2 \phi_3 \rangle = \mathcal{C}_{123}\, & z_{12}^{-(\Delta_1+\Delta_2-\Delta_3)} z_{23}^{-(\Delta_2+\Delta_3-\Delta_1)} z_{13}^{-(\Delta_1+\Delta_3-\Delta_2)} \\ & \times \bar{z}_{12}^{-(\overline{\Delta}_1+\overline{\Delta}_2-\overline{\Delta}_3)} \bar{z}_{23}^{-(\overline{\Delta}_2+\overline{\Delta}_3-\overline{\Delta}_1)} \bar{z}_{13}^{-(\overline{\Delta}_1+\overline{\Delta}_3-\overline{\Delta}_2)}\,, \end{aligned} \tag{2.161}$$

where we wrote $z_{ij} = z_i - z_j$. We point out that the value of the coefficient $\mathcal{C}_{123}$ is *not* arbitrary.[22] Although these results were specifically derived in two dimensions, they are readily generalised to $d > 2$ for scalar fields. This can be seen as follows. By a rotation the two points z_1 and z_2 can always be moved onto a predetermined line and the three points z_1, z_2, z_3 of the three-point function can always be made to fall into a given plane. Hence the restriction to $d = 2$ is enough to obtain the form of both two- and three-point functions.

5. Often it is useful to describe the critical behaviour of a statistical system through a field-theory with action $S[\phi]$. The standard procedure of the Hubbard-Stratonovich transformation allows one to find this action with a

[22] With the normalisation $\mathcal{C}_{ij} = \delta_{ij}$ in (2.159), the $\mathcal{C}_{ijk}$ are universal.

continuum field ϕ systematically for any given lattice model with discrete spin variables, see exercise 16 for an example. Then thermodynamic averages can be calculated from the functional integral

$$\langle \mathcal{A} \rangle = \frac{1}{Z} \int \mathcal{D}\phi \, \mathcal{A}[\phi] e^{-S[\phi]/T} \,, \tag{2.162}$$

where $Z = \int \mathcal{D}\phi e^{-S[\phi]/T}$ is the partition function. Here $\mathcal{D}\phi = \prod_{\boldsymbol{r}} \int \mathrm{d}\phi(\boldsymbol{r})$ is a shorthand for the functional integration over the values of the continuum field at all space points and which replaces the sum over all spin configurations in the lattice model. Symmetries of the model, such as conformal invariance, can be studied through the transformation properties of the action. At an RG fixed point, the action will be invariant under global scale-transformations. If one considers an arbitrary coordinate transformation (not necessarily conformal) $\boldsymbol{r} \mapsto \boldsymbol{r}' = \boldsymbol{r} + \boldsymbol{\epsilon}(\boldsymbol{r})$ with a 'small' $\boldsymbol{\epsilon}(\boldsymbol{r})$, the action should change according to

$$\delta S = \int \mathrm{d}\boldsymbol{r} \, T_{\mu\nu}(\boldsymbol{r}) \partial^{\mu} \epsilon^{\nu} \,, \tag{2.163}$$

where $\mu, \nu = 1, \ldots, d$ and Einstein's summation convention is used. Translation invariance is already implemented here and one has also used the fact that the action S is a scalar. This kind of expansion implicitly assumes sufficiently short-ranged interactions in order to be valid. Since the coordinate change $\boldsymbol{r} \mapsto \boldsymbol{r}'$ may be interpreted as the stretching of an elastic medium, $T_{\mu\nu}$ is referred to as **stress-energy tensor** or **energy-momentum tensor.**[23]

For a rotation-invariant theory, one easily sees that $T_{\mu\nu} = T_{\nu\mu}$ is a symmetric tensor. Moreover, for a system at an RG fixed point, the action S is invariant under dilatations, hence $T_{\mu}^{\mu} = 0$, i.e., the stress-energy tensor is traceless. Finally, for a special conformal transformation

$$\epsilon^{\mu}(\boldsymbol{r}) = \eta^{\mu} \boldsymbol{r}^2 - 2 r^{\mu} \boldsymbol{\eta} \cdot \boldsymbol{r} \tag{2.164}$$

where $\boldsymbol{\eta}$ is some constant infinitesimal vector, one has

$$T_{\mu\nu} \partial^{\mu} \epsilon^{\nu} = 2 T_{\mu\nu} \left(r^{\mu} \eta^{\nu} - \eta^{\mu} r^{\nu} \right) - 2 T_{\mu}^{\mu} \boldsymbol{\eta} \cdot \boldsymbol{r} = 0 \tag{2.165}$$

and the action (2.163) is invariant. Roughly speaking, the above arguments assert that [100]

$$\left.\begin{array}{l}\text{translation-invariance}\\ \text{rotation-invariance}\\ \text{scale-invariance}\\ \text{short-ranged interactions}\end{array}\right\} \Longrightarrow \text{conformal invariance.} \tag{2.166}$$

---

23 The standard construction of the *canonical* energy-momentum tensor from the Lagrangian density of a classical field-theory may omit divergence terms. In order to obtain continuity equations for the conserved quantities, it may be necessary to include those terms and one then arrives at an 'improved' energy-momentum tensor, see e.g. [168].

We point out that in $2D$, this argument can be generalised and ensures the invariance of S under the full infinite-dimensional conformal group. If one of the conditions in (2.166) is not met, full conformal invariance will not hold.[24]

6. The invariance of a critical two-dimensional equilibrium system under any analytic mapping of its coordinate space combined with an appropriate rescaling of local operators essentially determines the form of the scaling functions of all n-point correlation functions and the possible values of critical exponents. In fact, conformal invariance provides a partial classification scheme of phase transitions in two dimensions, see e.g. [51, 120, 122, 168, 270] for further information. This comes about from the inclusion of the effects of thermal fluctuations into the conformal generators, $\ell_n \mapsto L_n$, which modifies the conformal algebra to the so-called **Virasoro algebra**

$$[L_n, L_m] = (n-m)L_{n+m} + \frac{c}{12}(n^3 - n)\delta_{n+m,0} \tag{2.167}$$

(and similarly for $\bar{\ell}_n$). The Virasoro algebra is characterised by a real number c, the **central charge**. The mathematical representation theory of the Virasoro algebra leads, via the celebrated Kac table, to a classification of modular-invariant partition functions of equilibrium critical phenomena in two dimensions. The most simple models of equilibrium critical phenomena correspond to the unitary representations of the Virasoro algebra which are possible if either $c \geq 1$ or $c = 1 - 6/(m(m+1))$ with $m = 2, 3, 4\ldots$ which for each value of m defines a **conformal minimal model**. For example, the $2D$ Ising model at criticality corresponds to $c = 1/2$ ($m = 3$) and the allowed conformal weights are $\Delta = 0, \frac{1}{16}, \frac{1}{2}$. Since both magnetisation and energy densities are scalars, one has $x_\sigma = \Delta_\sigma + \overline{\Delta}_\sigma = \frac{1}{16} + \frac{1}{16} = \frac{1}{8}$ and $x_\varepsilon = \Delta_\varepsilon + \overline{\Delta}_\varepsilon = \frac{1}{2} + \frac{1}{2} = 1$. The conventional critical exponents of the Ising model can now be obtained from (2.145). Remarkably, all critical exponents derived from the conformal minimal models turn out to be *rational.*

This last result seems to be specific to two spatial dimensions. So far, there is no convincing evidence that the critical exponents in three dimensions, where an analogue of the Virasoro algebra does not exist, are rational.

7. Two-dimensional conformal invariance allows us to relate universal finite-size amplitudes to critical exponents [117]. To see this, consider the **logarithmic conformal transformation**

$$w = \frac{L}{2\pi} \ln z \tag{2.168}$$

[24] It can be checked explicitly (see exercise 17) that the improved energy-momentum tensor for free fields is symmetric and furthermore traceless for vanishing masses of the field ϕ, but we warn the reader that the above argument does contain hidden subtleties the existence of which have been exposed recently [526, 551, 14].

which maps the infinite complex z-plane onto an infinitely long strip of finite width L, with *periodic* boundary conditions. Under a finite transformation $w = w(z)$ and $\bar{w} = \bar{w}(\bar{z})$, the two-point function built from conformal primary scaling operator $\phi = \phi(z, \bar{z})$ with conformal weights $(\Delta, \overline{\Delta})$ transforms as

$$\langle \phi(z_1, \bar{z}_1)\phi(z_2, \bar{z}_2)\rangle_z = \tag{2.169}$$
$$\left(\frac{\mathrm{d}w}{\mathrm{d}z}(z_1)\frac{\mathrm{d}w}{\mathrm{d}z}(z_2)\right)^{\Delta}\left(\frac{\mathrm{d}\bar{w}}{\mathrm{d}\bar{z}}(\bar{z}_1)\frac{\mathrm{d}\bar{w}}{\mathrm{d}\bar{z}}(\bar{z}_2)\right)^{\overline{\Delta}}\langle \phi(w_1, \bar{w}_1)\phi(w_2, \bar{w}_2)\rangle_w .$$

Using the explicit form (2.159) of the two-point function, one obtains for the transformation (2.168), with $z = \exp(2\pi L^{-1} w) = \exp(2\pi L^{-1}(u + \mathrm{i}v))$

$$\begin{aligned}\langle \phi(w_1, \bar{w}_1)\phi(w_2, \bar{w}_2)\rangle_w &= \left(\frac{2\pi}{L}\right)^{2\Delta+2\overline{\Delta}}\left(\frac{z_1^{1/2} z_2^{1/2}}{z_1 - z_2}\right)^{2\Delta}\left(\frac{\bar{z}_1^{1/2} \bar{z}_2^{1/2}}{\bar{z}_1 - \bar{z}_2}\right)^{2\overline{\Delta}} \\
&= \left(\frac{2\pi}{L}\frac{\exp[\frac{\pi}{L}(w_1+w_2)]}{\exp(\frac{2\pi}{L}w_1) - \exp(\frac{2\pi}{L}w_2)}\right)^{2\Delta}\cdot\left(\frac{2\pi}{L}\frac{\exp[\frac{\pi}{L}(\bar{w}_1+\bar{w}_2)]}{\exp(\frac{2\pi}{L}\bar{w}_1) - \exp(\frac{2\pi}{L}\bar{w}_2)}\right)^{2\overline{\Delta}} \\
&= \left(\frac{\pi}{L}\frac{1}{\sinh[\frac{\pi}{L}(w_1 - w_2)]}\right)^{2\Delta}\cdot\left(\frac{\pi}{L}\frac{1}{\sinh[\frac{\pi}{L}(\bar{w}_1 - \bar{w}_2)]}\right)^{2\overline{\Delta}} ,\end{aligned} \tag{2.170}$$

where $w_1 - w_2 = (u_1 - u_2) + \mathrm{i}(v_1 - v_2)$. To understand the meaning of this result, consider the limits:

1. $u_1 - u_2 \ll L$. By a rotation of the coordinate system in the plane, one can always arrange for $v_1 = v_2$ and one recovers the bulk form.
2. $u_1 - u_2 \gg L$. One finds an asymptotic exponential decay

$$\begin{aligned}\langle \phi(u_1, v_1), \phi(u_2, v_2)\rangle_{\text{strip}} &\simeq \left(\frac{2\pi}{L}\right)^{2x} \\
&\times \exp\left[-\frac{2\pi}{L}(\Delta + \overline{\Delta})(u_1 - u_2) - \mathrm{i}\frac{2\pi}{L}(\Delta - \overline{\Delta})(v_1 - v_2)\right],\end{aligned} \tag{2.171}$$

which is the usual way to define a correlation length ξ via $\langle \phi(u, v)\phi(0, 0)\rangle \sim \exp(-u/\xi)$. Because of $x = \Delta + \overline{\Delta}$, one reads off [117]

$$\xi = L/(2\pi x). \tag{2.172}$$

Therefore, two-dimensional conformal invariance confirms the Privman-Fisher hypothesis (2.113), in the case of a slab geometry. The universal finite-size amplitudes (2.114) become [117, 70, 4]

$$S_i(0,0) = \begin{cases} 2\pi x_i & ; \text{ periodic} \\ \pi x_{i,\mathrm{s}} & ; \text{ free} \end{cases} , \quad Y(0,0) = \begin{cases} -\pi c/6 & ; \text{ periodic} \\ -\pi c/24 & ; \text{ free} \end{cases} \tag{2.173}$$

where x_i is the bulk scaling dimension related to the correlation length associated with the two-point function of the primary scaling operator ϕ_i and $x_{i,\mathrm{s}}$ is the corresponding surface scaling dimension. For the order parameter σ, one has $x_\sigma = \beta/\nu$ and $x_{\sigma,\mathrm{s}} = \beta_1/\nu$. The relations (2.173) permit a very efficient calculation of the central charge c and the scaling dimensions x_i, $x_{i,\mathrm{s}}$ in a given model.

The other universal amplitude combinations of (2.114) can also be found and read, for the slab geometry with periodic boundary conditions [119]

$$\begin{aligned}\lim_{\tau\to 0} g_{\mathrm{sing}}(\tau,0)\bar{\xi}_\varepsilon^2(\tau,0) &= \lim_{\mathfrak{z}\to\pm\infty} Y(\mathfrak{z},0)\, S_\varepsilon^{-2}(\mathfrak{z},0) = -\frac{c}{48\pi}\frac{2-\alpha}{1-\alpha}\\ \lim_{h\to 0} g_{\mathrm{sing}}(0,h)\bar{\xi}_\sigma^2(0,h) &= \lim_{\mu\to\pm\infty} Y(0,\mu)\, S_\sigma^{-2}(0,\mu) = -\frac{c}{48\pi}\frac{2-\gamma}{1-\gamma}\end{aligned} \quad (2.174)$$

where α and γ are the usual bulk critical exponents. Here, the correlation lengths $\bar{\xi}_i$ are defined from the normalised second moments

$$\bar{\xi}_i^2 := \frac{1}{4}\frac{\displaystyle\int_{\mathbb{R}^2} \mathrm{d}\boldsymbol{r}\, r^2 \langle\phi_i(\boldsymbol{r})\phi_i(\boldsymbol{0})\rangle}{\displaystyle\int_{\mathbb{R}^2} \mathrm{d}\boldsymbol{r}\, \langle\phi_i(\boldsymbol{r})\phi_i(\boldsymbol{0})\rangle} \quad (2.175)$$

It is an open problem if, or to what extent, ideas and methods of conformal invariance might be used in the description of non-equilibrium phase transitions. This question will be taken up in Vol. 2 of this book.

Problems

3. Consider the Ising model on a lattice with $\mathcal{N}$ sites. Write down the partition functions $Z = Z(T,h) = \exp(-G(T,h)/T)$ and $\tilde{Z} = \tilde{Z}(T,M) = \exp(-F(T,M)/T)$ for fixed field h and fixed magnetisation M, respectively. Show that in the limit $\mathcal{N}\to\infty$, one recovers the Legendre transformation $F = G + Mh$, together with $M = -\partial G/\partial h$.

4. Consider the Ising model in an external magnetic field on a one-dimensional ring with L sites. If $\mathfrak{T}$ is the transfer matrix, recall that the partition function $Z = \mathrm{tr}\ \mathfrak{T}^L$. Derive an explicit real-space renormalisation group transformation in terms of the model's parameters by defining the matrix $\mathfrak{T}_2 := \mathfrak{T}^2$ such that $Z = \mathrm{tr}\ {\mathfrak{T}_2}^{L/2}$ and study its fixed points.

5. For a system with a plane boundary, define for $\tau = 0$ the exponents $m_s \sim h^{1/\delta_s}$, $m_1 \sim h^{1/\delta_1}$ and $m_1 \sim h_1^{1/\delta_{11}}$ and show that [183]

$$\delta_s = \frac{\beta+\nu}{\beta_s} \quad , \quad \delta_1 = \frac{\beta+\nu}{\beta_1} \quad , \quad \delta_{11} = \frac{\nu}{2}\frac{d-\eta_{\|}}{\beta_1}$$

6. Fill in the details of the derivation of the two-scale-factor universality (2.113) in the finite-size scaling at equilibrium.

7. Consider the Brownian motion of a particle of mass m and submitted to an external force $F(t)$. The velocity (for simplicity, take $d=1$) satisfies the equation of motion

$$\frac{\mathrm{d}v(t)}{\mathrm{d}t} = -\gamma v(t) + \eta(t) + \frac{1}{m}F(t) \tag{2.176}$$

where $\eta(t)$ is a centred Gaussian noise with variance $\langle\eta(t)\eta(t')\rangle = 2B\delta(t-t')$. First, consider the case $F=0$ without an external force. From the condition that at equilibrium $\langle v^2\rangle_{\rm eq} = k_{\rm B}T/m$, derive the **Einstein relation** $B = \gamma k_{\rm B}T/m$. Next, at equilibrium, calculate the two-time correlation function $C = C(t-t') = \langle v(t)v(t')\rangle$ and the linear response function $R = R(t-t') = \delta\langle v(t)\rangle/\delta F(t')|_{F=0}$. Show that they satisfy the fluctuation-dissipation theorem $\Im\widehat{R}(\omega) = \big(\omega/(2\gamma k_{\rm B}T)\big)\widehat{C}(\omega)$ and verify the equivalence with (2.125). Explain the relationship with the Einstein relation.

8. Consider a classical system with interaction Hamiltonian $\mathcal{H}_0$ described by a canonical ensemble with temperature T. If $\phi(t)$ denotes the order-parameter field, add a time-dependent perturbation $\int \mathrm{d}t\, h(t)\phi(t)$ to the Hamiltonian and show directly that the linear response function is

$$\mathcal{R}_{\rm eq}(t,t') = \left.\frac{\delta\langle\phi(t)\rangle}{\delta h(t')}\right|_{h=0;\rm eq} = \frac{1}{T}\big(\langle\phi(t)\rangle\langle\phi(t')\rangle - \langle\phi(t)\phi(t')\rangle\big)$$

where all averages are taken at equilibrium.

9. Prove the following identity involving the n-fold commutator

$$[H,[H,\ldots,[H,A]\ldots]]_{\text{n times}} = \sum_{k=0}^{n}\binom{n}{k}(-1)^{n-k}H^k A H^{n-k} \tag{2.177}$$

and use this to prove the Heisenberg representation (2.122) by expanding both sides in powers of t and comparing coefficients.

10. The Liouville operator $\mathcal{L}:\mathcal{V}\to\mathcal{V}$ acts on the vector space

$$\mathcal{V} := \big\{A:\mathcal{H}\to\mathcal{H} | A \text{ linear and } \operatorname{tr} A^\dagger A < \infty\big\}$$

of linear operators A on a Hilbert space $\mathcal{H}$ such that $\operatorname{tr} A^\dagger A$ is finite. This vector space $\mathcal{V}$ has a Hilbert-space structure with the scalar product

$$S(A,B) := \operatorname{tr}\,\big(A^\dagger B\big)$$

Show that $\mathcal{L}$ as defined in (2.119) is Hermitean with respect to this scalar product and conclude that if A is Hermitean, then $A(0)e^{-\mathrm{i}\hbar t\mathcal{L}_0} = A(-t)$.

11. Show from the fluctuation-dissipation theorem that in the long-time limit

$$\begin{aligned}\lim_{t\to\infty} \mathcal{R}_{BA}(t) &= \mathcal{R}_{BA}^{stat} = \int_0^{1/T} \mathrm{d}\beta \left(\langle A(-\mathrm{i}\hbar\beta)B(0)\rangle_0 - \langle A\rangle_0 \langle B\rangle_0 \right) \\ &\stackrel{\hbar\to 0}{\simeq} \frac{1}{T}\left(\langle A(0)B(0)\rangle_0 - \langle A\rangle_0\langle B\rangle_0 \right) \qquad (2.178)\end{aligned}$$

which reproduces the well-known *static* fluctuation-dissipation theorem, since $\mathcal{R}_{BA}^{stat}$ is the static susceptibility of equilibrium statistical mechanics. Also check that in the limit $\mathcal{N} \to \infty$ the integral form (2.138) of the static fluctuation-dissipation theorem can be reproduced.

12. Prove the following form of the fluctuation-dissipation theorem

$$\chi''_{AA}(\omega) = \frac{\mathrm{i}}{\hbar}\int_0^\infty \mathrm{d}t\, \sin(\omega t) \langle [A(0), A(t)]\rangle_0 \qquad (2.179)$$

Can one write an analogous representation for the real part $\chi'_{AA}(\omega)$?

13. The dynamical behaviour at an equilibrium phase transition can be described in terms of **dynamical scaling**, which for the correlation function may be cast in the form

$$C(t, \boldsymbol{r}; \tau) = |\boldsymbol{r}|^{-(d-2+\eta)} \mathfrak{C}\left(|\boldsymbol{r}|t^{-1/z}; |\boldsymbol{r}||\tau|^{\nu}\right) \qquad (2.180)$$

with the usual equilibrium exponents and z is the **dynamical exponent**. For an infinitesimal periodic perturbation, with angular frequency ω, around the equilibrium state, derive (2.135).

14. Consider the infinitesimal conformal transformations as given by the complex generators (2.151). Rewrite them in terms of the infinitesimal transformations of the components of the $2D$ vector $\boldsymbol{r} = (r_1, r_2)$.

15. Verify the conformally co-variant three-point function (2.161). For scalar fields, extend to any dimension $d \geq 2$.

16. Consider a d-dimensional hypercubic lattice with $\mathcal{N}$ sites. The Ising model partition function can be written as

$$Z = \frac{1}{2^{\mathcal{N}}} \sum_{\{\sigma\}} \exp\left(\sum_{(a,b)} K_{ab}\, \sigma_a \sigma_b \right) \quad , \quad \sigma_a = \pm 1, \qquad (2.181)$$

where $\hat{K}$ is a matrix of coupling constants. Recalling the identity

$$\int_{-\infty}^{\infty} \mathrm{d}^{\mathcal{N}}\boldsymbol{y}\, \exp\left(-\frac{1}{2}\boldsymbol{y}\hat{Q}\boldsymbol{y}\right) = (2\pi)^{\mathcal{N}/2}(\det\hat{Q})^{-1/2}. \tag{2.182}$$

where $\hat{Q}$ is an $\mathcal{N} \times \mathcal{N}$ matrix, derive the **Hubbard-Stratonovich transformation** which relates the partition function $\mathcal{Z}$ to the functional integral of a scalar ϕ^4 theory.

17. Construct the energy-momentum tensor for a free scalar field. Does your construction agree with (2.166)?

18. Consider a two-dimensional critical system in a semi-infinite geometry, e.g. in the upper half-plane. Derive the form of the two-point function from conformal invariance [116].

19. Use a conformal transformation to derive the relationship between the universal finite-size scaling amplitude $S_i(0,0)$ and the surface scaling dimension $x_{i,\mathrm{s}}$ for a system defined on an infinitely long strip of width L with *free* boundary conditions [117].

Chapter 3
Directed Percolation

The term *percolation* (from the Latin *percolare* = to filter) means to make a liquid to pass through fine interstices and is often used in the context of filtering.[1] A filter is usually made of a porous substance such as cloth, paper, sand or charcoal, through which a liquid may be passed to cleanse it of the solid or impure matter held in suspension. The retained material accumulates, clogging the pores so that the filter becomes impermeable after some time and needs to be replaced. Therefore, it is important to understand how the transition from percolation to congestion takes place, and we shall use a particular formulation of this problem as a paradigm for absorbing phase transitions.

3.1 Directed Percolation at First Glance

Various simple models for percolation have been introduced and intensively studied. In such models the pores of the filter are represented by the sites of a lattice. Neighbouring pores are connected by small channels which are represented by the bonds of the lattice that connect adjacent sites. In order to mimic irregularities in the network, each of these channels is randomly open with probability p or closed otherwise. An important question would be how the **percolation probability** p, which controls the microscopic connectivity of the channels, influences the macroscopic permeability of the filter.

It turns out that in a sufficiently large system there is a *phase transition* from a macroscopically permeable phase to another phase, where the filter becomes clogged so that the penetration depth is finite. This transition is continuous and takes place at a well-defined **critical threshold** p_c. As in

[1] The french word *percolateur*=coffee machine makes this even more explicit and, because of the direction imposed by the gravitational field, refers to directed, rather than isotropic, percolation.

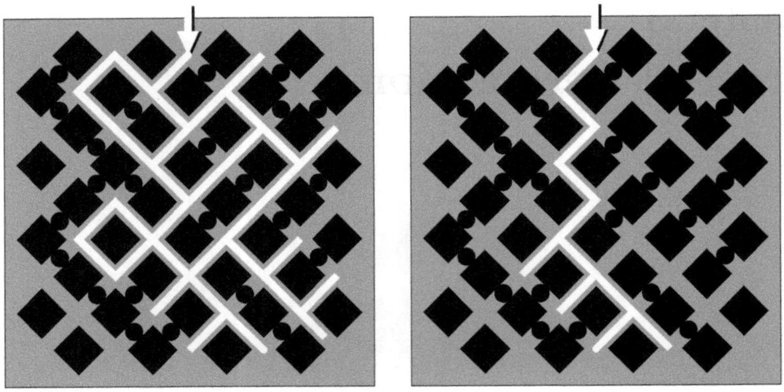

Fig. 3.1 Bond percolation: The figures show a finite lattice in a particular configuration of open and closed bonds. Water is injected from above in the middle, as indicated by the arrows. Left panel: In the case of isotropic (=undirected) percolation the water percolates through open channels in any direction, leading to a certain cluster of wet sites. Right panel: In the case of *directed* percolation the water can only propagate downwards, leading to a much smaller and more elongated cluster.

equilibrium statistical mechanics, the large-scale properties of percolation models close to the critical threshold turn out to be universal, i.e., they are determined by basic symmetries rather than microscopic details of the model. It is this universality which continues to fascinate theoretical physicists.

There are two fundamentally different versions of percolation. In *isotropic* (undirected) percolation, the agent can pass through open channels in any direction, while in **directed percolation** (DP) the water is restricted to flow along a preferred direction in space. For example, in a porous medium such a directed flow may be caused by a gravitational field, forcing the liquid to flow downwards. As shown in Fig. 3.1, the resulting clusters of wet sites are very different in both cases. Although both models exhibit percolation transitions, they take place at different critical thresholds and are characterised by different universal properties.

The two models also differ in so far as isotropic percolation in two space dimensions can be mapped exactly to the equilibrium q-state Potts model [529, 635] in the limit $q \to 1$ [224, 223] so that in two dimensions the critical exponents are known exactly [156, 487], see also appendix A. On the other hand, directed percolation is apparently not exactly solvable. An excellent introduction to isotropic percolation can be found in the textbook by Stauffer and Aharony [578]. In the present book, however, we are primarily concerned with directed percolation, which we shall from now on interpret as a stochastic many-particle process far from equilibrium.

The interpretation of DP as a stochastic process becomes possible by virtue of the strict order of cause and effect which allows one to interpret the preferred direction as a temporal coordinate. For example, in **bond directed**

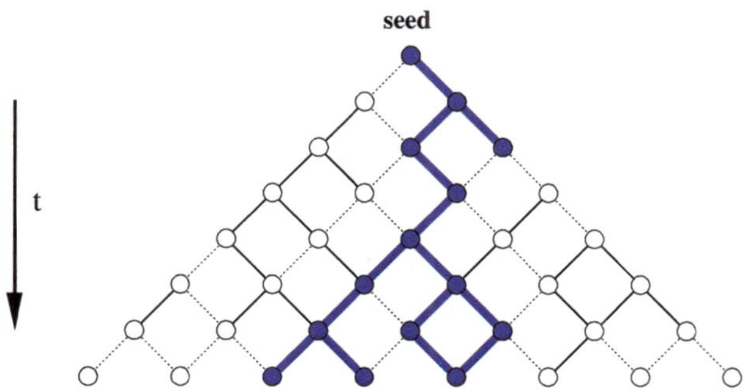

t	$N(t)$
0	1
1	1
2	2
3	1
4	2
5	2
6	3
7	3

Fig. 3.2 Bond directed percolation as a stochastic process evolving in time. The process shown here starts with a single wet site at the origin. It then evolves through a sequence of configurations along horizontal lines enumerated by a temporal index t. An important quantity to study is the number $N(t)$ of wet sites at time t.

percolation, where bonds are randomly blocked, we may enumerate horizontal rows by a temporal index t, commonly referred to as 'time', as shown in Fig. 3.2.[2] Knowing the configuration of wet sites at time t we can compute the next configuration at time $t+1$ by means of simple probabilistic rules. The same applies to **site directed percolation**, where sites instead of bonds are randomly blocked.[3]

Interpreting wet (active) sites as particles A and dry (inactive) sites as vacancies $\emptyset$ these probabilistic rules can be regarded as update rules of a **reaction-diffusion process**. For example, if both channels to the nearest neighbours at time $t+1$ are blocked, the trajectory of the particle terminates, meaning that the particle disappears by a death process $A \to \emptyset$. If only one channel is open the particle effectively diffuses to the left or right with equal probability. Finally, if both channels are open the particle duplicates creating an offspring, i.e., it undergoes the reaction $A \to 2A$. However, each site can be occupied by at most one particle. Therefore, if two particles happen to reach the same site, they merge to a single one by a coalescence process $2A \to A$. Therefore, DP can be interpreted as a reaction-diffusion process which involves

- particle removal (death) $A \to \emptyset$
- offspring production $A \to 2A$
- coalescence $2A \to A$

combined with single-particle diffusion, as demonstrated in Fig. 3.3. As will be discussed below, any stochastic particle process that effectively follows

[2] It is common to refer to a lattice with d directions perpendicular to the single 'time' direction as '$(d+1)$-*dimensional*'.

[3] In site percolation, the sites of a lattice are occupied with probability p and the occupied sites are linked by nearest-neighbour bonds to form clusters. In bond percolation, all sites are occupied and bonds are drawn between nearest-neighbour sites with probability p.

this reaction-diffusion scheme belongs generically to the universality class of DP.

Using this dynamical interpretation it is straightforward to generate explicit realisations of DP-clusters on a computer. To this end one simply iterates over all active sites at time t and activates their nearest neighbours at time $t+1$ independently with probability p. Usually a simple non-optimised source code takes less than a page. For example, the simple C-code shown in Fig. 3.4 measures the number of particles $N(t)$ in a bond DP process starting with a single seed, averaging the result over R independent runs.

To get a first impression of what DP clusters look like it is useful to plot the generated clusters for various percolation probabilities, as shown in Fig. 3.5. Obviously, if p is small, the generated clusters have the shape of a finite droplet, as shown in the left panel of the figure. On the other hand, if p is sufficiently large, activity may spread over the entire system within a cone-like region in space-time, generating an infinite cluster of active sites (see right panel of Fig. 3.5). However, even for large values of p it may happen that some of these clusters die out at an early stage.

Increasing the computational effort and analysing the cluster sizes statistically one observes the following phenomenology: Below a certain well-defined threshold $p < p_c$ all generated clusters remain finite while for $p > p_c$ some of the clusters (but not all of them) spread infinitely over the entire system. These two phases are separated by a sharp transition point at a specific critical threshold $p = p_c$ which in the case of bond DP in one space dimension is close to 0.6447. At this point finite clusters of all sizes are generated. These critical clusters are sparse and remind one of self-similar fractal structures. As will be explained below, the large-scale properties of critical clusters are universal, i.e., they do not depend on the microscopic details of the model under consideration.

At this point it is instructive to study the average number of active particles $\langle N(t) \rangle$ generated by the short C-program listed in Fig. 3.4. The result is shown in Fig. 3.6 in a double-logarithmic plot. Below the critical threshold the average number of particles $\langle N(t) \rangle$ first increases until it crosses over to an exponential decay. Above the critical point the increase accelerates until it crosses over to a linear increase. Precisely at the critical point $p = p_c$ the corresponding curve in the log-log plot appears to be straight, indicating power-law behaviour. In fact, it is found that $\langle N(t) \rangle$ measured at criticality increases for large t as t^{Θ}, where the exponent $\Theta \approx 0.302$ is just the slope of

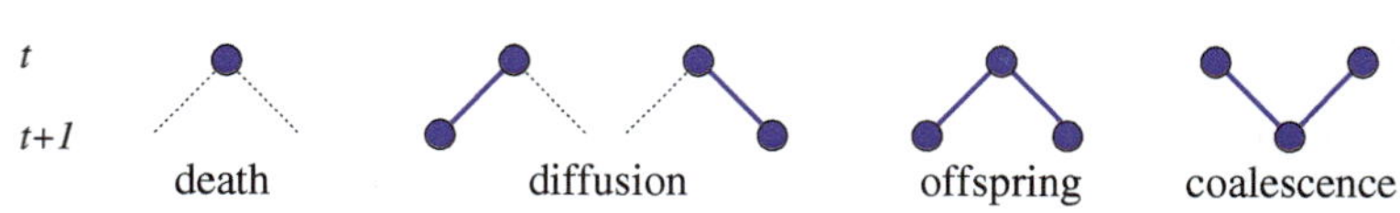

Fig. 3.3 Interpretation of bond directed percolation as a reaction-diffusion process of interacting particles (see text).

```
#include <fstream.h>
using namespace std;

const int T=1000;             // number of updates
const int R=10000;            // number of independent runs
const double p=0.6447;        // percolation probability

double RND(void) { return (double)rand()/0x7FFFFFFF; }

int main (void) {
int s[T][T],N[T],i,t,r;       // array of sites, N(t), indices
for (t=0; t<T; t++) N[t]=0; // clear particle counters
for (r=0; r<R; r++) {         // loop over R runs
    s[0][0]=1;                // place initial seed
    for (t=0; t<T-1; t++) { // temporal loop
        for (i=0; i<=t+1; ++i) s[t+1][i]=0;    // clear new config
        for (i=0; i<=t; ++i) if (s[t][i]==1) { // loop over wet sites
            N[t]++;                          // count wet sites
            if (RND()<p) s[t+1][i]=1;        // random activation left
            if (RND()<p) s[t+1][i+1]=1;      // random activation right
    }   }   }
ofstream os ("N.dat");        // write average N(t) to file
for (t=0; t<T-1; t++) os << t << ' ' << (double)N[t]/R << endl;
}
```

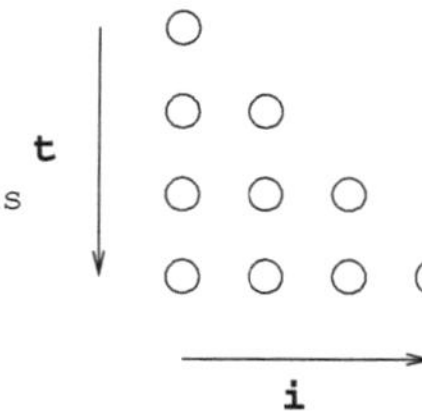

Fig. 3.4 Simple C-program generating a bond DP cluster, at p=0.6447.

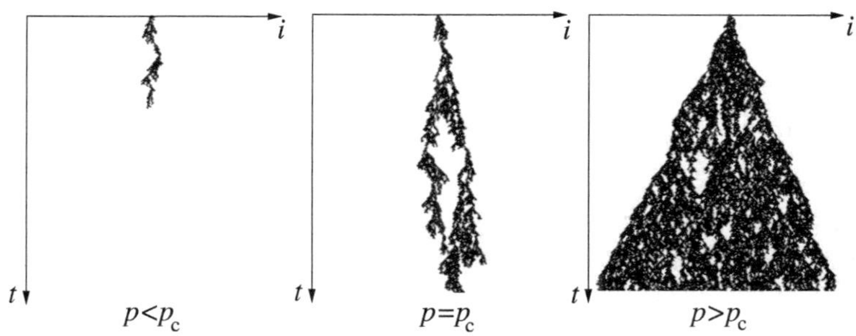

Fig. 3.5 Typical DP clusters grown from a single seed in 1+1 dimensions.

the straight line in the log-log plot. For small t, however, especially during the first few time steps, we observe deviations from the anticipated power-law behaviour. These so-called initial transients are caused by the discrete lattice structure and depend on non-universal details of the model. The asymptotic notation

$$\langle N(t)\rangle \sim t^{\Theta} \tag{3.1}$$

intentionally ignores such initial transients as well as the model-dependent proportionality constant and thus allows one to concentrate on the interesting universal behaviour valid in the limit of large times t and for large system sizes. The exponent Θ, however, is the same in all DP models, i.e., it is universal with respect to microscopic details of the model. It is this robust universality combined with its simplicity which makes DP so fascinating.

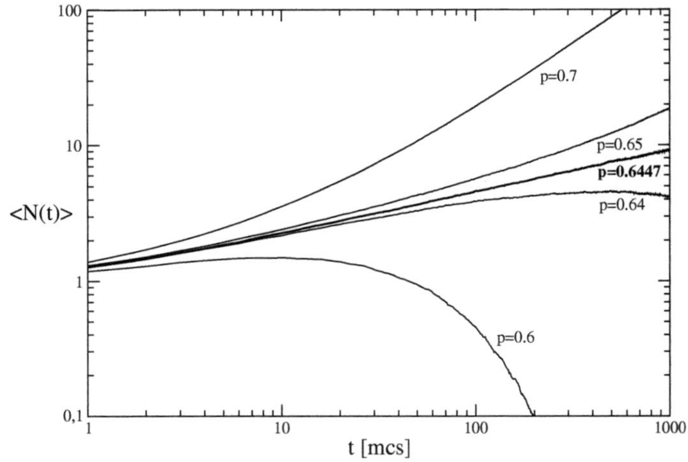

Fig. 3.6 Average number of particles $\langle N(t)\rangle$ as a function of time t (measured in units of Monte Carlo steps) for various percolation probabilities p.

3.2 Directed Percolation as a Stochastic Process

3.2.1 Basic Scaling Behaviour

As outlined in the previous section, DP can be interpreted as a stochastic process of diffusing particles which react as $A \to \emptyset$, $A \to 2A$. In addition, each site can be occupied by at most one particle, effectively introducing a coalescence process $2A \to A$. On a descriptive level the DP phase transition may be thought of as being caused by a competition of spontaneous reproduction and self-annihilation of particles. If the rate for offspring production $A \to 2A$ is sufficiently large, the (infinite) system is able to maintain a fluctuating active state characterised by a non-vanishing stationary particle density $\varrho > 0$. Contrarily, if the dynamics is dominated by death processes the system approaches a configuration without particles.

Such a configuration, from where the system cannot escape, is called **absorbing**. More specifically, an absorbing state is a configuration (or a set of several configurations) that can be reached by the dynamics but not be left by them. Obviously such absorbing states cannot obey detailed balance with any other active state, hence the system is by definition out of thermal equilibrium. Therefore, DP is said to display a **non-equilibrium phase transition** from a fluctuating phase into an absorbing state or, more concisely, an **absorbing phase transition**.

As in equilibrium statistical mechanics, it turns out that continuous phase transitions in systems out of equilibrium can be grouped into universality classes which are associated with certain critical exponents. Absorbing phase transitions are characterised by at least three independent critical exponents[4] β, $\nu_\perp$, and $\nu_\parallel$. The exponent β describes how the stationary density of particles in the active phase of an infinitely large system scales[5] close to criticality, i.e.,

$$\varrho \sim (p - p_c)^\beta . \tag{3.2}$$

Here p is the percolation probability that controls the effective rate for offspring production while p_c denotes its critical value where the transition takes place. Such a stationary state is characterised by a **correlation length** $\xi_\perp$ and a **correlation time** $\xi_\parallel$. They both diverge close to criticality as

$$\xi_\perp \sim (p - p_c)^{-\nu_\perp} , \qquad \xi_\parallel \sim (p - p_c)^{-\nu_\parallel} . \tag{3.3}$$

Since there is no symmetry between space and time, the two exponents $\nu_\perp$ and $\nu_\parallel$ are generally different (see Table 4.3 on page 159 for numerical values of these and other non-equilibrium exponents of DP).

In contrast to equilibrium statistical mechanics, stochastic models involve *time* as an independent degree of freedom on equal footing with the spatial degrees of freedom, allowing one to study dynamical properties such as relaxation phenomena. This requires one to specify an **initial state**. For example, starting with a fully occupied lattice at the transition $p = p_c$, the density of particles (active sites) decays algebraically as

$$\varrho(t) \sim t^{-\alpha} , \tag{3.4}$$

where $\alpha = \beta/\nu_\parallel$. Similarly, the spatial correlation length grows as

$$\xi_\perp(t) \sim t^{1/z} , \tag{3.5}$$

where $z = \nu_\parallel/\nu_\perp$ is the so-called **dynamical exponent**.

In finite systems one observes deviations from these asymptotic power laws. For example, on a finite lattice there is always a non-vanishing probability of reaching the absorbing configuration. Typically this leads to a breakdown of the power laws (3.4) and (3.5) when the correlation length $\xi_\perp$ becomes comparable with the lateral size (diameter) of the system. Generally such **finite-size effects** set in after a typical time that grows with the system size. For example, if L is the lateral size of the system (so that $\mathcal{N} \propto L^d$ is the total number of sites), the absorbing state is reached at a characteristic time t_f that scales as $t_f \sim L^z$.

---

[4] As will be shown in Sect. 4.1 there are actually two exponents β and β', meaning that absorbing phase transitions are generally described by *four* independent critical exponents. In the case of DP these two exponents coincide because of a special symmetry under time-reversal.

[5] See Table 4.1 on p. 102 for a collection of the definitions of the conventional critical exponents of absorbing phase transitions.

3.2.2 Universality and the DP Conjecture

As mentioned before there are a large variety of models which exhibit a DP phase transition. This means that the term 'directed percolation' does not refer to a particular model; rather it stands for a whole universality class of models which display the same type of transition, characterised by a certain set of critical exponents and scaling functions. The DP class comprises diverse types of systems, including models for epidemic spreading, forest fires, catalytic reactions, synchronisation of maps, and surface growth, to name only a few. In order to delineate this range of models, Janssen and Grassberger formulated their celebrated **DP-conjecture** [326, 240]. According to this conjecture, it is thought that a given model should generically belong to the DP universality class if

1. the model displays a continuous phase transition from a *fluctuating* active phase into a *unique* absorbing state,
2. the transition is characterised by a non-negative one-component order parameter,
3. the dynamic rules are short-ranged,
4. the system has no special attributes such as unconventional symmetries, conservation laws, or quenched randomness.

Apart from multicritical extensions of directed percolation (e.g. tricritical directed percolation, see Sect. 5.4) which violate the above hypothesis in the strict sense, no counterexamples have been found so far. In fact, the DP class seems to be even more general as it has been identified in models that violate some of the four conditions, for example in certain models with several [483, 482] or fluctuating absorbing states [470] or even in spreading processes with several particle species and multicomponent order parameters [254, 360, 11, 468, 505, 645, 310]. This means that the DP universality class is extremely robust.

In contrast to well-known equilibrium universality classes such as the $2D$ Ising model, even in 1+1 dimensions DP has not yet been solved exactly. Despite its simplicity and robustness all attempts to compute the critical exponents exactly have failed so far. In fact, numerical estimates suggest that the values of the critical exponents might turn out to be irrational numbers, rather than the simple rational values found in $2D$ integrable systems *at* equilibrium.

3.2.3 Simple Mean-Field Approximation

Most models that belong to the DP universality class can be interpreted directly as a stochastic particle processes. Let us now introduce effective reaction rates (probabilities per unit time) μ_r,μ_p, and μ_c for particle removal, offspring production, and coalescence:

$$\begin{aligned} A &\longrightarrow 2A &\quad& \text{with rate } \mu_{\mathrm{p}} \\ A &\longrightarrow \emptyset && \text{with rate } \mu_{\mathrm{r}} \\ 2A &\longrightarrow A && \text{with rate } \mu_{\mathrm{c}} \end{aligned}$$

Starting with such an effective reaction-diffusion scheme, it is straightforward to set up a simple **mean-field approximation** in terms of the mean particle density $\varrho(t)$, which neglects any spatial information (cf. appendix E and exercises 20,21,23,29). In this approximation the unary reactions $A \longrightarrow 2A$ and $A \longrightarrow \emptyset$ correspond to a linear term for particle gain and loss while the binary coalescence process $2A \longrightarrow A$ gives rise to a quadratic loss term, leading to the mean-field equation

$$\partial_t \varrho(t) = \tau \varrho(t) - g \varrho(t)^2 \,. \tag{3.6}$$

Here we introduced the so-called control parameter $\tau = \mu_{\mathrm{p}} - \mu_{\mathrm{r}}$ which, as we shall see below, is proportional to the distance from the critical point. Moreover we used the abbreviation $g = \mu_{\mathrm{c}}$ since in a field-theoretic context this parameter will play the role of a coupling constant.

In the asymptotic limit of large t the solution of the mean-field equation is given to leading order by

$$\varrho(t) \simeq \begin{cases} -\tau \left(g - \frac{\tau}{\varrho_0} \right)^{-1} \mathrm{e}^{\tau t} & \underset{t\to\infty}{\longrightarrow} 0 \quad \text{if } \tau < 0 \\ \left(\varrho_0^{-1} + gt \right)^{-1} & \underset{t\to\infty}{\longrightarrow} 0 \quad \text{if } \tau = 0 \\ \frac{\tau}{g} + \frac{\tau}{g^2} \left(g - \frac{\tau}{\varrho_0} \right) \mathrm{e}^{-\tau t} & \underset{t\to\infty}{\longrightarrow} \frac{\tau}{g} \quad \text{if } \tau > 0 \,, \end{cases} \tag{3.7}$$

where ϱ_0 denotes the initial density at $t = 0$. Obviously the steady-state solution $\varrho = 0$ corresponds to the absorbing state. For $\tau > 0$ the solution approaches a steady-state with a non-vanishing stationary density of active sites $\varrho = \tau/g$ while for $\tau = 0$ one obtains an algebraic decay. Therefore, within mean-field theory the transition takes place at the critical threshold $\tau_{\mathrm{c}} = 0$. Comparing these results with (3.2) and (3.4) one obtains the mean-field exponents $\beta_{\mathrm{MF}} = 1$ and $\alpha_{\mathrm{MF}} = 1$, respectively.

In order to obtain the third independent exponent, one has to include an additional term in the mean-field equation which provides spatial information. Since in DP active sites create offspring at nearest neighbours, the particles are subject to an effective diffuse motion. This can be accounted for by adding a term for diffusion in the mean-field equation

$$\partial_t \varrho(t, \boldsymbol{r}) \;=\; \tau \varrho(t, \boldsymbol{r}) - g \varrho(t, \boldsymbol{r})^2 + D \nabla^2 \varrho(t, \boldsymbol{r}) \,, \tag{3.8}$$

where D is the diffusion constant. Inserting the scaling ansatz $\varrho(t,\boldsymbol{r}) = (gt)^{-1}f(\boldsymbol{r}/t^{1/z})$, where $z = \nu_\parallel/\nu_\perp$ is the dynamical critical exponent, one obtains a consistent differential equation for f only if $z_{\mathrm{MF}} = 2$. Therefore, the mean-field exponents of directed percolation are given by

$$\beta_{\mathrm{MF}} = 1, \qquad \nu_{\perp,\mathrm{MF}} = 1/2, \qquad \nu_{\parallel,\mathrm{MF}} = 1. \tag{3.9}$$

3.2.4 Phenomenological Langevin Equation

The mean-field exponents are expected to be valid in sufficiently high space dimensions, where diffusive mixing is strong enough to suppress correlations. As we shall see now, the mean-field approximation is actually valid in space dimensions above the so-called **upper critical dimension** $d_c = 4$. To see this let us consider the **phenomenological stochastic Langevin equation** for DP which accounts for fluctuation effects on a coarse-grained level. This Langevin equation can be derived rigorously from the master equation of the contact process [326] and looks like the mean-field equation (3.8) extended by a suitable noise term:

$$\partial_t \varrho(t,\boldsymbol{r}) \;=\; \tau\varrho(t,\boldsymbol{r}) - g\varrho(t,\boldsymbol{r})^2 + D\nabla^2\varrho(t,\boldsymbol{r}) + \eta(t,\boldsymbol{r}) \,. \tag{3.10}$$

Here $\eta(t,\boldsymbol{r})$ stands for a density-dependent Gaussian noise field with the correlations

$$\langle \eta(t,\boldsymbol{r}) \rangle = 0 \,, \tag{3.11}$$

$$\langle \eta(t,\boldsymbol{r})\eta(t',\boldsymbol{r}') \rangle = \kappa\, \varrho(t,\boldsymbol{r})\, \delta(t-t')\, \delta^d(\boldsymbol{r}-\boldsymbol{r}') \,, \tag{3.12}$$

where κ controls the global noise amplitude. Note that the density appears on the right-hand side of the correlator, meaning that the amplitude of the noise is effectively proportional to $\sqrt{\varrho(t,\boldsymbol{r})}$. This ensures that the absorbing state $\varrho(t,\boldsymbol{r}) = 0$ does not fluctuate. The square-root behaviour is related to the fact that the noise describes **density fluctuations** on a coarse-grained scale, which can be viewed as the sum of individual noise contributions generated by each particle averaged over some mesoscopic box size. According to the **central limit theorem**, the distribution of the sum $\eta = \sum_{i=1}^{\mathcal{N}} \eta_i$ of $\mathcal{N}$ identically distributed random variables η_i (such that their average and variance are both finite) tends in the limit $\mathcal{N} \to \infty$ to a Gaussian. Hence one expects $\eta(t,\boldsymbol{r})$ to be drawn from a Gaussian distribution with an amplitude proportional to the square root of the number of active sites in the box.

By a dimensional analysis of the Langevin equation, one observes that the noise is irrelevant in $d > 4$, marginal in $d = 4$, and relevant in $d < 4$ dimensions (see exercise 43). This means that $d_c = 4$ is the **upper critical dimension** of directed percolation above which the mean-field exponents are correct. Below d_c fluctuation effects become relevant and the exponents have to be determined by a renormalisation group study of the corresponding field theory, as will be discussed in Sect. 4.3.4.

We note that (3.10) is the minimal Langevin equation needed to describe DP. It may also include higher order terms such as $\varrho^3(t,\boldsymbol{r})$, $\nabla^4\varrho(t,\boldsymbol{r})$, or higher-order contributions of the noise. The irrelevance of such higher-order terms can be understood from the renormalisation group (RG), as we shall explain later, and is the origin of universality in DP. Furthermore we note that Langevin equations for systems with absorbing states are difficult to iterate numerically. To our knowledge the most efficient method was proposed by Dornic et al. in [192].

3.2.5 Update Schemes and Evolution Equations

The dynamical rules of stochastic models can be defined in terms of various update schemes, the most important ones being **parallel** (synchronous) and **random-sequential** (asynchronous) updates.

3.2.5.1 Parallel Updates:

In models with parallel updates such as directed bond percolation (see Sect. 3.1) the time parameter t is discrete. For an update from time t to time $t+1$ the configuration c is mapped to a new configuration c' with probability $p_{c\to c'}$. Consequently the probability distribution $P_t(c)$ of finding the system at time t in the configuration c evolves as

$$P_{t+1}(c) = P_t(c) + \underbrace{\sum_{c'} p_{c'\to c} P_t(c')}_{\text{gain}} - \underbrace{\sum_{c'} p_{c\to c'} P_t(c)}_{\text{loss}} \,. \tag{3.13}$$

In this evolution equation gain and loss terms balance one another so that it conserves the total probability $\sum_c P_t(c) = 1$. Note that the evolution equation – unlike the process itself – is fully deterministic.

In a more compact form the evolution equation may be written as

$$|P_{t+1}\rangle = \mathcal{T}|P_t\rangle \,, \tag{3.14}$$

where $|P_t\rangle$ denotes a vector whose entries are the probabilities $P_t(c)$. This means that the corresponding vector space has a dimension equal to the number of possible configurations of the lattice. The so-called **transfer matrix** $\mathcal{T}$ is a linear operator that iterates the probability vector forward in time. Its matrix elements are given by

$$\langle c'|\mathcal{T}|c\rangle \;=\; \delta_{c,c'}\,\bigl(1-\sum_{c''} p_{c\to c''}\bigr) \;+\; (1-\delta_{c,c'})p_{c\to c'}\,. \tag{3.15}$$

The transfer matrix is real but in general non-symmetric, expressing the fact that stochastic processes are generally **irreversible**, i.e., they are not invariant under time-reversal. Moreover, the transfer matrix is constrained as it has to conserve probability. In order to express the condition for probability conservation in an elegant form let us define the bra vector as a sum over all possible configurations

$$\langle s| = \sum_c \langle c|\,, \tag{3.16}$$

where $\langle c|$ denotes the canonical basis of bra-vectors in terms of configurations c. Obviously, the system conserves probability if $\langle s|P_t\rangle = 1$ for all t. This implies the identity

$$\langle s|\mathcal{T} = \langle s|\,, \tag{3.17}$$

meaning that $\langle s|$ is a left eigenvector of $\mathcal{T}$ with eigenvalue 1. Consequently there exists also at least one right eigenvector $|s\rangle$ that obeys $\mathcal{T}|s\rangle = |s\rangle$. If the eigenvalue 1 is non-degenerate this state represents the stationary state of the stochastic process under consideration. However, in contrast to quantum mechanics, the components of $|s\rangle$ in the canonical representation may differ from those of $\langle s|$ and may be non-trivial since $\mathcal{T}$ is generally non-symmetric. In other words, ket and bra vectors are no longer related by a simple transposition.

Diagonalising $\mathcal{T}$, the real part of the eigenvalues is found to lie between 0 and 1 and its reciprocal negative logarithm gives the relaxation time of corresponding eigenmode. In principle, any stochastic system with parallel dynamics can be solved by diagonalising the transfer matrix, expanding the initial distribution $|P_0(c)\rangle = \sum_n c_n|n\rangle$ in terms of the eigenvectors $|n\rangle$. The formal solution is then given by

$$|P_t(c)\rangle \;=\; \sum_n c_n\, q_n^t\, |n\rangle\,, \tag{3.18}$$

where q_n are the eigenvalues corresponding to $|n\rangle$. However, in practise exact diagonalisation by solving the eigenvalue problem

$$\mathcal{T}|n\rangle = q_n|n\rangle \tag{3.19}$$

is a very challenging task and is usually restricted to small system sizes.

3.2.5.2 Random-Sequential Updates:

In models with random-sequential updates local transitions occur spontaneously just as in a radioactive decay. In this case the time parameter t is continuous and the probability distribution $P_t(c)$ evolves according to a **master equation**

$$\frac{\partial}{\partial t} P_t(c) = \underbrace{\sum_{c'} w_{c' \to c} P_t(c')}_{\text{gain}} - \underbrace{\sum_{c'} w_{c \to c'} P_t(c)}_{\text{loss}} \,. \tag{3.20}$$

Again the gain and loss terms balance one another so that the probability distribution remains normalised as time proceeds. It is important to note that the coefficients $w_{c \to c'}$ are *rates* rather than probabilities and carry the unit $[time]^{-1}$. Therefore, their numerical values may be larger than 1 and can be rescaled by changing the overall time scale.

Using again the vector notation this set of equations may be written as

$$\partial_t |P_t\rangle = -\mathcal{L} |P_t\rangle \,. \tag{3.21}$$

Here the **Liouville operator** $\mathcal{L}$, sometimes also called the **quantum Hamiltonian** because of its closeness to the Hamiltonians of magnetic quantum spin systems (see exercise 25), generates the temporal evolution. In the canonical representation it is defined by the matrix elements

$$\langle c' | \mathcal{L} | c \rangle = \delta_{c,c'} \sum_{c''} w_{c \to c''} - (1 - \delta_{c,c'}) w_{c \to c'} . \tag{3.22}$$

Conservation of probability $\langle s | P_t \rangle = 1$ now implies the existence of a left eigenvector with eigenvalue zero:

$$\langle s | \mathcal{L} = 0 \,. \tag{3.23}$$

This means that the sum over each column of the matrix $\mathcal{L}$ vanishes. Generally the corresponding right eigenvector $|s\rangle$ is a zero mode that represents the stationary state of the system. A formal solution of this first-order differential equation can be written in terms of a matrix exponential function as

$$|P_t\rangle = \sum_c \exp(-\mathcal{L}t)\, P_0(c) |c\rangle = \exp(-\mathcal{L}t) |P_0\rangle \,, \tag{3.24}$$

where $|P_0\rangle$ denotes the initial probability distribution at $t = 0$.

Like the transfer matrix $\mathcal{T}$, the Liouville operator $\mathcal{L}$ is generally non-symmetric and its form is restricted by probability conservation (3.23) and the positivity of the rates. In the mathematical literature such matrices are called *intensity matrices* or *generator matrices*, meaning that all diagonal (off-diagonal) entries are real and positive (negative) and that the sum over each column of the matrix vanishes. The eigenvalues of an intensity matrix may be complex, indicating oscillatory behaviour, but their real part is always non-negative. The non-vanishing eigenvalues represent relaxational eigenmodes which decay exponentially with time. A more formal statement may be given as follows:

Theorem: [315, 381, 558] *For a Liouville operator $\mathcal{L}$ which satisfies the master equation (3.21) and with $\langle s| = \sum_c \langle c|$ as a left eigenstate such that $\langle s| \mathcal{L} = 0$, the following holds. (i) There is a stationary state*

$$|s\rangle = \sum_c P_\infty(c)\,|c\rangle \tag{3.25}$$

such that $\mathcal{L}\,|s\rangle = 0$. (ii) Consider the eigenvalue problem $\mathcal{L}\,|n\rangle = E_n\,|n\rangle$, with $n = 0, 1, 2 \ldots$. Then

$$\Re E_n \geq E_0 = 0 \tag{3.26}$$

(iii) Let $|P_0\rangle = \sum_c P_0(c)|c\rangle$ be the normalised initial state such that the weights of the individual configurations satisfy $0 \leq P_0(c) \leq 1$ and $\langle s|P_0\rangle = 1$. Then for all times $t \geq 0$, one has

$$0 \leq P_t(c) \leq 1 \qquad \text{and} \quad \langle s|P_t\rangle = 1 \tag{3.27}$$

(iv) Let $\mathcal{L} : \mathbb{R}^n \to \mathbb{R}^n$ be a linear map such that for the elements $\mathcal{L}_{ij}$ of $\mathcal{L}$ holds

$$\mathcal{L}_{ij} \leq 0 \ \text{ for } \ i \neq j \ , \ \sum_{i=1}^{n} \mathcal{L}_{ij} = 0 \quad \forall j \in \{1, \ldots, n\} \tag{3.28}$$

Then $\mathcal{L}$ is a Liouville operator of a Markov process described by the master equation (3.21).

If all relaxational modes decay exponentially, how can the density of active sites in a DP process decay *algebraically* as $\varrho(t) \sim t^{-\alpha}$? In fact, such an algebraic decay is only possible in an *infinite* system with infinitely many eigenmodes and a quasi-continuous spectrum. The eigenmodes are then superposed in such a way that an algebraic decay is obtained. For example, an integral over exponential functions e^{-t/t_r} weighted by the amplitude $t_r^{-1-\alpha}$ gives

$$\int_0^\infty \mathrm{d}t_r\, t_r^{-1-\alpha}\, e^{-t/t_r} \sim t^{-\alpha}\,. \tag{3.29}$$

In a finite system, however, where the spectrum is discrete, such an algebraic behaviour can only be maintained within a certain temporal window until the system crosses over to an exponential decay towards the steady-state. We shall discuss such finite-size effects in more detail in Sect. 4.1.9.

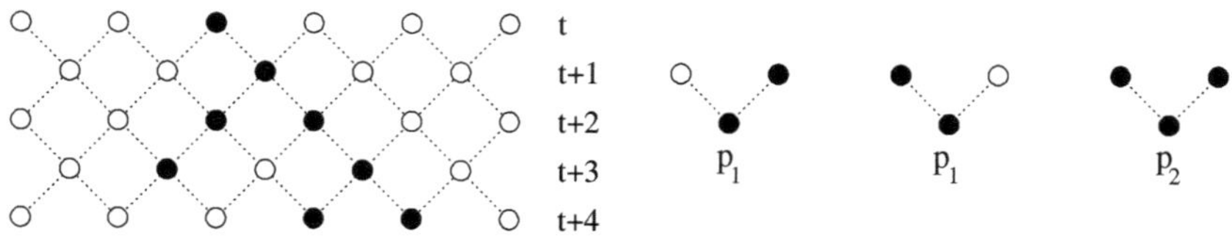

Fig. 3.7 The (1+1)-dimensional Domany-Kinzel automaton. Occupied sites are marked by full circles. The state of the system at time $t+1$ is obtained by an iteration of the dynamic rules according to the conditional probabilities $P[s_{i,t+1}|s_{i-1,t}, s_{i+1,t}]$. This means that the state $s_{i,t+1}$ of a given site i at time $t+1$ depends on the state of the left and right neighbours $(s_{i-1,t}, s_{i+1,t})$ at the previous time step t.

Another use of the Liouville operator is that it allows one to derive relations between different stochastic processes via stochastic similarity transformations, see exercises 26, 27.

3.3 Lattice Models of Directed Percolation

In this section we consider various lattice models that belong to the universality class of directed percolation. First, we discuss the Domany-Kinzel automaton which exhibits a non-trivial phase diagram controlled by two parameters. This model is very robust, easy to implement on a computer, and its phase diagram comprises a special point which allows one to study how the directed-percolation behaviour breaks down in presence of an additional symmetry. Secondly, we revisit the contact process which is a well-studied model in the mathematical literature. Thirdly, we consider the pair-contact process as an example of a model with infinitely many absorbing states. Finally, we also discuss the threshold transfer process as well as the Ziff-Gulari-Barshad model for heterogeneous catalysis, before briefly mentioning some further processes related to DP.

3.3.1 Domany-Kinzel Automaton

An important lattice model, which includes bond and site directed percolation as special cases, is the celebrated **Domany-Kinzel automaton** (DK) [188, 391]. The DK model is a stochastic **cellular automaton** defined on a tilted square lattice (see Fig. 3.7) whose sites s_i can be either occupied ($s_i = 1$) or empty ($s_i = 0$). Generally a cellular automaton is a discrete model consisting of a regular grid of cells, each in one of a finite number of states, evolving by *parallel* updates with a discrete time variable $t \in \mathbb{N}$ [634]. The DK model evolves stochastically according to certain conditional transition probabilities

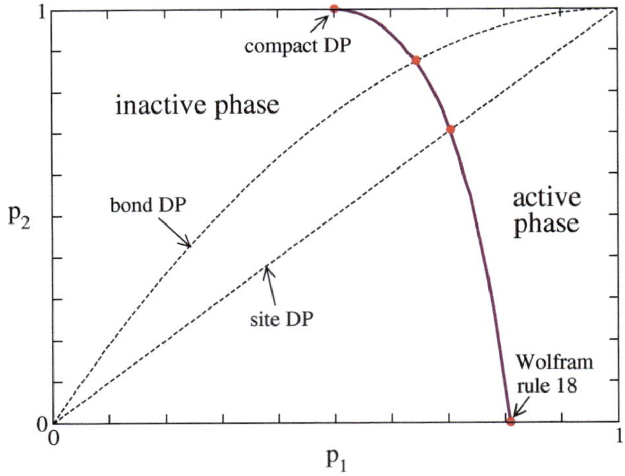

Fig. 3.8 Phase diagram of the (1+1)-dimensional Domany-Kinzel model. The inactive phase ($\varrho = 0$) is separated from the active phase ($0 < \varrho < 1$) by a line of second-order transitions (solid line). Bond directed percolation corresponds to the line $p_2 = p_1(2 - p_1)$ whereas site directed percolation (see Sect. 3.1) is obtained for $p_1 = p_2$. For $p_2 = 0$ the Domany-Kinzel automaton reduces to the cellular automata rule '18' of Wolfram's classification scheme [634]. For $p_2 = 1$ the system is characterised by a particle-hole symmetry, leading to a different universal scaling behaviour called *compact directed percolation*, see Sect. 5.3.

$P[s_{i,t+1}|s_{i-1,t}, s_{i+1,t}]$, which depend on *two* parameters p_1, p_2 and are defined by

$$P[1|0,0] = 0 \ , \quad P[1|0,1] = P[1|1,0] = p_1 \ , \quad P[1|1,1] = p_2 \ , \tag{3.30}$$

with $P[0|\cdot,\cdot] = 1 - P[1|\cdot,\cdot]$. The Domany-Kinzel model is widely used because it can be implemented efficiently on a computer. In order to determine the next configuration of active sites at time $t+1$, one simply has to iterate over all sites i which have at least one active nearest neighbour at time t. For those sites, one generates a random number $z_i \in (0,1)$ and performs the local update

$$s_{i,t+1} = \begin{cases} 1 & \text{if } s_{i-1,t} \neq s_{i+1,t} \quad\quad \text{and } z_i(t) < p_1 \ , \\ 1 & \text{if } s_{i-1,t} = s_{i+1,t} = 1 \text{ and } z_i(t) < p_2 \ , \\ 0 & \text{otherwise} \ . \end{cases} \tag{3.31}$$

This means that a site becomes active with probability p_2 if the two nearest neighbours at the previous time step were both active while it is activated with probability p_1 if only one of them was active. Thus the model is controlled by two parameters p_1 and p_2, giving rise to a two-dimensional phase diagram which is shown in Fig. 3.8. As can be seen, the active and the inactive phase are now separated by a *line* of phase transitions.

model	$p_{1,c}$	$p_{2,c}$	μ	Ref.
Wolfram rule 18	0.811(1)	0		[435]
	0.8015(4)	0.08015	1.5747(34)	[435]
	0.7894(3)	0.19735	1.4876(32)	[435]
	0.7668(2)	0.3834	1.3957(31)	[435]
	0.74515(7)	0.521605	1.3195(30)	[435]
	0.73300(10)	0.5864	1.2809(34)	[435]
site DP	0.70548515(20)	0.70548515	1.1902(30)	[355]
	0.67316(11)	0.807792	1.1046(26)	[435]
bond DP	0.644700185(5)	0.873762052	1	[355]
	0.62585(9)	0.9074825	0.93737(20)	[435]
	0.594305(15)	0.950888	0.82363(22)	[435]
	0.57870(8)	0.966429	0.76165(26)	[435]
	0.54865(7)	0.98757	0.60868(24)	[435]
	0.52469(6)	0.996911	0.44501(30)	[435]
	0.5124250(15)	0.99922875	0.31626(38)	[435]
	0.5024969(15)	0.99996903	0.14060(44)	[435]
compact DP	1/2	1	–	[390]

Table 3.1 Critical points of the (1+1)-dimensional Domany–Kinzel model. The column labelled μ gives the scaling factor between the particle density ϱ and the survival probability P_{sur}, as expected from rapidity-reversal-invariance (4.183), see Sect. 4.3.4.2.

The Domany-Kinzel model includes bond and site DP as special cases. For example, in bond directed percolation, the site $s_{i,t+1}$ is activated with probability p if only one of its nearest neighbours was active at time t, while it is activated with probability $1-(1-p)^2 = 2p-p^2$ if both of them were active. Comparing these probabilities with the DK update rule we retrieve bond directed percolation by setting $p_1 = p$ and $p_2 = p(2-p)$, shown as a dashed curve in the phase diagram Fig. 3.8. Similarly, the special case of directed site percolation [390], where sites instead of bonds are permeable with probability p and blocked otherwise, corresponds to the choice $p_1 = p_2 = p$. Finally, the special case $p_2 = 0$ is equivalent to the rule 'W18' of Wolfram's classification scheme of stochastic cellular automata [634]. Numerical estimates for the corresponding critical thresholds are listed in Table 3.1.

There is strong numerical evidence that the critical behaviour along the entire phase transition line (except for its upper terminal point) is that of DP, meaning that the model always exhibits the same type of long-range correlations. The short-range correlations, however, are non-universal and may change when moving along the phase transition line. They may even influence the visual appearance of the clusters, as illustrated in Fig. 3.9, where four typical snapshots of critical clusters are compared. As can be seen, the large-scale structure of the clusters in the first three cases is roughly the same although the microscopic texture seems to become bolder as one moves upwards along the phase transition line. As shown in [287], this visual impression can be traced back to an increase of the mean size of active droplets.

Approaching the upper terminal point the mean size of active droplets diverges and the generated clusters become compact (cf. Fig. 3.9), which explains why this special point is usually referred to as **compact directed percolation** (CDP). The exceptional behaviour at this point is due to an additional symmetry between active and inactive sites if $p_2 = 1$. Here the system has *two* symmetric absorbing states, namely, the empty and the fully occupied lattice. Moreover, the dynamic rules are invariant under the replacement $p_1 \leftrightarrow 1 - p_1$ so that the transition point is located exactly at $p_1 = 1/2$. It turns out that the dynamics at this point of the phase diagram is the same as in the (1+1)-dimensional Glauber-Ising model at zero temperature (or, equivalently, the **voter model** [422, 191]). As these models belong to a different universality class the term 'compact directed percolation' may be misleading. We shall come back to CDP in Sect. 5.3.

The transfer matrix of the (1+1)-dimensional Domany-Kinzel cellular automaton can be constructed as follows. Denoting by $c = \{s_1, s_2, \ldots, s_N\}$ the configuration of a row of N sites, the transfer matrix factorises into

$$\langle c'|\mathcal{T}|c\rangle = \prod_{j=1}^{N} \mathcal{T}_{s_{j-1},s_{j+1}}^{s'_j} \tag{3.32}$$

with appropriate (e.g. periodic) boundary conditions. According to the definition of the model the factors are given by

$$\mathcal{T}_{0,0}^{1} = 0\,, \quad \mathcal{T}_{0,1}^{1} = \mathcal{T}_{1,0}^{1} = p_1\,, \quad \mathcal{T}_{1,1}^{1} = p_2 \tag{3.33}$$

and conserve probability separately, i.e., $\mathcal{T}_{s_{j-1},s_{j+1}}^{0} + \mathcal{T}_{s_{j-1},s_{j+1}}^{1} = 1$. Note that for an even or infinite number of sites, the transfer matrix decomposes into two commuting factors on the even and the odd sublattices.

The Domany-Kinzel model can be generalised easily to higher spatial dimensions (see e.g. [241, 249, 445]). In the d+1-dimensional DK model the activation probability of site i at time $t+1$ depends on the number $n_i(t) = \sum_{j\in<i>} s_j(t)$ of active nearest neighbours at time t, i.e., the condi-

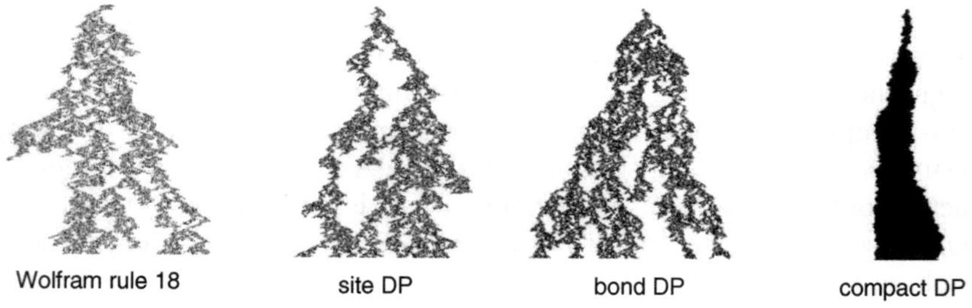

Fig. 3.9 Domany-Kinzel automaton: Critical clusters generated from a single active seed at different points of the phase transition line (cf. Fig. 3.8).

tional probabilities

$$P[1|0] = 0 ,$$
$$P[1|n] = p_n , \qquad (1 \le n \le 2d) \tag{3.34}$$

are now controlled by $2d$ parameters $p_1, \ldots, p_{2d}$. The special case of bond directed percolation corresponds to the choice $p_n = 1-(1-p)^n$ while for equal parameters $p_n = p$ one obtains site directed percolation in $d+1$ dimensions.

3.3.2 Contact Process

Another important model for directed percolation, frequently used in mathematically oriented communities, is the **contact process** (CP), sometimes also called the **Gribov process**. The contact process was originally introduced by Harris [267] as a model for epidemic spreading. It is defined on a d-dimensional square lattice whose sites can be either active ($s_i(t) = 1$) or inactive ($s_i(t) = 0$). Unlike the DK automaton, which uses parallel updates, the contact process evolves by *asynchronous* updates, meaning that the three processes (offspring production, on-site removal, and diffusion) occur spontaneously at certain *rates*. Although the contact process and the DK automaton differ significantly, they both display the same type of critical behaviour at the phase transition.

On a computer, the contact process can be implemented as follows. First, one defines an array with $\mathcal{N} = L^d$ sites and specifies the initial state. For each attempted update, a site i is randomly selected. Depending on its state $s_i(t)$ and the number of active neighbours $n_i(t) = \sum_{j\in\langle i\rangle} s_j(t)$, a new value $s_i(t+\mathrm{d}t) = 0, 1$ is assigned according to certain transition rates $w[s_i(t) \to s_i(t+\mathrm{d}t), n_i(t)]$. In the standard contact process these rates are defined by

$$w[0 \to 1, n] = (\lambda n)/(2d) \ , \quad w[1 \to 0, n] = 1 \ . \tag{3.35}$$

Here the parameter λ controls the infection rate and plays the same role as the percolation probability p in bond directed percolation. Its critical value λ_c depends on the dimension d and is listed in Table 3.2. On a computer this means that the transitions are carried out with the probabilities $\lambda n/(2d(1+\lambda))$ and $1/(1+\lambda)$, respectively, incrementing time by $\Delta t = 1/\mathcal{N}(1+\lambda)$ for each attempted update.[6]

In 1+1 dimensions, the dynamical rules of the contact process defined above can be recast as a two-site process as follows (see Fig. 3.10). For each

[6] An often-used alternative notation denotes empty sites by $\circ$ and occupied sites by $\bullet$. Then the rates of the (1+1)-dimensional contact process are $\bullet \xrightarrow{1} \circ$, $\bullet\circ\circ \xrightarrow{\lambda/2} \bullet\bullet\circ$ and $\bullet\circ\bullet \xrightarrow{\lambda} \bullet\bullet\bullet$.

model	d	1	2	3	4	5
CP	λ_c	3.297848(22)	1.64877(3)	1.31686(1)	1.19505(15)	1.13846(11)
PCP	p_c	0.077092(1)	0.20053(9)	0.25803(13)	–	0.29874(15)
sDP	p_c	0.70548515(20)	0.34457(1)	0.160958(6)	0.0755850(3)	0.0359725(2)
bDP	p_c	0.644700185(5)	0.287339(2)	–	–	–

Table 3.2 Critical points of the contact process [359, 171, 445] and the pair-contact process [151, 444] on a hypercubic lattice in d dimensions and for site and bond directed percolation on a bcc lattice [355, 249, 433] in d dimensions.

attempted update a pair of adjacent sites i and $i+1$ is randomly selected. If this pair is occupied by at least one particle they are updated according to the transition rates

$$\begin{aligned} w_{10\to 11} &= w_{01\to 11} = \lambda/2\,, \\ w_{10\to 00} &= w_{01\to 00} = w_{11\to 01} = w_{11\to 10} = 1/2\,, \end{aligned} \tag{3.36}$$

while the rates for all other transitions vanish. It is easy to verify that this definition of the dynamics is equivalent to the definition (3.35) in $d=1$ spatial dimensions.

The formulation in terms of two-site updates allows one to write down the master equation (see (3.20))

$$\frac{\partial}{\partial t} P_t(s'_1,\ldots,s'_N) = \sum_{i=1}^{\mathcal{N}} \Bigg[\sum_{s_i,s_{i+1},s'_i,s'_{i+1}} w_{s_i,s_{i+1}\to s'_i,s'_{i+1}} P_t(s_1,\ldots,s_i,s_{i+1},\ldots,s_N) - \sum_{s_i,s_{i+1},s'_i,s'_{i+1}} w_{s'_i,s'_{i+1}\to s_i,s_{i+1}} P_t(s_1,\ldots,s'_i,s'_{i+1},\ldots,s_N) \Bigg] \tag{3.37}$$

together with appropriate boundary conditions (e.g. periodic boundary conditions by identifying $s_{N+1}\equiv s_1$). Using the vector notation $\partial_t|P_t\rangle = -\mathcal{L}|P_t\rangle$ as in (3.21) the Liouville operator $\mathcal{L}$ can be written as a sum of two-site processes

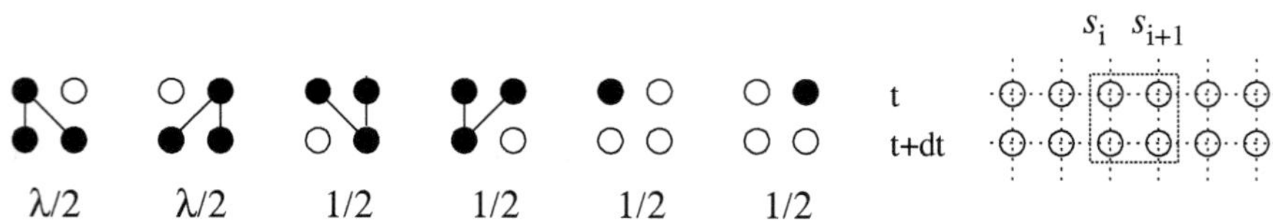

Fig. 3.10 Update rules for the $(1+1)$-dimensional contact process.

$$\mathcal{L} = \sum_{i=1}^{\mathcal{N}} \mathbf{1} \otimes \ldots \otimes \mathbf{1} \otimes \mathcal{L}_{i,i+1} \otimes \mathbf{1} \otimes \ldots \otimes \mathbf{1}\,, \tag{3.38}$$

where $\mathbf{1}$ is a 2×2 identity matrix and $\mathcal{L}_{i,i+1}$ denotes a 4×4 interaction matrix acting on sites i and $i+1$:

$$\mathcal{L}_{i,i+1} = \frac{1}{2} \begin{pmatrix} 0 & -1 & -1 & 0 \\ 0 & 1+\lambda & 0 & -1 \\ 0 & 0 & 1+\lambda & -1 \\ 0 & -\lambda & -\lambda & 2 \end{pmatrix} . \tag{3.39}$$

The spectrum of this operator will be analysed in Sect. 4.3.1.

Because of these compelling analytical properties the contact process is the preferred realisation of DP in the mathematical literature. Although an exact solution of the contact process is still lacking, a number of rigorous results could be derived. For example, it is possible to find lower and upper bounds for the critical parameter in the contact process [422], although they are useless for a precise numerical determination of the critical point λ_c.[7] Moreover, it was proven rigorously that a supercritical contact process starting from a single seed spreads within a well-defined light cone as long as it survives [198]. With the help of a limit theorem, it was also possible to prove that the supercritical phase of the contact process is ergodic [199]. More recently, it was proven in [85] that a contact process on a finite lattice in one dimension always terminates while a contact process in a slab ($[-K,K] \times \mathbb{R}^{d-1}$) in $d > 1$ dimensions has a critical threshold λ_c^K which converges to the usual value λ_c in d dimensions for $K \to \infty$. Although these results are not too surprising for physicists working in this field, it is quite hard to derive them rigorously.

3.3.3 Pair-Contact Process

The **pair-contact process** (PCP) was introduced by Jensen [350] and is one of the simplest models with infinitely many absorbing states showing a continuous phase transition. The process is defined on a d-dimensional hypercubic lattice with sites that may be either occupied or empty. The model evolves by asynchronous dynamics (random sequential updates) as follows. For each attempted update, a *pair* of adjacent sites is randomly selected. If both sites are occupied, they annihilate with probability p, otherwise another site adjacent to the selected pair is chosen and a particle is created, provided that the target site is empty (see Fig. 3.11). Solitary particles are not allowed to diffuse.

---

[7] For example, for $d = 1$, one has the rather loose bounds $3.078 < \lambda_c < 3.884$.

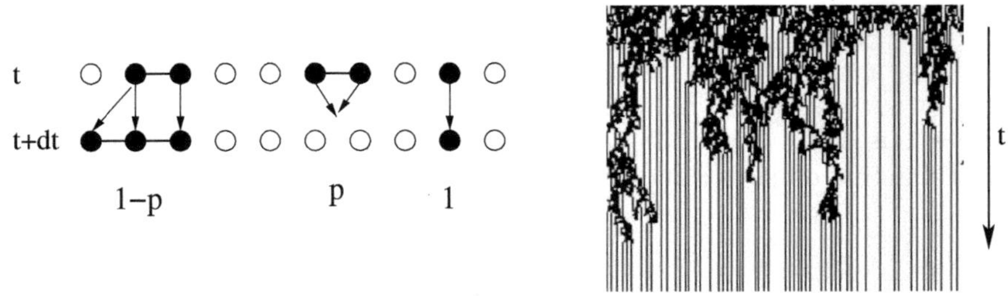

Fig. 3.11 The $(1+1)$-dimensional pair-contact process [350]. Left: Lattice sites may be either empty (open circles) or occupied by a particle (full circles). Pairs of occupied sites (solid bonds) are considered as active whereas isolated particles remain inactive. A pair is annihilated with rate p, otherwise an offspring is created at an empty neighbouring site selected at random. Right: Simulation of a subcritical PCP, terminating in an absorbing configuration of immobile solitary particles.

In this model only *pairs* of particles can annihilate and create offspring while isolated particles are completely frozen. Therefore any configuration without pairs of particles is absorbing. For example, on a one-dimensional chain with L sites and periodic boundary conditions the number N_a of absorbing states is asymptotically given by a power of the golden mean [126]

$$N_a \simeq \left(\frac{1+\sqrt{5}}{2}\right)^L \tag{3.40}$$

and diverges in the thermodynamic limit $L \to \infty$, see exercise 31. Therefore, the pair-contact process is said to have infinitely many absorbing states (in the limit $L \to \infty$).

The pair-contact process displays a continuous phase transition from a fluctuating active phase into a frozen state. In one spatial dimension, the current best estimate for the critical point is $p_c = 0.077092(1)$ [151]. However, in contrast to the DK model and the contact process, the order parameter ϱ of the PCP is the density of nearest-neighbour *pairs* of particles. As in the contact process and the Domany-Kinzel automaton, the conjugated field corresponds to spontaneous particle creation [444].

As demonstrated in Fig. 3.11 the PCP can be interpreted as a directed percolation process of pairs running on a disordered background of solitary particles. As the PCP has infinitely many absorbing states, the DP-conjecture by Janssen and Grassberger (see Sect. 3.2.2) cannot be applied. For this reason the critical behaviour of the pair-contact process was intensely investigated by simulations (including [357, 351, 175, 170, 177, 151]), showing that the critical scaling behaviour of the $(d+1)$-dimensional pair-contact process is indeed characterised by the same critical exponents as DP [350, 357, 433, 151]. Moreover, it was shown that the steady-state, dynamical and finite-size scaling functions coincide below and above the upper critical dimension $d_c = 4$ [444, 176, 446, 151], confirming a renormalisation-group study that predicts

DP universal behaviour [483] for the PCP in all dimensions. Thus, despite the different structure of the absorbing phase, the pair-contact process still belongs to the directed percolation universality class.

A modification of the pair-contact process, the so-called *pair-contact process with diffusion* (PCPD), has attracted a lot of attention in the last few years (see [274] for a recent review). The PCPD differs from the original PCP in so far as isolated particles are allowed to diffuse. In particular, the PCPD has only two absorbing states, namely, the empty lattice as well as a state with a single diffusing particle. This modification changes entirely the structure of the absorbing phase. The scaling behaviour of the PCPD is not yet understood and will be discussed in more detail in Sect. 5.8.

3.3.4 Threshold Transfer Process

The **threshold transfer process** (TTP) introduced in [470] is an example of a model with *fluctuating* absorbing states. Here, the lattice sites may be empty ($n = 0$), occupied by a single particle ($n = 1$), or by two particles ($n = 2$). The model evolves by random sequential updates, i.e., for each attempted update a site i is randomly selected and the following moves are carried out:

- If $n_i = 0$ a particle is created with probability r by setting $n_i := 1$
- If $n_i = 1$ the particle is removed with probability $1 - r$ by setting $n_i := 0$.
- If $n_i = 2$ the two particles may move to two different randomly selected neighbouring sites. The moves are carried out only if the target site hosts less than two particles.

In the TTP double-occupied sites are considered as active and their density is identified as the order parameter of the process. Consequently, the absorbing phase consists of all configurations devoid of double-occupied sites. This subspace is in fact absorbing, i.e., it can be reached by the dynamics but it cannot be left by them. Moreover, it is a fluctuating phase because each site switches randomly between $n_i = 0$ and $n_i = 1$. Therefore, the absorbing state is a fluctuating steady-state where the density of particles is controlled by the parameter r.

The phase transition in the TTP can be interpreted as a DP-like process of double-occupied sites running on top of the fluctuating absorbing state. Since the likelihood of such an active site generating another double-occupied site depends on the particle density in the absorbing state, this process is indirectly controlled by the parameter r. In fact, steady-state numerical simulations of the $(1 + 1)$-dimensional and $(2 + 1)$-dimensional TTP yield critical exponents and universal scaling functions that are in agreement

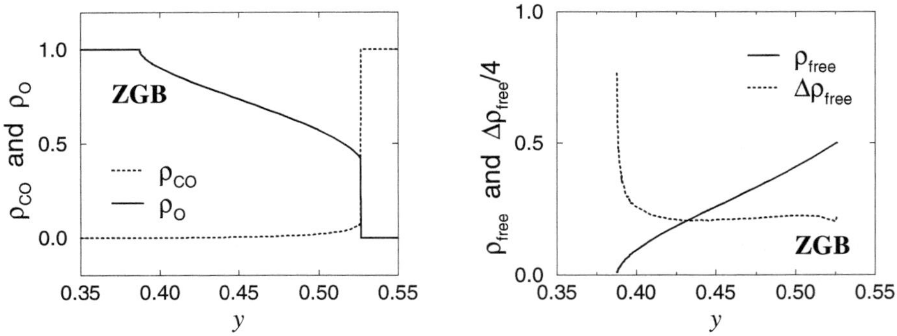

Fig. 3.12 Schematic phase diagram of the Ziff-Gulari-Barshad model for catalytic carbon-monoxide oxidation. The steady-state concentrations of CO and O_2 on the catalytic surface are plotted as functions of the CO concentration y in the gas phase. The system undergoes a second-order phase transition to the oxygen-poisoned state whereas a first order phase transition takes place if the CO-passivated phase is approached. The right figure shows the density of vacant lattice sites ρ_{free} and its fluctuations $\Delta\rho_{\text{free}}$. Reprinted with permission from [433]. Copyright (2004) World Scientific Publishing Company.

with the corresponding DP values [470, 446]. In $1D$, $r_c = 0.6894(3)$ and in $2D$, $r_c = 0.5876(5)$.

In order to obtain universal scaling functions, an external field has to be applied that is conjugated to the order parameter. In contrast to the previously discussed models the conjugated field cannot be implemented by simple particle creation at rate h, since particle creation would affect the particle density, i.e., the control parameter of the phase transition. Instead, the conjugated field has to be implemented as an additional term for particle diffusion [446], i.e., a particle moves to a randomly selected neighbour with probability h provided that $n < 2$.

3.3.5 Ziff-Gulari-Barshad Model

Many catalytic reactions such as the oxidation of carbon monoxide on a platinum surface exhibit transitions into so-called **catalytically poisoned states** where the system becomes trapped in a frozen configuration. If such a transition is continuous, it may mimic the reaction scheme of directed percolation. For this reason, catalytic reactions are promising candidates for a possible experimental realisation of DP (see Sect. 3.4).

A well-known model for heterogeneous catalysis which exhibits a DP-like absorbing phase transition is the **Ziff-Gulari-Barshad model** (ZGB) [647]. This model mimics catalysis of carbon monoxide oxidation

$$2\,\mathrm{CO} + \mathrm{O}_2 \longrightarrow 2\,\mathrm{CO}_2 \tag{3.41}$$

on a catalytic surface exposed to a gas composed of CO and O_2 molecules with fixed concentrations y and $1-y$, respectively. The catalytic surface is represented by a square lattice whose sites can either be empty, occupied by a CO molecule, or occupied by an O atom. The ZGB model evolves by random sequential updates according to the following probabilistic rules:

- O_2 molecules dissociate on the surface into two O atoms and fill pairs of adjacent vacant sites at rate $1-y$.
- CO molecules are adsorbed at empty sites at rate y.
- Adjacent CO molecules and O atoms recombine instantaneously to O_2 and desorb from the surface, leaving two empty sites behind.

Obviously, the system is trapped in catalytically poisoned states as soon as the lattice is completely covered either by carbon monoxide or by oxygen. These absorbing states attract the dynamics for sufficiently large CO and O_2 concentrations, which are controlled by the parameter y (see exercise 30 for a simple mean-field treatment). Numerical simulations show that catalytic activity occurs only in the range $0.390 < y < 0.525$ [360] (see Fig. 3.12) and that the system exhibits two different phase transitions. The transition into the oxygen-poisoned state is continuous while the other transition into the CO-passivated state is a first-order transition.

There is strong numerical evidence that the continuous phase transition belongs to the universality class of directed percolation [254, 360, 612]. At first glance, this might be surprising since the process involves two distinct chemical components, CO and O. However, in the ZGB model catalytic activity is related to the density of vacant sites, i.e., to a single component order parameter [254]. In this sense, the observed directed percolation behaviour is in full agreement with the universality hypothesis of Janssen and Grassberger.

Concerning possible experimental realisations we note that the ZGB model may be oversimplified. For example, a more realistic modelling of catalytic reactions has to incorporate mobility and desorption processes as well as a repulsive interaction between adsorbed oxygen molecules, see [647, 425]. These features affect the critical behaviour and drive the system away from the directed percolation universality class.

3.3.6 Further Non-equilibrium Phenomena Related to DP

Directed percolation plays an important role in various other models, where the connection to a spreading process is not immediately obvious. In the following, we consider five examples.

3.3.6.1 Roughening Transitions

DP-related roughening transitions can also be observed in certain solid-on-solid growth processes with random-sequential updates [17, 16]. As a key feature of these models, atoms are deposited everywhere but they can desorb only at the *edges* of existing layers. Moreover, the dynamics is constrained such that the heights at neighbouring lattice sites may differ at most by one unit. By varying the growth rate, such growth processes display a roughening transition from a non-moving smooth phase to a moving rough phase.

In this case the sites at zero height correspond to active sites A of a DP process. Deposition corresponds to the death process $A \to \emptyset$ while evaporation at the edges resembles offspring production $A \to 2A$. If the growth rate is small, the interface is effectively anchored to its bottom layer and a smooth phase is maintained. As the growth rate is increased, more islands on top of the bottom layer are produced until above a certain critical threshold they merge forming new layers at a finite rate, giving rise to a finite growth velocity. In an infinite system, the growth velocity scales near the transition as

$$v \sim \xi_{\|}^{-1} \sim (p - p_c)^{\nu_{\|}} , \tag{3.42}$$

where $\nu_{\|}$ is the temporal scaling exponent of DP.

3.3.6.2 Phase Coexistence in Non-Equilibrium Wetting

Recently there has been increasing interest in wetting phenomena far from thermal equilibrium. Such **non-equilibrium wetting processes** are modelled by a growth process on top of an inert substrate [296, 297, 149]. The substrate breaks translational invariance in the direction of growth, leading to a non-equilibrium wetting transition with an associated surface critical exponent. Imposing an additional attractive short-range force between substrate and interface, an additional phase emerges, where the bound and the moving phase coexist, becoming thermodynamically stable in the limit of infinite system size [298]. It was argued and confirmed by numerical simulations that one of the boundaries of the phase coexistence region, where the interface detaches from the bottom layer, is governed by a DP transition [233]. In this case the contact points, where the interface touches the substrate, correspond to the active sites of a DP process.

3.3.6.3 Synchronisation Transitions in Coupled Map Lattices

We briefly mention here that DP was also encountered in the study of coupled lattices of chaotic maps [42, 9, 234]. Roughly, a **chaotic map** is an iteration of real or complex numbers of the form $x_{n+1} = f(x_n)$ with $n \in \mathbb{N}$ which amplifies small perturbations.[8] Examples of discrete dynamical systems related to chaotic maps are the **logistic map** $x_{n+1} = rx_n(1-x_n)$ or the **tent map** $x_{n+1} = \frac{r}{4}(1-|2x-1|)$. A **map lattice** consists of many mutually coupled maps $x_n^{(j)}$ living on a given lattice with sites j, e.g. on a one-dimensional line. The idea is to couple two lattices $x_n^{(j)}$ and $y_n^{(j)}$, each of them evolving chaotically according to the same rules. Depending on the coupling constant these chains either evolve independently or they synchronise in a common chaotic state, i.e., the system displays a **synchronisation transition**. This transition was found to belong either to the universality class of multiplicative noise or to directed percolation. The order parameter is the so-called **Hamming distance** $h := \sum_j |x_n^{(j)} - y_n^{(j)}|$, which measures the difference between the states of the two chains.

3.3.6.4 Directed Lattice Animals

A randomly **branched polymer**, also called a **lattice animal**, is a connected graph on a lattice. The statistics of such graphs may be formulated in terms of a generating function $\mathcal{Z} = \sum_{n,b,k} a_{n,b,k} x^n y^b \tau^k$, where n is the number of occupied sites, b is the number of *bonds* between two occupied nearest-neighbour sites and k is the number of *contacts*, that is occupied nearest neighbour sites which are *not* linked by a bond and x, y, τ are fugacities. Finally, $a_{n,b,k}$ counts all animals with specified values of n, b, k. Depending

[8] For a discrete dynamical system $x_{n+1} = f(x_n)$, where the map $f : [0,1] \to [0,1]$ is three times differentiable, is symmetric, viz. $f(x) = f(1-x)$, is strictly increasing (i.e. $f'(x) > 0$) for $0 \leq x < \frac{1}{2}$ and satisfies $f\left(\frac{1}{2}\right) = 1$, there is a simple necessary and sufficient criterion for chaotic maps [35]: *Under the conditions stated above, a map f is chaotic if and only if the conditions (i) $f'(0) > 1$ and (ii) $f'(x)^2\{f,x\} < 0$ $\forall x \neq \frac{1}{2}$ are satisfied.* Here

$$\{f,z\} := \frac{f'''(z)}{f'(z)} - \frac{3}{2}\left(\frac{f''(z)}{f'(z)}\right)^2$$

is the **Schwarzian derivative**. The reader should check this on the logistic map with $r = 4$ or on maps like $f(x) = \sin(\pi x)$, $f(x) = 1 - (2x-1)^4$.

on the values of the fugacities, a large lattice animal may be preferably either in a spread-out or a collapsed state and one may become curious about the way the transition between the extended and collapsed states may occur, see e.g. [604] for an introduction.

Here, we exclusively consider *directed* lattice animals, which (i) must all 'start' from a single lattice site, (ii) all bonds must have a component in a preferred direction (for example the diagonal of a square lattice) and (iii) bonds and contacts are only counted if they have a component in the preferred direction. One speaks of the **strong embedding** of a lattice animal, if the only links between occupied sites are bonds (that is, contacts are absent). Dhar [163] showed that strongly embedded lattice animals are related to directed percolation. To understand his precise statement, consider a lattice gas with either empty or occupied sites labelled by $n_{i,j} = 0, 1$ and define the Hamiltonian

$$\mathcal{H} = -J_1 \sum_{i,j} n_{i,j}\big(n_{i+1,j} + n_{i,j+1}\big) - J_2 \sum_{i,j} n_{i,j} n_{i+1,j+1} \tag{3.43}$$

such that J_1 describes the interactions between nearest neighbour sites and J_2 is the interaction between two sites which have a common predecessor. Then one defines a partition function $Z = Z_{\mathcal{N}}(J_1, J_2) = \sum_{\{A\}} e^{-\mathcal{H}}$ where the sum runs over the strongly embedded directed lattice animal configurations. The remarkable observation is that the Boltzmann weights for any strongly embedded lattice animal, as generated by this partition function, are exactly the same as those found for a directed site-bond percolation problem, with the correspondence [163]

$$e^{J_1} = \frac{(2-p_B)(1-p_B p_S)}{1-2p_B p_S + p_B^2 p_S} \;,\quad e^{J_2} = \frac{1-2p_B p_S + p_B^2 p_S}{1-2p_B p_S + p_B^2 p_S^2} \tag{3.44}$$

where p_B and p_S are the directed percolation parameters for the bond and the site problem, respectively. Since there are quite a few exact results known for lattice animals, see [76] for a review, one might hope that some interesting lessons for directed percolation could be learnt. This correspondence has been checked numerically in detail [394] and furthermore can be extended to lattice animals without the constraint of strong embedding [282].

3.3.6.5 Damage Spreading

A few years ago, so-called **damage spreading** studies of non-equilibrium systems were very popular (see e.g. [294] and references therein). In these numerical studies, one simulates two copies of the same non-equilibrium system, using the same sequence of random numbers, but with slightly different initial conditions. Depending on the parameters, this initial 'damage' either disappears or it may spread over the entire system. The hope was that this would allow one to introduce the notion of chaos in the field of stochastic non-equilibrium systems. Grassberger was the first to notice that damage spreading transitions – irrespective of the model under investigation – belong generically to the DP universality class [242]. Later it was realised that the damage spreading technique is ambiguous in so far as it depends on the specific implementation of random numbers [301].

3.4 Experiments Related to Directed Percolation

Except for a very recent experiment by Takeuchi, Kuroda, Chaté and Sano [589], there are so far *no* experiments which reliably reproduce the critical behaviour of directed percolation, especially the set of critical exponents, on a quantitative level. In view of the simplicity of DP lattice models and the prominent role of DP as the basic universality class of absorbing-phase transitions, one might anticipate that experiments should not be too difficult to perform. The rareness of reliable experimental evidence is surprising, especially as various possible experimental realisations have been suggested in the past, and it is an open question why experiments of DP are so difficult to perform. Clearly, designing and performing further experiments in different settings is a challenging problem for the future.

In this section we describe various suggested experiments, discussing their particular advantages and problems. The aforementioned experiment by Takeuchi et al. [589], where DP exponents were clearly identified for the first time, will be discussed at the end in Sect. 3.4.3.3.

3.4.1 Experiments Resembling DP Dynamics

3.4.1.1 Catalytic Reactions

A necessary condition for any experiment related to DP is the existence of a clean absorbing state. As a candidate let us first discuss **catalytic reactions** on surfaces. Here the absorbing configuration corresponds to a so-called **catalytically poisoned state**, where the reactive surface is completely covered by one type of molecule, blocking any further dynamics.

An example is the catalytic reaction $2CO + O \to 2CO_2$ on a platinum surface. As discussed in Sect. 3.3.5, this process can be described in a simplified manner by the Ziff-Gulari-Barshad (ZGB) model. This model displays two transitions, namely, a discontinuous transitions into a CO-poisoned state

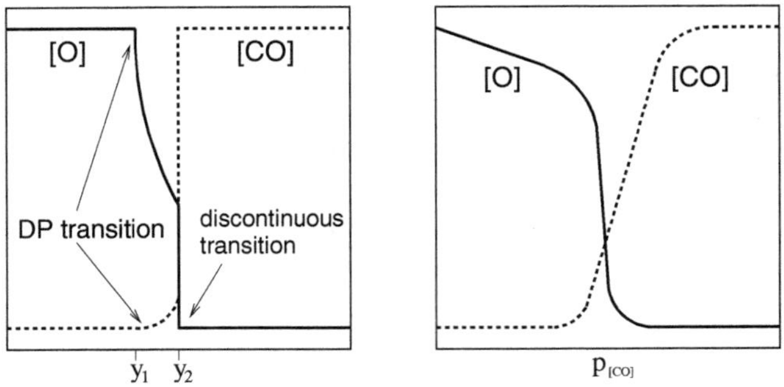

Fig. 3.13 Catalytic reactions in theory (left) and experiment (right). The schematic graphs show the concentrations of oxygen (solid line) and carbon-monoxide (dashed line) as a function of the CO adsorption rate.

and a DP transition into an O-poisoned state. However, in experiments such as the one on a Pt(210) surface in [202], only the discontinuous transition is seen. Similar results were obtained for Pt(111) and for other catalytic materials. Typically, the measured curves look like the one sketched on the right-hand side of Fig. 3.13, indicating that poisoning with oxygen does not occur. Instead the reactivity increases almost *linearly* with the CO pressure.

Why is the DP transition obscured or even destroyed under experimental conditions? As a likely explanation, the deviations may be caused by an **imperfect absorbing state**. For example, in the O-poisoned state a small amount of oxygen may still desorb from the surface, which in the language of DP would correspond to spontaneous creation of active sites by an external field h. The DP scaling behaviour may also be affected by inhomogeneities of the catalytic material, i.e. by spatially quenched disorder (see Sect. 5.11). Moreover, the effective reaction scheme may not be as simple as in the ZGB model. For example, it was observed that the catalytic reactions preferably take place at the perimeter of oxygen islands [633]. Furthermore, it was found that the adsorbed CO molecules on Pt(111) may form three different rotational patterns representing the c(4×2) structure of CO on platinum, leading to three competing absorbing states. Nevertheless, catalytic surface reactions remain a promising candidate for possible experimental realisations of DP in the future.

3.4.1.2 Percolation in Porous Media

Another possible realisation of DP is percolating water in a porous medium subjected to an external driving force. The medium could be a porous rock in a gravitational field where neighbouring pores are connected by channels with a certain probability. Depending on this probability, the penetration depth is either finite or the water may 'percolate' over infinitely long distances through the medium. Because of the gravitational field, the water can only flow downwards so that the percolation process is directed in space. Although this application seems to be quite natural and even gave DP its name, it is extremely difficult to realise experimentally. While the model assumes a non-conserved spreading agent, percolation in porous media is strongly influenced by the conserved volume of the fluid and the analysis is aggravated by various unwanted interactions such as capillary forces. Moreover, many materials have a broad distribution of pore sizes [283] and thus they are highly disordered.

3.4.1.3 Avalanches of Flowing Granular Matter

Some time ago it was proposed that simple systems of **flowing sand** on an inclined plane, e.g. the experiments performed by Douady and Daerr [194, 139], could serve as experimental realisations of DP [293, 294]. In these experiments glass beads are poured uniformly on to the top of an inclined plane covered by a rough velvet cloth. As the beads flow down a thin layer settles and remains immobile. Increasing the angle of inclination ϕ by $\Delta\phi$ the layer becomes dynamically unstable, i.e., by locally perturbing the system at the top of the plane an avalanche of flowing granular matter is released. In the experiment these avalanches have the shape of a fairly regular triangle with an opening angle Θ. As the increment $\Delta\varphi$ decreases, the value of Θ was found to scale as $\tan\Theta \sim (\Delta\varphi)^{x_a}$ with $x_a \approx 1$.

In order to explain the experimentally observed triangular form of the avalanches, Bouchaud et al. proposed a mean-field theory [75], predicting $x_a = 1/2$. As an alternative explanation, it was proposed that flowing sand may be understood as a nearest-neighbour spreading process [293, 294]. A corresponding lattice model showed a crossover from compact DP, where $x_a = 1$, to ordinary DP, where $x_a = \nu_\parallel - \nu_\perp \approx 0.6$. The crossover from CDP to DP is very slow and probably not accessible on laboratory scales. However, in principle the Douady-Daerr experiment, when performed on very large scales, may serve as a physical realisation of DP. As a precondition for the crossover, initially compact avalanches should thus be able to break up into several branches.

Such a breakup in several branches can be observed in the experiment in the limit of high angles of inclination [139]. Yet here the avalanches have no well-defined front, the propagation velocity of separate branches rather depends on their thickness. It is therefore no longer possible to interpret the vertical axis as a time coordinate. Also visually the branches are qualitatively different from DP clusters. Therefore, it remains an open question to what extent avalanches of flowing granular matter can be related to DP.

3.4.1.4 Calcium Dynamics in Living Cells

DP transitions may also occur in certain kinetic models for the dynamics of Calcium ions in living cells. Ca^{2+} ions play an important physiological role as second messenger for various purposes ranging from hormonal release to the activation of egg cells by fertilisation [62, 382]. The cell uses nonlinear propagation of increasing intra-cellular Ca^{2+} concentration, a so-called calcium wave, as a tool to transmit signals over distances which are much longer than the diffusion length. For example, propagating Ca^{2+} waves can be observed in the immature *Xenopus laevis* oocyte [412].

While most theoretical works on Calcium waves are based on mean-field methods [322], some authors have introduced models which also take the stochastic nature of Calcium release into account [383, 77]. In these models the transition between survival and extinction was found to belong to the DP universality class. However, inhomogeneities and finite-size effects of living cells make it practically impossible to identify the universal properties of such a transition.

3.4.1.5 Electron-Positron Collisions

There is a similar heuristic argument that directed percolation might be useful to describe some aspects of elementary particle fragmentation. For example, such a fragmentation takes place following the production of a quark-antiquark pair in a high-energy electron-positron collision $e^+e^- \to q\overline{q}$, where further quark-antiquark pairs are subsequently created out of the vacuum through the Schwinger mechanism, see [39]. In a confining theory such as quantum chromodynamics (QCD) one expects that $q\overline{q}$ pairs are linked by essentially one-dimensional flux tubes and the colour flux is modelled by the state of the lattice of directed bond percolation (the quarks and antiquarks are on the dual lattice). The creation of a new $q\overline{q}$ pair generates a supplementary colour field which compensates exactly for the one which generated the new pair and the creation probability can be expressed as a bond percolation problem [277].

3.4.2 Growth Processes Related to DP

3.4.2.1 Depinning Transitions

Some time ago Buldyrev et al. [94] reported measurements of an interface formed when a wet front propagates in paper by imbibition. Measuring the

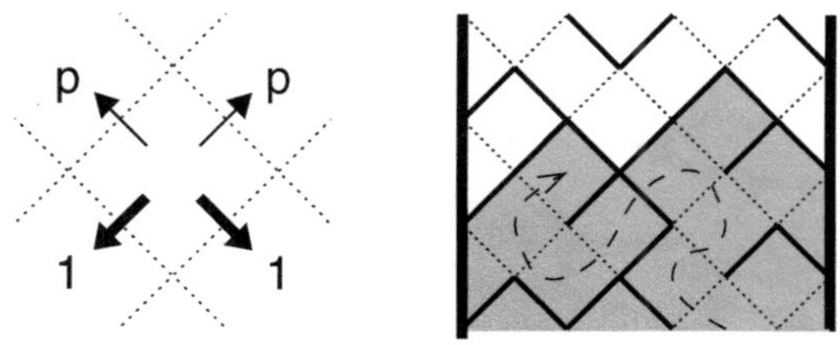

Fig. 3.14 Simple model exhibiting a depinning transition. The pores are represented by cells on a tilted square lattice. The permeability across the edges of the cells depends on the direction of the flow: If the fluid moves downwards all edges are permeable whereas in the upwards direction they are permeable with probability p and impermeable otherwise. The right panel of the figure shows a particular configuration of open (dashed) and closed (solid) edges. Water entering from below will be pinned along a directed path of solid lines, leading to a finite cluster of wet cells (shaded region). The dashed arrow is an example of an open path in order to illustrate how the water flows.

width w (the standard deviation of the heights) over a finite length ℓ of the interface, they found a power-law dependence $w \sim \ell^{\alpha_r}$ with a roughness exponent $\alpha_r = 0.63(4)$.

It was proposed that this process and in particular the value of α_r can be explained in terms of DP [591, 458]. To this end, one considers a simple model for interface pinning which is sketched in Fig. 3.14. In this model the pores of the paper are represented by cells of a tilted square lattice. The liquid can flow to neighbouring cells by crossing the edges of a cell. Depending on the direction of the flow these edges can either be open or closed. The symmetry in the vertical direction is broken by assuming that all edges are permeable in downwards direction, whereas in the upwards direction they can only be crossed with a certain probability p. Thus, starting with a horizontal row of wet cells at the bottom, one obtains a compact cluster of wet cells, as illustrated in the figure. Obviously, the size of the cluster (and thus the penetration depth of the liquid) depends on p. If p is large enough, the cluster is infinite, corresponding to a moving interface. If p is sufficiently small, the cluster is bound from above, i.e., the interface becomes pinned. Both regimes are separated by a depinning transition.

The depinning transition is related to DP as follows. As can be seen, a pinned interface is blocked along a path of impermeable edges running from one boundary of the system to the other. Obviously, the interface becomes pinned as soon as there is a directed path of impermeable bonds connecting the boundaries of the system. Hence the depinning transition is related to an underlying DP process oriented *perpendicular* to the direction of growth. Figure 3.15 shows a possible situation where the interface is pinned. The cluster's **backbone** (see below in Sect. 4.2.6), which consists of all bonds that connect the two boundaries, is indicated by bold dots. The shaded region denotes the resulting cluster of wet cells. As can be seen, the interface will

Fig. 3.15 Interface pinned along the backbone of a directed percolation cluster running from left to right.

be pinned at the lowest lying branch of the DP backbone. Therefore, the roughening exponent is expected to coincide with the meandering exponent

$$\alpha_r = \frac{\nu_\perp}{\nu_\parallel} \approx 0.63 \tag{3.45}$$

of the backbone. Comparing the prediction (3.45) with the experimental result $\alpha_r = 0.63(4)$ obtained by Buldyrev et al. [94] one finds an excellent coincidence, confirming the validity of the model introduced above. However, in subsequent experimental studies the relation to DP could not be confirmed. For example, Dougherty and Carle measured the dynamical avalanche distribution of an air/water interface moving through a porous medium made of glass beads [195]. Assuming an underlying DP process, the distribution $P(s)$ of avalanche sizes s is predicted to behave algebraically. In their experiment, however, they observed an essentially exponential behaviour $P(s) \sim s^{-b}e^{-s/L}$ and the estimates for the exponent b varied between -0.5 and 0.85. Moreover, Albert et al. [12] measured the propagation velocity of locally tilted parts of the interface, concluding that interfaces propagating in glass beads are related to the random-field Ising model rather than DP. As the experimental results are scattered over a wide range, most authors agree that depinning experiments should not yet be considered as a firmly established realisation of DP. Further experimental effort in this direction would be desirable.

Similar experiments were carried out in 2+1 dimensions with a spongy-like material used by florists, as well as fine-grained paper rolls [95]. In this case, however, the exponent α_r is not related to (2+1)-dimensional DP, instead it would correspond to the dynamic exponent of percolating directed interfaces in 2+1 dimensions. In experiments as well as in numerical simulations a roughness exponent $\alpha = 0.50(5)$ was obtained.

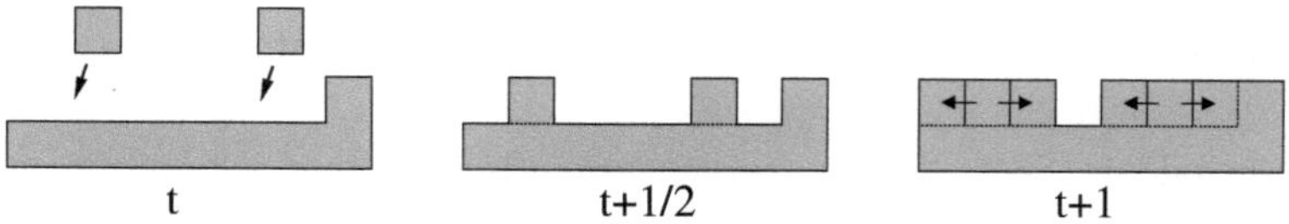

Fig. 3.16 Polynuclear growth model. In the first half time step atoms are deposited with probability p. In the second half time step islands grow deterministically by one step and coalesce.

3.4.2.2 Polynuclear Growth

DP also determines the critical behaviour of roughening transitions in certain growth processes. As an example let us consider so-called polynuclear growth (PNG) models [388, 418, 596], where particles first nucleate with probability p and then expand laterally (see Fig. 3.16). A key feature of PNG models is the use of *parallel updates*, giving rise to a maximal propagation velocity of one monolayer per time step. For a high adsorption rate the interface is smooth, propagating at maximal velocity $v = 1$. Decreasing the adsorption rate below a certain critical threshold, PNG models exhibit a roughening transition to a rough phase with $v < 1$. This transition can be observed even in one spatial dimension.

The relation to DP can be established as follows. Starting from a flat interface $h_i(0) = 0$, let us interpret sites at maximal height $h_i(t) = t$ as active sites of a DP process. Sites where particles are deposited remain active while all other sites become inactive. Similarly, lateral growth corresponds to offspring production of active sites. Therefore, the density of sites at maximal height

$$\varrho = \frac{1}{\mathcal{N}} \sum_i \delta_{h_i,t} \qquad (3.46)$$

corresponds to the density of active sites in a DP process and is expected to scale as $\varrho \sim (p_c - p)^\beta$ as the deposition probability p is varied. Although this mapping to DP is not exact, it is supported by numerical simulations.

Concerning experimental realisations of PNG models, the major problem is the use of parallel updates. In realistic experiments, atoms do not move synchronously, instead the adsorption events occur spontaneously, so that random sequential updates may be more appropriate to describe such experiments. However, using random-sequential updates in PNG models, the transition is lost since in this case there is no maximum velocity. It thus remains an open question to what extent PNG processes can be realised in experiments.

3.4.2.3 Growth with Evaporation at the Edges of Plateaus

As described in Sect. 3.3.6.1, certain growth processes, where evaporation is allowed only at the edges of plateaus, may exhibit a transition from a non-moving smooth phase to a moving rough phase which is driven by an underlying DP process [17, 16]. In particular, the growth velocity is found to scale as $v \sim (p - p_c)^{\nu_\parallel}$. Such a growth process may be interpreted as a PNG-process turned upside down. However, in contrast to the PNG model, which requires the use of parallel updates, the models introduced in [17, 16] work with random-sequential updates as well. Therefore, they may be suitable candidates for an experimental realisation. However, so far no experimental realisation of this type of growth process is known, probably because in experiments evaporation from the middle of a plateau (corresponding to spontaneous creation of activity in the language of DP) cannot be fully suppressed.

3.4.3 Intermittent Turbulence

3.4.3.1 Spiking Magnetic Fluids

A possible way of realising directed percolation in **intermittent turbulence** has been proposed by Pomeau [528]. The idea is that the transition to the turbulent state in a liquid should be similar to the transition of non-linear oscillators from a regular to an intermittent state. Coupling many of these oscillators, their interaction can be interpreted as a contamination process. The situation is similar to that of laminar convection rolls in Taylor-Couette flow which do not break up spontaneously into a turbulent state unless they are 'infected' by a turbulent neighbour.

An interesting experiment based on this idea has been carried out by Rupp et al. [554]. The experiment consists of a cylindrical electromagnet with a sharp edge at which a magnetic fluid is trapped in a ring. Due to the Rosensweig instability, the magnetic fluid forms spikes if the magnetic field is strong enough. Using a strong oscillating magnetic field, the spikes display an intermittent behaviour in space and time. Varying the field amplitude, Rupp et al. studied the onset of intermittency, finding the exponents β and $\nu_\perp$ in agreement with the predictions of DP while $\nu_\parallel$ differed significantly, see Table 3.3. Even the visual appearance of spatio-temporal clusters seemed to differ from those of critical DP. Therefore, it is unlikely that such an experiment can provide evidence for DP critical behaviour.

3.4.3.2 Intermittent Patterns of Falling Liquid Columns

A closely related experiment was carried out by Brunet and Limat [93], who studied spatiotemporal disorder in the patterns of falling liquid columns. In this experiment a liquid trickles down from a ring driven by gravity, forming small columns of liquid. The points from where these columns originate are not constant, instead they either oscillate around their rest position or they may even move to a different position. Moreover, sometimes columns disappear by coalescence and new columns may be born.

Depending on the flow rate, the authors find a continuous transition between transient and permanent chaos. Brunet and Limat raised the question whether this transition observed might be related to DP. However, as in the case of magnetic spikes there is a strong correlation between the sites and the resulting spatio-temporal patterns of turbulent sites do not look like typical DP clusters at criticality. In fact, Brunet and Limat find the exponents $\nu_\parallel = 0.80(25)$ and $\alpha = 0.60(15)$, which differ significantly from the DP values $\nu_\parallel \approx 1.734$ and $\alpha \approx 0.159$ in one dimension.

3.4.3.3 Turbulent Liquid Crystals

In a recent study, a new attempt to experimentally realise a steady-state transition in the DP class has been reported by Takeuchi et al. [589]. The system is a quasi-$2D$ thin layer (thickness 12μm) of the nematic liquid crystal MBBA in the electrohydrodynamic convection regime, subject to an external voltage in order to trigger the Carr-Helfrich instability. In this way, one obtains a transition between two different turbulent states, referred to as *dynamic scattering modes* DSM1 and DSM2. Close to the voltage threshold V_c for the appearance of DSM2, it is asserted in [589] that a regime of spatiotemporal intermittency is found where DSM2 plays the role of the active state and DSM1 corresponds to the absorbing state. It is also argued that the system under study should be less sensitive to two main problems of previous experiments:

(i) since it is macroscopic, sensitivity to quenched disorder is much less than in small systems on a molecular scale and
(ii) since it is turbulent in both states, it is less sensitive to long-range interactions [589].

As the order parameter is simply the density of DSM2, its observation gives an estimate of the non-universal threshold V_c and the exponent β for which Takeuchi et al. report the estimate $\beta = 0.59(4)$. Furthermore, the distributions of the size Δl and the duration Δt of the *inactive* regions are measured. From the expected power laws $e_\perp(\Delta l) \sim (\Delta l)^{-\epsilon_\perp/\nu_\perp}$ and $e_\parallel(\Delta t) \sim (\Delta t)^{-\epsilon_\parallel/\nu_\parallel}$ with the scaling relations $\epsilon_{\parallel,\perp}/\nu_{\parallel,\perp} = 2 - \beta/\nu_{\parallel,\perp}$

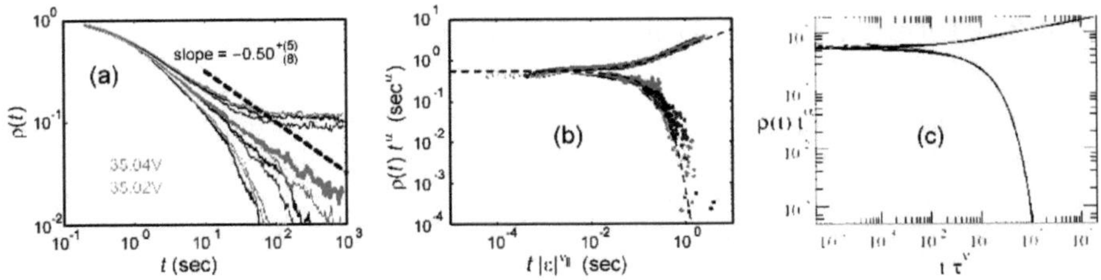

Fig. 3.17 Turbulent liquid crystals. (a) Experimental data obtained by Takeuchi et al., reprinted with permission from [589]. Copyright (2007) by the American Physical Society. (b) Data collapse of the experimental data. (c) Data collapse obtained by simulations.

(see p. 110) they find $\nu_{\perp,x} = 0.66(17)$, $\nu_{\perp,y} = 0.77(7)$ and $\nu_{\parallel} = 1.51(25)$, where the index x, y refers to the spatial direction in which the distribution is measured. According to [589], their results for β and $\nu_\perp$ compare quite well with the theoretical expectation for the DP class $\beta \approx 0.583$, $\nu_\perp \approx 0.733$, and $\nu_\parallel \approx 1.295$ (see Table 4.3), while the agreement for $\nu_\parallel$ is less satisfactory.

They hence carried out a different experiment by bringing the system first deep into the regime with a large DSM2 before ‘quenching’ the system to close to the voltage threshold. Measuring the relaxation of the DSM2 they extract a second estimate for $\beta/\nu_\parallel$. This leads to $\nu_\parallel = 1.181^{+(14)}_{-(21)}$, in better agreement with the theoretical prediction. Finally, they present their data in the form of a data collapse and find that their scaling function matches quite well with the corresponding scaling function of the contact process [589], as demonstrated in Fig. 3.17.

In this experiment, it has been possible for the first time to obtain estimates for *all* three independent exponents which are in fair agreement with the theoretical values for the DP class in 2+1 dimensions, as well as the entire scaling function. If these findings can be substantiated by further tests and improved estimates the turbulent liquid crystal studied by Takeuchi et al. would be the first experimental realisation of directed percolation.

3.4.4 Discussion

The results of various experiments to try to identify the critical behaviour of directed percolation in quasi-one-dimensional systems are listed in Table 3.3. In spite of considerable experimental effort, many of the measured exponents are not nearly precise enough to allow for a quantitative comparison with theoretical predictions.[9] It is remarkable that in various cases the largest de-

---

[9] In principle, one needs to measure four independent exponents, for example $\beta, \beta', \nu_\parallel, \nu_\perp$, in order to determine a universality class of absorbing phase transitions [470].

viations have been observed in the estimates for $\nu_\parallel$, hence future experiments should focus particularly on this exponent.

Why is it so difficult to perform experiments related to DP? It seems that the basic features of DP, which can be implemented so easily on a computer, are quite difficult to realise in nature. One of these assumptions is the existence of an absorbing state. In reality, such a non-fluctuating state is hard to realise. For example, a poisoned catalytic surface is never completely frozen, rather it will allow for small fluctuations which could be strong enough to 'soften' the transition, making it impossible to quantify the critical exponents. Another fundamental problem is **quenched disorder** due to microscopic inhomogeneities of the system. Depending on the type of disorder, even weak inhomogeneities might obscure or even destroy the DP transition.

In spite of all these difficulties, many physicists believe that DP should have a counterpart in reality, mostly because of its simplicity and robustness. Therefore, designing and performing such experiments remains one of the most important problems of non-equilibrium statistical physics. The recent liquid crystal experiment by Takeuchi et al. may be an important step in this direction.

experiment	geometry	β	$\nu_\perp$	$\nu_\parallel$	$\epsilon_\perp/\nu_\perp$	Ref.
RBC	annular		0.5			[109]
RBC	linear	0.30(5)	0.50(5)	0.50(5)	1.6(2)	[146, 145]
RBC	linear		0.5	0.5	1.7(1)	[146, 145]
viscous fingering	linear	0.45(5)	0.5		0.63(2)	[475]
line of vortices	linear			0.5		[628]
Taylor-Dean	linear	1.30(26)	≈ 0.64	≈ 0.73	1.67(14)	[153]
ferrofluidic spikes	annular	0.30(5)	1.2(1)	0.70(5)	1.70(5)	[554]
DP theory	$1D$	0.28	1.10	1.73	1.75	
Rayleigh-Taylor	$2D$	0.56(5)				[522]
turbulent liquid crystal MBBA	$2D$	0.59(4)	$0.66(17)^x$	1.51(25)	$1.10(22)^x$	[589]
			$0.77(7)^y$		$1.23(4)^y$	
				$1.18^{+(14)}_{-(21)}$		
DP theory	$2D$	0.58	0.73	1.30	1.20	

Table 3.3 Experimentally measured critical exponents in quasi-one-dimensional or two-dimensional systems. In $1D$ the geometry is either annular or linear. RBC stands for Rayleigh-Bénard convection. The exponent $\epsilon_\perp/\nu_\perp = 2 - \beta/\nu_\perp$ describes the distribution of spatial gaps. The numbers in brackets give the estimated error in the last digit(s) (most data are taken from [554]). In [589], indices x and y refer to the spatial direction in which the exponent was measured. The experimental data have to be compared to the theoretical values of DP, here rounded to two digits.

Problems

20. Consider the DP equation of motion on a mean-field level (also referred to as **Schlögl's first model**)

$$\lambda^{-1}\partial_t \varrho(t) = \tau\varrho(t) - g\varrho(t)^2 , \tag{3.47}$$

where we neglect the noise and spatial fluctuations of the order parameter. Derive how the order parameter relaxes in leading order to the steady state values $\varrho = \tau/g$ for $\tau > 0$ and $\varrho = 0$ for $\tau < 0$, respectively. Discuss the behaviour of the temporal correlation time $\xi_{\parallel}$ on which order parameter relaxations takes place. Consider in particular the case $\tau = 0$.

21. Taking spatial fluctuations of the order parameter into account, the Langevin equation of directed percolation is given by

$$\lambda^{-1}\partial_t \varrho(t,\boldsymbol{r}) = \left(\tau - g\,\varrho(t,\boldsymbol{r}) + \nabla^2\right)\varrho(t,\boldsymbol{r}) + h. \tag{3.48}$$

Consider small spatial deviations from the homogeneous steady-state order parameter $\varrho_{\rm s}(\tau,h)$ by introducing $\delta\varrho(t,\boldsymbol{r}) = \varrho(t,\boldsymbol{r}) - \varrho_{\rm s}(\tau,h)$. Derive within a linear approximation (Ornstein-Zernicke-like approach) in $\delta\varrho(t,\boldsymbol{r})$ the spatial correlation length $\xi_{\perp}(\tau,h)$ as well as the temporal correlation length $\xi_{\parallel}(\tau,h)$.

22. The relaxation of magnetic systems towards their steady-state is often described in terms of a stochastic Langevin equation for the coarse-grained time-dependent order-parameter (magnetisation) $m(t)$

$$\frac{\mathrm{d}m(t)}{\mathrm{d}t} = -\frac{\delta F[m]}{\delta m(t)} + \eta(t)$$

where $\eta(t)$ is a Gaussian centred noise with variance $\langle\eta(t)\eta(t')\rangle = 2T\delta(t-t')$ and $F[m]$ is the equilibrium free energy for which we take here the mean-field form $F[m] = -\frac{3}{2}\lambda^2 m^2 + \frac{1}{4}m^4$. Averaging over the noise, one has the mean-field equation $\dot{m} = 3\lambda^2\, m - m^3$. Find the stationary solutions and discuss the relaxation from an initial magnetisation $m(0) = m_0 \neq 0$ towards them.

23. Consider the diffusive contact process on a one-dimensional lattice, where in addition to the particle reactions admitted in the contact process, a particle may also hop to an empty nearest-neighbour site. Discuss the behaviour of the model within the site- as well as the pair-approximation and study the role of the diffusion rate D.

24. Consider **Schlögl's second model** defined by the reactions of particles of the single species A: $A + A + \emptyset \xrightarrow{\lambda} A + A + A$ and $A \xrightarrow{1} \emptyset$ where the spatially reflected reactions are understood to be included with the same rate. Analyse the model using the site-approximation.

25. Particles of a single species A can move on a chain of L sites such that each site is either empty (state $\emptyset$) or occupied by a single particle (state A). The following microscopic movements are admitted between any two nearest neighbours on the chain

$$A + \emptyset \xrightarrow{D_R} \emptyset + A \ , \quad \emptyset + A \xrightarrow{D_L} A + \emptyset$$

where $D_{R,L}$ are the diffusion rates to the right or the left, respectively. Find the Liouville operator $\mathcal{L}$ for that problem. Show that through a similarity transformation, it can be related to the quantum Hamiltonian of a Heisenberg ferromagnetic chain [15].

26. Consider the same system as in the previous exercise, with symmetric diffusion $D_R = D_L = D$. Furthermore, admit the following 'chemical reactions' between nearest-neighbour particles: (i) annihilation $A + A \xrightarrow{2\alpha} \emptyset + \emptyset$, (ii) coagulation $A + A \xrightarrow{\gamma} A + \emptyset$ and $A + A \xrightarrow{\gamma} \emptyset + A$ and (iii) death $A + \emptyset \xrightarrow{\delta} \emptyset + \emptyset$ and $\emptyset + A \xrightarrow{\delta} \emptyset + \emptyset$. If $\delta \neq 2\alpha + \gamma$, show through a similarity transformation of the Liouville operator that the n-particle correlator $c_n(t, \{r\}) = \langle a_{r_1} \ldots a_{r_n} \rangle(t)$, where a_r is the time-dependent density at site r, can be found either from the case without coagulation ($\gamma = 0$) or without annihilation ($\alpha = 0$) [399, 276, 573].

27. Analyse the diffusion-annihilation-coagulation process of exercises 25,26 by using non-equilibrium field-theory. Show how to re-derive the relationship between the correlation functions for different values of the reaction rates [31].

28. Consider a stochastic process, described by a master equation such that the detailed balance condition $w_{\alpha\beta} P_\beta^{\mathrm{eq}} = w_{\beta\alpha} P_\alpha^{\mathrm{eq}}$ holds and that the equilibrium probabilities $P_\alpha^{\mathrm{eq}} > 0$ for all configurations α. Show that the Liouville operator $\mathcal{L}$ is similar to a symmetric matrix $\mathcal{M}$.

29. Analyse the Domany-Kinzel automaton, using the site-approximation.

30. Study the phase-transition in the ZGB model in a simple mean-field approximation (for simplicity, consider first a one-dimensional version of the ZGB model before looking at the two-dimensional case). How many phase transitions do you find and are they of first or of second order?

31. Consider the pair-contact process on a chain with L sites and either open or periodic boundary conditions. Calculate the number $N_a(L)$ of stationary states for L large and prove in particular (3.40).

32. The **triplet-contact process** is characterised by the reactions of a single species A of particles, such that each site of the lattice is either empty or else occupied by exactly one particle and with the reactions

$$3A \longrightarrow 4A \ , \ 3A \longrightarrow \emptyset \tag{3.49}$$

For a one-dimensional lattice with L sites, show that for L large the number of steady-states is given by [274]

$$N_a(L) \simeq N_0 g^L \ , \ N_0 = \begin{cases} 1 & ; \text{ periodic boundary conditions} \\ 1.137\ldots & ; \text{ free boundary conditions} \end{cases} \tag{3.50}$$

and where $g \simeq 1.839\ldots$.

δ	z	β	$\nu_\parallel$	Ref.
0.32(1)	1.75(10)	–	2.5(2)	[507]
0.27(1)	1.8(1)	0.90(5)	–	[395]
0.33(1)	–	0.99(5)	–	[497]

Table 3.4 Estimates for the critical exponents of the (1+1)-dimensional diffusive triplet-contact process TCPD.

If one adds diffusion of single particles, one obtains the *triplet-contact process with diffusion* (TCPD) [507]. Table 3.4 lists some estimates for critical exponents along the line of second-order phase transitions. The upper critical dimension should be $d_c = 1$. We leave it to the reader to work out the cluster approximations for this model (why should one go beyond the pair-approximation here?).

33. Show that the generating function of a directed lattice animal can always be reduced to the case of strong embedding [282].

Chapter 4
Scaling Properties of Absorbing Phase Transitions

In this chapter, scaling properties of directed percolation will be generalised into a scaling theory for absorbing phase transitions. A large set of tools for their analysis will be presented.

4.1 Scaling in the Steady-State

As already outlined in Sect. 3.2.1, the continuous phase transition of DP exhibits universal scaling laws which can be described by a suitable phenomenological scaling theory. The most common way to characterise universality classes is in terms of critical exponents. For the convenience of the reader, we collect in Table 4.1 the definitions of the conventional critical exponents, as we shall use them in this book.[1] The various observables listed here will be introduced later in this chapter.

In fact, it turns out that the concept of scale-invariance, which proved to be so successful in equilibrium statistical mechanics (see Sect. 2.2), can be applied to non-equilibrium phase transitions as well.[2] The starting point is the assumption that the long-range properties of a critical many-particle process are scale-free and thus invariant under an appropriate multiplicative change of scales. Such scale-invariant systems are expected to exhibit power-law behaviour with certain universal critical exponents and scaling functions which are governed by symmetries and conservation laws rather than microscopic

---

[1] One should not confuse the control parameter τ with the exponent τ. In the older literature, $R^2(t) \sim t^{\tilde{z}}$ and $N(t) \sim t^{\eta}$ were frequently used. We warn the reader that the exponents of absorbing phase transitions are not always the direct analogues of equilibrium critical exponents.

[2] In the exercises for this chapter, the long-time behaviour of diffusion-annihilation-coagulation processes are studied, which provide another instance of statistical systems at an absorbing phase transition, including experimental illustrations.

observable	exponent	relation	conditions		
order parameter	β	$\varrho_s(\tau) \sim \tau^{\beta}$	$\tau > 0$	$h = 0$	$t \to \infty$
	β/σ	$\varrho_s(h) \sim h^{\beta/\sigma}$	$\tau = 0$	$h \neq 0$	$t \to \infty$
	α	$\varrho(t) \sim t^{-\alpha}$	$\tau = 0$	$h = 0$	
survival probability	β'	$P_s(\tau) \sim \tau^{\beta'}$	$\tau > 0$	$h = 0$	$t \to \infty$
	δ	$P(t) \sim t^{-\delta}$	$\tau = 0$	$h = 0$	
susceptibility	γ	$\chi = \partial\varrho_s/\partial h \sim \lvert\tau\rvert^{-\gamma}$	$\tau \neq 0$	$h = 0$	$t \to \infty$
variance	γ'	$\Delta_\varrho(\tau) \sim \lvert\tau\rvert^{-\gamma'}$	$\tau \neq 0$	$h = 0$	$t \to \infty$
correlation lengths	$\nu_\perp$	$\xi_\perp(\tau) \sim \lvert\tau\rvert^{-\nu_\perp}$	$\tau \neq 0$	$h \to 0$	$t \to \infty$
	$\nu_\parallel$	$\xi_\parallel(\tau) \sim \lvert\tau\rvert^{-\nu_\parallel}$	$\tau \neq 0$	$h = 0$	$t \to \infty$
autocorrelator		$c(t) \sim t^{\Theta - d/z}$	$\tau = 0$	$h = 0$	
temporal empty interval	$\epsilon_\parallel$	$e_\parallel(\Delta t) \sim (\Delta t)^{-\epsilon_\parallel/\nu_\parallel}$	$\tau = 0$	$h = 0$	$t \to \infty$
spatial empty interval	$\epsilon_\perp$	$e_\perp(\Delta r) \sim (\Delta r)^{-\epsilon_\perp/\nu_\perp}$	$\tau = 0$	$h = 0$	$t \to \infty$
number of active sites	Θ	$N(t) \sim t^{\Theta}$	$\tau = 0$	$h = 0$	
mean square spreading	z	$R^2(t) \sim t^{2/z}$	$\tau = 0$	$h = 0$	
mean mass of cluster	γ	$M(\tau) \sim (-\tau)^{-\gamma}$	$\tau < 0$	$h = 0$	
mean size of cluster	σ	$S(\tau) \sim (-\tau)^{-\sigma}$	$\tau < 0$	$h = 0$	
mean survival time	τ	$T(\tau) \sim (-\tau)^{-\tau}$	$\tau < 0$	$h = 0$	
mean spatial cluster volume	υ	$V(\tau) \sim (-\tau)^{-\upsilon}$	$\tau < 0$	$h = 0$	

Table 4.1 Definition of the conventional critical exponents of absorbing phase transitions, either in the stationary state depending on the control parameters τ and h, or for the dynamical behaviour *at* criticality ($\tau = h = 0$). Here $\tau < 0$ refers to the inactive/absorbing phase and $\tau > 0$ refers to the active phase. For the definition of the exponent α, the initial state should be the fully occupied lattice.

details of the model under consideration. This leads to numerous scaling relations between the exponents listed in Table 4.1, which are summarised in appendix B and will be derived in this chapter. Exponents for more specific kinds of critical behaviour (tricritical, surface,...) will be introduced when needed.

In the following, we introduce the reader to a generic scaling theory of absorbing phase transitions. We shall present this scaling theory on a phenomenological level in the context of DP, although it can be applied to other classes of continuous phase transitions into absorbing states as well. The essential difference compared to the equilibrium case lies in a separate treatment of space and time since in non-equilibrium statistical physics time plays the role of an independent degree of freedom. Moreover, there is no obvious symmetry between spatial and temporal degrees of freedom.

4.1.1 Order Parameters

We consider a generic many-particle system on a lattice that exhibits a continuous phase transition from a fluctuating active phase into one or several absorbing states. Unless stated otherwise, the system size is assumed to be infinite. The process is controlled by a control parameter τ which is zero at the transition and positive (negative) in the active (absorbing) phase. For example, in the contact process this parameter is proportional to the distance from criticality $\tau \propto \lambda - \lambda_c$. In addition, each site carries a local observable which tells us whether it is active or not. In the most simple examples of directed percolation, this observable is just the occupation number $s_i(t)$, while in the (1+1)-dimensional pair-contact process it would correspond to the product $s_i(t)s_{i+1}(t)$, which represents a pair density.

We are now interested in the properties of such a model in the active phase $\tau > 0$. Here the system relaxes towards a fluctuating steady-state whose space-time trajectory can be interpreted as an infinite cluster of active sites. So far the situation is similar to isotropic percolation (see Sect. 3.1 and Fig. 3.1), where an infinite cluster of connected sites emerges in the active phase. However, in contrast to the isotropic case, the cluster is now directed in time. This allows one to characterise such a cluster in two different ways:

(a) We may ask for the probability that a given site belongs to an infinite cluster that was generated in the past at $t = -\infty$ from a fully occupied lattice. Obviously, this probability is just the density ϱ of active sites in the stationary state.

(b) Alternatively we may ask for the probability that an isolated seed of activity (a single occupied site in an empty lattice) will generate an infinite cluster extending to $t = +\infty$. This quantity is related to the percolation probability $P_{\rm perc}$.

In general, these two quantities are different, although similar in character. They can both be used as equally legitimate order parameters, one of them probing the past and the other one probing the future. Both order parameters vanish in the absorbing phase and assume finite values in the active phase. Nevertheless, they may give different results since in general, the dynamic rules are not symmetric under time-reversal. In the active phase we therefore expect the two order parameters to scale algebraically as

$$\varrho \sim \tau^{\beta}, \qquad P_{\rm perc} \sim \tau^{\beta'} \tag{4.1}$$

with two, in general *different* exponents, β and β'.

A field-theoretic formulation of absorbing phase transitions (see Sect. 4.3.4) shows that the two order parameters are associated with a field ϕ and its associated response field $\widetilde{\phi}$ which can be interpreted as creation- and annihilation operators. Roughly speaking the exponent β emerges whenever the density of particles is measured. The exponent β', on the other hand, is related to the creation of particles, e.g. by applying an external field or by specifying the initial configuration. In correlation functions, which involve both creation

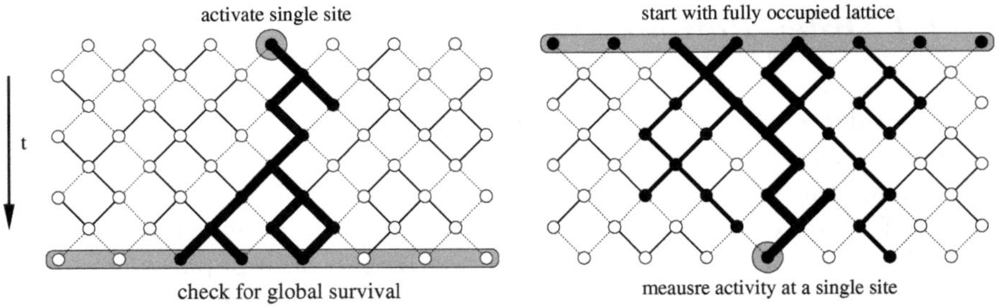

Fig. 4.1 Exact rapidity-reversal symmetry in bond directed percolation (see text).

and annihilation operators, both exponents are expected to appear. It should be emphasised that the emergence of two different order parameters with generally different exponents is a consequence of the directed nature of time and as such is a novel feature of non-equilibrium critical phenomena.

4.1.2 Rapidity-Reversal Symmetry of Directed Percolation

The emergence of two different order parameters (with two in general different exponents β and β') is a generic property of continuous phase transitions in models with a directed time-like dimension. In DP, however, a special invariance property under time-reversal, called **rapidity-reversal symmetry** [246], ensures that the two exponents coincide.[3] This means that DP is characterised by only three instead of four independent critical exponents.

In order to understand this symmetry heuristically, let us consider the special case of bond directed percolation (see Sect. 3.1), where this time-reversal symmetry is fulfilled *exactly* on the microscopic level. To see this, consider a particular realisation of open and closed bonds, as shown in the left panel of Fig. 4.1. Activating a single site at $t = 0$, a certain cluster represented by bold bonds is generated. At a given time t, we may ask for the so-called **survival probability** $P_{\text{sur}}(t)$ of finding at least one active site. In the active phase, there is a finite probability that the cluster survives forever, hence $P_{\text{sur}}(t)$ tends to a positive constant as $t \to \infty$. Clearly, this **ultimate survival probability** is just the percolation probability

$$P_{\text{perc}} = \lim_{t\to\infty} P_{\text{sur}}(t)\,. \tag{4.2}$$

which is non-zero in the active phase.

[3] The name rapidity-reversal comes from an application of directed percolation to high-energy collisions of elementary particles, in the context of Reggeon field-theory.

The right panel of Fig. 4.1 shows the same realisation of open and closed bonds turned upside down. *Provided* this time-reversed configuration has the same statistical weight as the original one, which is indeed the case for a *bond* DP process as we shall show later, a *bond* DP process running backward in time can be interpreted as a bond DP process evolving forward in time following exactly the same dynamical rules, the only difference being the exchange of initialisation and measurement.

In the time-reversed realisation of open and closed bonds we now consider the cluster which is generated by a fully occupied initial state at $t = 0$. Since the statistical weight of the time-reversed path in bond DP is the same as that of the original one, the probability of finding the central site at time t in the active state is exactly the same as the survival probability $P_{\rm sur}(t)$ in the original realisation. On the other hand, in a system with a fully occupied initial state this probability is just the density of active sites $\varrho(t)$. In other words, in bond DP the probability of finding a directed path from a single site at $t = 0$ to a horizontal line at time t coincides exactly with the probability of finding a directed path from a horizontal line at $t = 0$ to a particular site at time t. This leads to the remarkable identity [246]

$$P_{\rm sur}(t) = \varrho(t)\,. \tag{4.3}$$

Consequently, in the active phase the two order parameters saturate at the same value $P_{\rm perc} = \varrho$ which implies that the corresponding critical exponents $\beta = \beta'$ have to be identical ((4.3) is expressed by $\mu = 1$ for bond DP in Table 3.1).

It should be stressed that the exact rapidity-reversal symmetry in (4.3) is a *special property of bond DP*. In other realisations of DP, e.g. in site DP or the contact process, this symmetry is generally not exact. Instead one finds an *asymptotic* symmetry in the sense that $P_{\rm sur}(t)$ and $\varrho(t)$ become proportional in the limit $t \to \infty$. This proportionality implies that for all models belonging to the DP class the exponents β and β' are equal. We shall return to the rapidity-reversal symmetry in the context of a field-theoretical description of directed percolation in Sect. 4.3.4. In other models, which are not related to DP, this symmetry is not always fulfilled.

4.1.3 The Correlation Lengths $\xi_\perp$ and $\xi_\parallel$

In equilibrium statistical mechanics, continuous phase transitions are characterised by a single correlation length ξ that diverges at the critical temperature as $|T - T_c|^{-\nu}$. In the non-equilibrium case of an absorbing phase

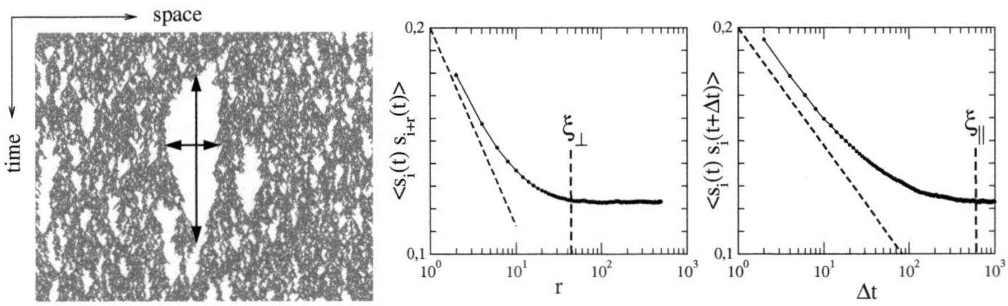

Fig. 4.2 Interpretation of the correlation lengths $\xi_\perp$ and $\xi_\parallel$ in an almost critical stationary (1+1)-dimensional bond DP process (see text). The dotted lines indicate the expected slopes $-\beta/\nu_\perp$ and $-\beta/\nu_\parallel$, respectively.

transition,[4] where 'space' and 'time' are different in character, there are instead *two* different correlation lengths $\xi_\perp$ and $\xi_\parallel$, where the indices $\perp$ and $\parallel$ denote spatial and temporal properties, respectively. Approaching the critical point, these correlation lengths are expected to diverge as

$$\xi_\perp \sim |\tau|^{-\nu_\perp}, \qquad \xi_\parallel \sim |\tau|^{-\nu_\parallel} \tag{4.4}$$

with generally different critical exponents $\nu_\perp$ and $\nu_\parallel$. Therefore, in the scaling regime, the two correlation lengths are related by

$$\xi_\parallel \sim \xi_\perp^z, \tag{4.5}$$

where $z = \nu_\parallel/\nu_\perp$ is the **dynamical exponent**. Roughly speaking this exponent tells us how fast a local perturbation spreads, ranging from $z = 1$ (deterministic 'light-cone' spreading) over $1 < z < 2$ (**superdiffusive spreading**) and $z = 2$ (diffusive spreading) to $z > 2$ (**subdiffusive spreading**). In many-particle models with short-range interactions one usually finds the dynamical exponent in the range $1 < z \leq 2$. Like the other critical exponents z is a universal quantity, i.e. its value is fully determined by the universality class and the dimensionality of the system.

Figure 4.2 illustrates the physical meaning of the correlation lengths $\xi_\perp$ and $\xi_\parallel$ visually. The left panel shows the space-time trajectory of a stationary (1+1)-dimensional DP process in the active phase. As can be seen, the cluster consists of active branches enclosing elongated inactive voids. The size of these voids is broadly distributed and extends up to certain maximal scales in space and time. These maximal scales may be interpreted as the correlation lengths $\xi_\perp$ and $\xi_\parallel$.

[4] In contrast, in *driven* diffusive systems in $d \geq 2$ spatial dimensions, *three* distinct length scales are expected: (i) $\xi_\perp \sim |\tau|^{-\nu_\perp}$ perpendicular to the direction of the external driving, (ii) $\xi_\parallel \sim |\tau|^{-\nu_\parallel}$ parallel to it and (iii) the correlation time $\tau_{\rm rel} \sim |\tau|^{-\nu_{\rm rel}}$. Hence, there is a dynamical exponent $z = \nu_{\rm rel}/\nu_\perp$ and an anisotropy exponent $\theta = \nu_\parallel/\nu_\perp$, which *a priori* are distinct. Systems of that kind will not be considered in this book.

As an alternative approach, the other two panels in Fig. 4.2 show the equal-time correlation function

$$c_\perp(r) = c_\perp(t,r) = \langle s_i(t) s_{i+r}(t) \rangle \tag{4.6}$$

and the autocorrelation function

$$c_\parallel(\Delta t) = \langle s_i(t) s_i(t+\Delta t) \rangle \tag{4.7}$$

measured numerically in a stationary DP process above criticality. In both cases the correlations first decay algebraically as $r^{-\beta/\nu_\perp}$ and $(\Delta t)^{-\beta/\nu_\parallel}$, respectively, until they saturate at the same constant value. In the saturated regime the two sites become uncorrelated so that this value is just equal to the squared stationary density ϱ^2. The crossovers take place at certain typical scales in space and time (indicated by vertical dashed lines in the figure) which correspond to the correlation lengths $\xi_\perp$ and $\xi_\parallel$.

4.1.4 Scale-Invariance

The phenomenological scaling theory for absorbing phase transitions assumes *simple scaling* (as opposed to multiscaling), meaning that the critical behaviour can be characterised in terms of only two diverging length scales $\xi_\perp$ and $\xi_\parallel$. In this case, all bulk exponents can be expressed in terms of the four independent critical exponents $\beta, \beta', \nu_\parallel, \nu_\perp$ by means of simple scaling relations.[5] The starting point is the hypothesis that a multiplicative change of the control parameter

$$\tau \mapsto \lambda\tau \tag{4.8}$$

rescales the order parameters and the correlation lengths by

$$\varrho \mapsto \lambda^\beta \varrho \tag{4.9}$$

$$P_{\text{perc}} \mapsto \lambda^{\beta'} P_{\text{perc}} \tag{4.10}$$

$$\xi_\perp \mapsto \lambda^{-\nu_\perp} \xi_\perp \tag{4.11}$$

$$\xi_\parallel \mapsto \lambda^{-\nu_\parallel} \xi_\parallel \tag{4.12}$$

Likewise, all other measurable quantities and parameters have to be multiplied by λ raised to suitable powers, the so-called **scaling powers**. For example, all spatial and temporal quantities such as distances r and time intervals Δt have to be rescaled in the same way as the correlation lengths, i.e.,

$$r \mapsto \lambda^{-\nu_\perp} r\,, \qquad \Delta t \mapsto \lambda^{-\nu_\parallel} \Delta t\,. \tag{4.13}$$

[5] As explained above in Sect. 4.1.2, the exponents β and β' coincide in systems with a rapidity-reversal symmetry such as DP.

Scale-invariance implies that any function with n parameters can be expressed as a leading power law times a scaling function that depends on only $n-1$ parameters (cf. Sect. 2.2.2). In many cases, scaling functions are universal in the same sense as the critical exponents, i.e., they are determined by a small number of fundamental parameters, while they do not depend on microscopic 'details' of the dynamics.

In order to demonstrate how this mechanism works, let us first consider the density of active sites $\varrho_s(\tau)$ in the stationary state with $\tau > 0$. Scale-invariance implies that the order parameter ϱ_s can be written as a generalised homogeneous function

$$\lambda^{\beta}\varrho_s(\tau) = \tilde{r}(\lambda\tau)\,. \tag{4.14}$$

Choosing $\lambda = \tau^{-1}$ one obtains the well-known power law

$$\varrho_s(\tau) \;=\; \tau^{\beta}\tilde{r}(1)\,. \tag{4.15}$$

Thus the function $\varrho_s(\tau)$, which depends on only one parameter, is fully determined (up to a prefactor) by scale-invariance.

4.1.5 Two-Point Correlation Function in the Steady-State

As another example we consider the two-point correlation function

$$c(t_1, t_2; \boldsymbol{r}_1, \boldsymbol{r}_2) \;=\; \langle s(t_1, \boldsymbol{r}_1)s(t_2, \boldsymbol{r}_2)\rangle \tag{4.16}$$

in the stationary active state with $\tau > 0$. Because of translational invariance in space and time this function depends on three parameters

$$c(\Delta t, r; \tau) = \langle s(t_1, \boldsymbol{r}_1)s(t_2, \boldsymbol{r}_2)\rangle\,, \tag{4.17}$$

where $\Delta t = t_2 - t_1$ and $r = |\boldsymbol{r}_2 - \boldsymbol{r}_1|$. Scale-invariance implies that a scale transformation according to (4.8) and (4.13) changes the correlation function by a factor λ^{κ} with some unknown exponent κ, i.e., sufficiently close to the critical point the correlation function can be described in terms of a generalised homogeneous function

$$c(\,\lambda^{-\nu_\parallel}\Delta t, \lambda^{-\nu_\perp}r;\, \lambda\tau) \;\simeq\; \lambda^{\kappa}c(\Delta t, r; \tau)\,. \tag{4.18}$$

This relation holds for arbitrary $\lambda > 0$. Thus, choosing λ in such a way that one of the arguments on the left-hand side becomes constant, we may express $c(r, \Delta t, \tau)$ as a leading power law times a scaling function which depends on only two parameters. For example, setting $\lambda = \tau^{-1}$ the above relation reduces to

$$c(\Delta t, r; \tau) = \tau^{\kappa} c(\tau^{\nu_{\|}} \Delta t, \tau^{\nu_{\perp}} r;\, 1) = \tau^{\kappa} f(\Delta t/\xi_{\|}, r/\xi_{\perp}), \tag{4.19}$$

where f is a scaling function that depends on only two arguments.

The unknown exponent κ in the leading power law can be determined by considering the limit $r \to \infty$ in the stationary active phase. In this case, the two points become uncorrelated so that $c(\Delta t, \infty; \tau) = \varrho^2$. Since ϱ scales as τ^{β} we obtain

$$\kappa = 2\beta. \tag{4.20}$$

As in equilibrium statistical mechanics, it is often useful to introduce *metric factors* (cf. Sect. 2.2), by which the scaling function can be expressed in terms of dimensionless and appropriately normalised arguments. For any parameter x we shall denote the corresponding metric factor by a_x. Moreover, we shall denote the normalised scaling functions by the corresponding capital letters with a tilde symbol. For example, in the case of the two-point correlation function we use the notation

$$c(\Delta t, r; \tau) \simeq \lambda^{-2\beta}\, \tilde{C}(\lambda^{-\nu_{\|}} a_t \Delta t, \lambda^{-\nu_{\perp}} a_r r;\, \lambda a_{\tau} \tau). \tag{4.21}$$

The metric factors contain all non-universal system-dependent features of the scaling form and may depend e.g. on the lattice structure, the interaction range, and the update scheme. With a suitable convention for the choice of the metric factors, the scaling function $\tilde{C}$ is fully universal, hence it should be the same for all systems within a given universality class. For example, for the two-point correlation function given above, a possible normalisation would be

$$\tilde{C}(0,1;0) = \tilde{C}(1,0;0) = \tilde{C}(0,0;1) = 1. \tag{4.22}$$

With this convention the non-universal metric factors are just the amplitudes (proportionality factors) of the corresponding power laws. For example, at criticality the equal-time correlation function decays as

$$\begin{aligned} c(\Delta t, r; \tau)\Big|_{\Delta t=0,\tau=0} &\simeq \lambda^{-2\beta}\, \tilde{C}(0, \lambda^{-\nu_{\perp}} a_r r; 0)\Big|_{\lambda^{-\nu_{\perp}} a_r r=1} \\ &= (a_r r)^{-2\beta/\nu_{\perp}}\, \tilde{C}(0,1;0) \;=\; (a_r r)^{-2\beta/\nu_{\perp}}. \end{aligned}$$

4.1.6 Empty-Interval Probabilities in the Steady-State

Let us now consider the stationary active phase of a system with an absorbing phase transition and analyse the time series of subsequent activations at a particular site (see Fig. 4.3). What is the size distribution $e_{\|}(\Delta t, \tau)$ of empty (=inactive) intervals Δt between subsequent activations at this particular site?

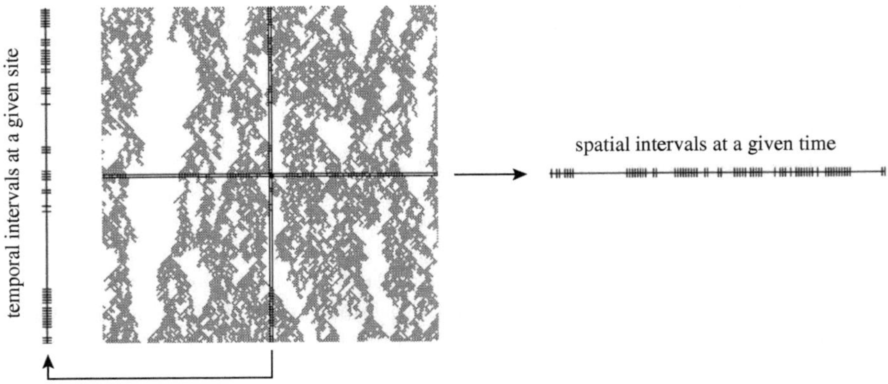

Fig. 4.3 Temporal and spatial empty-interval distribution in a stationary (1+1)-dimensional DP process.

The distribution $e_{\|}(\Delta t, \tau)$ is an example of a so-called **empty-interval probability**. It is important to note that empty-interval distribution functions differ significantly from the ordinary two-point correlation function discussed above. While two-point functions probe correlated activity of two points ignoring intermediate configurations, empty-interval functions require the *absence* of activity along a line that connects the two points. In this sense such quantities may be regarded as multipoint correlation functions. While two-point correlation functions saturate at $\varrho^2 > 0$ in the stationary state, empty-interval functions tend to zero if the distance between the two points increases.

4.1.6.1 Scaling of the Temporal Empty-Interval Distribution Functions

We now *postulate* that empty-interval distribution functions have scaling properties analogous to those of two-point correlation functions, although with different asymptotic limits and different scaling indices. This means that under a scale-transformation (4.8) and (4.13) we expect them to scale as $e_{\|} \mapsto \lambda^{\epsilon_{\|}} e_{\|}$ with some unknown exponent $\epsilon_{\|}$, leading to the scaling form

$$e_{\|}(\Delta t; \tau) = \lambda^{-\epsilon_{\|}} \tilde{e}_{\|}(\lambda^{-\nu_{\|}} \Delta t; \lambda \tau) . \tag{4.23}$$

Choosing $\lambda = \Delta t^{1/\nu_{\|}}$ one obtains (where spatial translation-invariance was also assumed)

$$e_{\|}(\Delta t; \tau) = \Delta t^{-\epsilon_{\|}/\nu_{\|}} \tilde{e}_{\|}(\Delta t^{1/\nu_{\|}} \tau) \tag{4.24}$$

with some unnormalised scaling function $\tilde{e}_{\|}$. The exponent $\epsilon_{\|}$ can be determined as follows. On the one hand, the temporal interval distribution has to be normalised, i.e.

$$\sum_{\Delta t=1}^{\infty} e_{\parallel}(\Delta t;\tau) = 1 \tag{4.25}$$

for all $\tau > 0$, including the critical limit $\tau \to 0^+$. This means that for small arguments the scaling function $\tilde{e}_{\parallel}$ tends to a constant and the exponent $\epsilon_{\parallel}/\nu_{\parallel}$ has to be larger than 1. On the other hand, it is easy to see that the reciprocal of the density of active sites ϱ^{-1} is by definition exactly equal to the mean interval-size

$$\overline{\Delta t} = \sum_{\Delta t=1}^{\infty} \Delta t\, e_{\parallel}(\Delta t;\,\tau). \tag{4.26}$$

In the continuum limit this sum can be approximated by an integral

$$\overline{\Delta t} \approx \int_a^{\infty} \mathrm{d}\Delta t\ \Delta t\ e_{\parallel}(\Delta t,\tau) \tag{4.27}$$

with a suitable lower cutoff a that accounts for the discreteness of the sum for small Δt. Assuming that $1 < \epsilon_{\parallel}/\nu_{\parallel} < 2$ and inserting the scaling form (4.24) one finds by substituting $\Delta t \mapsto \tau^{-\nu_{\parallel}}\,\Delta t$ that the average interval size scales as

$$\overline{\Delta t} \approx \int_a^{\infty} \mathrm{d}\Delta t\ \Delta t^{1-\epsilon_{\parallel}}\ \tilde{e}_{\parallel}(\Delta t^{1/\nu_{\parallel}}\tau) \ \sim\ \tau^{\epsilon_{\parallel}-2\nu_{\parallel}}\,, \tag{4.28}$$

where we substituted $\Delta t \to \tau^{-\nu_{\parallel}}\,\Delta t$ in the integrand. Comparing this expression with $\overline{\Delta t} = \varrho^{-1} \sim \tau^{-\beta}$, we arrive at the scaling relation

$$\epsilon_{\parallel} = 2\nu_{\parallel} - \beta\,. \tag{4.29}$$

Hence, as the main result, the distribution of temporal empty intervals is found to scale as

$$e_{\parallel}(\Delta t;\tau) \ \simeq\ \Delta t^{-(2-\beta/\nu_{\parallel})}\,\tilde{e}_{\parallel}(\Delta t^{1/\nu_{\parallel}}\tau)\,. \tag{4.30}$$

In particular, at criticality the distribution decays algebraically, i.e.,

$$e_{\parallel}(\Delta t,0) \simeq \Delta t^{-(2-\beta/\nu_{\parallel})}. \tag{4.31}$$

Consequently, at criticality the time-series of subsequent activations forms a fractal set with the (box) fractal dimension (see appendix H and exercise 46)

$$d_{f,\parallel} = 1 - \frac{\beta}{\nu_{\parallel}}. \tag{4.32}$$

Note that this scaling *ansatz* requires $\epsilon_{\parallel}/\nu_{\parallel}$ to be larger than 1, which in turn implies that $\beta < 1$. For all absorbing phase transitions known so far, this is indeed the case, at least to a very good approximation.

4.1.6.2 Scaling of the Spatial Empty-Interval Distribution Functions

Similarly, we may study the *spatial* distribution $e_\perp(r;\tau)$ of active sites along a line at a given time t in a (1+1)-dimensional system (see Fig. 4.3). Denoting by Δr the spatial size of empty intervals (see Fig. 4.3) and repeating the calculation, one arrives at the analogous result

$$e_\perp(\Delta r;\tau) \simeq (\Delta r)^{-(2-\beta/\nu_\perp)}\, \tilde{e}_\perp(\Delta r^{1/\nu_\perp}\tau) \tag{4.33}$$

with the scaling relation

$$\epsilon_\perp = 2\nu_\perp - \beta\,. \tag{4.34}$$

At criticality the spatial empty-interval distribution decays as

$$e_\perp(\Delta r;0) \sim (\Delta r)^{-(2-\beta/\nu_\perp)} \tag{4.35}$$

and thus the fractal dimension along a one-dimensional line in space is $d_{f,\perp} = 1-\beta/\nu_\perp$. More generally, in a $d+1$-dimensional system the set of active sites at a given time, i.e. a spatial cut of the cluster, has according to this simple scaling argument a fractal structure with the fractal dimension[6]

$$d_{f,\perp} = d - \frac{\beta}{\nu_\perp}\,. \tag{4.36}$$

4.1.7 The External Field h

In equilibrium statistical mechanics, order parameters are associated with certain conjugated fields. If such a field is applied externally, it causes a response of the corresponding order parameter. Off-criticality, this response grows to lowest order *linearly* with the external field. The corresponding proportionality constant, called **susceptibility**, depends on the temperature and diverges algebraically at the critical point (cf. Sect. 2.2).

In absorbing phase transitions the situation is similar. Here the conjugated field h of the order paramater ϱ_s corresponds to spontaneous creation of

[6] Precise calculations of the fractal dimensions in $(1+1)D$ directed percolation [269] found a small *non-zero* value for the Renyi codimension $\bar{\mathcal{D}} = \beta/\nu_\perp - \bar{d} = 0.018(5)$, where $\bar{d} = d - d_{f,\perp}$ is the fractal *co*dimension, whereas (4.36) implies $\bar{\mathcal{D}} = 0$. This finding was supported by an analysis of the scaling behaviour of the **factorial moments** [64, 65] $F_q := \langle n(n-1)\cdots(n-q+1)\rangle/\langle n\rangle^q \sim \delta^{-(q-1)\bar{\mathcal{D}}}$ as a function of the binning size $\delta = \Delta r$, which is in agreement with simple scaling and led to $\bar{\mathcal{D}} = 0.022(4)$ [277]. This suggests that intermittency effects might modify the scaling arguments used to derive (4.36). The same calculations also indicate that the calculation of $d_{f,\parallel}$, see (4.32), might be affected by so-called *multifractal* generalisations of the simple scaling examined in this book.

activity. In ordinary DP, this field can simply be implemented as spontaneous particle creation at rate h. In other realisations of DP, e.g. in the threshold transfer process (see Sect. 3.3.4), the definition of h is more subtle. Being a conjugated field, the application of h destroys the absorbing state and causes a linear response of ϱ in the off-critical regime. The corresponding susceptibility

$$\chi = \frac{\partial}{\partial h} \varrho_s(\tau, h) \tag{4.37}$$

turns out to diverge at zero field as

$$\chi \sim |\tau|^{-\gamma} \tag{4.38}$$

with the susceptibility exponent γ. At criticality $\tau = 0$, one finds that the stationary density varies with the external field as

$$\varrho_s \sim h^{\beta/\sigma} \tag{4.39}$$

which defines the so-called **gap exponent** or **field exponent** σ. This equation is analogous to the one for the magnetisation at the critical isotherm of equilibrium systems, cf. Sect. 2.2.2.

4.1.7.1 General Scaling Properties of the External Field

Dimensional analysis of the asymptotic power laws in (4.38) and (4.39) suggest that χ and h have to be rescaled as

$$\chi \mapsto \lambda^{-\gamma}\chi\,, \qquad h \mapsto \lambda^{\sigma} h\,, \tag{4.40}$$

where γ and σ are the exponents associated with χ and h. Scaling theory predicts that in presence of an external field the order parameter and the susceptibility can be expressed in terms of generalised homogeneous functions as

$$\varrho_s(\tau, h) \simeq \lambda^{-\beta}\, \tilde{R}(a_\tau \tau\lambda, a_h h\lambda^{\sigma})\,, \tag{4.41}$$

$$a_\chi\, \chi(\tau, h) \simeq \lambda^{\gamma}\, \tilde{X}(a_\tau \tau\lambda, a_h h\lambda^{\sigma})\,. \tag{4.42}$$

In what follows, we shall normalise the universal scaling function of the order parameter by

$$\tilde{R}(1,0) = \tilde{R}(0,1) = 1\,. \tag{4.43}$$

Thus, the metric factors a_τ and a_h can be determined for each model from the amplitudes of the power laws

$$\varrho_s(\tau, 0) \simeq (a_\tau \tau)^{\beta} \quad \text{and} \quad \varrho_s(0, h) \simeq (a_h h)^{\beta/\sigma}\,. \tag{4.44}$$

Since the susceptibility is defined as the derivative of the order parameter with respect to the conjugated field, the scaling functions and amplitudes are related by

$$\tilde{X}(x,y) = \partial_y \tilde{R}(x,y)\,, \qquad a_\chi = a_h^{-1}\,, \tag{4.45}$$

leading to the scaling relation

$$\gamma = \sigma - \beta\,. \tag{4.46}$$

This scaling law corresponds to the well-known **Widom law** of equilibrium phase transitions (see Sect. 2.2 and (2.16)). Moreover, an additional hyperscaling law, to be derived in Sect. 4.2.10, can be used to relate the exponents γ and σ to the standard exponents $\beta, \nu_\parallel$, and $\nu_\perp$ by

$$\begin{aligned} \gamma &= \nu_\parallel + d\nu_\perp - \beta - \beta'\,, \\ \sigma &= \nu_\parallel + d\nu_\perp - \beta'\,. \end{aligned} \tag{4.47}$$

We note that special values of scaling function are often useful for checking the consistency of numerical results. For example, comparing (4.42) for $\tau = 0$

$$a_\chi\, \chi(0,h) \simeq (a_h h)^{-\gamma/\sigma}\, \tilde{X}(0,1)\,, \tag{4.48}$$

with the definition of the susceptibility

$$\chi = \partial_h\, \varrho_s = \partial_h\, (a_h h)^{\beta/\sigma} \tag{4.49}$$

one finds that the special value $\tilde{X}(0,1)$ of the scaling function for the susceptibility is equal to the quotient β/σ. This relation can be used as a consistency check for numerical estimates of the susceptibility. It also illustrates that critical exponents may sometimes be expressed as special values of scaling functions.

4.1.7.2 The External Field Within Mean-Field Theory

We now demonstrate that the scaling relations derived above can be applied consistently within the mean-field approximation of directed percolation. Extending the mean-field equation (3.6) by a term for spontaneous particle creation at rate h, one obtains

$$\partial_t \varrho(t) = \tau \varrho(t) - g\varrho(t)^2 + (1 - \varrho(t))h\,. \tag{4.50}$$

The stationary solution reads

$$\varrho_s(\tau, h) = \frac{\tau - h}{2g} \pm \sqrt{\left(\frac{\tau - h}{2g}\right)^2 + \frac{h}{g}}\,. \tag{4.51}$$

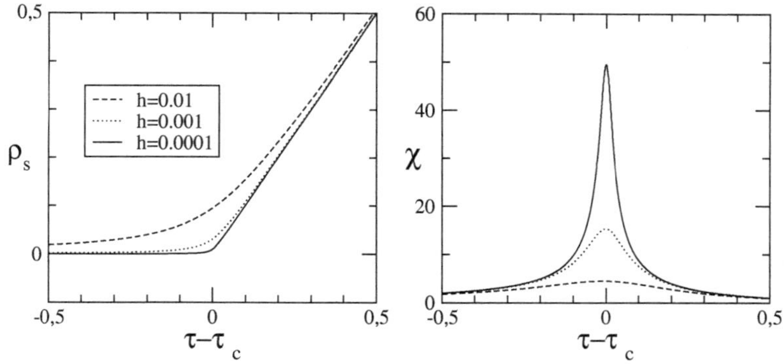

Fig. 4.4 Left: The order parameter $\varrho_{\rm s}$ of DP as a function of τ for various external fields h as predicted by mean-field theory. Right: The corresponding susceptibility $\chi = \partial\varrho_{\rm s}/\partial h$. For $h > 0$ the susceptibility displays a finite peak which diverges in the limit $h \to 0$ at the critical point.

The solution with the positive sign is the physical one (the other one yields negative densities) and is plotted in the left panel of Fig. 4.4. As expected, the state without particles ($\varrho_{\rm s} = 0$) is no longer absorbing. At criticality ($\tau = 0$) this solution reduces to

$$\varrho_{\rm s}(0,h) = \frac{1}{2g}\left(\sqrt{h^2+4gh}-h\right) = \sqrt{\frac{h}{g}} + \mathrm{O}\left(h\right). \tag{4.52}$$

Comparing this result to (4.39) one finds that the mean-field value of the gap exponent is given by

$$\sigma_{\rm MF} = 2\,. \tag{4.53}$$

Next, we examine the order parameter close to the critical point. To this end we perform in (4.51) the limits $\varrho_{\rm s} \to 0$, $\tau \to 0$, and $h \to 0$ with the constraints that $\varrho_{\rm s}/\sqrt{h}$ and $\varrho_{\rm s}/\tau$ are finite, giving

$$\varrho_{\rm s}(\tau,h) \simeq \frac{\tau}{2g} + \sqrt{\left(\frac{\tau}{2g}\right)^2 + \frac{h}{g}}\,. \tag{4.54}$$

The corresponding order parameter susceptibility

$$\chi(\tau,h) = \frac{\partial\varrho_{\rm s}(\tau,h)}{\partial h} \simeq \frac{1}{2g}\left[\left(\frac{\tau}{2g}\right)^2 + \frac{h}{g}\right]^{-1/2} \tag{4.55}$$

diverges at the transition point as

$$\chi(\tau,0) \simeq |\tau|^{-1}\,, \qquad \chi(0,h) \simeq \frac{1}{2}\,(gh)^{-1/2}. \tag{4.56}$$

Comparing this result with (4.38), we obtain the mean-field susceptibility exponent

$$\gamma_{\mathrm{MF}} = 1\,. \tag{4.57}$$

These results are consistent with the scaling relation (4.46).

4.1.8 Fluctuations of the Order-Parameter in the Steady-State

The steady-state of an *infinite* system in the active phase is characterised by a stationary density of active sites with a constant value $\varrho_{\mathrm{s}} > 0$ which does not fluctuate. In numerical simulations, however, where the system size is *finite*, the density is a fluctuating quantity. Depending on the system size, two different situations have to be distinguished:

(a) If the lateral system size L is of the same order or smaller than the correlation length $\xi_\perp$, the critical properties are affected by so-called **finite-size effects**. In particular, the system may enter the absorbing state after some time. Finite-size effects can be described in terms of scaling laws and will be discussed in the following subsection.

(b) If the lateral system size L is finite but much larger than $\xi_\perp$, the active state is practically stable and the order parameter fluctuates around a stationary value.

The fluctuations in the latter case can be explained as follows. For system sizes $L \gg \xi_\perp$ the system effectively consists of $(L/\xi_\perp)^d$ uncorrelated subsystems that contribute independently to fluctuations of ϱ_{s}. According to the central limit theorem, the sum of these fluctuations is expected to be Gaussian in the limit $L \to \infty$ with a statistical spread decreasing as $\sqrt{L^d}$. This allows one to define a variance per unit volume by

$$\Delta_\varrho = \Delta_\varrho(\tau) = \lim_{L\to\infty} L^d \left(\langle\varrho^2\rangle - \langle\varrho\rangle^2\right)\,, \tag{4.58}$$

where $\langle\ldots\rangle$ denotes the temporal average

$$\langle\varrho^k\rangle := \lim_{T\to\infty} \frac{1}{T}\sum_{t=1}^{T} \varrho(t)^k\,. \tag{4.59}$$

To approximate this limit in a numerical simulation, the averaging time has to be much larger than the correlation time $\xi_{||}$. Especially close to criticality, where $\xi_{||}$ is large, the estimation of Δ_ϱ is thus a numerically challenging task.

The decay of $\varrho(t)$ towards a stationary value ϱ_{s}, as well as its fluctuations in the steady-state of a (1+1)-dimensional bond DP process are demonstrated

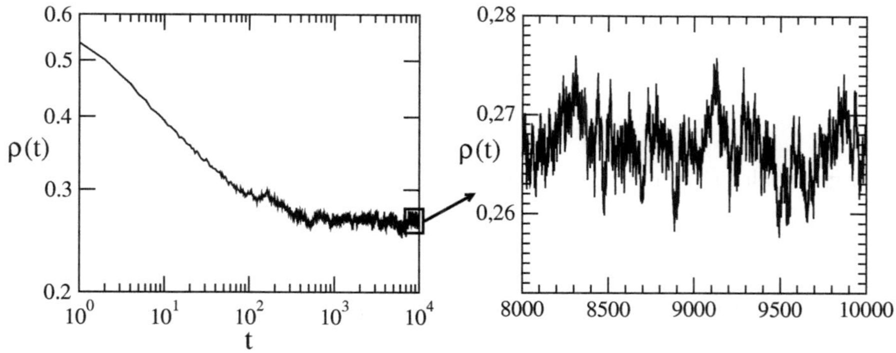

Fig. 4.5 Order-parameter fluctuations in the steady-state of a $(1+1)$-dimensional bond DP process with $L = 10^5$ sites slightly above criticality ($p - p_c = \tau = 0.005$). The right panel shows a zoom of the steady-state fluctuations for large t.

in Fig. 4.5. Performing simulations for different values of τ one finds that the order parameter variance grows algebraically as

$$\Delta_\varrho(\tau) \sim |\tau|^{-\gamma'} . \tag{4.60}$$

In order to determine the exponent γ', we apply the scale transformation(4.9)-(4.13) to (4.58):

$$\begin{aligned} L^d \left(\langle \varrho^2 \rangle - \langle \varrho \rangle^2\right) &\sim |\tau|^{-\gamma'} \\ \Rightarrow \quad (\lambda^{-\nu_\parallel} L)^d \left(\langle \lambda^{2\beta} \varrho^2 \rangle - \langle \lambda^{2\beta} \varrho \rangle^2\right) &\sim |\lambda \tau|^{-\gamma'} \end{aligned} \tag{4.61}$$

Obviously, scale-invariance requires that

$$\gamma' = d\nu_\perp - 2\beta . \tag{4.62}$$

Scale-invariance also allows one to express the variance of steady-state fluctuations in terms of a scaling form as

$$a_\Delta \, \Delta_\varrho(\tau, h) \simeq \lambda^{\gamma'} \, \tilde{D}(a_\tau \tau \lambda, a_h h \lambda^\sigma) . \tag{4.63}$$

Normalising $\tilde{D}(0,1) = 1$, the metric factor a_Δ can be determined from the amplitude of the power-law

$$\Delta_\varrho(0, h) \simeq (a_\Delta \, h)^{-\gamma'/\sigma} . \tag{4.64}$$

For an explicit illustration, we mention that steady-state fluctuations can be studied by adding an appropriate noise term in (4.50) representing rapidly-varying degrees of freedom. In that case, the mean-field fluctuations are given by [480]

$$\Delta_{\varrho} \propto \frac{\frac{\tau}{2g} + \sqrt{\left(\frac{\tau}{2g}\right)^2 + \frac{h}{g}}}{\sqrt{\left(\frac{\tau}{2g}\right)^2 + \frac{h}{g}}} . \tag{4.65}$$

For zero field, the fluctuations vanish in the absorbing phase ($\tau < 0$) whereas they remain constant in the active phase ($\tau > 0$). In other words, the fluctuations do not diverge at the critical point, they rather exhibit a finite jump. As in equilibrium phase transitions, a finite jump of a quantity corresponds to a vanishing exponent, i.e., $\gamma'_{\rm MF} = 0$. However, in the present case a vanishing exponent is obtained only within the mean-field approximation.

We note that in equilibrium statistical mechanics, the intensity of order parameter fluctuations is known to scale in the same way as the susceptibility, meaning that γ' equals γ. This relationship is at the core of the so-called **fluctuation-dissipation theorem** (see Sect. 2.4). In the non-equilibrium case, however, comparison of (4.47) and (4.62) shows that the exponents γ and γ' are in general distinct and therefore the fluctuation-dissipation theorem cannot be valid.

4.1.9 Finite-Size Scaling in the Steady-State

Most of the scaling laws derived so far are valid in infinite systems. However, numerical methods such as Monte Carlo simulations or series expansions are restricted to finite systems. It is therefore important to understand how far finite-size effects influence the properties of the system. Generally finite-size effects are expected when the spatial correlation length becomes comparable with the lateral size of the system. As known from equilibrium statistical mechanics (see Chap. 2), this is the reason why finite-size effects are particularly strong close to the critical point [220].

4.1.9.1 Finite-Size Scaling Forms

For absorbing phase transitions, finite-size effects emerge in two different ways:

(a) Approaching criticality the spatial correlation length $\xi_\perp$ increases but the finite lateral size L prevents it from becoming infinite. As a result, the phase transition is destroyed and its singularities become rounded and shifted.

(b) In a finite system, there is always a small but non-zero probability of reaching the absorbing state. Consequently the system *always* reaches the absorbing state within a certain typical time depending on L, even if $\tau > 0$, meaning that the steady-state in the active state is no longer stable. However, applying an external conjugated field h, which destroys the absorbing state by spontaneous creation of activity, the steady-state is non-trivial and depends on τ, h, as well as on the system size L [437].

As in equilibrium, finite-size effects can be controlled by finite-size scaling laws. To this end one has to include the system size as an additional parameter in the corresponding scaling functions. Being a length, the parameter L changes under rescaling as

$$L \mapsto \lambda^{-\nu_\perp} L \,. \tag{4.66}$$

Therefore, the appropriate scaling form for the stationary density reads

$$\varrho_{\rm s}(\tau, h, L) \simeq \lambda^{-\beta}\, \tilde{R}_{\rm pbc}(\lambda a_\tau \tau,\, \lambda^\sigma a_h h,\, \lambda^{-\nu_\perp} a_L L)\,. \tag{4.67}$$

Note that the universal finite-size scaling function $\tilde{R}$ depends on the particular choice of the boundary conditions, as well as on the shape of the finite system (see e.g. [311, 312, 375, 374, 22]) while all information concerning the structure of the lattice is contained in the metric factor a_L. In this book, we are primarily concerned with hypercubic lattices of size L^d with aspect ratio 1 and periodic boundary conditions (pbc).

In the so-called **thermodynamic limit** $L \to \infty$, all finite-size effects disappear and the universal scaling function (4.41) is recovered, i.e.,

$$\tilde{R}_{\rm pbc}(x, y, \infty) = \tilde{R}(x, y)\,. \tag{4.68}$$

Finite-size effects can be included in any scaling function by adding an L-dependent argument in the same way as in (4.67). For example, it is straightforward to set up finite-size scaling forms for the fluctuations Δ_ϱ and the susceptibility χ [433].

4.1.9.2 Finite-Size Cumulants

In equilibrium statistical mechanics, it is particularly useful to consider ratios or combinations of order parameter moments. The most important one is the fourth order cumulant Q, also called the **Binder cumulant**, which is defined as [67]

$$Q = 1 - \frac{\langle \varrho_{\rm s}^4 \rangle}{3\langle \varrho_{\rm s}^2 \rangle^2}\,. \tag{4.69}$$

If the order parameter is characterised by a centred Gaussian distribution (as e.g. in the Ising model for $T > T_{\rm c}$), this cumulant tends to zero at the critical point. In a finite system, the Binder cumulant obeys the scaling form

$$Q(\tau,h,L) \simeq \tilde{Q}_{\text{pbc}}(a_\tau \tau\lambda,\, a_h h\lambda^\sigma,\, a_L L\lambda^{-\nu_\perp})\,. \tag{4.70}$$

Notice that the Binder cumulant is dimensionless, so that no metric factor a_Q has to be introduced. Setting $a_L L\,\lambda^{-\nu_\perp} = 1$, one has for zero field at criticality

$$\begin{aligned} Q(0,0,L) &= Q(\tau,0,L)\Big|_{\tau=0} \\ &\simeq \tilde{Q}_{\text{pbc}}(a_\tau \tau\,(a_L L)^{-\nu_\perp},0,1)\Big|_{\tau=0} \\ &= \tilde{Q}_{\text{pbc}}(0,0,1) \end{aligned} \tag{4.71}$$

which is L-independent and universal for given boundary conditions and a fixed shape of the system. At equilibrium, the universal value $\tilde{Q}_{\text{pbc}}(0,0,1)$ would correspond to an intersection point if one plotted Q as a function of the control parameter τ for various system sizes L at zero field. Thus it is possible to determine the transition point from the common intersection point. This **cumulant intersection method** is very useful and has been applied in numerous works.

Cumulants have also been investigated for several absorbing phase transitions [175, 137, 176, 279]. However, in this case the order parameter in the active phase has a non-zero mean and therefore the cumulant tends to $Q = 2/3$ in the thermodynamic limit. Moreover, in *finite* systems with absorbing states the order parameter and its moments $\langle \varrho^k \rangle$ vanish as soon as $\mathrm{O}(\xi_\perp) = L$ even in the active phase. Thus the powerful cumulant intersection method cannot be applied to absorbing phase transitions.

In order to circumvent this problem, one could keep the system at $\tau = 0$ artificially active by means of an external field $h > 0$, studying the limit $h \to 0$. In this case it is convenient to normalise the universal scaling function $\tilde{Q}_{\text{pbc}}$ by the condition

$$\tilde{Q}_{\text{pbc}}(0,1,1) = 0 \tag{4.72}$$

which fixes the metric factor a_L. However, carrying out the limit it turns out that Q diverges at the critical point, reflecting the non-fluctuating absorbing state.

A ratio that remains finite at criticality is obtained via [440]

$$U = \frac{\langle \varrho_s^2 \rangle \langle \varrho_s^3 \rangle - \langle \varrho_s \rangle \langle \varrho_s^2 \rangle^2}{\langle \varrho_s \rangle \langle \varrho_s^4 \rangle - \langle \varrho_s \rangle \langle \varrho_s^2 \rangle^2}\,. \tag{4.73}$$

This ratio is as useful for absorbing phase transitions as the Binder cumulant Q is for equilibrium, i.e., its value at criticality characterises the universality class.

4.2 Dynamical Scaling Behaviour

So far, we have studied the steady-state properties of a system with an absorbing phase transition in the active phase $\tau > 0$. Let us now address the question how a system prepared in a certain initial configuration evolves in time. It turns out that the time-dependent order parameters exhibit **dynamical scaling**. In the corresponding dynamical scaling forms, the time parameter t appears as an additional variable. Obviously, t has to scale in the same way as the temporal correlation length $\xi_{\|}$, i.e.,

$$t \mapsto \lambda^{-\nu_{\|}} t \,. \tag{4.74}$$

As we shall see in the following, dynamical scaling implies certain scaling forms which do depend on the initial configuration.

4.2.1 Homogeneously Active Initial State

Let us first consider the decay of the particle-density $\varrho(t;\tau)$ in an infinite system starting with a fully active initial configuration. For zero external field $h = 0$ scale-invariance implies the scaling form

$$\varrho(t;\tau) \simeq \lambda^{-\beta} \tilde{R}_{\text{full}}(\lambda^{-\nu_{\|}} a_t t; \lambda a_\tau \tau). \tag{4.75}$$

Here the subscript 'full' is used to indicate that the process starts with a fully active initial configuration. Setting $\lambda^{-\nu_{\|}} a_t t = 1$, the scaling form turns into

$$\varrho(t;\tau) \simeq (a_t t)^{-\beta/\nu_{\|}} \, \tilde{R}_{\text{full}}(1; a_\tau \tau (a_t t)^{1/\nu_{\|}}) \,. \tag{4.76}$$

In particular, at criticality ($\tau = 0$) the density decays according to a power law

$$\varrho(t) \sim t^{-\alpha} \,, \tag{4.77}$$

where $\alpha = \beta/\nu_{\|}$. Normalising the scaling function by $R_{\text{full}}(0,1) = 1$ the non-universal metric factor a_t can be determined from the amplitude of this power law.

The scaling form (4.76) can be verified by plotting $\varrho(t;\tau)\, t^{\alpha}$ versus $t|\tau|^{\nu_{\|}}$ for different values of τ near criticality. As shown in the right panel of Fig. 4.6, all data sets collapse onto two different curves depending on the sign of τ. As will be discussed in Sect. 4.3.6, such a data collapse can be used to determine critical exponents numerically.

In numerical simulations with a homogeneously active initial state, one usually starts with a fully occupied lattice $\varrho_0 = 1$. However, the scaling forms given above can be applied to any type of homogeneous initial configuration

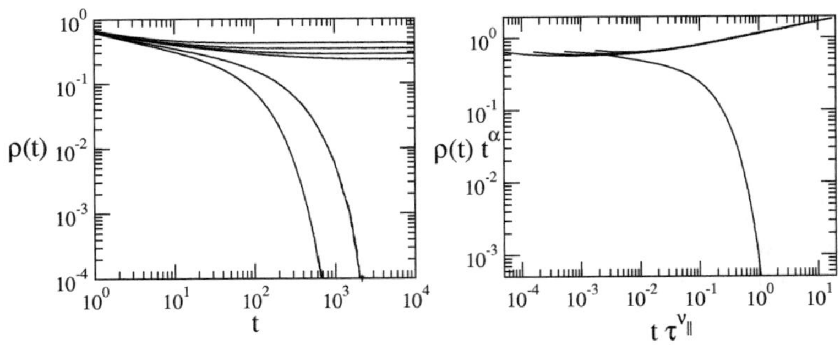

Fig. 4.6 The order parameter decay of (1+1)D bond DP close to the the critical point. The left graph shows simulation data starting with a fully occupied lattice for $\tau = \pm 0.0256, 0.0128, \dots 0.0016$. The corresponding data collapse is shown in the right panel, where the data is rescaled according to (4.76).

without in-built long-range correlations. For example, in simulations starting with a finite-density random initial state, where each site is occupied with probability $p_0 < 1$, one observes an initial increase of the order parameter followed by a crossover to the predicted scaling behaviour. This so-called **critical initial slip** will be discussed in more detail at the end of this subsection.

4.2.2 Pair-Connectedness Function, I

The dynamical scaling behaviour of systems starting with non-trivial initial configurations can be derived from the so-called **pair-connectedness function** $\Upsilon(t_1, t_2; \boldsymbol{r}_1, \boldsymbol{r}_2; \tau)$ that probes the existence of a causal connection between two points in space-time. In DP the pair-connectedness function is defined as the probability of finding a conducting directed path of open bonds from site $(t_1, \boldsymbol{r}_1)$ to site $(t_2, \boldsymbol{r}_2)$. More generally, the pair-connectedness function is defined as the probability that a cluster generated at site $\boldsymbol{r}_1$ at time t_1 in an otherwise empty system activates site $\boldsymbol{r}_2$ at time t_2. Because of translational invariance in space and time the pair-connectedness function depends only on the differences $r := |\boldsymbol{r}_2 - \boldsymbol{r}_1|$ and $t := t_2 - t_1 \geq 0$ so that we may write $\Upsilon(t, r; \tau)$.

Note that the pair-connectedness function differs from ordinary two-point correlation functions and empty-interval functions (see (4.17) and (4.23)) in so far as it probes the actual existence of a causal path between the two points. For example, in the case of equal times ($t = t_2 - t_1 = 0$) the pair connectedness function in a directed process with short-range interactions is always zero while the corresponding two-point or empty-interval function measured in a DP process with a positive density of active sites may be non-zero.

In the steady-state limit $t \to \infty$, the asymptotic behaviour of the pair-connectedness function can be determined as follows. On the one hand, the probability that the seed placed at $(t_1, \boldsymbol{r}_1)$ generates an infinite cluster is just the ultimate survival probability. On the other hand, in the limit $t_2 \to \infty$ the probability that a randomly chosen site belongs to the infinite cluster is just the steady-state density ϱ_s, hence

$$\lim_{t\to\infty} \Upsilon(t,r;\tau) = P_{\rm perc}(\tau)\varrho_s(\tau) \sim \tau^{\beta+\beta'} . \tag{4.78}$$

This suggest that the pair-connectedness function has to be rescaled as

$$\Upsilon \mapsto \lambda^{\beta+\beta'}\Upsilon \tag{4.79}$$

leading to the scaling form

$$\Upsilon(t,r;\tau) \simeq \lambda^{-\beta-\beta'}\, \tilde{\Upsilon}(\, a_t\lambda^{-\nu_\|}t, a_r\lambda^{-\nu_\perp}r;\, \lambda\tau) \,. \tag{4.80}$$

Note that in this expression the two exponents β and β' play a symmetric role.

4.2.3 Spreading Profile at Criticality

The pair-connectedness function provides information on how fast a cluster generated from a single seed in an otherwise empty system spreads in space and time. Monitoring the cluster at time t after generation at the origin one obtains a certain density profile $\varrho(t,r;\tau) = \Upsilon(t,r;\tau)$.

Figure 4.7 shows the numerically determined spreading profile of $\varrho(t,r)$ in a $(1+1)$-dimensional contact process at criticality $\tau = 0$ for various times. According to the scaling form (4.80), these profiles are expected to scale as

$$\varrho(t,r;0) \simeq (a_t t)^{-\alpha-\delta}\, \tilde{\Upsilon}\Big(1, a_r r (a_t t)^{-1/z}; 0\Big) \tag{4.81}$$

with $\alpha = \beta/\nu_\|$ and $\delta = \beta'/\nu_\|$. As demonstrated in the central panel of Fig. 4.7, one obtains an excellent data collapse by plotting $\varrho(t,r)t^{\alpha+\delta}$ as a function of $r/t^{1/z}$. The right panel shows the same data in a semi-logarithmic representation, demonstrating that the relative deviations are most pronounced in the tails of the distribution. Moreover, one finds that the asymptotic scaling function $\tilde{\Upsilon}(1,r;0)$ is not a Gaussian, reflecting that critical DP differs significantly from a simple random walk. To illustrate the non-Gaussian behaviour, the dashed line indicates a normal distribution fitted to the data in the centre, which turns out to deviate significantly in the tails. Analysing the asymptotic

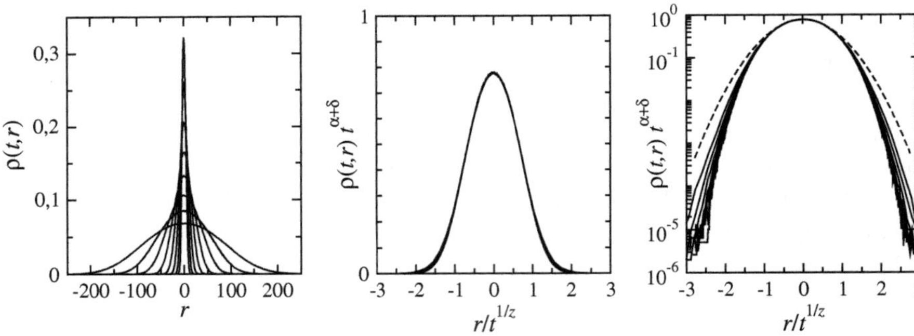

Fig. 4.7 Left: Spreading profile of a cluster generated from a single seed in a (1+1)-dimensional contact process at criticality for $t = 16, 32, 64, \ldots, 2048$. Center: Data collapse according to (4.81). Right: Same data collapse in semi-logarithmic plot. The dashed line indicates a Gaussian distribution fitted at the origin.

profile for large r in more detail it seems to decay as $\tilde{\Upsilon}(1, r; 0) \sim \exp(-a\, r^{\phi})$ with an exponent $\phi \approx 2.4$. The exact form of $\tilde{\Upsilon}$, however, is still unknown.

4.2.4 Clusters Generated from a Single Seed

The generation of clusters from a localised active seed in an otherwise inactive system is an important numerical technique for the study of absorbing phase transitions and the estimation of critical exponents. This technique was first introduced by Grassberger and de la Torre [246] and exploits the scaling properties of the pair-connectedness function.

In the absorbing phase $\tau < 0$, all clusters are finite while in the active phase $\tau > 0$, there is a finite probability of generating an infinite cluster. To characterise the growth of the clusters in both phases, one usually measures

- the survival probability $P_{\rm sur}(t)$ averaged over many clusters,
- the number of active sites $N_{\rm a}(t)$ at time t averaged over all clusters, and
- the mean square spreading (e.g. the squared radius of gyration) from the origin $R_{\rm s}^2(t)$ averaged over surviving clusters.

At criticality these quantities display asymptotic power laws of the form

$$P_{\rm sur}(t) \sim t^{-\delta}\,, \qquad N_{\rm a}(t) \sim t^{\Theta}\,, \qquad R_{\rm s}^2(t) \sim t^{\tilde{z}} \tag{4.82}$$

with the **survival probability exponent** δ, the so-called **slip exponent** Θ, and the **spreading exponent** $\tilde{z}$ which is different from the usual dynamical exponent z (see below). Sufficiently close to criticality, these quantities obey the scaling forms

$$a_P P_{\rm sur}(t;\tau) \simeq \lambda^{-\delta\nu_\parallel}\,\tilde{P}_{\rm seed}(a_t\lambda^{-\nu_\parallel}t;\,a_\tau\lambda\tau) \tag{4.83}$$
$$a_N N_{\rm a}(t;\tau) \simeq \lambda^{\Theta\nu_\parallel}\,\tilde{N}_{\rm seed}(a_t\lambda^{-\nu_\parallel}t;\,a_\tau\lambda\tau) \tag{4.84}$$
$$a_R R_{\rm s}(t;\tau) \simeq \lambda^{\nu_\perp}\,\tilde{\mathcal{R}}_{\rm seed}(a_t\lambda^{-\nu_\parallel}t;\,a_\tau\lambda\tau)\,. \tag{4.85}$$

By definition the percolation probability equals the ultimate survival probability, i.e., $P_{\rm perc}(\tau) = \lim_{t\to\infty} P_{\rm sur}(t,\tau) \sim \tau^{\beta'}$ in the active phase, hence

$$\delta = \beta'/\nu_\parallel\,. \tag{4.86}$$

In order to determine Θ, we express the average number of particles $N_{\rm a}(t,\tau)$ in terms of the pair-connectedness function by

$$\begin{aligned} a_N N_{\rm a}(t;\tau) &= \int \mathrm{d}\boldsymbol{r}\,\Upsilon(t,|\boldsymbol{r}|;\tau) \\ &= \int \mathrm{d}\boldsymbol{r}\,\lambda^{-\beta-\beta'}\,\tilde{\Upsilon}(a_t\lambda^{-\nu_\parallel}t,\,a_r\lambda^{-\nu_\perp}r;\,a_\tau\lambda\tau) \\ &\simeq \lambda^{d\nu_\perp-\beta-\beta'}\int \mathrm{d}\boldsymbol{r}\,\tilde{\Upsilon}(a_t\lambda^{-\nu_\parallel}t,\,a_r r;\,a_\tau\lambda\tau)\,, \end{aligned} \tag{4.87}$$

where we used (4.80) and substituted $r \mapsto \lambda^{\nu_\perp}r$. Carrying out the integral one obtains a function depending on two arguments

$$a_N N_{\rm a}(t;\tau) \;=\; \lambda^{d\nu_\perp-\beta-\beta'}\tilde{N}_{\rm seed}(a_t\lambda^{-\nu_\parallel}t;\,a_\tau\lambda\tau)\,. \tag{4.88}$$

Comparison with (4.84) yields the so-called **generalised hyperscaling relation** [470]

$$\frac{d}{z} = \Theta + \frac{\beta}{\nu_\parallel} + \frac{\beta'}{\nu_\parallel} \tag{4.89}$$

which holds for almost all universality classes of absorbing phase transitions below their upper critical dimension. With $\alpha = \beta/\nu_\parallel$ and $\delta = \beta'/\nu_\parallel$ the hyperscaling relation is often written as

$$\Theta = \frac{d}{z} - \alpha - \delta. \tag{4.90}$$

In DP, where the rapidity-reversal symmetry (see Sect. 4.1.2) implies $\alpha = \delta$, the hyperscaling relation reduces to

$$\Theta = \frac{d}{z} - 2\delta. \tag{4.91}$$

We shall come back to the rapidity-reversal symmetry in a field-theoretic context in Sect. 4.3.4.

Finally, we can compute the exponent $\tilde{z}$ by evaluating the integral (see also exercise 45)

$$\begin{aligned}R^2(t;\tau) = \langle |\boldsymbol{r}|^2 \rangle &= \frac{1}{N(t)} \int \mathrm{d}\boldsymbol{r}\, r^2\, \Upsilon(t, |\boldsymbol{r}|; \tau) \\ &\simeq \frac{1}{N(t)} \int \mathrm{d}\boldsymbol{r}\, r^2\, \lambda^{-\beta-\beta'}\, \tilde{\Upsilon}(a_t\lambda^{-\nu_\parallel} t,\, a_r\lambda^{-\nu_\perp} r;\, a_\tau\lambda\tau) \\ &= \frac{1}{N(t)} \lambda^{(d+2)\nu_\perp - \beta - \beta'} \int \mathrm{d}\boldsymbol{r}\, r^2\, \tilde{\Upsilon}(a_t\lambda^{-\nu_\parallel} t,\, a_r r;\, a_\tau\lambda\tau) \\ &= \lambda^{2\nu_\perp}\, \tilde{R}^2(a_t\lambda^{-\nu_\parallel} t;\, a_\tau\lambda\tau)\,, \end{aligned} \tag{4.92}$$

leading to the relation[7]

$$\tilde{z} = \frac{2}{z} = \frac{2\nu_\perp}{\nu_\parallel}. \tag{4.93}$$

This result can also be obtained from dimensional analysis of $R_s^2(t) \sim t^{\tilde{z}}$.

4.2.5 Properties of Clusters in the Absorbing Phase

The pair-connectedness function can also be used to derive the scaling behaviour of various other quantities which characterise a cluster generated from a single seed in the absorbing phase. For example, the mean cluster mass M is given by the pair-connectedness function integrated over space and time

$$M(\tau) = \int \mathrm{d}\boldsymbol{r} \int_0^\infty \mathrm{d}t\, \Upsilon(t, \boldsymbol{r}; \tau)\ . \tag{4.94}$$

Inserting the scaling relation (4.80) and substituting the scaling variables we obtain a scaling law for the average cluster mass measured in an infinite system below criticality:

$$\begin{aligned} M &\simeq \int \mathrm{d}\boldsymbol{r} \int_0^\infty \mathrm{d}t\, \lambda^{-\beta-\beta'}\, \tilde{\Upsilon}(a_t\lambda^{-\nu_\parallel} t,\, a_r\lambda^{-\nu_\perp} r;\, a_\tau\lambda\tau) \\ &= \int \mathrm{d}\boldsymbol{r} \int_0^\infty \mathrm{d}t\, t^{-\alpha-\delta}\, \tilde{\Upsilon}(a_t,\, a_r t^{-\nu_\perp/\nu_\parallel} r;\, a_\tau t^{1/\nu_\parallel}\tau) \\ &\sim |\tau|^{-\nu_\parallel(1+\Theta)}\,. \end{aligned} \tag{4.95}$$

Similarly, the mean survival time T, the mean spatial volume V, and the mean cluster size S of a cluster in the *in*active phase read (exercise 45)

$$T = \int \mathrm{d}t\, P_{\mathrm{sur}}(t) \sim |\tau|^{-\nu_\parallel(1-\delta)}\,, \tag{4.96}$$

[7] In the original work by Grassberger and de la Torre [246], and various other papers, z and $\tilde{z}$ have been denoted as Z and z, which in the past led to frequent misunderstandings with respect to the definition of z. We strongly encourage the reader to use the by now commonly accepted convention that z stands exclusively for the dynamical exponent $\nu_\parallel/\nu_\perp = z$.

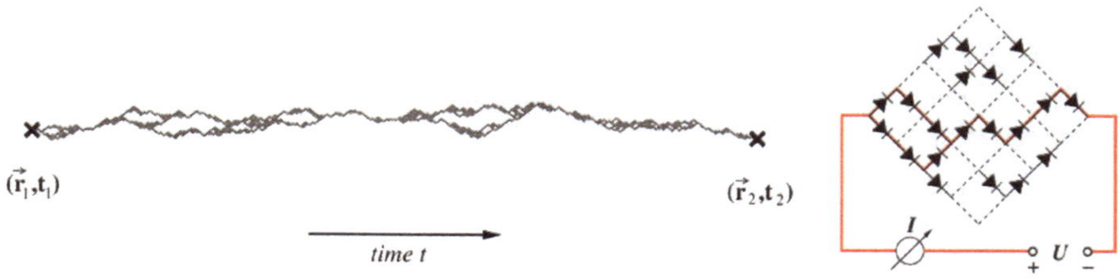

Fig. 4.8 Left: Backbone (bold dots) of a two-point correlation function in a critical DP process. Right: In a random diode network the backbone consists of all elements carrying a non-zero electric current.

$$V = \int \mathrm{d}t\, P_{\mathrm{sur}}(t) t^{d/z-1} \sim |\tau|^{-\nu_\parallel (d/z-\delta)}\,, \tag{4.97}$$

$$S = \int \mathrm{d}t\, P_{\mathrm{sur}}(t) t^{d/z} \sim |\tau|^{-\nu_\parallel (d/z+1-\delta)}\,. \tag{4.98}$$

For these quantities, we obtain the following scaling relations[8]

$$M \sim |\tau|^{-\gamma}\,, \qquad \gamma = \nu_\parallel(1+\Theta) = \nu_\parallel + d\nu_\perp - \beta - \beta'\,, \tag{4.99}$$

$$T \sim |\tau|^{-\tau}\,, \qquad \tau = \nu_\parallel(1-\delta) = \nu_\parallel - \beta'\,, \tag{4.100}$$

$$V \sim |\tau|^{-v}\,, \qquad v = \nu_\parallel(d/z-\delta) = d\nu_\perp - \beta'\,, \tag{4.101}$$

$$S \sim |\tau|^{-\sigma}\,, \qquad \sigma = \nu_\parallel(d/z+1-\delta) = \nu_\parallel + d\nu_\perp - \beta'\,. \tag{4.102}$$

4.2.6 Pair-Connectedness Function, II

The pair-connectedness function $\Upsilon(t_2-t_1, \boldsymbol{r}_2-\boldsymbol{r}_1;\tau)$ can be determined numerically by generating a cluster at the point $(t_1,\boldsymbol{r}_1)$ in an otherwise inactive system and measuring the response at the point $(t_2,\boldsymbol{r}_2)$, averaging over many realisations. However, as illustrated in Fig. 4.8, only a subset of the cluster actually contributes to a response at this point. This subset of sites and bonds connecting the two points $(t_1,\boldsymbol{r}_1)$ and $(t_2,\boldsymbol{r}_2)$ is called the **backbone** of the pair-connectedness function. Roughly speaking the backbone consists of all sites and bonds that are connected with the sites $(t_1,\boldsymbol{r}_1)$ and $(t_2,\boldsymbol{r}_2)$ by directed paths, cutting off all dangling ends of the cluster. For example, in a random resistor-diode network [339, 580, 581, 300] as shown in Fig. 4.8 the backbone would consist of all diodes which carry a non-zero current.

The probability that a given site $(t,\boldsymbol{r})$ with $t_1 < t < t_2$ belongs to the backbone is proportional to the product of the probabilities $\Upsilon(t-t_1,|\boldsymbol{r}-\boldsymbol{r}_1|;\tau)$ and $\Upsilon(t-t_2,|\boldsymbol{r}_2-\boldsymbol{r}|;\tau)$ of finding both a directed path from the origin to

[8] On should not confuse here the control parameter τ with the critical exponent τ of the mean survival time T.

this intermediate site *and* another directed path from this site to the terminal point, divided by the probability $\Upsilon(t_2 - t_1, |\boldsymbol{r}_2 - \boldsymbol{r}_1|, ; \tau)$ that the backbone exists. Hence the average number of backbone sites $N_b(t)$ at time t can be expressed as

$$N_b(t) \simeq \int \mathrm{d}\boldsymbol{r}\, \frac{\Upsilon(t - t_1, |\boldsymbol{r} - \boldsymbol{r}_1|; \tau)\, \Upsilon(t_2 - t_1, |\boldsymbol{r}_2 - \boldsymbol{r}|; \tau)}{\Upsilon(|\boldsymbol{r}_2 - \boldsymbol{r}_1|, t_2 - t_1; \tau)} . \tag{4.103}$$

At criticality $\tau = 0$ the scaling form (4.80) implies

$$N_b(t) \sim \left(\frac{(t - t_1)\,(t_2 - t)}{t_2 - t_1} \right)^{\Theta} \tag{4.104}$$

and therefore the total mass of a critical backbone scales as

$$N_b^{\mathrm{tot}} = \int_{t_1}^{t_2} \mathrm{d}t\, N_b(t) \sim (t_2 - t_1)^{1+\Theta} . \tag{4.105}$$

Interpreting the backbone as the subset of a random-diode network which actually contributes to the conductivity between the two points (see Fig. 4.8), one observes that for most of the time the electric current is distributed among *several* conducting branches. Sometimes, however, the full current passes a single site. These so-called **red sites** (or **red bonds**) are particularly important since they are the weakest links of the backbone and the first to break when the control parameter is varied.

The total number of red bonds can be calculated by varying the control parameter τ infinitesimally and monitoring the linear response of the pair-connectedness function. For example, decreasing the percolation threshold by $p \to (1 - \epsilon)p$ in a bond DP process a fraction ϵ of all bonds will be removed. Therefore, to lowest order in ϵ the probability that the backbone is cut into two pieces is $\epsilon N_b^{\mathrm{red}}$, where N_b^{red} denotes the number of red bonds. Consequently the number of red bonds is given by

$$N_b^{\mathrm{red}} = \frac{\partial_\tau \Upsilon(t_2 - t_1, |\boldsymbol{r}_2 - \boldsymbol{r}_1|; \tau)}{\Upsilon(t_2 - t_1, |\boldsymbol{r}_2 - \boldsymbol{r}_1|; \tau)} = \partial_\tau \ln \Upsilon(t_2 - t_1, |\boldsymbol{r}_2 - \boldsymbol{r}_1|; \tau) . \tag{4.106}$$

Inserting the scaling form for the pair-connectedness function (4.80) one is led to

$$N_b^{\mathrm{red}} \simeq \lambda\, \tilde{N}_b^{\mathrm{red}}(a_t \lambda^{-\nu_\parallel} t,\, a_r \lambda^{-\nu_\perp} r;\, \lambda\tau). \tag{4.107}$$

For critical backbones probing the same location in space ($\boldsymbol{r}_1 = \boldsymbol{r}_2, \tau = 0$) this scaling form implies that the number of red bonds increases with time as

$$N_b^{\mathrm{red}} = (t_2 - t_1)^{1/\nu_\parallel} . \tag{4.108}$$

Note that in bond DP the backbone is statistically invariant under time-reversal, reflecting the rapidity-reversal symmetry (4.3) discussed previously

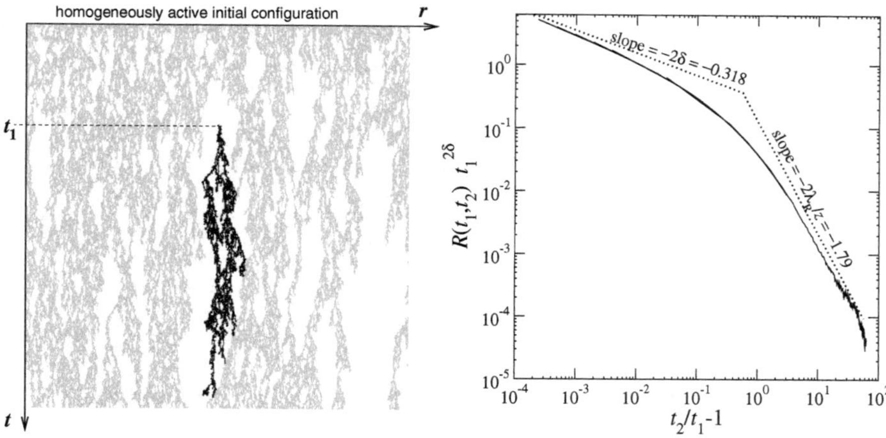

Fig. 4.9 Response to an external perturbation in the contact process at criticality. Left: The creation of a particle at time t_1 in a DP process starting with a fully occupied lattice increases the average density of active sites for $t > t_1$. Right: Data collapse of the autoresponse function $R(t_2, t_1)$ according to the scaling form (4.113) for various values of $t_1 = 2^6, \dots, 2^{16}$. The dashed lines indicate the expected asymptotic slopes of the scaling function.

in Sect. 4.1.2. In other realisations of DP this symmetry is not exact but emerges asymptotically for large distances in space and time.

4.2.7 Response Function

A similar quantity is the **response function** $R(t_2, t_1; \boldsymbol{r}_2, \boldsymbol{r}_1; \tau)$ which is defined as the average response measured at site $(t_2, \boldsymbol{r}_2)$ to a local perturbation by an external field at site $(t_1, \boldsymbol{r}_1)$ in a system starting with a certain initial configuration:

$$R(t_2, t_1; \boldsymbol{r}_2, \boldsymbol{r}_1; \tau) := \left. \frac{\partial \varrho(t_2, \boldsymbol{r}_2)}{\partial h(t_1, \boldsymbol{r}_1)} \right|_{h(t_1, \boldsymbol{r}_1)=0} . \tag{4.109}$$

Usually it is assumed that the system starts with a homogeneously active state (fully occupied lattice) and that the process is perturbed by creating activity at site $(t_1, \boldsymbol{r}_1)$. The response function is translationally invariant in space and thus depends only on $r = |\boldsymbol{r}_2 - \boldsymbol{r}_1|$. However, in contrast to the pair-connectedness function, the response function is *not* translationally invariant in time, since time-translation-symmetry is broken by the presence of a non-absorbing initial state.

In a simulation the response function can be determined by measuring the difference between $\varrho(t_2, \boldsymbol{r}_2)$ with and without perturbation at site $(t_1, \boldsymbol{r}_1)$

averaged over many realisations. Such simulations are numerically challenging and can be optimised by parallel processing of several perturbations, see appendix G.

Since the response function is the connected two-point correlation function of the conjugated field and the order parameter, it has to be rescaled as

$$R \mapsto \lambda^{\beta+\beta'} R\,, \tag{4.110}$$

leading to the scaling form

$$R(t_2,t_1;r;\tau) \;\simeq\; \lambda^{-\beta-\beta'}\,\tilde{R}(a_t\lambda^{-\nu_\parallel}t_2,\, a_t\lambda^{-\nu_\parallel}t_1,\, a_r\lambda^{-\nu_\perp}r;\, \lambda\tau)\,. \tag{4.111}$$

At criticality $\tau = 0$, choosing $\lambda = (a_t t_1)^{-\nu_\parallel}$, this scaling form reduces to

$$R(t_2,t_1;r) \;\simeq\; (a_t t_1)^{-\alpha-\delta}\,\tilde{R}\Big(1,\, t_2/t_1,\, a_r r(a_t t)^{-1/z};\, 0\Big)\,. \tag{4.112}$$

In particular, the **autoresponse function** at criticality can be expressed as

$$R(t_2,t_1) \;\sim\; t_1^{-\alpha-\delta}\,\tilde{R}(1,t_2/t_1,0;0) =: t_1^{-\alpha-\delta}\,f_R(t_2/t_1)\,. \tag{4.113}$$

The asymptotic behaviour of this scaling function at criticality can be determined as follows. Keeping $t_2 - t_1$ fixed and taking $t_1 \to \infty$ one measures the response to perturbation in an empty system which should be given by the pair-connectedness function

$$\lim_{\substack{t_2-t_1=\Delta t\\ t_1\to\infty}} R(t_2,t_1) \;=\; \Upsilon(t_2-t_1) \;\sim\; (t_2-t_1)^{-\alpha-\delta} \tag{4.114}$$

hence

$$f_R(1+\epsilon) \sim \epsilon^{-\alpha-\delta} \quad \text{for } 0<\epsilon\ll 1\,. \tag{4.115}$$

However, keeping t_1 fixed and taking $t_2 \to \infty$ one finds an algebraic decay

$$f_R(\xi) \sim \xi^{-\lambda_R/z} \tag{4.116}$$

with a new exponent λ_R, the so-called **autoresponse exponent**. Recently it was shown that in DP this exponent is related to the standard exponents [45]

$$\lambda_R = d + z + \beta/\nu_\perp\,. \tag{4.117}$$

A proof of this relation will be outlined in the second volume of this book.

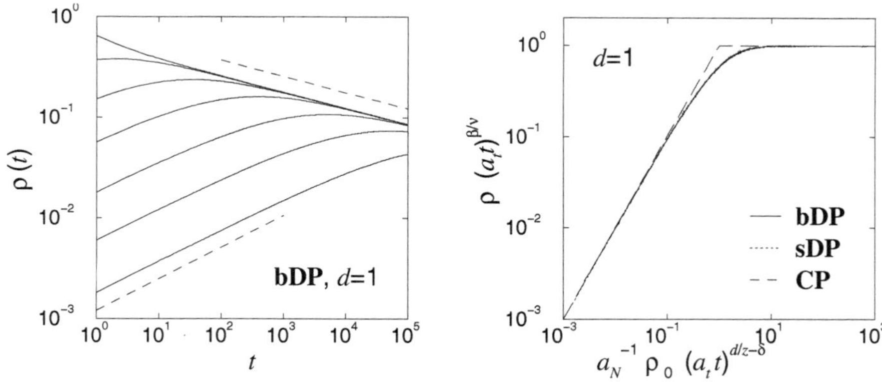

Fig. 4.10 Critical initial slip of the particle density measured in three different (1+1)-dimensional DP processes at criticality. The left figure shows the particle density $\varrho(t)$ in bond-DP for various initial densities $\varrho_0 = 0.001, 0.0033, 0.01, 0.033, 0.1, 0.33, 1$ as a function of time. The dashed lines indicate the slopes $+\Theta$ and $-\alpha$. The right figure shows a collapse of the universal scaling function $\tilde{R}_{\rm slip}(1, x, 0)$ measured in site and bond DP as well as in the contact process. The data sets are restricted to the scaling regime of the models and rescaled according to (4.119). The dashed lines indicate the asymptotic scaling behaviour (4.121).

4.2.8 Early-Time Behaviour and Critical Initial Slip

So far, we have considered two types of initial configurations, namely, the entirely active state (fully occupied lattice) and a localised active seed (a single particle) at the origin. Now we turn to homogeneous initial conditions with a density of active sites $\varrho_0 < 1$ in which the lattice sites are uncorrelated.

While for a fully occupied lattice ($\varrho_0 = 1$), the particle density in a critical system decreases as $\varrho(t) \sim t^{-\beta/\nu_\parallel}$, one observes that the average number of particles in simulations starting with a single seed *increases* as $N(t) \sim t^{\Theta}$. For low-density initial states, where active sites are separated by a certain typical distance, we therefore expect a crossover from an initial increase caused by the growth of individual clusters to a subsequent decay when the clusters begin to interact.

Figure 4.10 shows the density of active sites in a bond-DP process for various initial densities. As can be seen, the density first increases until it crosses over to the usual asymptotic decay. This crossover is known as the **critical initial slip** of non-equilibrium systems [338].

To describe the scaling properties of the initial slip we assume that the initial particle density ϱ_0 appears as an additional scaling field in the scaling forms. Since an uncorrelated homogeneous state with a finite density of particles may be created by applying the conjugated field at $t = 0$ homogeneously in the whole system, this additional scaling field should be rescaled in the same way as $\int \mathrm{d}^d\boldsymbol{r}\, h(t,\boldsymbol{r})$, hence

$$\varrho_0 \;\mapsto\; \lambda^{d\nu_\perp - \beta'}\,\varrho_0\,. \tag{4.118}$$

The corresponding scaling form is then given by

$$\varrho(t;\varrho_0,\tau) \;\simeq\; \lambda^{-\beta}\,\tilde{R}_{\text{slip}}(a_t t\,\lambda^{-\nu_\parallel};\, a_0\varrho_0\,\lambda^{d\nu_\perp-\nu_\parallel\delta},\, a_\tau\tau\lambda)\,. \tag{4.119}$$

Choosing $\lambda = (a_t t)^{1/\nu_\parallel}$, this scaling form reduces at criticality to

$$\varrho(t;\varrho_0) \;\simeq\; (a_t t)^{-\alpha}\,\tilde{R}_{\text{slip}}(1;\, a_0\varrho_0\,(a_t t)^{dz-\delta}\,,0)\,. \tag{4.120}$$

In order to recover $\varrho(t) \sim t^{-\alpha}$ for $\varrho_0 = 1$ and $N_{\text{a}}(t) \sim t^{\Theta}$ for $\varrho_0 \to 0$ (see equations (4.77) and (4.82)) this scaling function behaves asymptotically as

$$\tilde{R}_{\text{slip}}(1;x,0) \;\simeq\; \begin{cases} x & \text{for} \quad x \ll 1 \\ \text{const} & \text{for} \quad x \gg 1\,. \end{cases} \tag{4.121}$$

Furthermore, the scaling function $\tilde{R}_{\text{slip}}$ is related to the functions $\tilde{N}_{\text{seed}}$ and $\tilde{R}_{\text{full}}$ via, see (4.84) and (4.75)

$$\tilde{R}_{\text{full}} = \tilde{R}_{\text{slip}}\Big|_{\varrho_0=1} \tag{4.122}$$

$$\tilde{N}_{\text{seed}} = L^d\,\tilde{R}_{\text{slip}}\Big|_{\varrho_0=L^{-d}}\,. \tag{4.123}$$

On a more descriptive level, the initial slip can be interpreted as follows. In a low-density initial state, the active sites are separated by a certain average distance $r_0 = \varrho_0^{-1/d}$. Initially, each of them generates an independent cluster so that the density of active sites first increases as $\varrho \sim \varrho_0 t^{\Theta}$. However, only a fraction $\propto t^{-\delta}$ of these clusters survives, each of them spanning a volume of $\xi_\perp^d \sim t^{d/z}$. The surviving clusters start touching each other when this volume becomes comparable with $(\varrho_0 t^{-\delta})^{-1}$, i.e.

$$\mathrm{O}(a_0\varrho_0\,(a_t t_{\text{c}})^{d/z-\delta}) \;=\; 1 \tag{4.124}$$

or, ignoring the amplitudes,

$$t_{\text{c}} \sim \varrho_0^{1/(\delta-d/z)}\,. \tag{4.125}$$

Note that this condition is in agreement with the assumed scaling power in (4.118). Right at the crossover, the density of active sites scales as

$$\varrho(t_{\text{c}}) \;\sim\; t_{\text{c}}^{\delta-d/z+\Theta} \;\sim\; t_{\text{c}}^{-\alpha}\,, \tag{4.126}$$

where we used the hyperscaling relation (4.89). Therefore, the system crosses over to the expected asymptotic decay $\varrho \sim t^{-\alpha}$ with an amplitude *independent* of ϱ_0. In fact, as shown in the left panel of Fig. 4.10, all curves are not only asymptotically parallel but even collapse onto each other. This means that t_c *marks the time at which the memory of the initial conditions is lost.*

In the right panel of Fig. 4.10 we plot the order parameter for three different lattice models as a function of time. The data are rescaled according to the universal scaling form (4.120). The excellent data collapse confirms the validity of the scaling form and demonstrates that the observed crossover is indeed a *universal* phenomenon. It is worth mentioning that the models use different update schemes, namely, random-sequential and parallel updates. Thus the data-collapse affirms that non-universal features caused by different update schemes are completely absorbed in the metric factors a_t and do not affect the form of the universal scaling function.

4.2.9 Fractal Initial Conditions

In the last section, we have considered how homogeneous initial states with a finite density ϱ_0 of *uncorrelated* particles influence the scaling behaviour. This raises the question how a critical system starting with a *correlated* configuration evolves in time. As an example, we consider initial states which are characterised by a fractal dimension $d_\mathrm{f} \in [0, d]$. To simplify matters, we assume the fractal to be *simple* (opposed to multifractality[9]) and furthermore require that the different fractal dimensions which one may define (see appendix H) coincide.

As an example, consider an artificially generated fractal distribution on a one-dimensional line. As described in [299] this can be done by successively placing active sites separated by uncorrelated empty intervals of size ℓ. These interval sizes are algebraically distributed as $P(\ell) \sim \ell^{-1-d_\mathrm{f}}$, where the fractal dimension $d_\mathrm{f} > 0$ plays the role of a control parameter, see exercise 46.

Since the density of active sites would be zero for $d_\mathrm{f} < 1$, an appropriate cutoff in form of a maximal interval size has to be introduced. In numerical simulations, a natural upper cutoff is the system size. Below this cutoff the two-point correlation function in the initial configuration decays as

$$\langle s_i s_{i+r} \rangle \sim r^{d_\mathrm{f}-1} \,, \tag{4.127}$$

We can now use such a fractal initial state to study the temporal evolution of a *critical* DP process, interpolating continuously between the homogeneous and localised case. In fact, for $d_\mathrm{f} = 1$ the particles are homogeneously distributed, leading to the usual long-time behaviour $\varrho(t) \sim t^{-\alpha}$, while for $d_\mathrm{f} \to 0$, the fractal dimension tends to zero, corresponding to the localised case of a single seed, where we expect the density to increase as $\varrho(t) \sim t^{\Theta}$.

---

[9] A **multifractal** is characterised by a continuous distribution of dimensions, rather than a single one. In this book, multifractals will not be considered.

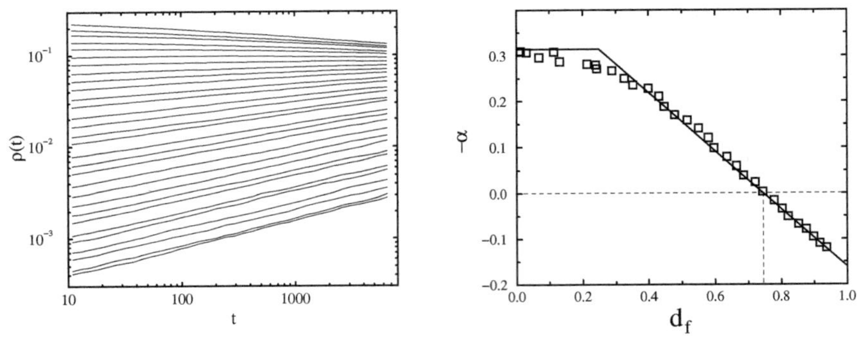

Fig. 4.11 Fractal initial conditions in a critical (1+1)-dimensional DP process. Left: Decay of $\varrho(t)$ for various values of $d_\mathrm{f} \in [0,1]$. Right: Numerical estimates for the exponents $-\alpha(d_\mathrm{f})$. The theoretical prediction (4.129) is shown by the solid line. The dashed line indicates the 'natural' correlations of DP (see text). Reprinted with permission from [299]. Copyright (1998) by the American Physical Society.

In between we find a power-law decay

$$\varrho(t) \sim t^{-\alpha(d_\mathrm{f})} \tag{4.128}$$

with a continuously varying exponent

$$\alpha(d_\mathrm{f}) = \begin{cases} -\Theta & \text{for } d_\mathrm{f} \le d_\mathrm{f}^* \\ -\frac{1}{z}(d - d_\mathrm{f} - \beta/\nu_\perp) & \text{for } d_\mathrm{f} > d_\mathrm{f}^*, \end{cases} . \tag{4.129}$$

As can be seen in Fig. 4.11, this power law extends over the whole dynamical range and thus differs significantly from the critical initial slip discussed above. For very low values of d_f the exponent $\alpha(d_\mathrm{f})$ saturates at the value $-\Theta$. In this limit, the initial configuration is so sparse that it behaves in the same way as a single seed. For $d_\mathrm{f} > d_\mathrm{f}^* = \beta/\nu_\perp$, however, correlations in the initial state become relevant and modify the entire temporal evolution.

As shown in [603] the relation (4.129) can be proved by a field-theoretic calculation. Interestingly, this calculation does not depend on the specific form of correlations in the initial state but only on its scaling dimension d_f. This has an interesting consequence: No matter how the particles are distributed – as long as the initial distribution exhibits simple scaling with a certain fractal dimension, the particle-density decays according to the scaling form (4.128) and (4.129).

It is interesting to note that a critical DP process itself generates two-point correlations of the form $\langle s_i s_{i+r} \rangle \sim r^{-\beta/\nu_\perp}$, corresponding to the 'natural' fractal dimension $d_{\mathrm{f},\perp} = d - \beta/\nu_\perp$ of DP (cf. (4.36)). In other words, DP itself creates a fractal distribution of active sites, and it would be interesting to know what happens if a state with the same fractal dimension is used as

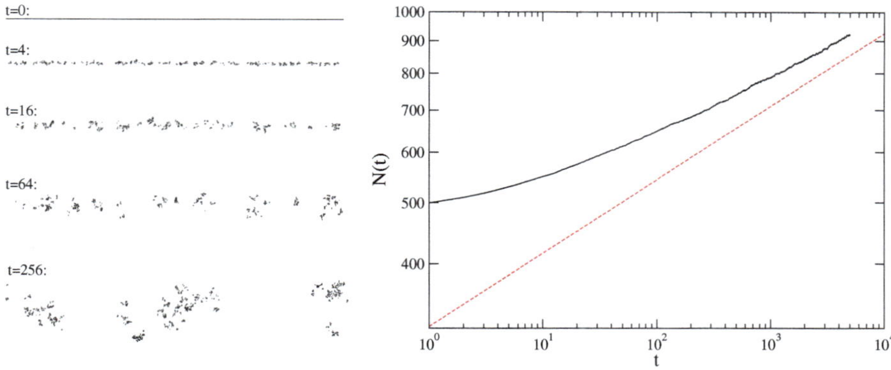

Fig. 4.12 Directed percolation in 2+1 dimensions starting with a one-dimensional line of active sites. Left: Snapshots of the process taken at certain times. Right: Average number of active sites $N(t)$ versus time. The dashed line indicates the expected slope $\alpha(d_f) = (1-\beta/\nu_\perp)/z \approx 0.116$.

initial configuration. In fact, starting a (1+1)-dimensional DP process with an artificial initial configuration of uncorrelated intervals with dimension $d_{f,\perp} = 1 - \beta/\nu_\perp$ one observes that the density of particles remains almost constant (see dashed lines in Fig. 4.11). Nevertheless the process is non-stationary as it generates higher-order correlations between the intervals.

The formula (4.129) works nicely in higher dimensions and is even valid for an initially active d_h-dimensional hyperplane embedded in $d > d_h$ spatial dimensions. For example, Fig. 4.12 shows the evolution of the number of particles in a two-dimensional contact process starting with a one-dimensional line of active sites. With $d = 2$ and $d_h = 1$ (4.129) yields

$$\alpha(d_f) = \frac{2 - 1 - \beta/\nu_\perp}{z} \approx 0.116 . \tag{4.130}$$

This prediction is in convincing agreement with the numerical simulation.

4.2.10 Influence of an External Field

Let us now investigate how an external field h influences the dynamics. As discussed in Sect. 4.1.7, such an external field h can be introduced by adding a process for spontaneous creation of activity in the bulk. This field is conjugated to the order parameter and destroys the absorbing state.

In (4.40) it was postulated that the external field scales as $h \mapsto \lambda^\sigma h$ with a certain exponent σ. This exponent can be determined as follows. A homogeneous initial condition with density ϱ_0 can be imposed by applying the external field

$$h = \varrho_0 \, \delta(t) \,. \tag{4.131}$$

Since the δ-function scales as $\delta(t) \mapsto \lambda^{\nu_\parallel} \delta(t)$ while ϱ_0 scales according to (4.118), dimensional counting allows one to express the field and susceptibility exponents in terms of the four standard exponents (β, β', $\nu_\parallel$, $\nu_\perp$) as

$$\sigma = d\nu_\perp + \nu_\parallel - \beta' \,, \tag{4.132}$$

$$\gamma = d\nu_\perp + \nu_\parallel - \beta' - \beta \,, \tag{4.133}$$

proving the relations (4.47) in Sect. 4.1.7. In the case of the DP universality class ($\beta = \beta'$), the field and susceptibility exponents are given by β, $\nu_\parallel$, and $\nu_\perp$. Furthermore, the average mass of a cluster generated from a single seed can be considered as the response of the system to a perturbation such as (4.131). Thus both, the mean cluster mass and the susceptibility diverge at the critical point in the same way, see (4.38) and (4.99).

Since the external field scales as $h \mapsto \lambda^\sigma h$ it can be included by adding a scale-invariant combination $a_h \lambda^\sigma h$ to the list of arguments in the scaling form (4.63), where a_h is the corresponding scale factor:

$$\varrho(t; \tau, h) \;\simeq\; \lambda^{-\beta} \, \tilde{R}(a_t \lambda^{-\nu_\parallel} t; a_\tau \tau \, \lambda, a_h h \, \lambda^\sigma) \,. \tag{4.134}$$

Therefore, in a critical system ($\tau = 0$) with a non-vanishing external field the density first decays and then saturates at a positive constant according to

$$\varrho(t; h) \sim t^{-\alpha} \, \tilde{R}(a_h t^{\sigma/\nu_\parallel} h) \,. \tag{4.135}$$

Hence, by plotting $\varrho(t;h) t^\alpha$ over against $t^{\sigma/\nu_\parallel} h$, it is possible to estimate the exponent σ by a data collapse.

4.2.11 Finite-Size Scaling

As in the static case, the dynamical scaling behaviour may be affected by **finite-size effects**. Typically such finite-size effects set in when the spatial correlation length becomes comparable with the system size, allowing the system to terminate in the absorbing state.

In order to take finite-size effects into account, one may extend the scaling forms by an additional argument. For example, in a finite system with $\mathcal{N} = L^d$ sites, the order parameter obeys the finite-size scaling form

$$\varrho(t; \tau; L) \simeq \lambda^{-\beta} \, \tilde{R}_{\text{full,pbc}}(\lambda^{-\nu_\parallel} a_t t; \; \lambda^{-\nu_\perp} a_L L, \; \lambda a_\tau \tau) \,, \tag{4.136}$$

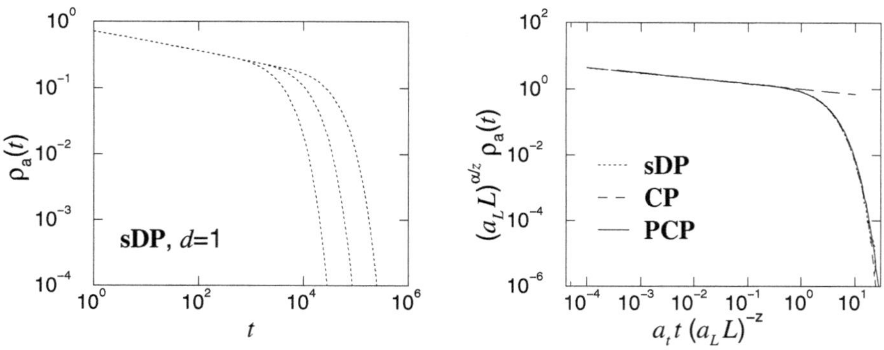

Fig. 4.13 The order parameter decay at the critical point for $d = 1$. The left figure shows the raw data of one-dimensional site DP for $L = 64, 128, 256$ (from left to right). The rescaled data, i.e., the universal scaling function $\tilde{R}_{\text{full,pbc}}(x; 1, 0)$ [see (4.138)] is presented in the right figure for site DP, the contact process and the pair-contact process, all collapsing onto a single curve. The long-dashed line corresponds to the power-law behaviour of the infinite system [(4.77)]. Reprinted with permission from [433]. Copyright (2004) World Scientific Publishing Company.

The indices 'full' and 'pbc' indicate that the scaling function $\tilde{R}$ depends on both the initial state, the shape of the system, and the particular choice of the boundary conditions (here periodic boundaries).

Initially, the spatial correlation length $\xi_\perp$ is smaller than the system size L and finite-size effects can be neglected, i.e., $\tilde{R}_{\text{full,pbc}} \approx \tilde{R}_{\text{full}}$. Finite-size effects set in when $\mathrm{O}(\xi_\perp) = L$ which happens at a typical time t_{FSS} with

$$t_{\text{FSS}} \;=\; a_t^{-1}\,(a_L L)^z \,. \tag{4.137}$$

For example, setting $\lambda^{-\nu_\perp} a_L L \;=\; 1$ in (4.136), one obtains at the critical point

$$\begin{aligned} \varrho(t;0;L) &\simeq (a_L L)^{-\alpha/z}\,\tilde{R}_{\text{full,pbc}}(a_t t (a_L L)^{-z};1,0) \\ &= (a_L L)^{-\alpha/z}\,\tilde{R}_{\text{full,pbc}}(t/t_{\text{FSS}};1,0)\,. \end{aligned} \tag{4.138}$$

For $t \ll t_{\text{FSS}}$ the scaling function obeys the power-law $\tilde{R}_{\text{pbc}}(x;1,0) \sim x^{-\alpha}$ whereas $\tilde{R}_{\text{pbc}}(x;1,0)$ decays exponentially in the finite-size scaling regime $t \gg t_{\text{FSS}}$. This is illustrated in Fig. 4.13, where the raw and rescaled data are shown for a finite DP process in one spatial dimension. Again the data of models with different update schemes collapse onto a unique universal curve $\tilde{R}_{\text{full,pbc}}(x;1,0)$. A similar finite-size scaling analysis can be performed for the survival probability P_{sur} or the number of active sites N_{a}. We refer the interested reader to [433].

In numerical simulations, it may sometimes be necessary to avoid finite-size effects. A simple but very reliable method is to simulate a series of finite-size systems with $L = 2, 4, 8, 16, \ldots$ growing in powers of 2 and to monitor how $t_{\rm FSS}$ scales with L. This allows one to extrapolate how large the system should be so that for a given simulation time, finite-size effects can be neglected. If finite-size effects cannot be ignored, it may be best to turn them into a tool, along the lines described in appendix F.

4.2.12 Universality of Finite-Size Amplitudes

We now describe an extension of the Privman-Fisher hypothesis, which we discussed in Chap. 2 for equilibrium systems, to absorbing phase transitions and especially to directed percolation, following [279]. Since the derivation requires us to keep trace of many metric factors, we begin by considering first the phenomenological scaling behaviour for the infinite system and shall only introduce the specific finite-size scaling properties when really needed.

Assuming spatial translation-invariance, the order parameter ϱ, the survival probability P and the pair-connectedness function Υ are expected to scale as (in a notation close to that of Chap. 2, see also exercise 47)

$$\begin{aligned} \varrho(t,\boldsymbol{r};\tau,h) &= D_{1\varrho}\,\xi_\perp^{-\beta/\nu_\perp}\,\mathcal{E}^\pm\left(D_0\frac{t}{\xi_\perp^z},\frac{\boldsymbol{r}}{\xi_\perp};D_2h|\tau|^{-\beta-\gamma}\right)\\ P(t,\boldsymbol{r};\tau,h) &= D_{1P}\,\xi_\perp^{-\beta'/\nu_\perp}\,\mathcal{F}^\pm\left(D_0\frac{t}{\xi_\perp^z},\frac{\boldsymbol{r}}{\xi_\perp};D_2h|\tau|^{-\beta-\gamma}\right)\\ \Upsilon(t,\boldsymbol{r};\tau,h) &= D_{1\Upsilon}\,\xi_\perp^{\Theta z-d}\,\mathcal{U}^\pm\left(D_0\frac{t}{\xi_\perp^z},\frac{\boldsymbol{r}}{\xi_\perp};D_2h|\tau|^{-\beta-\gamma}\right) \end{aligned} \tag{4.139}$$

where the D's are non-universal metric factors, $\mathcal{E},\mathcal{F},\mathcal{U}$ are universal scaling functions where the index distinguishes between the cases $\tau > 0$ and $\tau < 0$, $\xi_\perp = \xi_0|\tau|^{-\nu_\perp}$ is the spatial correlation length, $\xi_\| = \xi_\perp^z/D_0$ is the temporal correlation length and z is the dynamical exponent.

In the steady-state, and for $h = 0$, one expects $\varrho_{\rm s} \sim \tau^\beta$ and $P_{\rm perc} \sim \tau^{\beta'}$. Then the scaling forms of $\varrho_{\rm s}$, of $P_{\rm perc}$ and of the autoconnectedness are

$$\begin{aligned} \varrho_{\rm s}(\tau,h) &= D_{1\varrho}\,\xi_0^{-\beta/\nu_\perp}\,\widetilde{\mathcal{E}}^\pm\left(D_2h|\tau|^{-\beta-\gamma}\right)|\tau|^\beta\\ P_{\rm perc}(\tau,h) &= D_{1P}\,\xi_0^{-\beta'/\nu_\perp}\,\widetilde{\mathcal{F}}^\pm\left(D_2h|\tau|^{-\beta-\gamma}\right)|\tau|^{\beta'}\\ \Upsilon_{\rm s}(0;\tau,h) =: \Upsilon_{\rm s}(\tau,h) &= D_{1P}\,\xi_0^{\Theta z-d}\,\widetilde{\mathcal{U}}^\pm\left(D_2h|\tau|^{-\beta-\gamma}\right)|\tau|^{\beta+\beta'}, \end{aligned} \tag{4.140}$$

where $\widetilde{\mathcal{E}}^\pm = \lim_{r_\|\to\infty}\mathcal{E}^\pm$ and similarly for $\mathcal{F}$ and $\mathcal{U}$. In the active phase ($\tau > 0$), the surviving clusters will create an average density $\sim \tau^\beta$ in the interior of the spreading cone. Therefore, the auto-connectedness should in the steady-state saturate at the value $\Upsilon_{\rm s}(\tau,h) = \varrho_{\rm s}(\tau,h)P_{\rm perc}(\tau,h)$. Comparison of the scaling forms then yields, setting $h = 0$,

$$D_{1\Upsilon} = D_{1\varrho} D_{1P} \frac{\widetilde{\mathcal{E}}^{\pm}(0)\widetilde{\mathcal{F}}^{\pm}(0)}{\widetilde{\mathcal{U}}^{\pm}(0)} \tag{4.141}$$

The mean cluster mass (4.94) scales as

$$M(\tau,h) = \frac{D_{1\Upsilon}}{D_0} \xi_{\perp}^{\gamma/\nu_{\perp}} \overline{\mathcal{U}}^{\pm}\left(D_2 h|\tau|^{-\beta-\gamma}\right) \tag{4.142}$$

where (4.139) was used and $\overline{\mathcal{U}}^{\pm}$ is a new universal function related to $\mathcal{U}^{\pm}$.

So far the discussion has been completely general. For systems in the DP class, we consider (i) the effect of a weak external field h (related to the rate of a particle creation process $\circ \to \bullet$). A site at a given time becomes active if it was connected with at least one active site in the past, where a particle was created by the field. The number of such sites is equal to the cluster size, the probability of becoming active is given by the density $\varrho_{\rm s}(\tau,h) \simeq 1-(1-h)^{M(\tau,h)} \simeq hM(\tau,h)$ for h small. Therefore,

$$M(\tau,0) = \left.\frac{\partial \varrho_{\rm s}(\tau,h)}{\partial h}\right|_{h\to 0} \tag{4.143}$$

Comparison with the scaling forms for $\varrho_{\rm s}$ and M leads to

$$D_{1P} = D_0 D_2 \xi_0^{-(\beta+\gamma)/\nu_{\perp}} \mathcal{A}^{\pm} \tag{4.144}$$

where $\mathcal{A}^{\pm}$ is an universal amplitude. Recall that (ii) directed (bond) percolation is rapidity-reversal-invariant, hence

$$\varrho_{\rm s}(\tau,h) = P_{\rm perc}(\tau,h) \tag{4.145}$$

As a consequence, $\beta = \beta'$ and $D_{1\varrho} = D_{1P}$. Therefore, combining the equations (4.140,4.142,4.144) and using the hyperscaling relation (4.133), we have

$$\begin{aligned} \varrho_{\rm s}(\tau,h) &= D_0 D_2 \xi_0^{-d-z} |\tau|^{\beta} \hat{\mathcal{M}}_1^{\pm}\left(D_2 h|\tau|^{-\beta-\gamma}\right) \\ M(\tau,h) &= D_0 D_2^2 \xi_0^{-d-z} |\tau|^{-\gamma} \hat{\mathcal{M}}_2^{\pm}\left(D_2 h|\tau|^{-\beta-\gamma}\right) \end{aligned} \tag{4.146}$$

with universal functions $\hat{\mathcal{M}}_n^{\pm}(x) := {\rm d}^n \hat{\mathcal{M}}^{\pm}(x)/{\rm d}x^n$. Finally, we define a new function $\mu = \mu(\tau,h)$ by $\varrho_{\rm s}(\tau,h) = \partial\mu(\tau,h)/\partial h$, which implies

$$\mu(\tau,h) = D_0 \xi_0^{-d-z} |\tau|^{(d+z)\nu_{\perp}} \hat{\mathcal{M}}^{\pm}\left(D_2 h|\tau|^{-\beta-\gamma}\right) \tag{4.147}$$

In particular, because of $\xi_{\|} = \xi_{\perp}^{z}/D_0$ we obtain the extension of the Privman-Fisher hyperuniversality (2.112) [535] to the DP class in the form [279]

$$\mu(\tau,0)\xi_{\perp}^{d}(\tau,0)\xi_{\|}(\tau,0) \underset{\tau\to 0}{\longrightarrow} \text{universal constant} \tag{4.148}$$

We can now include finite-size effects in the discussion. To be specific, consider a rod geometry, such that the lattice is infinite in the 'time' direction, but is hypercubic with finite width L in the 'space' directions. Adopting the usual expectations of finite-size scaling, we expect that the scaling functions should now be written as follows:

$$\hat{\mathcal{M}}_n^{\pm} = \hat{\mathcal{M}}_n^{\pm}\left(D_2 h|t|^{-\beta-\gamma}; L\xi_{\perp}^{-1}\right) \tag{4.149}$$

and without introducing any further metric factor. Scaling out L and using (4.148) we have for the DP class the following finite-size scaling forms for the temporal and spatial correlation lengths, generalising the equilibrium form (2.113) [279]

$$\xi_{\|,i}^{-1} = L^{-z} D_0 R_i\left(C_1 \tau L^{1/\nu_{\perp}}\right) \quad , \quad \xi_{\perp,i}^{-1} = L^{-1} S_i\left(C_1 \tau L^{1/\nu_{\perp}}\right), \tag{4.150}$$

where (for given boundary conditions) R_i and S_i are *universal* scaling functions, the index i refers to the observable of which the correlators are studied and the entire non-universality can be absorbed into the non-universal metric factors C_1 and D_0 (since in finite systems, the only stationary state is the absorbing state, there is no non-trivial analogue of the equilibrium Gibbs functional).

Therefore, the finite-size scaling amplitudes $S_i(0)$ of the spatial correlation lengths $\xi_{\perp}^{-1}$ are *universal*, as are *ratios* of temporal correlation lengths

$$\xi_{\|,i}^{-1}/\xi_{\|,j}^{-1} = R_i(0)/R_j(0).$$

The latter ones are easily calculated from the spectrum of the Liouvillian/quantum Hamiltonian $\mathcal{L}$, see appendix C and Sect. 4.3.1.

A different consequence of the above argument now concerns the universality of certain moments. Since

$$\begin{aligned}
\langle |\boldsymbol{r}|^n \rangle &= \frac{\displaystyle\int_{\Omega(L)} \mathrm{d}\boldsymbol{r} \int_0^\infty \mathrm{d}t\, |\boldsymbol{r}|^n \Upsilon(t,\boldsymbol{r}; L/\xi_{\perp})}{\displaystyle\int_{\Omega(L)} \mathrm{d}\boldsymbol{r} \int_0^\infty \mathrm{d}t\, \Upsilon(t,\boldsymbol{r}; L/\xi_{\perp})} \\
&= \xi_{\perp}^n \frac{\displaystyle\int_{\Omega(L/\xi_{\perp})} \mathrm{d}\boldsymbol{r} \int_0^\infty \mathrm{d}t\, |\boldsymbol{r}|^{n-(\beta+\beta')/\nu_{\perp}} \mathcal{U}^{\pm}(t,\boldsymbol{r}; L/\xi_{\perp})}{\displaystyle\int_{\Omega(L/\xi_{\perp})} \mathrm{d}\boldsymbol{r} \int_0^\infty \mathrm{d}t\, |\boldsymbol{r}|^{-(\beta+\beta')/\nu_{\perp}} \mathcal{U}^{\pm}(t,\boldsymbol{r}; L/\xi_{\perp})} \\
&= \xi_{\perp}^n \bar{\Xi}_n(L/\xi_{\perp}) \;=\; L^n \Xi_n(L/\xi_{\perp})
\end{aligned} \tag{4.151}$$

where $\Omega(L)$ is a d-dimensional hypercube of linear extent L, the scaling functions $\bar{\Xi}_n$ and Ξ_n are indeed universal functions, at least for the DP universality class [279].

It still is an open problem whether the above argument can be extended to situations where the two auxiliary assumptions (i,ii) made above (see p. 139) and which are specific to directed percolation, no longer hold true. However, explicit results in some models indicate that the results derived here in a relatively restricted context, might be of a more general validity.

1. The diffusive pair-contact process (PCPD) is described by the reactions, together with their rates

$$\bullet\bullet\circ \xrightarrow{(1-p)(1-D)/2} \bullet\bullet\bullet \;,\; \bullet\bullet \xrightarrow{p(1-D)} \circ\circ \;,\; \bullet\circ \xleftrightarrow{D} \circ\bullet \,. \tag{4.152}$$

(here and below, the spatially symmetric counterparts are understood to occur with the same rates). The phase diagram is shown in Fig. S.5. We concentrate on the region where pair-annihilation dominates, such that the mean particle-density $\varrho \sim t^{-1/2}$ in $1D$. On a finite lattice, the inverse relaxation time $\xi_{\|,L}^{-1} = \Gamma_L$ is given by the lowest energy gap of the Liouville operator $\mathcal{L}$. Through a combination of Bethe-ansatz solutions and numerical diagonalisation and extrapolation, one finds in $1D$ [279]

$$\Gamma_L = \frac{aD}{L^2}\left(1 + \mathrm{O}(L^{-2})\right) \;,\; a = \begin{cases} 2\pi^2 & ; \text{ periodic boundary conditions} \\ \pi^2 & ; \text{ free boundary conditions} \end{cases} \tag{4.153}$$

throughout the entire absorbing phase $p \leq p_c(D)$ (see Sect. 5.8 for a discussion of the behaviour *at* the critical line). Up to a normalisation, this indicates the universality of the finite-size scaling amplitude.

2. The diffusion-annihilation-coagulation process is given by the reactions

$$\bullet\bullet \xrightarrow{D\gamma} \bullet\circ \;,\; \bullet\bullet \xrightarrow{2D\alpha} \circ\circ \;,\; \bullet\circ \xleftrightarrow{D} \circ\bullet \;, \tag{4.154}$$

If $\alpha + \gamma = 1$, the $1D$ mean steady-state particle density can be found from free-fermion methods, see exercise 37. One has $\lim_{L\to\infty} L\varrho_{L,s} = (1 + \alpha/\gamma)/(1 + 2\alpha/\gamma)$ which depends on the branching ratio α/γ. Remarkably, the finite-size amplitude of the inverse relaxation time $\xi_{\|,L}^{-1} = \Gamma_L$ for a chain with L sites can be found from Bethe ansatz methods, for both periodic and free boundary conditions, and turns out to be *independent* of α and γ [271]. This nicely confirms hyperuniversality.

3. The diffusion-coagulation-production process is given by the reactions

$$\bullet\circ \xleftrightarrow{D} \circ\bullet \;,\; \bullet\bullet \xrightarrow{D} \bullet\circ \;,\; \bullet\circ\bullet \xrightarrow{2D\lambda} \bullet\bullet\bullet \,. \tag{4.155}$$

It is known that the long-time behaviour of the $1D$ mean particle-density is, independently of λ, $\varrho \sim t^{-1/2}$ [273], see also exercise 39. Again, for a periodic chain with L sites, the exact inverse relaxation time $\xi_{\|,L}^{-1} = 2\pi^2 DL^{-2}(1 + \mathrm{O}(L^{-2}))$ is λ-independent and hence universal, up to normalisation.

An extension of studies of this kind to different universality classes would be of interest and could become of diagnostic value for the identification of new universality classes.

4.3 Methods of Analysis

Despite the simplicity of directed percolation as a lattice model, an exact treatment, let alone a rigorous solution that explains its critical behaviour quantitatively, is still lacking. As at equilibrium, the mean-field theory provides correct results only above the upper critical dimension d_c. Below the upper critical dimension, however, fluctuation effects and correlations change the critical behaviour entirely, in particular the values of the critical exponents and the form of scaling functions. For this reason, directed percolation has been studied by a variety of approximate methods, as will be discussed in the following.

4.3.1 Exact Diagonalisation

In principle, any stochastic model defined on a *finite* lattice can be solved exactly by diagonalisation of the time evolution operator. For example, in models with parallel updates, which evolve by the linear evolution equation

$$|P_{t+1}\rangle = \mathcal{T}|P_t\rangle\,, \tag{4.156}$$

the general solution can be expanded in terms of the eigenfunctions of $\mathcal{T}$, see (3.18). Similarly, in models with random sequential updates evolving by a master equation (3.21)

$$\partial_t|P_t\rangle = -\mathcal{L}|P_t\rangle\,. \tag{4.157}$$

the probability distribution $|P_t\rangle$ can be computed by diagonalising the Liouville operator $\mathcal{L}$. In the case of two-state models such as DP with $\mathcal{N} = L^d$ sites the evolution operator $\mathcal{T}$ (or $\mathcal{L}$) is a $2^{\mathcal{N}} \times 2^{\mathcal{N}}$ matrix.

Since it is often very difficult to determine eigenvalues and eigenvectors exactly by analytical methods, such matrices are usually diagonalised numerically on a computer. However, depending on the computational capabilities the system size is strongly limited. For example, directly solving a (2+1)-dimensional contact process on a 4×4 lattice means to diagonalise a $65\,536 \times 65\,536$ matrix which requires about 17 gigabytes of memory, not to mention the CPU time needed for its diagonalisation, although the memory requirements are the limiting factor in this kind of calculation. Even when using optimised algorithms (exploiting symmetries and simplifications

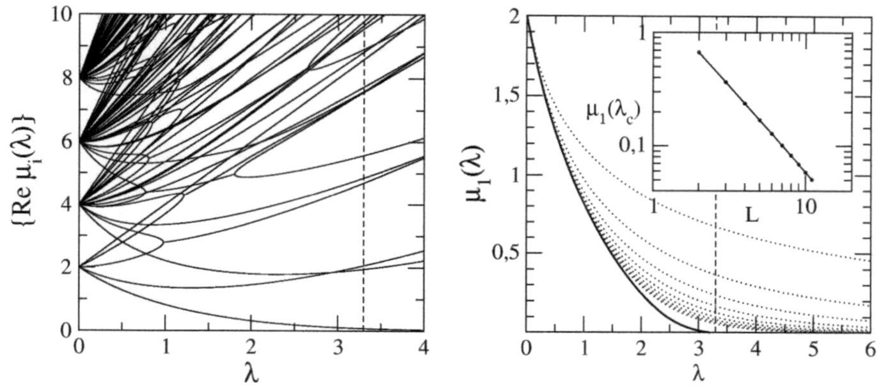

Fig. 4.14 Real part of the eigenvalue spectrum of the (1+1)-dimensional contact process with periodic boundary conditions. Left: Low-lying eigenvalues for system size $L = 8$ as a function of λ. The dashed line marks the critical point $\lambda_c \simeq 3.2978$ of an infinite system. Right: Eigenvalue μ_1 as a function of λ for system sizes $L = [2, 3, 4, \ldots, 11]$ from top to bottom (dotted lines) extrapolated to $L \to \infty$ (solid line). The inset demonstrates how the eigenvalue μ_1 varies with L at the critical point $\lambda = \lambda_c$.

for sparsely occupied matrices) the numerical effort grows exponentially, limiting the manageable system size.[10] Nevertheless, the method of exact diagonalisation can be used as an approximation. Moreover, it is of conceptual interest.

In appendix C, we show how the matrix $\mathcal{L}$ of a (1+1)-dimensional contact process with periodic boundary conditions can be constructed explicitly by using a symbolic programming language. Solving the eigenvalue problem

$$\mathcal{L}(\lambda) \, |\psi_i\rangle = \mu_i \, |\psi_i\rangle \tag{4.158}$$

numerically by standard numerical diagonalisation techniques [531], one obtains a spectrum of eigenvalues $\{\mu_i\}$. The real part of the spectrum for a chain with $L = 8$ sites is shown in Fig. 4.14. As in all reaction-diffusion models, the lowest eigenvalue μ_0 is zero, representing the stationary state. In the contact process, this zero mode is simply the absorbing configuration while the other eigenvectors represent the relaxational modes of the system. As can be seen, all relaxational eigenmodes have a short lifetime, except for the first excited state $|\psi_1\rangle$, whose eigenvalue μ_1 tends to zero as λ increases. This eigenvector represents the active state of the system shortly before it becomes trapped in the absorbing state. As the smallest eigenvalues determine the life time of the process we may identify μ_1^{-1} with the temporal correlation length $\xi_\parallel$.

In finite systems, there is always a finite probability of reaching the absorbing state, hence $\mu_1 > 0$. In infinite systems, however, this eigenvalue decreases

---

[10] It is one of the attractive features of the density-matrix renormalisation group [625, 626] that it solves this difficulty.

with λ and eventually vanishes at the critical point (see extrapolated curve in the right panel of Fig. 4.14). At criticality $\lambda = \lambda_c$, where $\xi_{\parallel} \sim L^z$, one expects that

$$\mu_1(L) \sim L^{-z}\,, \tag{4.159}$$

where $z = \nu_{\parallel}/\nu_{\perp}$ is the dynamical exponent. In fact, plotting $\mu_1(L)$ versus L in a double-logarithmic plot and measuring the slope (see inset of Fig. 4.14)) for $L = 2$ to $L = 11$ one obtains the estimate $z \approx 1.55$, which is in fair agreement with the DP value $z = 1.580745(10)$, see Table 4.3.

Near criticality, the first gap of the spectrum $\mu_1(\lambda, L)$ is expected to obey the finite-size scaling form

$$\mu_1(\lambda, L) = L^{-z}\, h(L\tau^{\nu_{\perp}})\ , \tag{4.160}$$

where $\tau = \lambda - \lambda_c$. Thus, by plotting $\mu_1 L^z$ over against $L\tau^{\nu_{\perp}}$, the exponents z and $\nu_{\perp}$ can be determined by data collapse. In order to determine the third exponent β, it would be necessary to analyse the eigenvector $|\psi_1\rangle$ and to compute the average density of active sites which obeys a similar scaling form.[11]

Larger system sizes can be reached by using sparse-matrix methods, see appendix C and especially Sect. 4.3.8 on the density-matrix renormalisation group (DMRG). The specific techniques of finite-size scaling combined with sequence extrapolation, which go beyond simple visual collapse, are explained in appendix F and further illustrations can be found again in Sect. 4.3.8 on the DMRG. These methods allow a considerable increase in numerical precision, although for DP they are by now inferior to the series expansion methods (see Sect. 4.3.3).

For studies of the *stationary* states only, an interesting alternative using the **quasi-stationary distribution** has been proposed recently [173]. If $p_c(t)$ is the probability that a non-absorbing configuration c occurs at time t and if $P(t)$ is the survival probability of the system until time t, the quasi-stationary distribution is defined by

$$\bar{p}_c := \lim_{t\to\infty} \frac{p_c(t)}{P(t)} \tag{4.161}$$

such that $\bar{p}_A = 0$ for any *absorbing* configuration A and one has the normalisation $\sum_{c\neq A} \bar{p}_c = 1$. Denote by $r_c := \sum_{c'} w_{c'\to c}\bar{p}_{c'}$ the quasi-stationary probability flux into configuration c, according to the master equation (3.20), and by r_A the flux into the absorbing state A. Further, let $w_c = \sum_{c'} w_{c\to c'}$ denote the total rate of transitions out of configuration c. Then, starting from some initial guess for $\bar{p}_c$, the quasi-stationary distribution is found from the iteration scheme [173]

---

[11] Computationally, it is considerably more efficient to adapt Yang's technique (1952) for the calculation of the spontaneous magnetisation in the $2D$ Ising model and to look at the matrix element of the density operator between the two lowest eigenstates of $\mathcal{L}$, see [270, 125] and Sect. 4.3.8.

$$\bar{p}'_c = a\bar{p}_c + \frac{(1-a)r_c}{w_c - r_A} \tag{4.162}$$

and the free parameter $a \in [0,1]$ is chosen for optimal convergence (values of $a \approx 0.1$ are recommended [173]). For example, the order parameter on a lattice of linear extent L is estimated as $\varrho_s = L^{-1}\sum_c n_c\bar{p}_c$, where n_c is the number of occupied sites of the configuration c. Analysing these data by the finite-size scaling methods described in appendix F, practical experience suggests that the method outlined is a useful general purpose-method, with an accuracy comparable to other general methods. For example, considering the (1+1)-dimensional contact process on rings up to $L = 23$ sites, an estimate for the critical point $\lambda_c = 3.29791(1)$ is obtained [173], after extrapolation with the BST algorithm, see appendix F.2. Similarly, numerical estimates of several exponents and universal moment ratios, for both the $1D$ contact process and the (non-diffusive) pair-contact process, are in good agreement with each other, often to more than three or four digits (in agreement with the expected universality), as well as with the precise estimates collected in Table 4.3 [173].

Therefore, absorbing phase transitions are described by the same *mathematical mechanism* as equilibrium critical phenomena: in the absorbing phase, the largest eigenvalue of the transfer matrix (or, respectively, the lowest eigenvalue of the Liouville operator/quantum Hamiltonian) is non-degenerate, but becomes degenerate at the critical point. On a finite lattice, this degeneracy is lifted such that *at* criticality, the lifting of the degeneracy (lowest gap) is of order $O(L^{-z})$ in the linear size of the system, while the lifting is exponentially small throughout the ordered phase.

4.3.2 Yang-Lee and Fisher Zeros

In equilibrium statistical mechanics, a large variety of continuous phase transitions has been analysed by studying the distribution of so-called *Fisher zeros* (obtained when varying the control parameter τ) and *Yang-Lee zeros* (obtained when varying the external field h) [639, 638, 159]. The starting point is the observation that a second-order phase transition of an infinite system is characterised by a non-analyticity (a discontinuous derivative) of various measurable quantities at the critical point. Hence, approximating them by a power series in the control parameter, the zeroes in the complex plane should approach the transition point in a forcipate manner.

At equilibrium, one usually expresses the partition function of a finite system as a polynomial in a control parameter and investigates the roots of this polynomial in the complex plane. If the control parameter is related to the temperature T, these roots are called **Fisher zeros**. For example, in the case of the Ising model, the Fisher zeros lie on a circle in the complex plane and approach the real line from both sides in the vicinity of the phase transition as the system size increases. In the thermodynamic limit, any neighbourhood of the critical point captures infinitely many of these zeroes, causing non-analytic behaviour at the transition. Alternatively, the control parameter may be related to the external field h, in which case the roots are called **Yang-Lee zeros**.

Recently, it has been shown that the concept of Yang-Lee zeros can also be applied to non-equilibrium systems [24, 72]. For example, in the case of DP [25, 142], one has to consider the survival probability $P_{\text{sur}}(t)$ (see (4.82)) of a finite percolation tree as a function of the percolation probability p in the complex plane. The survival probability of DP and the partition sum of an equilibrium system are similar in many respects. They are both positive in the physically accessible regime and can be expressed as polynomials in finite systems. As the system size tends to infinity, infinitely many Yang-Lee zeros approach the real line at the critical point, in both cases giving rise to a non-analytic behaviour at the transition. In different models, such as the *asymmetric exclusion process*, one tries to define a non-equilibrium analogue of the partition function and then proceeds to analyse its zeros, see [72].

In bond DP, the survival probability $P_{\text{sur}}(t)$ is given by the sum over the weights of all possible configurations of bonds, where each conducting bond contributes to the weight with a factor p, while each non-conducting bond contributes with a factor $1-p$. As shown in [142], the polynomial for the survival probability can be expressed as a sum over all cluster configurations c reaching the horizontal row at time t. Therefore, the polynomial is of the form

$$P_{\text{sur}}(t) = \sum_c p^n (1-p)^m \,, \tag{4.163}$$

where n denotes the number of bonds in the interior of the cluster while m is the number of bonds belonging to its hull. Summing up all weights in (4.163), one obtains a polynomial of degree t^2+t. For example, for bond DP the first five polynomials in one spatial dimension are given by

$$\begin{aligned}
P_{\text{sur}}(0) &= 1 \\
P_{\text{sur}}(1) &= 2p - p^2 \\
P_{\text{sur}}(2) &= 4p^2 - 2p^3 - 4p^4 + 4p^5 - p^6 \\
P_{\text{sur}}(3) &= 8p^3 - 4p^4 - 10p^5 - 3p^6 + 18p^7 + 5p^8 - 30p^9 + 24p^{10} - 8p^{11} + p^{12} \\
P_{\text{sur}}(4) &= 16p^4 - 8p^5 - 24p^6 - 8p^7 + 6p^8 + 84p^9 - 29p^{10} - 62p^{11} - 120p^{12} \\
&\quad + 244p^{13} + 75p^{14} - 470p^{15} + 495p^{16} - 268p^{17} + 83p^{18} - 14p^{19} + p^{20} \,.
\end{aligned} \tag{4.164}$$

As t increases, the number of possible configurations grows rapidly, leading to complicated polynomials with very large coefficients.

The distribution of zeros in the complex plane for $t = 15$ is shown in Fig. 4.15. As can be seen, the zeros are not arranged regularly on a circle, instead their distribution reminds one of a fractal, perhaps being a signature of the non-integrable nature of DP. As expected, the zeros approach the phase transition point from above and below. Their distance from the transition point is found to scale as $t^{-1/\nu_\parallel}$ in agreement with basic scaling arguments.

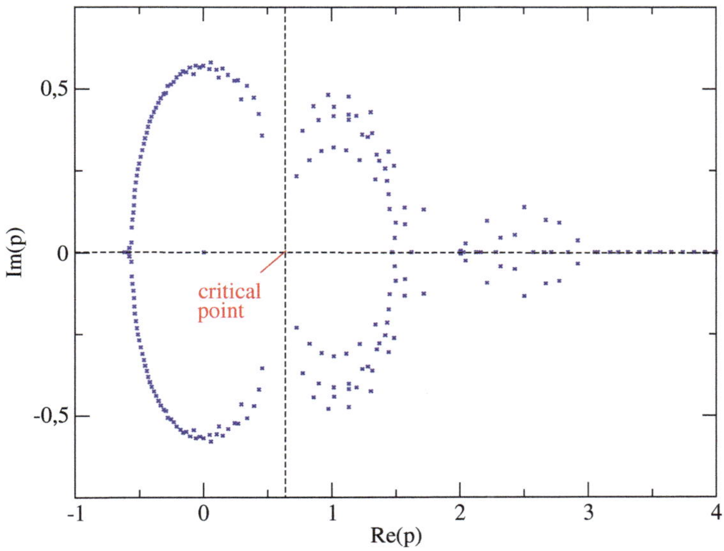

Fig. 4.15 Distribution of Yang–Lee zeros of the polynomial $P_{\rm sur}(15)$ in the complex plane. The critical point is marked by an arrow. After [142].

4.3.3 Series Expansion

The most precise estimates of the critical exponents of DP in 1+1 dimensions have been obtained by series expansions combined with Padé approximants. This technique is very similar to low- or high-temperature expansions in equilibrium statistical physics, i.e., the order parameter is expressed as a power series in terms of the control parameter. In what follows we demonstrate how this method works in the example of the (1+1)-dimensional contact process, following [359].

Firstly, the Liouville operator (3.38) of the contact process is separated into two parts

$$\mathcal{L}(\lambda) = \mathcal{L}_0 + \lambda \mathcal{L}_1 \,, \tag{4.165}$$

where $\mathcal{L}_0$ describes spontaneous removal of particles $A \to \emptyset$ while $\mathcal{L}_1$ describes offspring production $A \to 2A$ at neighbouring sites. The starting point of a series expansion in the inactive phase is to consider λ as a small parameter and to expand the probability distribution $|P_t\rangle$ in terms of λ. To this end it is convenient to introduce the Laplace transform of $|P_t\rangle$ and to expand it as a power series in λ:

$$|\overline{P}(\mu)\rangle = \int_0^\infty \mathrm{d}t\, e^{-\mu t}\, |P_t\rangle = \sum_{n=0}^{\infty} \lambda^n\, |\overline{P}_n(\mu)\rangle\ . \tag{4.166}$$

Applying $\mu + \mathcal{L}(\lambda)$ from the left, then using the master equation (3.21), and finally integrating by parts, this equation turns into

$$(\mu + \mathcal{L}_0 + \lambda\mathcal{L}_1) \sum_{n=0}^{\infty} \lambda^n |\overline{P}_n(\mu)\rangle = |P_0\rangle\,, \tag{4.167}$$

where $|P_0\rangle$ denotes the initial particle configuration. Comparing the coefficients, one may readily derive the recursion relation

$$(\mu + \mathcal{L}_0)\,|\overline{P}_n(\mu)\rangle = \begin{cases} |P_0\rangle & \text{if } n = 0 \\ -\mathcal{L}_1 |\overline{P}_{n-1}(\mu)\rangle & \text{if } n \geq 1 \end{cases}. \tag{4.168}$$

Let us further assume that the process starts with a single particle at the origin. Following [359], we denote this initial state as $|P_0\rangle = |\bullet\rangle$, where the bullet indicates a single active site. Since $\mathcal{L}_0$ can only destroy a particle, the vector $|\overline{P}_0(\mu)\rangle$ is a superposition of configurations with at most one particle. Applying (4.168), it is then easy to show that

$$|\overline{P}_0(\mu)\rangle = \frac{1}{\mu(\mu+1)}|0\rangle + \frac{1}{1+\mu}|\bullet\rangle. \tag{4.169}$$

Similarly, the vector $|\overline{P}_1(\mu)\rangle$ is a superposition of configurations with at most two particles:

$$|\overline{P}_1(\mu)\rangle = -\frac{1}{(\mu+1)^2(\mu+2)}|0\rangle + \frac{\mu}{(\mu+1)^2(\mu+2)}|\bullet\rangle + \frac{1}{(\mu+1)(\mu+2)}|\bullet\bullet\rangle. \tag{4.170}$$

Repeating this procedure the vectors $|\overline{P}_n(\mu)\rangle$ can be constructed iteratively. As n increases, the expressions for the coefficients become very complex and have to be generated by symbolic computer programs up to a certain order.

Having determined the vectors $|\overline{P}_n(\mu)\rangle$, one can express the temporal behaviour of any observable $X(t) = \langle s|X|P_t\rangle$ as a power series in λ, where $\langle s| = (1,1,1,\ldots,1)$ denotes the sum over all configurations introduced in (3.16). In particular, the temporal integral over such an observable is given by

$$\int_0^\infty \mathrm{d}t\, X(t) = \lim_{\mu\to 0} \langle s|X|\overline{P}(\mu)\rangle = \sum_{n=0}^{\infty} \lambda^n \lim_{\mu\to 0} \langle s|X|\overline{P}_n(\mu)\rangle \,. \tag{4.171}$$

As an example let us consider the survival probability $P_{\mathrm{sur}}(t)$ that the system has not yet reached the absorbing state at time t. This quantity may be expressed as

$$P_{\mathrm{sur}}(t) = 1 - \langle 0|P_t\rangle = \langle s|P_t\rangle - \langle 0|P_t\rangle \,, \tag{4.172}$$

where $\langle 0| = (1,0,0,\ldots,0)$ projects onto the absorbing state. The mean survival time T of clusters (4.96) in the inactive phase can then be expanded in powers of λ by

$$T = \int_0^\infty \mathrm{d}t\, P_{\mathrm{sur}}(t) = \lim_{\mu\to 0} \overline{P}_a(\mu) = \sum_{n=0}^{\infty} \lambda^n \lim_{\mu\to 0}\Big(\langle s|\overline{P}_n(\mu)\rangle - \langle 0|\overline{P}_n(\mu)\rangle \Big). \tag{4.173}$$

Since $T \sim (-\tau)^{\beta'-\nu_\parallel}$ we can estimate λ_c and $\beta - \nu_\parallel$ by determining the location and the amplitude of the singularity in the logarithmic derivative

$$\frac{\mathrm{d}}{\mathrm{d}\lambda} \ln T \simeq \frac{\nu_\parallel - \beta'}{\lambda_c - \lambda} + \mathrm{const} \tag{4.174}$$

by using Padé approximants [259, 367].

A general review on series expansion can be found in [259]. Series expansions were applied to (1+1)-dimensional DP for the first time in [209], where the critical exponents could be determined with a relative accuracy of about 10^{-3}. Refined simulations [211] led the authors to the conjecture that the DP exponents might be given by the *rational* values $\beta = 199/720$, $\nu_\perp = 26/15$, and $\nu_\parallel = 79/72$. In a sequence of papers [359, 358, 361, 352, 353] the error margins could be further reduced down to $10^{-4} \ldots 10^{-5}$. These improved estimates showed that the conjectured rational values were incorrect, suggesting that the critical exponents of DP are probably *irrational* numbers. In addition, the exponents were found to be independent of the type of lattice under consideration. Currently, the most precise estimates are given in [355, 356]. Series expansions for DP were also performed in two spatial dimensions [362], see Table 4.3 on p. 159 for the numerical values.

4.3.4 Field-Theoretical Methods

Field-theoretic Renormalisation group theory is probably the most fascinating analytical approach to directed percolation. While the predictive power concerning the values of exponents is limited, field-theory offers a deep insight into the origin of universality, the justification of scaling relations, and the crossover to mean-field behaviour in high dimensions.

4.3.4.1 Field-Theoretic Action

As we have seen in Sect. 3.2.4, the critical behaviour of the DP universality class is described by a minimal stochastic Markovian process represented by the Langevin equation [326, 330]

$$\lambda^{-1}\partial_t \varrho(t,\boldsymbol{r}) = \left(\tau - g\,\varrho(t,\boldsymbol{r}) + \nabla^2\right)\varrho(t,\boldsymbol{r}) + h + \eta(t,\boldsymbol{r}) \qquad (4.175)$$

with the field-dependent Gaussian noise

$$\langle \eta(t,\boldsymbol{r})\eta(t',\boldsymbol{r}')\rangle \;=\; \lambda^{-1}\,\kappa\,\varrho(t,\boldsymbol{r})\,\delta(t-t')\,\delta(\boldsymbol{r}-\boldsymbol{r}')\,, \qquad (4.176)$$

where we have absorbed the diffusion constant D in a parameter λ on the left-hand side for later convenience. Higher-order terms such as ϱ^3, ϱ^4, or $\nabla^4\varrho$ turn out to be irrelevant under renormalisation as long as $g > 0$. Negative values of g give rise to a first-order transition, whereas the special case $g = 0$ is associated with a tricritical point, which will be discussed in Sect. 5.4.

Simple dimensional counting gives the mean-field exponents (see exercise 43)

$$\beta_{\mathrm{MF}} = \beta'_{\mathrm{MF}} = 1\;\;,\;\;\nu_{\perp,\mathrm{MF}} = \frac{1}{2}\;\;,\;\;\nu_{\parallel,\mathrm{MF}} = 1 \qquad (4.177)$$

and further shows that the noise is irrelevant for $d > 4$, while it is relevant for $d < 4$. This confirms that $d_c = 4$ is the upper critical dimension of directed percolation. Above the upper critical dimension d_c mean-field theories are valid and present instructive insight into the critical behaviour (cf. Sect. 3.2.3, Sect. 4.1 as well as [480]).

In dimensions $d < d_c$, fluctuation effects become important and lead to a behaviour that deviates from the mean-field prediction. In this case, the Langevin equation (4.175) in its bare form is no longer valid, since a continuum description in the form of a partial differential equation cannot account for the fractal structure of the emerging percolation clusters. However, field-theoretic renormalisation group methods provide a way to restore the discreteness of the underlying lattice through the back-door, either in form of a cutoff in momentum space or by some other regularisation procedure. In this case path integral representations are more adequate than the Langevin equation approach (the connection between Langevin equations and path integrals is outlined in appendix D).

In the past two decades, field-theoretic renormalisation-group techniques have been applied to estimate the critical exponents and universal scaling functions (see [342, 593, 592] for comprehensive reviews of the field-theoretical treatment of DP). Stationary correlation functions as well as response functions can be determined by calculating path integrals with weight $\exp(-\mathcal{J})$, where the dynamic functional $\mathcal{J}$ describes the considered stochastic process. Up to higher-order terms which are irrelevant in the RG sense, the dynamic functional describing DP is given by [325, 190, 326]

$$\mathcal{J}[\widetilde{\varrho},\varrho] = \lambda \int \mathrm{d}t\,\mathrm{d}\boldsymbol{r}\; \widetilde{\varrho}\left[\lambda^{-1}\partial_t\varrho - (\tau+\nabla^2)\varrho - \left(\frac{\kappa}{2}\widetilde{\varrho} - g\,\varrho\right)\varrho - h\right] \tag{4.178}$$

where $\widetilde{\varrho}(t,\boldsymbol{r})$ denotes the **response field** conjugated to the Langevin noise field [463]. Remarkably, the above functional is well-known from high energy physics and corresponds to the Lagrangian of **Reggeon field-theory** [1, 123]. Since DP represents the simplest realisation of a non-equilibrium phase transition, its field-theory is often regarded as the non-equilibrium counterpart of the famous ϕ^4-theory of equilibrium critical phenomena (which by a Hubbard-Stratonovich transformation is equivalent to the Ising model, see exercise 16).

4.3.4.2 Rapidity-Reversal Symmetry

Rescaling the fields in (4.178) by

$$\widetilde{\varrho}(t,\boldsymbol{r}) = \mu\,\widetilde{n}(t,\boldsymbol{r}), \qquad \varrho(t,\boldsymbol{r}) = \mu^{-1}\,n(t,\boldsymbol{r}), \qquad h = \mu^{-1}\,\bar{h} \tag{4.179}$$

with $\mu^2 = 2g/\kappa$, turns the functional $\mathcal{J}$ into

$$\mathcal{J}[\widetilde{n},n] = \lambda \int \mathrm{d}t\,\mathrm{d}\boldsymbol{r}\; \widetilde{n}\left[\lambda^{-1}\partial_t n - (\tau+\nabla^2)n - \sqrt{\frac{g\kappa}{2}}\,(\widetilde{n}-n)n - \bar{h}\right]. \tag{4.180}$$

In this form, the cubic terms $\widetilde{n}^2 n$ and $\widetilde{n}n^2$ have the same coefficient and thus play a symmetric role. In fact, for $h = 0$ the action (4.180) is now invariant under the **duality transformation**

$$\widetilde{n}(t,\boldsymbol{r}) \longleftrightarrow -n(-t,\boldsymbol{r})\,. \tag{4.181}$$

Note that this duality symmetry, also called **rapidity-reversal symmetry** in Reggeon field-theory, is a simple consequence of the path integral formulation while it is considerably less evident in the Langevin equation.

Usually, a duality transformation defines a dual stochastic process that might differ from the original one [422], relating dual pairs of order parameters. For example, the survival probability $P_{\mathrm{sur}}(t)$ that a cluster generated by a single seed is still active after t time steps, is mapped by the duality transformation to the density of active sites $\varrho^{\mathrm{dual}}(t)$ in the dual process starting with a fully occupied lattice [331] (see exercise 48)

$$P_{\mathrm{sur}}(t) \simeq \mu^2\,\varrho^{\mathrm{dual}}(t)\,. \tag{4.182}$$

The invariance of the action (4.180) under the transformation (4.181) tells us that the dual process is again a DP process running backwards in time,

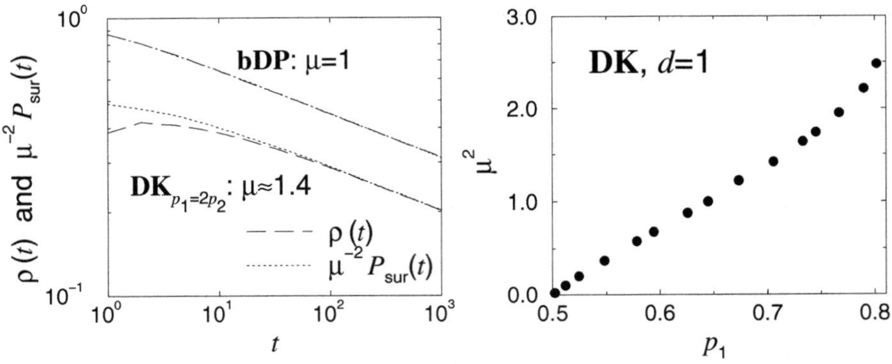

Fig. 4.16 Self-duality of DP: The order parameter decay $\varrho(t)$ (starting with a fully occupied lattice) and the survival probability $P_{\rm sur}(t)$ (starting with a single seed) for two distinct critical points of the Domany-Kinzel automaton (left figure). For bond-DP the duality symmetry is fulfilled microscopically, leading to $\varrho(t) = P_{\rm sur}(t)$. In general, this equivalence is only obeyed asymptotically, i.e., $\varrho(t) \simeq \mu^2 P_{\rm sur}(t)$. Here, the curves for $p_2 = p_1/2$ are shown. Moving along the critical line of the phase diagram, the parameter μ decreases (see right figure) until it vanishes at the termination point which reflects the violation of the rapidity reversal for compact DP. Thus μ parametrises the critical line of the one-dimensional DK automaton. Reprinted with permission from [435].

meaning that directed percolation is **self-dual**. Therefore, $\varrho(t)$ and $\varrho^{\rm dual}(t)$ coincide, leading to the relation

$$P_{\rm sur}(t) \simeq \mu^2 \varrho(t) . \tag{4.183}$$

Thus, the duality symmetry can be identified as the field-theoretic equivalent of the rapidity-reversal symmetry discussed in Sect. 4.1.2. However, the field-theoretic treatment shows that this duality symmetry is not just a special property of bond-DP, rather it is a characteristic symmetry of the universality class of DP. This means that all models belonging to the DP universality class obey, at least asymptotically, the rapidity-reversal symmetry after suitable coarse-graining. In other words, the duality is not necessarily an exact symmetry of the microscopic dynamics [326]. For example, in bond-DP the duality symmetry is exact with $\mu = 1$, while for contact process and the pair-contact process (see below in Sect. 5.5) the symmetry holds only asymptotically with $\mu \neq 1$. Numerical values of μ are listed in Table 3.1 for the (1+1)-dimensional Domany-Kinzel automaton, and the asymptotic nature of self-duality is shown in Fig. 4.16.

Irrespective of the particular value of μ, the asymptotic equivalence leads to the relation [246]

$$\delta = \alpha . \tag{4.184}$$

or by means of $\alpha = \beta/\nu_{\|}$ and $\delta = \beta'/\nu_{\|}$ to the scaling law

$$\beta = \beta' . \tag{4.185}$$

Thus, compared to general absorbing phase transitions, DP is characterised by three instead of four independent critical exponents.

4.3.4.3 Field-Theoretic ϵ-Expansion of Critical Exponents

By a field-theoretic renormalisation group calculation it is possible to calculate the critical exponents as a power series (loop expansion) in $\epsilon = d_c - d$. For example, the standard exponents β, $\nu_\perp$, $\nu_\|$, and z are given in two loop order by [89, 88, 90, 326]

$$\beta = 1 - \frac{\epsilon}{6}\left[1 - \left(\frac{11}{288} - \frac{53}{144}\ln\frac{4}{3}\right)\epsilon + \mathrm{O}(\epsilon^2)\right], \tag{4.186}$$

$$\nu_\perp = \frac{1}{2} + \frac{\epsilon}{16}\left[1 + \left(\frac{107}{288} - \frac{17}{144}\ln\frac{4}{3}\right)\epsilon + \mathrm{O}(\epsilon^2)\right], \tag{4.187}$$

$$\nu_\| = 1 + \frac{\epsilon}{12}\left[1 + \left(\frac{109}{288} - \frac{55}{144}\ln\frac{4}{3}\right)\epsilon + \mathrm{O}(\epsilon^2)\right], \tag{4.188}$$

$$z = 2 - \frac{\epsilon}{12}\left[1 + \left(\frac{67}{288} - \frac{59}{144}\ln\frac{4}{3}\right)\epsilon + \mathrm{O}(\epsilon^2)\right]. \tag{4.189}$$

4.3.4.4 Equation of State

Field-theoretic methods allow one to calculate scaling functions as a power series in ϵ. The Widom-Griffiths scaling form of the equation of state close to the transition point (to be specific for $x < 0$ and $|x| \ll 1$) is given by [332]

$$\begin{aligned}\tilde{H}(x,1) = 1 - x &+ \frac{\epsilon}{6} K\left[(2-x)\ln(2-x) - 2(1-x)\ln 2\right] \\ &+ \frac{\epsilon^2}{72}\left[(4-x)\ln^2(2-x) - 4(1-x)\ln^2 2\right] + \mathrm{O}(\epsilon^3),\end{aligned} \tag{4.190}$$

where

$$K = 1 + \epsilon\left(\frac{85}{288} + \frac{29}{72}\ln 2 - \frac{53}{144}\ln 3\right). \tag{4.191}$$

As in equilibrium statistical physics, the equation of state may be expressed in a much simpler form if one chooses a suitable parametric representation [559, 365]. For example, introducing the parameters $a_\tau \tau = R(1-\theta)$ and $\varrho = R^\beta \theta/2$ the equation of state given above turns into [332]

$$a_h \, h \;=\; \left(\frac{R^\beta}{2}\right)^{\sigma/\beta} \; H(\theta) \tag{4.192}$$

with metric factors a_τ, a_h and the scaling function

$$H(\theta) \;=\; \theta\,(2-\theta) + \mathrm{O}(\epsilon^3)\,. \tag{4.193}$$

In this parametrisation, the whole phase diagram lies in the range $R \geq 0$ and $\theta \in [0,2]$. It is remarkable that the scaling forms for DP are less complicated than e.g. those of the Ising model [8].

It is straightforward to calculate the susceptibility from the above results. In particular, the universal amplitude ratio $\tilde{\mathrm{X}}(+1,0)/\,\tilde{\mathrm{X}}(-1,0)$ can be estimated at two-loop order to be [332]

$$\frac{\tilde{\mathrm{X}}(+1,0)}{\tilde{\mathrm{X}}(-1,0)} = 1 - 2\beta + \mathrm{O}(\epsilon^3) \tag{4.194}$$

$$= 1 - \frac{\epsilon}{3}\left[1 - \left(\frac{11}{288} - \frac{53}{144}\ln\frac{4}{3}\right)\,\epsilon\, + \mathrm{O}(\epsilon^2)\right].$$

As we shall see below, critical exponents and universal amplitude combinations are approximated by the ϵ-expansion with a similar accuracy (see Sect. 4.3.6). This is different with respect to equilibrium, where the exponents are usually estimated more accurately by ϵ-expansion than amplitude combinations [536].

Quite recently, a new approach to renormalised field-theory, applied to absorbing phase transitions, was developed, which is not based on a perturbative improvement around the Gaussian fixed point - and is therefore called the **non-perturbative renormalisation group** [107, 104, 105, 106, 108, 102]. A detailed description of the method is beyond the scope of this book, but we emphasise that the non-perturbative character of this technique, which comes about through an exact integral equation for the correlators, provides a complementary and useful point of view on the collective behaviour close to an absorbing phase transition. See e.g. [103] for a pedagogical introduction.

4.3.4.5 Breakdown of Finite-Size Scaling

The field-theoretic renormalisation group allows us to estimate not only critical exponents, the equation of state and the related susceptibility but also correlation functions and response functions, the survival and percolation probabilities, as well as finite size-scaling functions (see e.g. [326, 336, 331, 440, 333] and references therein). Moreover, the field-theoretic analysis reveals an interesting anomaly of finite-size scaling above the upper critical dimension, the so-called **breakdown of finite-size scaling** which we shall discuss in the following.

As an example, let us consider the so-called **zero-momentum approximation**. Introducing the spatial Fourier transforms of the fields n and $\widetilde{n}$

$$n(t,\boldsymbol{r}) = \sum_{\boldsymbol{q}} \mathrm{e}^{i\boldsymbol{q}\cdot\boldsymbol{r}} \widehat{n}(\boldsymbol{q},t) \tag{4.195}$$

in a system of linear size L and periodic boundary conditions, a dynamic functional for the homogeneous zero mode $\boldsymbol{q} = 0$ can be constructed as follows. First the fields $n, \tilde{n}$ are decomposed into homogeneous modes $\widetilde{\Phi}(t)$, $\Phi(t)$ and inhomogeneous orthogonal complements $\widetilde{\Psi}(t,\boldsymbol{r})$, $\Psi(t,\boldsymbol{r})$, i.e.

$$n(t,\boldsymbol{r}) = \Phi(t) + \Psi(t,\boldsymbol{r}) \tag{4.196}$$

with $\Phi(t) = L^{-d} \int \mathrm{d}^d\boldsymbol{r}\, n(t,\boldsymbol{r})$. This leads to a decomposition of the response functional into $\mathcal{J} = \mathcal{J}_0[\Phi,\widetilde{\Phi}] + \mathcal{J}_1[\Phi,\widetilde{\Phi};\Psi,\widetilde{\Psi}]$ with

$$\mathcal{J}_0 \;=\; \lambda L^d \int \mathrm{d}t\, \widetilde{\Phi} \left[\lambda^{-1}\partial_t \Phi - \tau\Phi \;-\; \sqrt{\frac{g\kappa}{2}}\,(\widetilde{\Phi} - \Phi)\Phi \;-\; \bar{h} \right] . \tag{4.197}$$

The inhomogeneous modes $\widetilde{\Psi}$ and Ψ can be eliminated by a functional integration and it turns out that $\mathcal{J}_1$ contributes to the leading scaling behaviour for $d \le d_c$ (see [336, 342] for a detailed analysis). Here, we follow [440] and focus our attention to the mean-field regime. For $d > d_c$, $\mathcal{J}_1$ can be neglected, because it provides only corrections to the leading asymptotic scaling behaviour. In order to analyse $\mathcal{J}_0$, it is convenient to define the rescaled quantities

$$\varphi(s) \;=\; L^{d/2}\Phi(t) \qquad \text{and} \quad s \;=\; \sqrt{2g\kappa}\, L^{-d/2}\lambda\, t\, . \tag{4.198}$$

Then, introducing the rescaled control parameter and the rescaled field

$$T \;=\; \sqrt{2g\kappa}^{\,-1}\, L^{d/2}\tau \quad \text{and} \quad H \;=\; \sqrt{2g\kappa}^{\,-1}\, L^{d}\, \bar{h} \tag{4.199}$$

the functional (4.197) reads

$$\mathcal{J}_0 \;=\; \int \mathrm{d}s\, \widetilde{\varphi} \left[\partial_s\varphi - T\varphi \;-\; \frac{1}{2}(\widetilde{\varphi} - \varphi)\varphi \;-\; H \right] . \tag{4.200}$$

Note that the whole dynamic functional depends on the rescaled parameters T and H only. Furthermore, the rescaled parameters contain the irrelevant variable g in a dangerous, i.e., singular way. Comparing (4.199) to the finite-size scaling form [(4.67) with $\beta_{\mathrm{MF}} = 1$ and $\sigma_{\mathrm{MF}} = 2$] one obtains the finite-size scaling exponent for periodic boundary conditions

$$\nu_\perp^* = \frac{2}{d}. \tag{4.201}$$

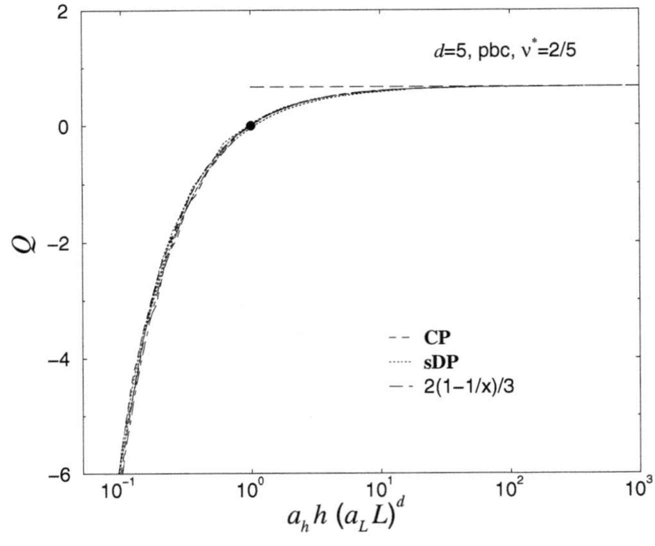

Fig. 4.17 The universal fourth-order cumulant $\tilde{Q}_{\mathrm{pbc}}(0, x, 1)$ as a function of the rescaled field $a_h h(a_L L)^d$ at criticality for $d > d_c$. The field-theoretical results are in perfect agreement with numerical data of the five-dimensional contact process (CP) and of the five-dimensional site-directed percolation process (sDP) (see [440] for details). The bold circle corresponds to the normalisation of the universal scaling function (4.72) which specifies the metric factor a_L. Reprinted with permission from [433]. Copyright (2004) World Scientific Publishing Company.

The symbol * indicates that the scaling exponent differs from the mean-field value of the correlation length exponent $\nu_{\perp,\mathrm{MF}} = 1/2$ for $d > 4$. Thus within the mean-field regime, the finite-size scaling forms are not controlled by the ratio $L/\xi_\perp \sim L|\tau|^{\nu_\perp}$ but by the ratio $L|\tau|^{\nu_\perp^*}$. This scaling anomaly is often termed *breakdown of finite-size scaling* and is related to the appearance of an additional length scale l_∞ [69, 201], which diverges as $l_\infty \sim |\tau|^{-\nu_\perp^*}$. Analogous to equilibrium, this length scale coincides with ξ below d_c, meaning that $\nu_\perp = \nu_\perp^*$. Thus, the exponent $\nu_\perp^*$ fulfils the hyperscaling relation $\nu_\perp^* d = 2\beta + \gamma'$ in all dimensions.

In addition to the finite-size scaling exponent $\nu_\perp^*$, it is also possible to derive universal scaling functions. The dynamic functional (4.197) is associated with a Fokker-Planck equation [336] which leads to the following moments of the order parameter, for $\tau = 0$ [440]

$$\langle \varphi^k \rangle = 2^{k/2}\, \Gamma(H + k/2) \,/\, \Gamma(H). \tag{4.202}$$

In this way, the universal finite-size scaling functions of the order parameter, the order parameter fluctuations as well as certain ratios can be derived. For example, the ratios U and Q ((4.73) and (4.69)) are given by

$$U = \frac{1}{2} \quad \text{and} \quad Q = \frac{2}{3}\left(1 - \frac{1}{2H}\right). \tag{4.203}$$

Note that the ratio Q is not finite at the critical point ($H \to 0$). As already mentioned, this reflects the different nature of the phases with vanishing order parameter in equilibrium and in absorbing phase transitions. This scaling scenario is also observed in numerical investigations. Corresponding data of the five-dimensional contact process as well as of five-dimensional site-DP are shown in Fig. 4.17. As can be seen, the numerical data fits perfectly to (4.203), an impressive demonstration of the usefulness of field-theory in non-equilibrium critical phenomena.[12]

4.3.5 Methods for Exact Solution

Although it has not yet been possible to find an exact solution for a model in the DP universality class, we briefly comment on methods to solve systems with an absorbing phase transition exactly. Indeed, since there are many integrable two-dimensional classical spin systems and their one-dimensional quantum counter parts [50], one might hope to identify stochastic processes with an integrable Liouville operator $\mathcal{L}$ and which might then be solved by Bethe ansatz techniques. For example, the master equation for simple symmetric diffusion can be exactly mapped onto the Schrödinger equation of the XXZ Heisenberg ferromagnet [15]. A systematic investigation of integrable stochastic systems was started in [13, 563, 566]. However, although very interesting mathematical structures related to Hecke algebras, quantum groups and their quotients have been discovered, it turned out that the number of integrable non-equilibrium models, even in one spatial dimension, is rather limited. A detailed discussion of integrable models and Bethe-ansatz methods is beyond the scope of this book, see e.g. [565, 271] and references therein.[13]

---

[12] The breakdown of finite-size scaling discussed in this section follows the lines of dangerously irrelevant variables such that the scaling functions may become *singular* for $d > d_c$. However, it is also possible that scaling functions may *vanish*. An example occurs in the **surface tension** $\sigma(\tau)$ of Reggeon field-theory, defined by $\xi_{\|,L}(\tau) \sim \exp\left(\sigma(\tau) L^d\right)$ in the active phase $\tau > 0$, in close analogy with Ising models for $T < T_c$. From standard scaling arguments including hyperscaling, one would expect $\sigma(\tau) \sim \tau^{d\nu_\perp}$. However, from the Liouville operator associated to $(1+1)D$ Reggeon field-theory, one obtains $\sigma(\tau) \sim \tau^{\mu}$, with $\mu = 1.74(6)$, considerably larger than the expectation $\mu \approx 1.1$ [278]. Qualitatively, this means that the so-called 'node-link-blob picture' of undirected percolation, which states that percolation clusters should be described as collections of nodes at distances $\sim \xi_{\text{bulk}}$, separated by tenuous links, should not be valid for directed percolation, since it implies $\mu = d\nu_\perp$ [278]. Similar effects also arise in certain first-order transitions [317].

[13] `http://www.iop.org/EJ/journal/-page=extra.topical2/jstat` gives a list of recent topical articles on the Bethe ansatz.

model	reactions			conditions
I	$\bullet\circ \stackrel{D}{\longleftrightarrow} \circ\bullet$	$\circ\circ \stackrel{\nu}{\longrightarrow} \circ\bullet$	$\bullet\;\circ \longrightarrow \circ\;\circ$	$D = \nu$
II	$\bullet\circ \stackrel{D}{\longleftrightarrow} \circ\bullet$	$\bullet\bullet \stackrel{\gamma}{\longrightarrow} \bullet\circ$	$\bullet\;\circ \longrightarrow \bullet\;\bullet$	$D = \gamma$
III		$\bullet\circ \longrightarrow \bullet\bullet$	$\bullet\circ \longrightarrow \circ\circ$	
IV	$\bullet\circ \stackrel{D}{\longleftrightarrow} \circ\bullet$	$\bullet\bullet \stackrel{2\alpha}{\longrightarrow} \circ\circ$	$\circ\circ \stackrel{2\sigma}{\longrightarrow} \bullet\bullet$	$\alpha + \sigma = D$
V	$\bullet\circ \stackrel{D}{\longleftrightarrow} \circ\bullet$	$\bullet\bullet \stackrel{2\alpha}{\longrightarrow} \circ\circ$	$\bullet\bullet \stackrel{\gamma}{\longrightarrow} \circ\circ$	$\alpha + \gamma = D$
VI	$\bullet\circ \stackrel{D}{\longleftrightarrow} \circ\bullet$	$\circ\circ \stackrel{2\sigma}{\longrightarrow} \bullet\bullet$	$\circ\circ \stackrel{\nu}{\longrightarrow} \bullet\circ$	$\sigma + \nu = D$
VII		$\bullet\circ \longleftrightarrow \circ\circ$	$\bullet\bullet \longleftrightarrow \bullet\circ$	
VIII		$\bullet \longrightarrow \circ$	$\circ \longrightarrow \bullet$	

Table 4.2 Schematic list of the parity-symmetric reaction schemes of stochastic systems whose Liouville operator can be reduced to a free-fermion form by a local stochastic similarity transformation. For reversible reactions, the forward and backwards rates are assumed to be equal.

On the other hand, a formal discussion of integrable Liouville operators sometimes allows one to use stochastic similarity transformations in order to obtain an overview of at least certain classes of integrable models. For example, if the Liouville operator can be brought into the form

$$\mathcal{L} = -\sum_{\ell=1}^{L} \left[D\left(\sigma_\ell^- \sigma_{\ell+1}^+ + \sigma_\ell^+ \sigma_{\ell+1}^-\right) + \eta\left(\sigma_\ell^+ \sigma_{\ell+1}^+ + \sigma_\ell^- \sigma_{\ell+1}^-\right) + h\sigma_\ell^z \right] \tag{4.204}$$

then standard free-fermion methods can be applied [276, 565] to diagonalise $\mathcal{L}$ exactly and to find explicit time-dependent solutions for the observables (see exercises). In Table 4.2 we list those stochastic systems, in one spatial dimension, which describe the evolution of a single species of particles which can be reduced to the Liouville operator (4.204) by a *local* similarity transformation $\mathcal{L}' = \mathcal{B}\mathcal{L}\mathcal{B}^{-1}$ with $\mathcal{B} = \bigotimes_{\ell=1}^{L} B_\ell$, see [276, 275, 565, 271].[14] In writing these schematic reaction schemes, we assume that all reactions are symmetric under a global left-right inversion (parity symmetry) and only specify one of the reactions involved. Reactions rates are only indicated in order to spell out relations between various reaction rates (within a given model), unspecified reaction rates remain arbitrary. These correspondences remain valid if *biased* diffusion and reaction rates are admitted.

The models in Table 4.2 can be identified as follows. Models I/II are the diffusion-coagulation-decoagulation process and its conjugate under a particle-hole exchange. Model III is the *voter model*, see Sect. 5.2. Model IV is related to the $1D$ kinetic Ising model with Glauber dynamics [276] (see Vol. 2 of this book) and models V/VI are the diffusion-annihilation-coagulation process and its conjugate under a particle-hole exchange. Model

[14] A very interesting idea [565] is to consider **enantiodromy** transformations $\mathcal{L}' = \mathcal{B}\mathcal{L}^{\mathrm{T}}\mathcal{B}^{-1}$, where T denotes the transpose, but this does not yet seem to have been explored systematically.

	$d=1$ [353]	$d=2$ [612, 249]	$d=3$ [349]	Mean-field
$\beta=\beta'$	0.276486(8)	0.5834(30)	0.813(9)	1
$\nu_\perp$	1.096854(4)	0.7333(75)	0.584(5)	1/2
$\nu_\parallel$	1.733847(6)	1.2950(60)	1.110(10)	1
σ	2.554216(13)	2.1782(171)	2.049(26)	2
γ'	0.543882(16)	0.2998(162)	0.126(23)	0
γ	2.277730(5)	1.5948(184)	1.237(23)	1
$\eta_\perp$	1.504144(19)	1.5912(148)	1.783(16)	2
$\delta=\alpha$	0.159464(6)	0.4505(10)	0.732(4)	1
Θ	0.313686(8)	0.2295(10)	0.114(4)	0
z	1.580745(10)	1.7660(16)	1.901(5)	2
$\tilde{D}(1,0)$ [445]	1.46(12)	1.65(9)	1.83(11)	2
$\frac{\tilde{X}(+1,0)}{\tilde{X}(-1,0)}$ [445]	0.033(4)	0.25(1)	0.65(3)	1
R_χ [445]	0.60(4)	0.72(4)	0.86(8)	1
U [440]	0.833(11)	0.704(13)	0.61(2)	1/2

Table 4.3 The critical exponents and various universal amplitude combinations of the directed percolation universality class. For $d=1$, the exponents γ, $\nu_\perp$, and $\nu_\parallel$ are obtained from a series expansion by Jensen [351]. For $d=2$ and $d=3$ activity spreading simulations are performed yielding δ, Θ, as well as the dynamical exponent z [612, 349]. In addition, the exponent $\nu_\parallel$ is determined [249, 349] in order to estimate the full set of exponents via scaling laws. The numbers in brackets give the estimated uncertainty in the last digit(s).

VII is a zero-temperature Glauber-Ising model in a weak magnetic field and domain-wall driving [565] and model VIII describes particles which exhibit free decay and creation.

4.3.6 Monte Carlo Simulations

Non-equilibrium phase transitions of non-integrable models such as DP are often investigated by Monte Carlo simulations. In contrast to equilibrium, where one has to construct a dynamics for a given model (see appendix G), the simulation of a non-equilibrium process is much easier because the model is already defined as a time-dependent stochastic process.

One of the aims of numerical simulations is to find precise estimates of the critical point, the critical exponents and of the scaling functions. For systems with absorbing states, there are two main simulation techniques, which differ by their initial condition, namely, simulations with homogeneous initial states and simulations starting from a single active seed. In the following, we summarise these methods, using directed percolation as the paradigmatic example. For reference, Table 4.3 gives a list of recent estimates of critical exponents and universal amplitudes of the DP universality class.

4.3.6.1 Homogeneous Initial States

The first task is to find the critical point of the system. To this end, one first simulates the system starting with a fully occupied lattice, measuring the density of active sites as a function of time. To avoid finite-size effects in this type of simulation, the system has to be sufficiently large, typically larger than $T^{1/z}$, where T is the total simulation time and $z = \nu_\parallel/\nu_\perp$ is the expected dynamical critical exponent. Moreover, the density has to be averaged over many runs in order to reduce the statistical error. Equivalently, one may perform a single run using a very large system.

The typical scenario is shown in Fig. 4.6 in the example of bond DP. If the system is in the absorbing phase ($p < p_c$), the curves veer down, whereas they veer up and saturate if the system is in the active phase. By a successive adjustment of the critical parameter (most efficiently by halving the remaining interval) one can estimate the critical point within a certain margin, depending on the simulation time and the quality of the statistical average. To find out whether a curve veers up or down, just divide by the expected slope and trust your eyes, which are often more reliable in detecting a curvature than any other tool.

Having determined the critical parameter, one can estimate the exponent $\alpha = \beta/\nu_\parallel$ by measuring the slope of the straight line in a double logarithmic plot. As mentioned before, never use the χ^2 error displayed by a regression algorithm, instead try different slopes and check their compatibility with the numerical results visually. As a rough rule of thumb, one expects the relative error in the estimate of a critical exponent to be at least 10 times larger than the error in the critical parameter.

Next, one can determine the exponent $\nu_\parallel$ by plotting different data sets in the vicinity of the critical point rescaled according to the scaling form (4.76) by plotting $\varrho(t)t^\alpha$ versus $t(p-p_c)^{\nu_\parallel}$, as shown in the right panel of Fig. 4.6. This can be done manually but it is advisable to write a short program that carries out the data collapse and displays the graph directly on the screen, which allows one to adjust α and $\nu_\parallel$ quite easily. The error bars are then determined by varying the exponents until the data collapse breaks down.

Finally, the third exponent is determined by a finite-size scaling analysis. Here it is advisable to increase the lateral system size in powers of 2 ($L = 2, 4, 8, 16, \ldots$) and to measure the decay of the density $\varrho(t)$. Plotting $\varrho(t)t^\alpha$ versus t/L^z the dynamical exponent z has to be adjusted in such a way that all curves collapse. Note that for small systems, lattice effects may lead to severe corrections, so it may be impossible to collapse all curves. In this case these data sets have to be discarded.

4.3.6.2 Seed Simulations

Seed simulations provide an alternative method of analysing absorbing phase transitions. Here one starts with a single active site in an otherwise inactive system. This site serves as the seed of a growing cluster whose scaling properties were already discussed in Sect. 4.2.4. In seed simulations one usually measures three time-dependent quantities, namely the survival probability $P_{\text{sur}}(t)$, the average number of active sites $N_{\text{a}}(t)$, and their mean square spreading from the origin $R_{\text{s}}^2(t)$ averaged over surviving runs. At criticality these quantities are expected to obey power laws of the form

$$P_{\text{sur}}(t) \sim t^{-\delta}\,, \qquad N_{\text{a}}(t) \sim t^{\Theta}\,, \qquad R_{\text{s}}^2(t) \sim t^{2/z} \tag{4.205}$$

The process is first simulated at criticality. Empirically, it turned out that $N_{\text{a}}(t)$ is the most sensitive quantity with respect to deviations from criticality and should be used to determine or verify the critical point. Simulations at criticality permit one to estimate two independent exponents ($\delta = \beta'/\nu_{\parallel}$ and $z = \nu_{\parallel}/\nu_{\perp}$), while the exponent Θ allows one to verify the generalised hyperscaling relation $\Theta = d/z - \alpha - \delta$. The third exponent $\nu_{\parallel}$ can be determined by off-critical simulations, plotting e.g. $N_{\text{a}}(t)t^{-\Theta}$ versus $t(p-p_c)^{\nu_{\parallel}}$ and adjusting $\nu_{\parallel}$ in such a way that the curves collapse.

Seed simulations can be optimised in various ways. The most efficient one is to keep the positions of active sites in a dynamically generated list. Such list structures are provided by standard software packages such as the C++ Standard Template Library (STL).[15] This method virtually eliminates finite-size effects because the possible values of the coordinates are only limited by the range of integer numbers.

4.3.6.3 Comparison of Scaling Functions

For the identification of a universality class, the analysis of scaling functions is often more reliable than a simple comparison of critical exponents. While for the latter the numerical differences between different universality classes are often small, the scaling functions may differ significantly. Thus the agreement of universal scaling functions provides not only additional but also independent and more convincing evidence in favour of a conjecture that the phase transitions of two models belong to the same universality class. For example, let us consider the scaling behaviour of the equation of state. Setting $a_h h \lambda^{\sigma} = 1$ in (4.41) we obtain the scaling form

$$\varrho_{\text{s}}(\tau, h) \simeq (a_h h)^{\beta/\sigma}\, \tilde{R}(a_\tau \tau\, (a_h h)^{-1/\sigma}, 1)\,. \tag{4.206}$$

---

[15] See e.g. `http://www.sgi.com/tech/stl/` for more information.

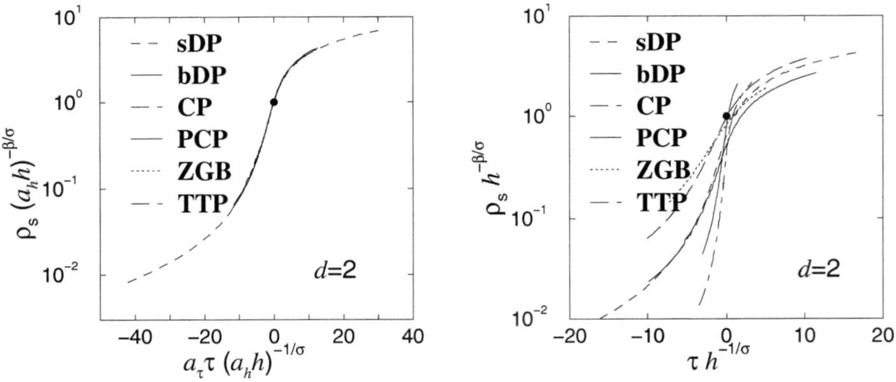

Fig. 4.18 The universal scaling function $\tilde{R}(x,1)$ of the order parameter of the two-dimensional directed percolation universality class (left panel). The data of all considered models collapse onto to the same scaling function, which is an impressive manifestation of the robustness of the directed percolation universality class with respect to variations of the microscopic interactions. If the metric factors a_τ and a_h are not taken into account, each model is characterised by its own scaling function (right panel). The bold circles mark the condition $\tilde{R}(0,1) = 1$. Reprinted with permission from [433]. Copyright (2004) World Scientific Publishing Company.

Plotting the rescaled stationary order parameter $\varrho_s(\tau, h)\,(a_h h)^{-\beta/\sigma}$ as a function of the rescaled control parameter $a_\tau \tau (a_h h)^{-1/\sigma}$, the data sets of different models belonging to the DP universality class have to collapse onto a single universal curve $\tilde{R}(x,1)$. This is shown in Fig. 4.18 for six different lattice models. Despite the different interaction details of e.g. bond-DP (bDP) and site-DP (sDP), despite the fact that the activity is coupled to a frozen (PCP) or fluctuating (TTP) background of inactive particles, despite the different nature of the absorbing phase (a unique state for e.g. the CP and infinitely many states for the PCP), despite the fact that the ZGB model contains two distinct components in contrast to the other models, despite the different implementation of the conjugated field (particle creation or diffusion), and despite the different lattice structures and update schemes used, – all rescaled data collapse onto a unique universal scaling function! This data collapse is an impressive manifestation of the robustness of the DP universality class. Furthermore, it allows one to identify irrelevant parameters, i.e., those parameters and interaction details which do not affect the scaling behaviour.

A similar analysis can be performed in various dimensions and the results are displayed in Fig. 4.19. The universal scaling functions depend on the dimension d below the upper critical dimension $d_c = 4$. In this regime it is worth comparing the numerical results with those of analytical approximations such as renormalisation group results derived by the powerful and versatile ϵ-expansion. As expected, the differences increase with increasing perturbation parameter ϵ, i.e., with increasing distance to the upper critical

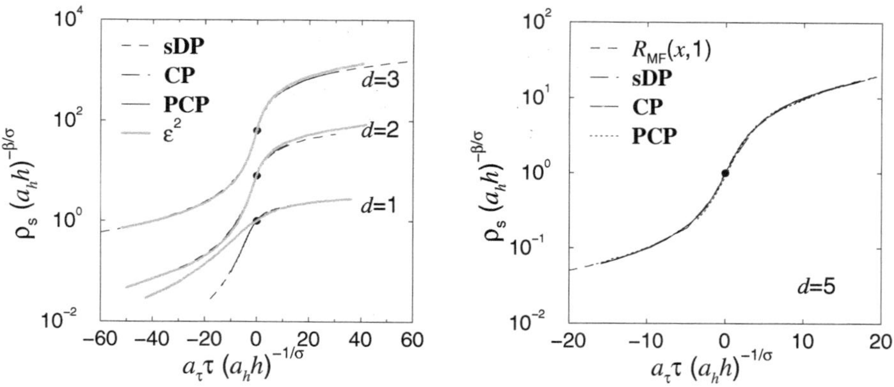

Fig. 4.19 The universal scaling function $\tilde{R}(x,1)$ of the directed percolation universality class below (left) and above (right) the upper critical dimension $d_c = 4$. The two- and three-dimensional data are vertically shifted in order to avoid overlaps. The bold circles mark the condition $\tilde{R}(0,1) = 1$. For $d < d_c$ we compare the numerical data to those of an ϵ-expansion (grey lines, see (4.193) and (4.192) and [332]). For $d > d_c$ the analytically known mean-field scaling function (4.207) fits perfectly to the numerical data. Reprinted with permission from [433]. Copyright (2004) World Scientific Publishing Company.

dimension. The deviations are particularly strong in $d = 1$, indicating that higher orders of the expansion than ϵ^2 are necessary to describe the scaling behaviour in low dimensions.

Above the upper critical dimension, the scaling behaviour of the lattice models has to equal the analytically known mean-field scaling function, e.g. for the equation of state (see (4.54))

$$\tilde{R}_{\mathrm{MF}}(x,y) = \frac{x}{2} + \sqrt{\left(\frac{x}{2}\right)^2 + y}\ . \tag{4.207}$$

Rescaled data of three different five-dimensional models are shown in Fig. 4.19. The agreement between the numerical and the analytical results confirm that all considered models exhibit the same type of simple mean-field behaviour. In other words, the value of the upper critical dimension is less than five. In this way, a reliable upper bound on d_c can be obtained just by checking whether the numerical or experimental data are consistent with the analytically known mean-field scaling function. This method was also applied successfully in universality classes different from DP where the value of d_c is not known exactly [438].

Right at the upper critical dimension, the scaling behaviour is characterised by mean-field power laws modified by logarithmic corrections. For example, the order parameter at zero field behaves as

$$\varrho_{\mathrm{s}} \sim \tau\,|\ln\tau|^{1/3}\,. \tag{4.208}$$

Recent numerical investigations [10, 438, 446, 255] as well as analytical results [341] have shown that the concept of universal scaling functions can also be applied to systems at the upper critical dimension. Instead of going into details we refer the interested reader to the review [433].

As in the equation of state, the universality also manifests itself in the scaling functions of the fluctuations and the susceptibility, respectively. For example, Fig. 4.20 shows the universal susceptibility scaling function $\tilde{X}(x,1)$, see (4.42). The susceptibility is obtained by computing the numerical derivative of the order parameter with respect to the conjugated field. All scaling functions exhibit a clear maximum signalling the divergence of χ at the critical point.

As already discussed in Sect. 4.1.9, a universality class is not only characterised by critical exponents and universal scaling functions but also by various **universal amplitude combinations** [536, 279]. These amplitude combinations emerge from the universality of the scaling functions since universal amplitude combinations are just particular values of the scaling functions. In particular, the measurement of amplitude combinations in experiments or simulations provides a reliable test for theoretical predictions. For example, approximation schemes of the renormalisation group, such as ϵ- or $1/n$-expansions, are widely used to obtain explicit and systematic estimates of the amplitude combinations (see Sect. 4.3.4). Here, we shall focus on the amplitude ratio of the susceptibility. Other universal amplitude combinations of DP are discussed in [433], see also Table 4.3.

Consider the singular behaviour of the susceptibility below and above the transition for $h = 0$

$$\chi(\tau, 0) \simeq \begin{cases} a_{\chi,+}\,\tau^{-\gamma} & ;\ \text{if } \tau > 0 \\ a_{\chi,-}\,(-\tau)^{-\gamma} & ;\ \text{if } \tau < 0 \end{cases} . \tag{4.209}$$

Using the scaling form (4.42), the susceptibility ratio

$$\frac{\chi(\tau > 0, h)}{\chi(\tau < 0, h)} = \left.\frac{\tilde{X}(\ a_\tau \tau\,\lambda, a_h h\,\lambda^{\sigma})}{\tilde{X}(-a_\tau \tau\,\lambda, a_h h\,\lambda^{\sigma})}\right|_{a_\tau|\tau|\lambda=1} = \frac{\tilde{X}(+1, x)}{\tilde{X}(-1, x)} \tag{4.210}$$

is clearly a universal quantity for all values of the scaling argument $x = a_h h |a_\tau \tau|^{-\sigma}$. In particular it equals the ratio $a_{\chi,+}/a_{\chi,-}$ for a vanishing field

$$\frac{a_{\chi,+}}{a_{\chi,-}} = \lim_{h\to 0} \frac{\tilde{X}(+1, x)}{\tilde{X}(-1, x)} = \frac{\tilde{X}(+1, 0)}{\tilde{X}(-1, 0)} . \tag{4.211}$$

Within the mean-field approximation (4.55) leads to the result

$$\frac{\tilde{X}_{\mathrm{MF}}(+1, x)}{\tilde{X}_{\mathrm{MF}}(-1, x)} = 1 \tag{4.212}$$

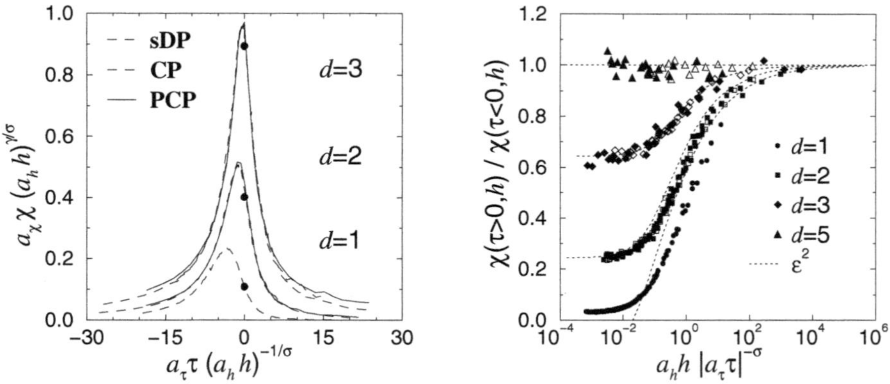

Fig. 4.20 The universal scaling function $\tilde{X}(x, 1)$ of the directed percolation universality class below the upper critical dimension (left figure). The two- and three-dimensional data are vertically shifted in order to avoid overlaps. The bold circles mark the condition $\tilde{X}(0, 1) = \beta/\sigma$ for each d, reflecting the accuracy of the numerical analysis. The right figure displays the universal scaling function $\tilde{X}(1, x)/\tilde{X}(-1, x)$ for various dimensions and models (closed symbols site-DP, open symbols PCP and CP). The dashed lines correspond to an ϵ-expansion of a renormalisation group approach (see (4.195) and [332]). The universal amplitude $\tilde{X}(1, 0)/\tilde{X}(-1, 0)$ is obtained from the extrapolation $a_h h |a_\tau \tau|^{-\sigma} \to 0$. Reprinted with permission from [433]. Copyright (2004) World Scientific Publishing Company.

for all x. The susceptibility ratio $\tilde{X}(+1, x)/\tilde{X}(-1, x)$ is shown in Fig. 4.20 for various models and different dimensions. The five-dimensional data agrees well with the mean-field prediction. Far away from the transition point, the critical fluctuations are suppressed and the behaviour of all systems is well described by the mean-field solution. Approaching criticality, the critical fluctuations increase and a crossover to the d-dimensional behaviour takes place.

It is instructive to compare these results with field-theoretic findings [332] since the theoretical curve reflects the accuracy of the RG estimates of all involved quantities, namely the exponent, the scaling functions, as well as the non-universal metric factors. The perfect agreement between the numerical data and the RG curve for $d = 3$ indicates that all quantities are approximated well. In the two-dimensional case, a horizontal shift is observed between the numerical data and the RG-estimates. Thus the RG-approach yields good estimates for the exponents and the scaling functions, but the metric factors are less precise. For $d = 1$, the ϵ^2-approximation does not render appropriate estimates of the DP scaling behaviour, similarly for the equation of state (see Fig. 4.19).

The investigation of the universal amplitude ratio $\tilde{X}(+1, 0)/\tilde{X}(-1, 0)$ for a vanishing field offers a more quantitative check of the renormalisation group results. Numerically this ratio is obtained by an extrapolation of the susceptibility ratio $\tilde{X}(+1, x)/\tilde{X}(-1, x)$ to $x \to 0$. The estimated values listed in Table 4.3 may be compared to (4.195). Despite the negative and therefore unphysical results for $d = 1$ the RG results agree well with the numerical data (see also Fig. 4.20). For example, the two-dimensional values differ only by 3%. That has to be compared to the difference of the critical exponents. The RG estimate of e.g. the order parameter exponent (4.186) differs for $d = 2$ from the best known numerical value by 6% (see Table 4.3). The comparable accuracy of the critical exponent and of the amplitude combination is contrary to what is observed in critical equilibrium systems, where the exponents are usually calculated more accurately by ϵ-expansions (e.g. 6% difference for γ) than universal amplitude combinations (e.g. 115% for the susceptibility ratio). A possible explanation is that the ϵ^2-approximation yields for $\epsilon = 1, 2$ much better results for DP than e.g. for the Ising model. This is also reflected by the remarkably simple form of the equation of state within the ϵ-expansion (see (4.193)).

So far we have considered universal scaling functions and amplitude combinations of infinite systems in the steady-state. This analysis can be extended easily to steady-state and dynamical finite-size scaling functions and amplitude combinations. For example, we already have presented the universal scaling behaviour of the fourth-order cumulant Q above d_c in Fig. 4.17 and the dynamical order parameter behaviour in Fig. 4.13. We refer the interested reader to [433] for a systematic review on the analysis of these universal quantities.

4.3.7 Universal Moment Ratios

A different set of quantities, which may be useful for diagnostic purposes, considers universal ratios of moments of the order parameter [175, 176, 172, 151]. Let

$$m_n := \langle \varrho_s^n \rangle \tag{4.213}$$

denote the nth moment of the order parameter. From the usual scaling arguments, one expects that ratios such as m_2/m_1^2, $m_3/(m_1 m_2)$ or m_4/m_2^2 should be universal. Similarly, if one considers the cumulants

$$K_2 := m_2 - m_1^2 \ , \quad K_4 := m_4 - 4m_3 m_1 - 3m_2^2 + 12 m_2 m_1^2 - 6m_1^4, \tag{4.214}$$

the ratio K_4/K_2^2 should be universal. In Table 4.4, results for several moments ratios as estimated by quasi-stationary simulations are listed for the contact process and the non-diffusive pair-contact process, which are thought to be in the same universality class. Also included are the pair-contact process with diffusion, the parity-conserved universality class and compact directed percolation, which will be defined in Chap. 5.

A comparison of these numbers allows one to appreciate quantitatively a few points. It is satisfying to see that the estimates for the contact process

d	model	m_2/m_1^2	$m_3/(m_1m_2)$	m_4/m_2	K_4/K_2^2	Ref.
1	DP	1.1736(1)	1.301(3)	1.554(2)	-0.505(3)	[175]
	PCP	1.1738(2)	1.303(3)	1.558(2)	-0.493(3)	[175]
	$D = 0.1$	1.140(15)	1.27(2)	1.55(3)	0.1(2)	[151]
	$D = 0.5$	1.166(8)	1.310(15)	1.61(2)	0.0(1)	[151]
	$D = 0.85$	1.170(6)	1.301(3)	1.61(4)	-0.1(1)	[151]
	PC	1.3340(4)				[176]
	CDP	1.142(8)				[172]
2	DP	1.3257(5)	1.569(1)	2.093(8)	-0.088(4)	[175]
	PCP	1.323(3)	1.56(1)	2.07(1)		[175]

Table 4.4 Universal moment ratios for several systems in $d = 1$ and 2 spatial dimensions. The moments are calculated from the particle- or pair-density, respectively, according to which is related to the physical order parameter [151]. The models are: contact process/directed percolation (DP), pair-contact process (PCP), diffusive pair-contact process (PCPD), labelled by the value of the diffusion constant $D \in [0, 1]$, parity-conserving class (PC) and compact directed percolation (CDP).

and the pair-contact process, for both dimensions $d = 1, 2$, are practically identical. We also see that distinct universality classes (PC, CDP) may be distinguished by different values of the moment ratios or the cumulants. These observations confirm what one finds by studying the critical exponents in these systems. With respect to the diffusive pair-contact process, the situation is not so clear. If the diffusion constant D is large enough, most of the moment ratios in $d = 1$ appear to be close to the DP-values, although the (rather imprecise) cumulants are quite different. On the other hand, the data for D small might be seen as an indication of a critical behaviour different from DP, or else as an example of especially large corrections to scaling.[16] This illustrates the difficulties which may sometimes arise in interpreting purely numerical data. We shall come back to the PCPD in Sect. 5.8. Certainly, it will be useful to build up a database of universal moment ratios, in order to create a tool for future analysis.

The form of the distribution of the order parameter and its scaling behaviour is also analysed in [176, 151].

4.3.8 Density-Matrix Renormalisation-Group Methods

Diagonalising large matrices such as Liouville operators and transfer matrices can in certain cases be greatly helped through a relatively recent method invented by S. White and called the **density-matrix renormalisation-group** (DMRG). As compared with exact diagonalisation, it allows one to

[16] Interestingly, the study of the finite-size scaling of the universal ratios of correlation times rather suggests particularly strong corrections to scaling for large values of D [279].

treat considerably larger systems, with a numerical precision which may even exceed what can be achieved by Monte Carlo methods. On the other hand, the technique is only efficient for systems with extremely short-ranged interactions. In non-equilibrium systems, efficient implementations only exist in one spatial dimension. Since the method is well-presented in a lecture-notes volume [518] and recent developments are brilliantly covered in two recent reviews [560, 561], we limit ourselves here to a brief illustration on how the DMRG might be used to study steady-states or time-dependent averages. Adopting the language used in DMRG studies, we use in this section the terms 'Liouville operator $\mathcal{L}$' and 'quantum Hamiltonian H' synonymously.

In the following we shall consider two models as illustration for the application of the DMRG to non-equilibrium systems. Both models describe particles of a single species A on a chain with N sites such that each site is either empty ($\emptyset$) or occupied by a single particle (A). Throughout, we shall use open (free) boundary conditions, since they facilitate the application of the DMRG. The two models are defined by the following dynamical rules:

1. **Diffusion-annihilation model**: Particles hop to neighbouring sites with unit rate. If the target site is empty the particle diffuses by $A\emptyset \leftrightarrow \emptyset A$. Otherwise, if the target site is occupied, the two particles annihilate pairwise ($2A \to \emptyset\emptyset$). This model has two stationary states, namely, the empty lattice and the state with only one diffusing particle. The smallest non-vanishing eigenvalue of the Liouville operator $\mathcal{L}$ can be found exactly from the Bethe ansatz [279]
$$E_2(N) = 2p\left(1 - \cos\frac{\pi}{N+1}\right) \sim N^{-2} \tag{4.215}$$
which implies a dynamical exponent $z = 2$.
2. **Branching-fusing process**: Here the reactions $2A \to \emptyset\emptyset$, $A \to \emptyset$ and $A\emptyset \leftrightarrow \emptyset A$ occur with rate p and the reactions $A\emptyset, \emptyset A \to AA$ occur with rate $1-p$. This model exhibits a transition between an empty state and a fluctuating active steady-state at some critical threshold p_c. This transition is expected to be in the DP universality class, confirmed by estimates of both bulk and surface exponents, as shown in Table 4.5. Quantitatively, the results indicate that the DMRG is a viable general-purpose method, which produces results comparable in precision to other well-established techniques.

4.3.8.1 Density Matrix Renormalisation Group Algorithm

The DMRG algorithm [626] finds selected approximate eigenvalues and eigenvectors of a given (symmetric) time evolution operator $\mathcal{L}$. That desired eigen-

exponent	series	simulation	DMRG	best DP estimate
$\beta/\nu_\perp$	0.2520(1)	0.25208(4)	0.249(3)	0.2520
$\beta_1/\nu_\perp$	0.6690(1)	0.664(7)	0.667(2)	0.6689
$\nu_\perp$	1.0969(1)	1.09684(1)	1.08(2)	1.0968
$z = \nu_\parallel/\nu_\perp$	1.5806(2)	1.58074(4)	1.580(1)	1.5807

Table 4.5 Comparison of some critical exponents (bulk and surface) in 1+1 dimensions as found for the branching-fusion model by the DMRG [125] with results from directed percolation as obtained from series data [212] and Monte Carlo simulations [482, 409]. The numbers in brackets give the estimated uncertainty. The best estimates for DP are taken from Table 4.3 and (4.238), rounded to four digits.

vector $|\psi\rangle$ is called a **target state** and the process of selecting $|\psi\rangle$ is referred to as **targeting** (it is possible to target several states, see below). One assumes that $\mathcal{L}$ is defined on an open chain with L sites and has the local structure

$$\mathcal{L} = \sum_{i=1}^{N-1} \mathcal{L}_{i,i+1} \tag{4.216}$$

where $\mathcal{L}_{i,i+1}$ is a local time evolution operator acting on two neighbouring sites. The method is iterative and proceeds in two steps. The first one is the so-called **infinite-system method** (ISM). Suppose we are interested in the ground state of $\mathcal{L}$. As the starting point, consider a chain of four lattice sites which can be represented as $B_l^{(1)} \bullet \bullet B_r^{(1)}$, where $\bullet$ denotes a single site and $B_{r,l}^{(1)}$ are blocks at the left and right side of the chain. Initially, they contain only one spin, that is $B_{r,l}^{(1)} = \bullet$. At this point, the main loop begins. The time evolution operator $\mathcal{L}$ is easily written down and its ground state wave function $\psi_0(\alpha_l, i_l, j_r, \beta_r)$ can be found via standard diagonalisation routines [531], where α_l and β_r denote degrees of freedom of the blocks $B_r^{(1)}$ and $B_l^{(1)}$ and the indices i_l, j_r refer to the spin degrees of freedom of the single lattice points in the middle of the chain. The density matrix for the left part of the system is defined as

$$\rho^{(l)}(\alpha_l, i_l; \gamma_l, k_l) = \sum_{j_r,\beta_r} \psi_0(\alpha_l, i_l, j_r, \beta_r)\psi_0(\gamma_l, k_l, j_r, \beta_r)\,. \tag{4.217}$$

As a shortcut, one may write instead $\rho^{(l)} = \mathrm{tr}_l(|\psi_0\rangle\langle\psi_0|)$, where $\mathrm{tr}_{l,r}$ denotes a partial trace either in the left or right part of the system. Next, one solves the non-sparse eigenvalue problem $\rho|\Omega_i\rangle = \omega_i|\Omega_i\rangle$ via some standard routine. The eigenvalues of the density matrix are non-negative and can be ordered according to $\omega_1 \geq \omega_2 \geq \omega_3 \geq \ldots \geq 0$. Furthermore, if the ground state vector of $\mathcal{L}$ is normalised according to $\langle\psi_0|\psi_0\rangle = 1$, one has $\sum_i \omega_i = 1$. Each eigenvalue ω_i is equal to the probability of finding the left part of the chain in the corresponding density matrix eigenvector $|\Omega_i\rangle$ when the whole system is in the ground state $|\psi_0\rangle$. The configurational space reduction is then obtained by keeping only the first m dominant density matrix eigenvectors $|\Omega_i\rangle$ with

$i = 1, 2, \ldots m$, corresponding to the m largest ω_i. Formally, the truncation can be represented by

$$O_m^T \left(B_l^{(1)} \bullet \right) O_m = B_l^{(2)} , \tag{4.218}$$

where $O_m = [|\Omega_1\rangle, \ldots, |\Omega_m\rangle]$. The accuracy of the projection operation can be described by the **truncation error** $\epsilon = 1 - \sum_{i=1}^{m} \omega_i$. This projection operation is repeated for the right part as well to obtain $B_r^{(2)}$. Combining the two blocks with new sites one gets $B_l^{(2)} \bullet \bullet B_r^{(2)}$, e.g. a chain of $N = 6$ sites after the first pass through the main loop. The next pass through the main loop begins by writing down $\mathcal{L}$ for this longer chain.

Applying this procedure repeatedly at the left and right part of the system, one generates larger and larger systems. At each iteration step, two new sites are added in the middle of the chain and the boundaries are pushed further away from each other. Schematically, this may be illustrated as

$$B_l^{(1)} \bullet \bullet B_r^{(1)} \to B_l^{(2)} \bullet \bullet B_r^{(2)} \to \ldots \to B_l^{(N/2-1)} \bullet \bullet B_r^{(N/2-1)} .$$

This ISM method is repeated, often until $N \approx 10^3$. However, for systems close to criticality, the numerical precision achieved is not sufficient and a second step, the so-called **finite-size method** (FSM), is needed.

The starting point of the FSM is the target vector $|\psi\rangle$ for a chain of given length N, as generated by the ISM described above. At this point another sequence of iterations is started. First, one calculates better approximations for the blocks on the left part representing more than $N/2 - 1$ sites, using as before (4.218), while for the blocks on the right part, one uses blocks generated in previous iterations in order to keep the total length of the system fixed at N. Schematically this looks as follows:

$$B_l^{(N/2-1)} \bullet \bullet B_r^{(N/2-1)} \to B_l^{(N/2)} \bullet \bullet B_r^{(N/2-2)} \to \ldots \to B_l^{(N-3)} \bullet \bullet B_r^{(1)} .$$

Second, this procedure is reversed and the larger blocks on right part of the system are refined. Schematically, this may be written as

$$B_l^{(N-3)} \bullet \bullet B_r^{(1)} \to B_l^{(N-2)} \bullet \bullet B_r^{(2)} \to \ldots \to B_l^{(1)} \bullet \bullet B_r^{(N-3)} .$$

In these steps, the B_r are updated according to (4.218), while the B_l are taken from the blocks calculated previously. Finally, the B_l are updated again through the sequence

$$B_l^{(1)} \bullet \bullet B_r^{(N-3)} \to B_l^{(2)} \bullet \bullet B_r^{(N-4)} \to \ldots \to B_l^{(N/2-1)} \bullet \bullet B_r^{(N/2-1)}$$

until one is back at the left-right symmetric partition. This completes a so-called 'sweep' of the renormalisation procedure. The target vector extracted at this stage can be used as starting point for the next FSM iteration. Usually, two or three such sweeps achieve convergence.

In practise, critical exponents are found from combining the DMRG with finite-size scaling so that estimates for several values of N are required. For better efficiency, one uses the FSM to calculate from the same run quantities for chains of different lengths simultaneously as follows. The blocks generated at the end of the FSM for a chain of length N_0 are used as starting point for further DMRG calculation: first, one uses the ISM to enlarge the system symmetrically until a length $N_1 > N_0$ is reached and then one switches back to the FSM.

In carrying out these calculations for reaction-diffusion systems, one has the additional difficulty that the time evolution operator $\mathcal{L}$ to be diagonalised is non-symmetric. This can create serious problems concerning the numerical stability of the algorithm. A good way to find both eigenvalues and eigenvectors of $\mathcal{L}$ is through the **Arnoldi algorithm** for which efficient implementations are available.[17] But we warn the reader that even the convergence of the extreme eigenvalues is not always completely regular, meaning that subsequent estimates do not always form Sturm chains.

The Arnoldi algorithm iteratively generates one sequence of orthonormal vectors $|q_i\rangle$ forming the columns of a matrix $Q_n = [|q_1\rangle, \ldots, |q_n\rangle]$ and a Hessenberg matrix $\mathcal{L}_n = Q_n^T \mathcal{L} Q_n$ with elements $\mathcal{L}_{ij}$. One starts from a random vector $|q_0\rangle$ of unit length and $h_{10} = 1$ and iterates using

$$h_{k+1,k}|q_{k+1}\rangle = \mathcal{L}|q_k\rangle - \sum_{i=1}^{k} h_{ik}|q_i\rangle \tag{4.219}$$

with $h_{ij} = \langle q_i|\mathcal{L}|q_j\rangle$, where $\langle q_i| = |q_i\rangle^T$ and $h_{k+1,k}$ is determined by enforcing unit length for $|q_{k+1}\rangle$.

The non-symmetric Hessenberg matrix is then diagonalised using the standard QR algorithm. Some of the eigenvalues of $\mathcal{L}_n$ will converge towards some of those of $\mathcal{L}$, and associated eigenvectors of $\mathcal{L}_n$ can be transformed into those of $\mathcal{L}$ using Q_n. The subtleties in assuring the convergence are dealt with in the available packages.[18]

4.3.8.2 The Choice of the Density-Matrix

The choice of the density-matrix is an important issue in DMRG studies. For the reaction-diffusion type models at hand, it turned out that

---

[17] See `ftp://caam.rice.edu` in the directory `pub/people/software/ARPACK`

[18] An alternative is to use a non-symmetric Lanczos method, see [125] for details.

N_l	N_r	usual precision	high precision
6	6	0.0211852795111	0.0211852795111
7	7	0.0173940538620	0.0173940538302
8	8	0.0146003960454	0.0146003961355
9	7	0.0146003960889	0.0146003961161
10	6	0.0146003961644	0.0146003961865
9	7	0.0146003961642	0.0146003961866
8	8	0.0146003962502	0.0146003962379
7	9	0.0146003963897	0.0146003962378
6	10	0.0146003962420	0.0146003962377
7	9	0.0146003961645	0.0146003962378
8	8	0.0146003961420	0.0146003962379
9	9	0.0124729862108	0.0124729862559
10	8	0.0124729861363	0.0124729862333
11	7	0.0124729861847	0.0124729862262
12	6	0.0124729861234	0.0124729862244
11	7	0.0124729861816	0.0124729862251
10	8	0.0124729861765	0.0124729862258
9	9	0.0124729869859	0.0124729861961

Table 4.6 Energy gap $\Gamma = E_1(p, N)$ for the branching-fusing model with $p = 0.8403578$ and with $m = 32$ states retained, as determined by the FSM of the DMRG. N_l and N_r indicate the lengths of the left and right parts, which vary during the application of the finite-system method. 'Usual precision' refers to 14-digit precision arithmetics and 'high precision' to 30 digits. After [125].

$$\rho_i := \frac{1}{2} \operatorname{tr} \left\{ |\psi_i^{(l)}\rangle\langle\psi_i^{(l)}| + |\psi_i^{(r)}\rangle\langle\psi_i^{(r)}| \right\} \tag{4.220}$$

appears to give the best convergence, where $|\psi_i^{(l)}\rangle$ and $|\psi_i^{(r)}\rangle$ denote the (normalised) left and right eigenvectors corresponding to the i-th eigenvalue of the non-symmetric time evolution operator $\mathcal{L}$. It is the optimal choice in the sense that it minimises simultaneously the distance of the trial vectors from the exact right and left eigenstates $|\psi_i^{(l)}\rangle$ and $|\psi_i^{(r)}\rangle$, see [125].[19]

When applied to the first excited state of the branching-fusing model, one finds a convergence for the energy up to the 11th digit. Even the ground-state energy is estimated as $E_0 \simeq 10^{-8}$, very close indeed to the exact value of zero, although the first excited state was used as the target state, see [125] for details. We remark that targeting simultaneously the ground-state and the first excited state may improve the precision of the ground-state energy but leads to a loss of accuracy in the excited state.

In Tables 4.6 and 4.7 the convergence of the non-symmetric DMRG with the FSM included is illustrated. We stress that for critical systems the use of the FSM is necessary in order to achieve data of sufficient accuracy. At least for the cases considered here, an increase in m did not lead to an increase of accuracy. On the other hand, for chains with more than $N \approx 50$ sites the

[19] Still, the choice (4.220) may not be the best in all situations. For example, in the q-symmetric quantum Heisenberg chain a non-symmetric choice for the density-matrix was preferable [380].

N_l	N_r	DMRG	exact
6	6	0.0581163651488	0.0581163651479
7	7	0.0437047728086	0.0437047985324
8	8	0.0340537551881	0.0340538006322
9	7	0.0340537521644	
10	6	0.0340537779114	
9	7	0.0340537779110	
8	8	0.0340538006303	0.0340538006322
7	9	0.0340538006306	
6	10	0.0340538006308	
7	9	0.0340538006312	
8	8	0.0340538006323	0.0340538006322
9	9	0.0272774426890	0.0272773931946
9	9	0.0272773931864	0.0272773931946
9	9	0.0272773931999	0.0272773931946

Table 4.7 Lowest gap $\Gamma = E_2(p, N)$ for the diffusion-annihilation model with $p = 1$ and with $m = 32$ states, found from the FSM. N_l and N_r indicate the lengths of the left and right parts. The last column gives the exact result (4.215). For total length 18, only the three results for equal lengths are given. After [125].

DMRG lost its numerical stability, presumably because of an accumulation of small errors in the diagonalisation processes.

4.3.8.3 Exponents from Finite-Size Scaling

Close to criticality, one expects for the energy gap the following finite-size scaling form

$$\Gamma(p, N) = N^{-\Theta} G\left(|p - p_c| N^{1/\nu_\perp}\right) , \tag{4.221}$$

where $G(x)$ is a scaling function. A finite-size estimate $p_c(N)$ for the critical point, along with an estimate $z(N)$ is found by solving the equation

$$\frac{\ln\left[\Gamma(p, N+2)/\Gamma(p, N)\right]}{\ln[N/(N+2)]} = \frac{\ln\left[\Gamma(p, N)/\Gamma(p, N-2)\right]}{\ln[(N-2)/N]} =: z(N) . \tag{4.222}$$

In addition, density profiles can be found. If $\hat{n}(\ell)$ is the density operator at position ℓ, the density profile is naturally calculated from the ground states $|s\rangle = |\psi_0^{(r)}\rangle$ and $\langle s| = \langle \psi_0^{(l)}|$, normalised as $\langle s|s\rangle = 1$, by

$$n(\ell) = \langle \psi_0^{(l)} | \hat{n}(\ell) | \psi_0^{(r)} \rangle = \ell^{-\beta/\nu_\perp} F_1(\ell/N) , \tag{4.223}$$

where F_1 is a scaling function. The quantity $n(\ell)$ will be non-vanishing only if the boundary reaction $\emptyset \to A$ is included with rate p' at the two sites at the edges. Of course, one must then analyse the behaviour of the function in p' (see below). According to (4.223) the quantity $N^{\beta/\nu_\perp} n(\ell)$ depends on ℓ only through the scaling variable ℓ/N, see the left panel in Fig. 4.21.

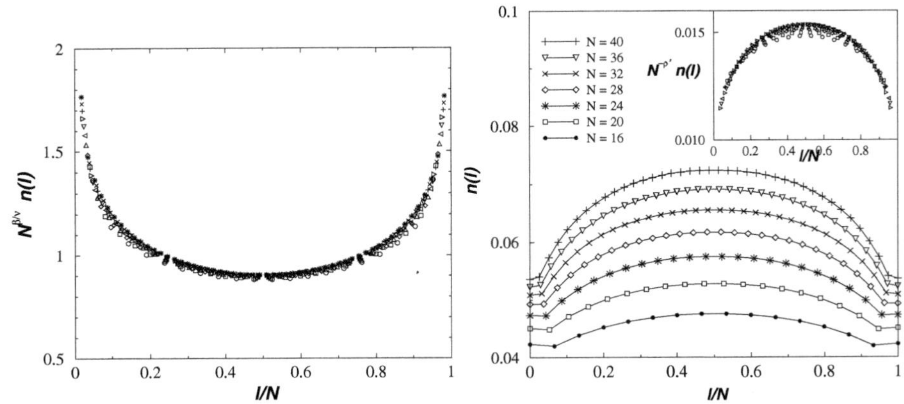

Fig. 4.21 Finite-size scaling of the particle-density profile in the critical branching-fusing model with $p = 0.84036$ and $p' = 0.3$ (left panel) and $p' = 0.002$ (right panel). The inset shows the collapse of the scaled densities onto a single curve as expected from (4.225)), with $\rho' = (\beta_1 - \beta)/\nu_\perp$. Reproduced from [125] with kind permission of Springer Science and Business Media.

In particular, for the particle density taken in the centre at $\ell = N/2$ one expects the scaling form

$$n(N/2;p) = N^{-\beta/\nu_\perp} F_2 \left(|p - p_c| N^{1/\nu_\perp} \right) \tag{4.224}$$

and one can again obtain estimates for the critical point $p_c(N)$ and the exponent $\beta/\nu_\perp$.

The boundary rate p' leads, below the upper critical dimension, to non-monotonic profiles. This counter-intuitive effect is analogous to short-time critical dynamics, giving rise to the **slip exponent** [343] (see also Sect. 4.2.4). In analysing the role of p' for the scaling behaviour, it can be shown that [125]

$$n(\ell, p') \;=\; \ell^{-\beta/\nu_\perp} F_3(\ell/N, p'\ell^{x_1}) \;\sim\; \ell^{(\beta_1-\beta)/\nu_\perp} F_4(\ell/N)\,, \tag{4.225}$$

where the last relation is expected to hold for $p' \to 0$. Here the scaling function F_3 should first satisfy $\lim_{\lambda\to\infty} F_3(\ell/N, \lambda) = F_1(\ell/N)$ and second $F_3(\ell/N, \lambda) \sim \lambda$ in analogy to what is known [549, 550] in equilibrium surface critical phenomena where furthermore $x_1 = \beta_1/\nu_\perp$ is related to the surface exponent β_1 of the order parameter.[20] The change in the finite-size scaling behaviour as a function of p' is illustrated in Fig. 4.21.

Finally, profiles may also be found from matrix elements which allow one to avoid the cumbersome double limit $p' \to 0$ followed by $N \to \infty$ altogether. If $|\psi_1^{(r)}\rangle$ and $\langle\psi_1^{(l)}|$ are the first excited eigenstates of $\mathcal{L}$, consider

[20] Surface critical phenomena of DP, where an analogous exponent β_1 appears, will be discussed below in Sect. 4.4.1

	from Γ			from $n(\ell)$		from $\mathfrak{n}(\ell)$	
N	$p_c(N)$	$z(N)$	$(z-1/\nu_\perp)(N)$	$p_c(N)$	$\beta/\nu_\perp(N)$	$\beta/\nu_\perp(N)$	$\beta_1/\nu_\perp(N)$
10	0.815486295	0.830071389	0.177917024				
12	0.822241704	0.923515450	0.248868030			0.211498060	0.524156106
14	0.826556808	0.996672190	0.303005003			0.214641534	0.540022433
16	0.829477408	1.055258740	0.345372694	0.844595690	0.174469664	0.217449273	0.552833181
18	0.831547147	1.103159519	0.379339662			0.219928645	0.563377349
20	0.833068754	1.143030157	0.406998719	0.843578941	0.183190533	0.222113618	0.572197559
22	0.834221223	1.176727177	0.430122954			0.224042959	0.579677788
24	0.835115836	1.205580740	0.449620922	0.842813911	0.191175959	0.225753163	0.586096418
26	0.835824726	1.230565614	0.466327849			0.227274607	0.591673398
28	0.836396350	1.252411806	0.480376056	0.842276687	0.197772094	0.228648854	0.596559909
30						0.229893751	0.600879333
32				0.841894149	0.203168874	0.230988495	0.604833883
34						0.231983300	0.608060181
36				0.841617080	0.207584161	0.232890344	0.611490572
40				0.841415905	0.211176020		
∞	0.84036(1)	1.580(1)	0.66(2)	0.8406(3)	0.24(1)	0.249(3)	0.667(2)

Table 4.8 Finite-size estimates of the critical point p_c and of various exponents for the branching-fusing model, obtained from the gap Γ, from the density profile $n(\ell)$, as well as from the matrix element $\mathfrak{n}(\ell)$. The last row shows the $N \to \infty$ limit obtained by the BST algorithm. The numbers in brackets give the estimated uncertainties in the last digit. After [125].

$$\mathfrak{n}(\ell) := \langle \psi_1^{(l)} | \hat{n}(\ell) | \psi_1^{(r)} \rangle \sim \begin{cases} N^{-\beta/\nu_\perp} & ; \text{ if } \ell/N \approx 1/2 \\ N^{-\beta_1/\nu_\perp} & ; \text{ if } \ell/N \approx 0, 1 \end{cases} , \tag{4.226}$$

where one may simply set $p' = 0$ and which again gives access to bulk and surface exponents, depending on whether one measures deep inside the bulk or close to a surface.

The results from these three different methods are compared in Table 4.8. We see that the finite-size data themselves for the exponents still move considerably with the number of sites N, and a precise sequence extrapolation towards $N \to \infty$ is needed. In this kind of application, the so-called **BST algorithm** [96, 280, 270] (see appendix F) has proved to give very reliable results. Because of the free boundary conditions used, the final results are quite far from the finite-N data, but it is very satisfying to see that results obtained from different quantities are in agreement with each other and also with the literature, see also Table 4.5.

4.3.8.4 Light-Cone Transfer Matrix Renormalisation-Group

In order to make the $1D$ contact process suitable for a DMRG-treatment, it is useful to slightly modify the rules, as illustrated in Fig. 4.22. It is also common to parametrise the critical parameter of the contact process by $p := (1+\lambda)^{-1}$. With this parametrisation the dynamics can be recast as follows: If a lattice site i is occupied, the particle is removed with probability p or a particle is created at a randomly chosen neighbouring site of i with probability $1-p$.

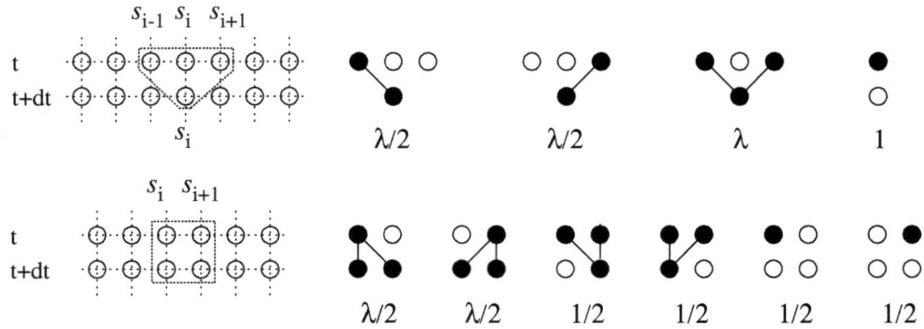

Fig. 4.22 Dynamical rules for the contact process on a $1D$ lattice. Upper line: updating rules for a Monte-Carlo simulation. Lower line: updating rules for the LCTMRG. After [287].

A variant of the DMRG, the so-called **light-cone transfer-matrix density matrix renormalisation group** (LCTMRG) [384], provides an alternative route to the calculation of time-dependent quantities. The algorithm we describe here is an improvement on earlier DMRG approaches [385, 205] and has the following advantages:

- There is no need for random numbers and ensemble averages since *all* relevant ensembles and correlations (in the sense explained below) are taken into account. One LCTMRG run takes a few minutes while 1000 MC runs may take days. The resulting correlation functions are very smooth and no repeated runs, e.g. for improved statistics, are needed.
- The transfer matrix enables us to take the thermodynamic limit $N \to \infty$ *exactly*.

However, the LCTMRG is still plagued by numerical instabilities whose exact origin is unclear, restricting the calculation to about 1000 time steps. The LCTMRG is not very useful for models where each site may have many different states ($n \gg 2$), or where the interaction spans more than two or three sites.

We now give a brief description of the algorithm as devised by Kemper et al. [384] and with the extensions [204] needed to measure two-time quantities, as we will be required to do in Vol. 2 of this book in the studies of ageing. The dynamics of the one-dimensional stochastic process can be mapped by a Trotter-Suzuki checkerboard decomposition onto a two-dimensional classical model: This is the geometric interpretation of DP in two spatial dimensions, directed along one of the two axes.

The checkerboard is made up of plaquettes τ ("local transfer-matrices") encoding the local interaction according to the rules in Fig. 4.22:

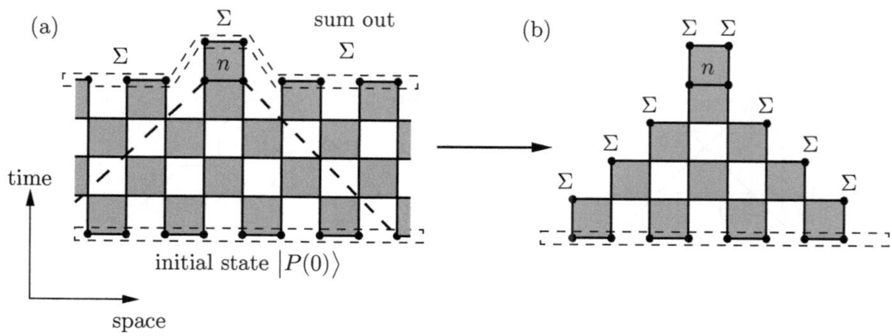

Fig. 4.23 (a) Trotter-Suzuki decomposition of $2\Delta t$ time steps. The resulting $2D$ lattice consists of local plaquette interactions τ and is infinitely extended in the space direction. The dimension of the time direction is finite and the boundary conditions are fixed by $\langle 1|$ and $|P(0)\rangle$. (b) Reduction of the $2D$ lattice to a triangle structure. All other plaquettes trivialise, i.e. do not contribute to the state of the top of the triangle. Reprinted with permission from [384].

$$(\tau)^{l_1 l_2}_{r_1 r_2} = \langle l_2 r_2 | e^{-\Delta t \cdot h} | l_1 r_1 \rangle = \quad \text{with} \quad l_i, r_i \in \{0,1\} \tag{4.227}$$

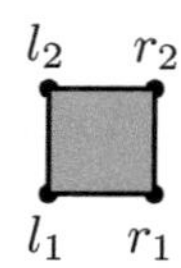

where h is the local transition-rate matrix from two neighbouring sites l_1, r_1 at time t to the same sites at time $t + \Delta t$. The time step $\Delta t \ll 1$ should be chosen sufficiently small.

One determines the thermodynamic properties of the system by a transfer matrix: this ensures that the system is truly infinite in space, while we can follow the short-time dynamics for a certain number of time steps.

Because of probability conservation in (4.228) and causality (at each time step, only a neighbouring site may be affected by the local interaction), the measurement of a local observable $n_i(t)$ at time step t and site i depends only on the 'past light-cone' of this site on the classical $2D$ lattice [205] (see Fig. 4.23):

$$\forall l_1, r_1 : \sum_{l_2 r_2} (\tau)^{l_1 l_2}_{r_1 r_2} = 1, \qquad = 1. \tag{4.228}$$

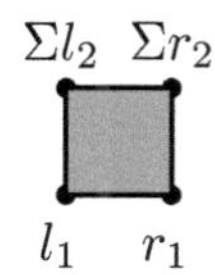

As the dimension of the exact transfer-matrix grows exponentially with the number of time steps, we use the density-matrix renormalisation group (DMRG) idea to decimate the state space by splitting the system into two strongly correlated parts, called the 'system' and 'environment'. Kemper et

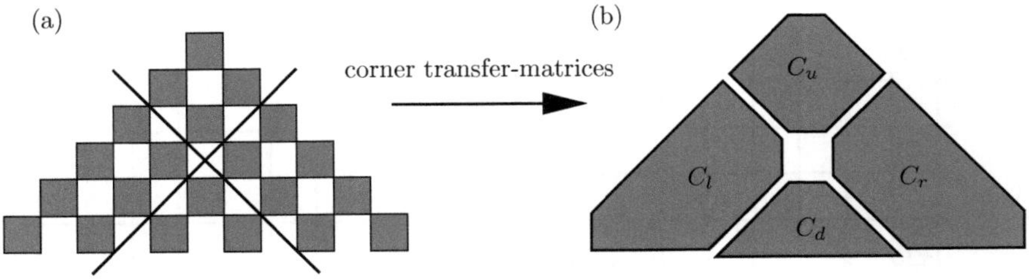

Fig. 4.24 Splitting the light-cone into four corner transfer-matrices by diagonal cuts. Reprinted with permission from [384].

al. [384] have proposed an efficient realisation of the DMRG algorithm applied to corner transfer-matrices. These are obtained by diagonal cuts through the checkerboard: the light cone is split into four parts diagonally along the future and past light cone of the centre point of the triangle (see Fig. 4.24).

A modification of the algorithm is required [204] in order to be able to compute the autocorrelation and autoresponse functions necessary for investigation of the ageing behaviour, which will be discussed in the second volume of this book. Usually for a local observable $n_i(t)$ the expectation value $N(t) := \langle n_i(t)\rangle$ is obtained by multiplying the local transfer-matrix at site i and time t with $n_i(t)$ before applying the initial and final conditions and taking the trace over temporal indices. But for studies on ageing, one is also interested in the two-time autocorrelation $\langle n_i(t)n_i(s)\rangle$ and its connected counterpart $C(t,s)$, along with the autoresponse function

$$C(t,s) = \langle n_i(t)n_i(s)\rangle - \langle n_i(t)\rangle\langle n_i(s)\rangle \ , \ \ R(t,s) = \left.\frac{\delta\langle n_i(t)\rangle}{\delta h_i(s)}\right|_{h=0} \ , \tag{4.229}$$

where the external field h is realised as a rate of spontaneous creation of particles on the site i. In order to find the autocorrelator $\langle n_i(t)n_i(s)\rangle$, the algorithm has been modified to multiply the local transfer matrices τ adjacent to site i with n_i both at time steps s and t before the trace. Then the connected autocorrelation $C(t,s)$ is computed via (4.229), and the derivative of the connected autocorrelation function is computed from a symmetric difference, i.e.

$$\frac{\partial C(t,s)}{\partial s} := \frac{C(t,s+\Delta t/2) - C(t,s-\Delta t/2)}{\Delta t} \tag{4.230}$$

which is sufficiently accurate (i.e. independent of Δt for $\Delta t = 0.01\ldots0.05$).

Likewise, when applying an external field h_i in order to compute $R(t,s)$, the local τ adjacent to site i at time step $t = s$ is modified to include particle production at rate h_i. However, as we are interested in the derivative with respect to the external field, it is better to compute this derivative analytically: the time evolution operator in the presence of an external field h_i on

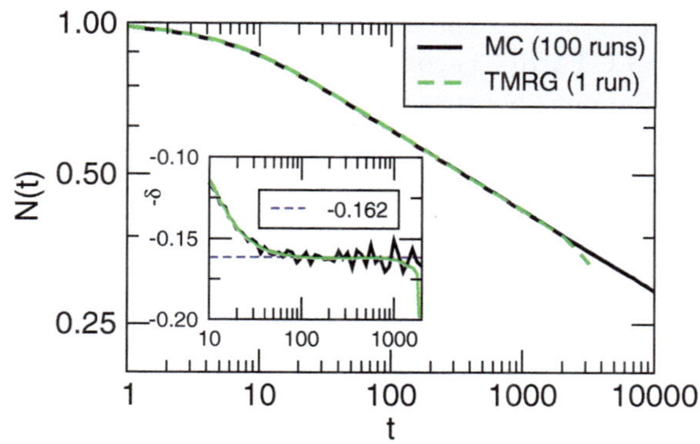

Fig. 4.25 Comparison of the mean particle-density $N(t)$ found by Monte Carlo (MC) and by the LCTMRG for the critical $1D$ contact process. Direct estimates for $-\delta$ are shown in the inset. The LCTMRG becomes numerically unstable around $t = 1000$. Times are measured in units of $\Delta t = 0.02$. Reprinted with permission from [204].

site i is

$$\mathcal{L}_{h_i} = \mathcal{L} + \mathrm{id}_{1,\dots,i-1} \otimes h_i \begin{pmatrix} -1 & 0 \\ 1 & 0 \end{pmatrix}_i \otimes \mathrm{id}_{i+1,\dots,N}$$

where $\mathcal{L}$ is the Liouville operator (or Hamiltonian) for the CP. Then using the state at time $t = s$, $|P(s)\rangle = e^{-\mathcal{L}s}\,|P(t=0)\rangle$, and the final state $\langle s|$, see (3.16)

$$\begin{aligned} R(t,s) &= \lim_{h_i,\Delta t' \to 0} \langle s|\, e^{-\mathcal{L}(t-s-\Delta t')} \left(\frac{e^{-\mathcal{L}_{h_i}\Delta t'} - e^{-\mathcal{L}\Delta t'}}{h_i \Delta t'} \right) |P(s)\rangle \qquad (4.231) \\ &= \lim_{h_i,\Delta t' \to 0} \langle s|\, e^{-\mathcal{L}(t-s-\Delta t')} \left(\frac{(\mathcal{L} - \mathcal{L}_{h_i})\Delta t' + \mathrm{O}((\Delta t')^2)}{h_i \Delta t'} \right) |P(s)\rangle \\ &= \langle s|\, e^{-\mathcal{L}(t-s)} \left(\mathrm{id}_{\dots,i-1} \otimes - \begin{pmatrix} -1 & 0 \\ 1 & 0 \end{pmatrix}_i \otimes \mathrm{id}_{i+1,\dots} \right) |P(s)\rangle \;, \end{aligned}$$

where the matrix is written in terms the local basis $(0, A)$ on site i. This has two advantages:

1. The limit $h_i \to 0$ is taken *exactly*, thus there is no danger of triggering a phase transition by inserting extra particles into the system.
2. No numerical derivative is necessary which would have included the difference of two very similar quantities, so this method is numerically more accurate.

In Fig. 4.25, we compare the results for $N(t) \sim t^{-\delta}$ of the critical contact process in $1D$ obtained from Monte Carlo (MC) or the LCTMRG, respectively. First, we observe that the LCTMRG data are fairly smooth, as is especially evident when studying the exponent δ directly (see inset). However, we also see that the LCTMRG becomes numerically unstable around $t = 1000$ time steps. This happens because the basis vectors of the reduced state space offered during the renormalisation by the LCTMRG method step become inadequate: the expectation value of the identity operator $\langle 1 \rangle$ is around 1 (as it should be) only for the first several hundred time steps, but it then decreases to below 0.1. However, the onset of instability can in practice always be identified very reliably. The reason for this instability is that DMRG works best if system and environment are quite strongly entangled, which is not the case here.

4.4 Other Critical Properties

4.4.1 Surface Critical Behaviour

In a critical system that generates scale-free long range correlations as time proceeds a local defect may cause a non-local response of the order parameter. This applies in particular to boundary conditions, which usually affect the entire critical behaviour of a system. In equilibrium, for example, fixing the spins of a critical semi-infinite Ising model at the boundary, one observes a non-local response of the magnetisation which decays algebraically with the distance from the boundary. In equilibrium statistical physics, the critical behaviour at surfaces induces an additional **surface exponent** β_1 for the order parameter field in the vicinity of the wall [68, 115, 183, 316, 523]. This exponent is generally independent of the usual bulk exponents and describes how deviations of the order parameter caused by the boundary decay in the bulk of the system, see also Sect. 2.3.6.

A similar picture emerges in non-equilibrium statistical physics. However, as non-equilibrium systems involve time as an extra dimension, one has to distinguish between spatial, temporal, and mixed boundary conditions. For example, in non-equilibrium statistical physics a ubiquitous type of *temporal* boundary condition is the initial configuration of a dynamical process. On the other hand, the simplest example of a *spatial* boundary is a static $(d-1)$-dimensional hyperplane, on which we shall focus in the following.

As a $(d-1)$-dimensional hyperplane breaks translational invariance in space the order parameter $\varrho(t, \boldsymbol{r}; \tau)$ will depend on the perpendicular distance $r_\perp = |\boldsymbol{r}_\perp|$ from the hyperplane. Assuming that $r_\perp$ scales like a length and invoking usual scaling arguments the order parameter in a semi-infinite system is expected to obey the scaling form

$$\varrho(t, r_\perp; \tau) \simeq t^{-\alpha}\, \tilde{w}(r_\perp / t^{1/z},\, r_\perp \tau^{\nu_\perp})\,, \tag{4.232}$$

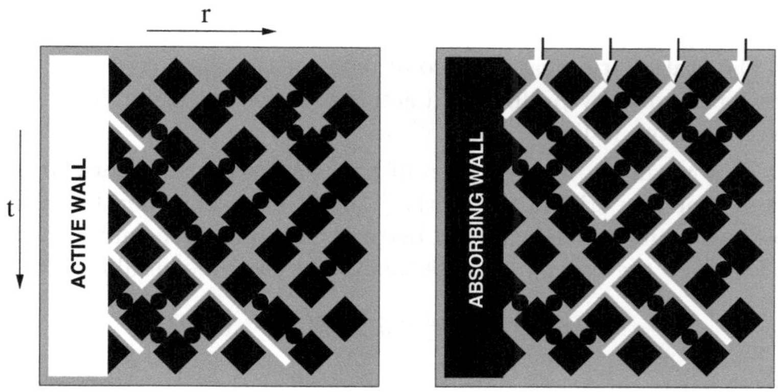

Fig. 4.26 Directed percolation with a wall. The figure shows a configuration of a $(1+1)$-dimensional bond DP process. Left: Initially inactive system with an active wall. Right: Initially active system with an absorbing wall.

where we have suppressed the metric factors for simplicity.

In models with an absorbing phase transition two important types of boundary conditions have been studied (see Fig. 4.26). One possibility is to keep all sites on the hyperplane in the active state. Such a boundary is usually referred to as an **active wall**. The other possibility is to keep the boundary sites inactive, imposing a so-called **absorbing wall**. As we shall see below, these two types of boundary conditions correspond to different scaling functions $\tilde{w}_{\mathrm{act}}$ and $\tilde{w}_{\mathrm{abs}}$.

4.4.1.1 DP with an Active Wall

Let us first consider an active wall (see left panel of Fig. 4.26) and study the corresponding mean-field theory. This will be the directed percolation analogue of the *extraordinary* transition, see Sect. 2.3.6. In the continuum limit an active wall can be considered as a hyperplane at which the particle density diverges. The wall acts as a permanent source of activity, destroys the absorbing state, and leads to a finite response of the order parameter field which becomes stationary in the limit $t \to \infty$.

In order to compute the stationary profile as a function of $r_\perp$ we have to solve the stationary mean-field equation (cf. (3.8))

$$\lambda^{-1}\partial_t \varrho(t,\boldsymbol{r}) \;=\; \tau\varrho(t,\boldsymbol{r}) - g\varrho^2(t,\boldsymbol{r}) + \nabla^2\varrho(t,\boldsymbol{r}) \;=\; 0 \qquad (4.233)$$

together with the boundary condition

$$\varrho(t,\boldsymbol{r})\Big|_{r_\perp=0} = \infty\,, \qquad (4.234)$$

where $r_\perp$ denotes the coordinate perpendicular to the wall. At criticality $\tau = 0$ the stationary mean-field solution is of the form $\varrho_s(\boldsymbol{r}) \sim r_\perp^{-2}$. Thus the stationary density profile of an active wall decays algebraically with the distance from the wall.

Let us now turn to the question of how such a profile is formed over the course of time in a critical and initially inactive system. Clearly, the density in close proximity to the wall will become constant after only a short time. Since for $\tau = 0$ the order parameter obeys the simplified scaling form

$$\varrho(t, r_\perp) \simeq t^{-\alpha}\, \tilde{w}_{\rm act}(r_\perp/t^{1/z}, 0) \tag{4.235}$$

the stationarity of the profile nearby implies that the scaling function $\tilde{w}_{\rm act}(y, 0)$ behaves as $\tilde{w}_{\rm act}(y, 0) \sim y^{-\beta/\nu_\perp}$ for small arguments $y \ll 1$. Hence in the limit $t \to \infty$ the profile becomes stationary and has the form

$$\varrho_s(\boldsymbol{r}) \sim r_\perp^{-\beta/\nu_\perp} \tag{4.236}$$

in agreement with the previous mean-field result. For finite t, however, this power-law profile extends only over a finite range until it crosses over to an exponential decay when $y \approx 1$. This crossover takes place at a distance $r_\perp \sim t^{1/z}$ and moves away from the wall as time advances.

The left panel of Fig. 4.27 shows snapshots of density profiles taken at certain times in a critical (1+1)-dimensional contact process with a permanently active site at the origin starting with an empty lattice. Plotting $\varrho(t, r_\perp)\, t^\alpha$ as a function of $r_\perp/t^{1/z}$ one obtains an excellent data collapse, confirming the scaling form (4.235) as well as the expected crossover from an initial algebraic decrease $\tilde{w}_{\rm act}(r, 0) \sim r^{-\beta/\nu_\perp}$ to an exponential decay.

Including the control parameter τ the above scaling form allows one to examine the off-critical behaviour at the surface. For example, in the active phase the profile becomes stationary within the correlation time $\xi_\parallel \sim \tau^{-\nu_\parallel}$ and the influence of the wall extends over a range proportional to the correlation length $\xi_\perp \sim \tau^{-\nu_\perp}$.

Note that the surface critical behaviour near an active wall involves only the usual bulk exponents, i.e., in DP an active wall does not induce an independent surface exponent.

4.4.1.2 DP with an Absorbing Wall

In the case of an *absorbing* wall the situation is different (see [337, 212, 409, 408, 228] as well as [307, 229] for recent reviews). Here the order parameter is set to zero at the wall and the system still has an absorbing state. This is analogous to the *ordinary* transition of equilibrium systems described in Sect. 2.3.6. In the active phase of a semi-infinite system, we therefore expect

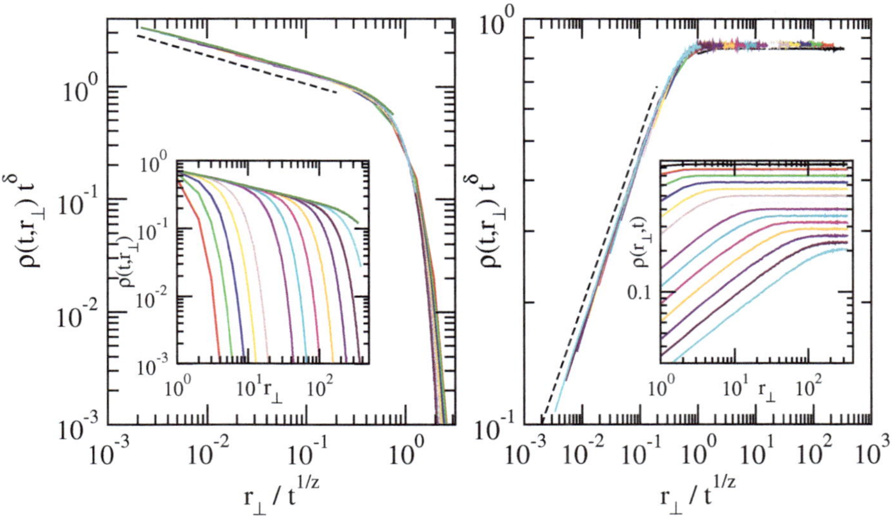

Fig. 4.27 Directed percolation with a wall. The left (right) panel shows a data collapse of the averaged density profiles near an active (absorbing) wall measured in a critical (1+1)-dimensional contact process at $t = 1, 2, 4, \ldots, 16384$. The dashed lines indicate the expected slopes $-\beta/\nu_\perp$ and $(\beta_1 - \beta)/\nu_\perp$, respectively. The insets display the corresponding raw data used for the data collapses.

to find the usual stationary density ϱ_s in the bulk and a **depletion zone** next to the wall whose thickness is related to the correlation length $\xi_\perp \sim \tau^{-\nu_\perp}$.

A suitable order parameter that characterises the depth of the depletion zone is the average activity ϱ_s of lattice sites adjacent to the wall. In numerical simulations this surface activity is found to scale as

$$\varrho_1 \sim \tau^{\beta_1} \tag{4.237}$$

with a new exponent β_1. This surface critical exponent is larger than the bulk exponent β and takes the values [354, 229]

$$\beta_1 = \begin{cases} 0.73371(2) & \text{in } d = 1\,, \\ 1.07(5) & \text{in } d = 2\,. \end{cases} \tag{4.238}$$

Again it is instructive to study the mean-field limit. In the continuum formulation the surface density $\varrho_s(t)$ may be expressed as the average activity

$$\varrho_1(t) = \varrho(t, \boldsymbol{r})\Big|_{r_\perp = a} \tag{4.239}$$

measured at a fixed microscopically small distance away from the wall which may be interpreted as the lattice spacing a. In order to determine $\beta_{\mathrm{MF},1}$ within a mean-field approximation, we now have to solve (3.8) with the Dirichlet boundary condition

$$\varrho(t,\boldsymbol{r})\Big|_{r_\perp=0} = 0\,, \tag{4.240}$$

i.e., the order parameter vanishes along the hyperplane. Next to the wall, where the density is small and the non-linear term of the mean-field equation can be neglected, $\varrho(t,\boldsymbol{r})$ has to increase linearly with $r_\perp$. Therefore, the surface density is proportional to the first derivative of the density profile

$$\varrho_1(t) \sim \frac{\partial}{\partial r_\perp}\varrho(t,\boldsymbol{r})\Big|_{r_\perp=0} . \tag{4.241}$$

Multiplying the one-dimensional stationary mean-field equation

$$\tau\varrho_{\rm s}(r_\perp) - g\varrho_{\rm s}^2(r_\perp) + \varrho_{\rm s}''(r_\perp) = 0 \tag{4.242}$$

by $\varrho_{\rm s}'(r_\perp)$ and integrating over $r_\perp$ one obtains the equation

$$\frac{1}{2}\tau\varrho_{\rm s}^2(r_\perp) - \frac{1}{3}g\varrho_{\rm s}^3(r_\perp) + \left[\varrho_{\rm s}'(r_\perp)\right]^2 = C\,, \tag{4.243}$$

where C is an integration constant. Since $\varrho_{\rm s} = \tau/g$ is the stationary solution in the bulk, this constant takes the value $C = \tau^3/6g^2$. Hence the derivative of the order parameter at the wall (and therewith the surface density) is given by

$$\varrho_1 \sim \varrho_{\rm s}'(0) = \sqrt{\frac{\tau^3}{3g^2}} \tag{4.244}$$

and thus scales as $\tau^{3/2}$, i.e., the surface exponent in the mean-field limit is

$$\beta_{\rm MF,1} = 3/2\,. \tag{4.245}$$

This value has to be compared with the bulk exponent $\beta_{\rm MF} = 1$, confirming the inequality $\beta_{\rm MF,1} > \beta_{\rm MF}$ in the mean-field case.

To describe the scaling properties of an absorbing wall in the general case, we can again use the scaling form (4.235), although with a different scaling function $\tilde{w}_{\rm abs}$. For example, starting with a fully occupied lattice at criticality the growing depletion zone is described by the scaling form

$$\varrho(t,r_\perp) \simeq t^{-\alpha}\,\tilde{w}_{\rm abs}(r_\perp/t^{1/z},0)\,. \tag{4.246}$$

This scaling function differs from the one considered in (4.235) and has the following asymptotic properties. For large arguments, i.e., deep in the bulk, the process does not feel the influence of the absorbing wall so the density of active sites decays as $t^{-\alpha}$, hence $\tilde{w}_{\rm abs}(r,0) \simeq$ const for $r \gg 1$. Similarly, in the vicinity of the wall the order parameter is expected to decay as $\varrho_1(t) \sim t^{-\alpha_1}$, where $\alpha_1 = \beta_1/\nu_\parallel$, meaning that the scaling function has to increase as $\tilde{w}_{\rm abs}(r,0) = r^{(\beta_1-\beta)/\nu_\perp}$ for $r \ll 1$. As demonstrated in the right panel

	$d=1$ [354]	$d=2$ [229]	mean-field	one-loop [228]
$\beta=\beta'$	0.276486(8)	0.583(3)	1	$1-\epsilon/6$
$\nu_\perp$	1.096854(4)	0.733(8)	1/2	$1/2+\epsilon/16$
$\nu_\parallel$	1.733847(6)	1.295(6)	1	$1+\epsilon/12$
$\beta_1=\beta_1'$	0.73371(2)	1.07(5)	3/2	$3/2-7\epsilon/48$

Table 4.9 Bulk and surface exponents of directed percolation with an absorbing wall (ordinary transition).

of Fig. 4.27 for DP in 1+1 dimensions, this is indeed the case. A similar observation can also be made within the context of finite-size scaling, as was already shown in Fig. 4.21.

In early papers, the numerical value of the surface exponent $\beta_1 \approx 0.734$ compared to the DP bulk exponent $\nu_\parallel \approx 1.734$ in 1 + 1-dimensions led to speculation about a possible scaling relation $\nu_\parallel \stackrel{?}{=} 1+\beta_1$. However, more recent estimates (see Table 4.9) rather indicate that this relation does not hold [354]. In fact, by dimensional analysis it seems to be unlikely that β_1 and $\nu_\parallel$ are related by a simple linear scaling relation. Moreover, in 2+1 dimensions, the numerical value $\beta_1 = 1.07(5)$ cannot be simply related to the other exponents. Similarly, a field-theoretic one-loop approximation [228]

$$\beta_1 = \frac{3}{2} - \frac{7}{48}\epsilon + \mathrm{O}(\epsilon^2) \qquad (\epsilon = 4-d) \tag{4.247}$$

indicates that the surface exponent is generally independent of the other exponents. See Table 4.5 for a comparison of the results found for $\beta_1/\nu_\perp$, $\beta/\nu_\perp$, $\nu_\perp$ and z from different numerical techniques.

As discussed in Sect. 4.1.1, other universality classes of absorbing phase transitions are generally described in terms of *two* independent bulk exponents β and β'. Similarly, the scaling theory of surface critical behaviour next to an absorbing wall requires us to introduce two surface exponents β_1 and β_1' which may be different. In DP, however, the rapidity-reversal symmetry (which is still valid in presence of an absorbing wall) forces them to be equal.

4.4.1.3 Critical Behaviour at Edges

In $d \geq 2$ dimensions the intersection of two $d-1$-dimensional absorbing walls defines an **absorbing edge** with opening angle ψ. Near the edge the order parameter is known to scale with an exponent $\beta_{\mathrm{ed}}(\psi)$ which varies continuously with ψ (see [228, 229] for details). For $\psi = \pi$ the two walls coincide, forming a single absorbing wall, hence $\beta_{\mathrm{ed}} = \beta_1$. Other numerical values are shown in Table 4.10.

Interestingly, in the limit $\psi \to 2\pi$ the bulk exponent $\beta_{\mathrm{MF,ed}} = \beta$ is recovered. This case corresponds to a semi-infinite absorbing half-plane. For example, in 2+1 dimensions such a boundary would be an absorbing semi-infinite line. Although the critical behaviour is changed in the vicinity of the line, it is apparently not affected at the tip.

4.4.1.4 Moving Absorbing Wall

The simplest type of a time-dependent boundary condition is a **moving absorbing wall.** As an example let us consider an absorbing hyperplane in a DP process that moves away with constant velocity $v = \frac{\mathrm{d}}{\mathrm{d}t} r_\perp$, where $r_\perp$ again denotes the coordinate perpendicular to the wall (see Fig. 4.28). Remarkably, in such a system one observes a continuous surface transition in the active phase, where the bulk-process is off-critical.

The transition is caused by two competing velocities, namely, the bulk propagation velocity $v_0 \sim \xi_\perp/\xi_\parallel$ of the DP process in the active phase, and the velocity v of the moving wall. Obviously, a non-zero density next to the surface can only be maintained if the DP process expands faster than the wall moves away, i.e., $v > v_0$. As shown in Fig. 4.28, the stationary surface density $\varrho_{s,1}$ measured at the co-moving boundary sites decreases continuously and eventually vanishes when $v = v_0$. Although the numerical results suggest a linear decrease, a scaling theory for this transition is not yet known. Even the mean-field equation subjected to a linearly moving Dirichlet boundary condition does not provide a simple scaling solution.

4.4.1.5 DP Confined by an Absorbing Parabola

As a critical DP cluster is known to spread in a typical region $r \sim t^{1/z}$ it is interesting to study the influence of an absorbing boundary condition in the form of a generalised parabola $r = \pm ct^{\psi}$, where ψ is a control exponent. This problem was investigated by Kaiser and Turban in [371, 372]. As expected, it turns out that the boundary is relevant for $\psi > 1/z$ and irrelevant if $\psi < 1/z$. Since under scaling transformations (4.8) and (4.13) the amplitude c scales as $c \mapsto \lambda^{\nu_\parallel \psi - \nu_\perp} c$, the impact of such a boundary can be taken into account by adding the scale-invariant argument $c/t^{\psi - 1/z}$ to the parameters of the corresponding scaling functions. For example, the survival probability of DP confined by a generalised parabola scales as

ψ	$\pi/2$	$3\pi/4$	π	$5\pi/4$	2π (bulk)
$\beta_{\mathrm{ed}}(\psi)$	1.6(1)	1.23(7)	1.07(5)	0.98(5)	0.583(4)
$\beta_{\mathrm{MF,ed}}(\psi)$	2	5/3	3/2	7/5	1

Table 4.10 Edge exponent β_{ed} in DP for various opening angles ψ in 2+1 dimensions and the corresponding mean-field values. Values taken from [229].

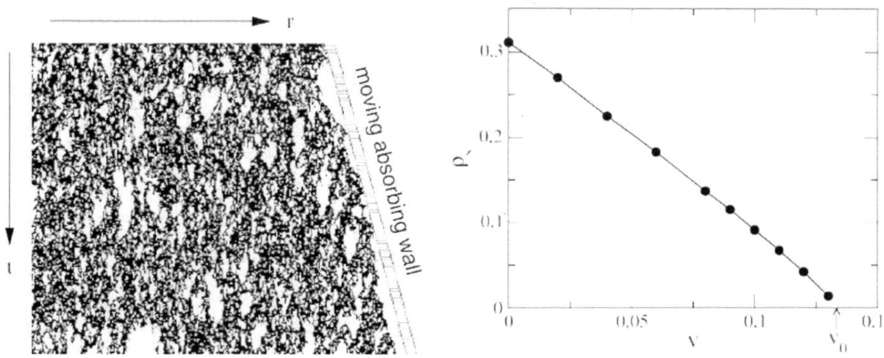

Fig. 4.28 Directed percolation with a moving absorbing wall. A contact process in the active phase $\lambda > \lambda_c$ expands in space, bounded by an absorbing wall that moves at constant velocity $v = \mathrm{d}r/\mathrm{d}t$. In the vicinity of the wall the surface density ϱ_s decreases with increasing v and finally vanishes when reaching the bulk velocity v_0.

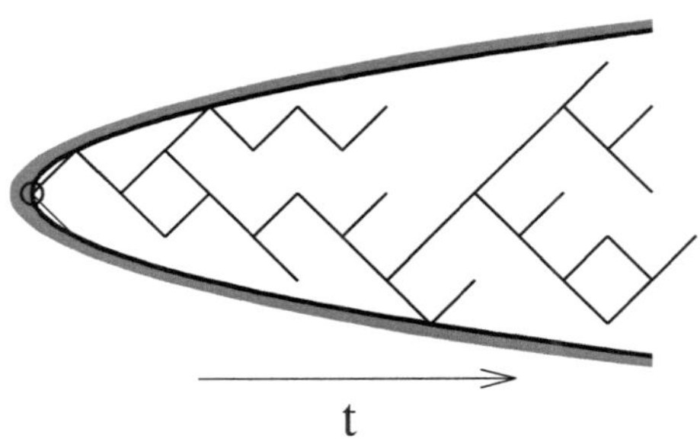

Fig. 4.29 Illustration of bond directed percolation confined to a generalised parabola.

$$P_{\text{sur}}(t) \simeq t^{-\delta}\, \tilde{p}_{\text{seed}}(\tau t^{1/\nu_\|},\, c/t^{\psi-1/z}) \,. \tag{4.248}$$

Instead of going further into the details we refer the interested reader to the reviews [307, 229].

4.4.2 Persistence Exponents

Having looked at length into the behaviour of densities and correlators, we now consider a completely different class of observables, related to the the study of so-called **first-passage problems** [543]. Rather than looking at averages, one rather looks for the probability that a certain event happens at time t for the first time. For example, one may ask for the time at which the magnetisation of an Ising ferromagnet first changes its sign. More generally, one may ask for the probability that a certain observable in a random process did *not* cross its expectation value until time t. At a critical point, the study of such quantities may shed a different light on their fluctuation properties. Such **persistence probabilities** have a surprisingly rich behaviour.

We shall first briefly state some results from equilibrium critical systems before considering the persistence properties of absorbing phase transition, again for the paradigmatic case of directed percolation.

4.4.2.1 Global and Local Persistence Probabilities

There are mainly two types of persistence studies. For *global* observables such as the total magnetisation or the average particle density, one studies the **global persistence probability** $P_g(t)$ that an order parameter does not cross its expectation value up to time t [454, 413, 474, 562, 193, 499, 455, 567]. In *critical* dynamical systems, it is observed that the global persistence probability decays algebraically as $P_g(t) \sim t^{-\Theta_g}$. For Markovian dynamics one can show that the **global persistence exponent** can be expressed by the scaling relation $\Theta_g z = \lambda_C - d + 1 + \eta/2$, where η and λ_C are the static and the **autocorrelation exponent**,[21] respectively.

The other category of persistence studies, on which we shall focus here, deals with *local* observables such as the occupation number or spin orientation at individual sites. Here one is interested in the **local persistence probability** $p_\ell(t, \boldsymbol{r})$ that the local state at a given position $\boldsymbol{r}$ in space has not changed up to time t. For example, in spin models $p_\ell(t, \boldsymbol{r})$ may be defined as the probability that a spin at a given site does not flip until time t. If the system is translationally invariant, the persistence probability does not depend on $\boldsymbol{r}$ so that $p_\ell(t)$ is just the fraction of spins that did not flip until time t.

In many systems evolving towards a scale-free state the local persistence probability is found to decay algebraically as

$$p_\ell(t) \sim t^{-\Theta_\ell}\,, \tag{4.249}$$

where Θ_ℓ is the so-called **local persistence exponent** [81, 158, 121, 160, 161, 162, 157, 308, 295, 555, 143, 567, 200]. This exponent is generally different from Θ_g and seems to have certain universal features, i.e., its numerical value may coincide in various models belonging to the same universality class. However, in contrast to Θ_g, which can be expressed in terms of the bulk exponents by a scaling relation, Θ_ℓ is believed to be an independent exponent. One of the most important exact results was obtained by Derrida, Hakim and Pasquier [160, 161], who were able to compute the persistence exponent

---

[21] In the context of ageing phenomena, we shall see in Vol. 2 that the exponent λ_C describes the decay of two-time autocorrelation functions.

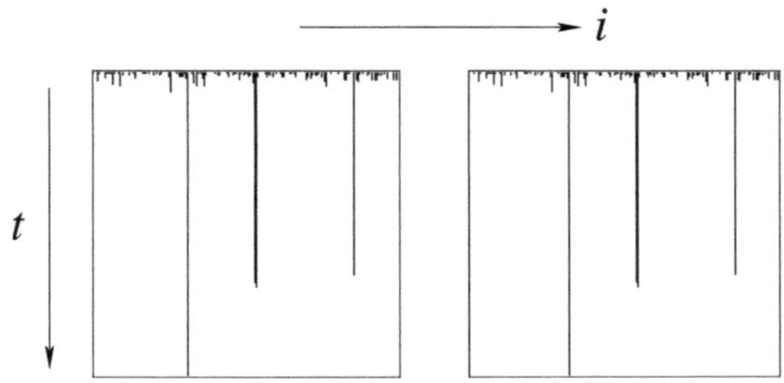

Fig. 4.30 Local persistence in DP. The figure shows a (1+1)-dimensional critical DP process starting with a fully occupied lattice. The persistent sites are marked by black vertical lines. The right panel shows the persistent sites of the same run without the active sites.

$$\Theta_\ell(q) = -\frac{1}{8} + \frac{2}{\pi^2}\left[\arccos\left(\frac{2-q}{\sqrt{2}\,q}\right)\right]^2 \tag{4.250}$$

in the one-dimensional zero-temperature Ising and Potts-q models with heat-bath dynamics, see appendix G. This exact solution maps the Glauber-Ising dynamics to a time-reversed coagulation process with a special boundary condition and relates Θ_ℓ to a universal amplitude close to the boundary.

4.4.2.2 Local Persistence in Directed Percolation

In models with absorbing states, local persistence may be defined as the probability that a given site – monitored from a moment shortly after the initial state up to time t – remains locally absorbing (inactive). For example, in a DP process starting with a fully occupied infinite lattice, $p_\ell(t)$ would be the density of sites which have been never activated up to time t, excluding the fully occupied initial state at $t = 0$. A typical run of a critical DP process in 1+1 dimensions is shown in Fig. 4.30, where locally persistent sites are represented as vertical lines.

In [295] the local persistence exponent of (1+1)-dimensional DP was determined by simulating a Domany-Kinzel cellular automaton, which led to the estimate $\Theta_\ell = 1.50(2)$. This value suggested that Θ_ℓ is probably unrelated to the other bulk exponents of DP, $\beta, \nu_\parallel$ and $\nu_\perp$. Moreover, it was speculated whether Θ_ℓ is exactly equal to the rational value 3/2 or not. However, extensive Monte-Carlo simulations of the (1+1)-dimensional contact process shown in Fig. 4.31 led to the improved estimate [230]

$$\Theta_\ell = 1.512(6) \tag{4.251}$$

which appears to be inconsistent with the possibility of a rational value 3/2.

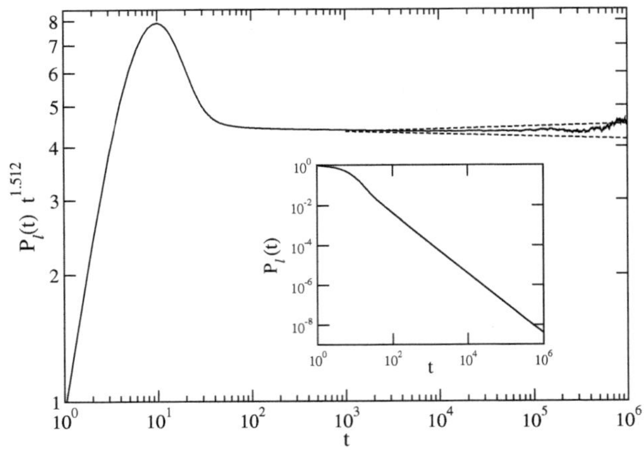

Fig. 4.31 High-precision measurement of local persistence in the 1d contact process. The plot shows $p_\ell(t)$ multiplied by t^{Θ_ℓ} with $\Theta_\ell = 1.512$. The dotted lines indicate the error margin of the estimated exponent ± 0.006. The inset shows the raw data produced by simulation.

The measurement of the local persistence exponent in higher dimensions is extremely difficult because of strong finite-size effects and very long transients. So far no conclusive results exist.

4.4.2.3 Off-Critical and Finite-Size Scaling of the Persistence Probability

If the local persistence probability is a well-behaved order parameter, it should exhibit similar scaling laws as e.g. the density of active sites. This

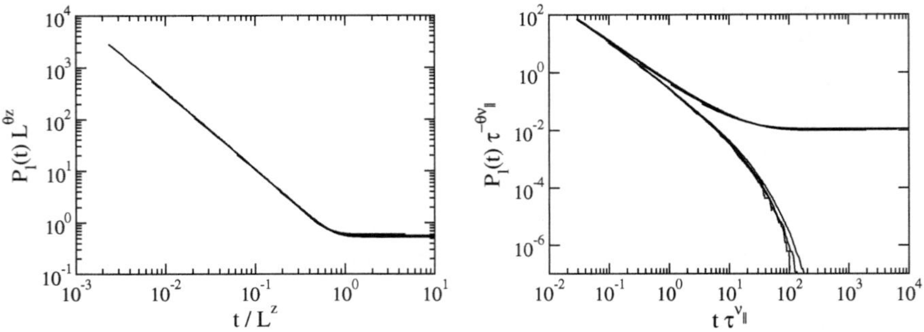

Fig. 4.32 Left: Local persistence probability in finite-size simulations of a (1+1)-dimensional contact process with $L = 8, 16, 32, 64, 128$ sites, plotted as a data collapse according to the scaling form (4.252). Right: Corresponding graph for off-critical simulations with a distance from criticality $\tau = \lambda - \lambda_c = 0.0512,\ 0.1024,\ \ldots, 0.8192$.

means that one expects a scaling law of the form

$$p_\ell(t) = t^{-\Theta_\ell}\, \tilde{F}\left(\frac{t}{L^z}, \frac{t}{\tau^{-\nu_\parallel}}\right), \tag{4.252}$$

where $\tilde{F}$ is a scaling function. In order to demonstrate the validity of this scaling form numerically, we present in Fig. 4.32 numerical data for two special cases:

(a) *Finite-size scaling:* A finite (1+1)-dimensional contact process at criticality reaches the absorbing state within finite time such that there is a finite probability of persistent sites surviving forever. Therefore, the persistence probability saturates at a constant value. Hence, plotting $p_\ell(1,t)L^{\Theta_\ell z}$ over against t/L^z all curves collapse onto a single one, as demonstrated in Fig. 4.32.

(b) *Off-critical simulations:* For an infinite (1+1)-dimensional contact process below (above) the percolation threshold, we expect the persistence probability to saturate (decay exponentially). According to the scaling form (4.252) the curves should collapse onto two different curves for either case when $p_\ell(1,t)\tau^{-\Theta_\ell \nu_\parallel}$ is plotted over against $t\tau^{\nu_\parallel}$. As shown in the right panel of Fig. 4.32, this is indeed the case.

4.4.2.4 Local Persistence as a Quantity Depending on Two Time Parameters

In models with absorbing states, local persistence is usually introduced as the probability that a given site is not activated until time t, excluding the initial configuration at $t = 0$ where all sites are active. This definition suggests that the persistence probability $p_\ell(t)$ depends only on a *single* time parameter t. However, with only one parameter t, it is impossible to define a meaningful continuum limit of the persistence probability. For example, a fully occupied initial state in the lattice model translates into an infinite δ-peak-like density of active sites at $t = 0$ in the continuum description so that the probability for a given site to remain inactive in the interval $0 < \tau \leq t$ would be zero. Therefore, the only way to introduce persistence consistently is to define p_ℓ as the probability that a given site remains inactive in some interval $[t_0, t]$. This means that local persistence should actually be defined as a quantity $p_\ell(t, t_0)$ that depends on *two* time parameters t_0 and t.

The inset of Fig. 4.33 shows the temporal decay of the persistence probability $p_\ell(t_0, t)$ in a critical (1+1)-dimensional contact process as a function of t for various values of t_0. As can be seen, by increasing t_0 the initial transient becomes more pronounced. Moreover, the curves are shifted to the right. However, the asymptotic slope, i.e., the persistence exponent Θ_ℓ, is not affected.

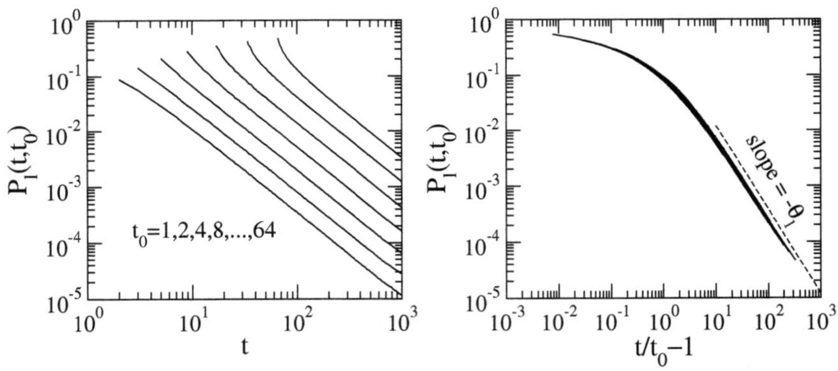

Fig. 4.33 Measurement of the local persistence probability $p_\ell(t, t_0)$ in a (1+1)-dimensional critical contact process as a function of two variables t and t_0. The inset shows the original data while the main graph demonstrates a data collapse according to (4.253).

For an infinite system at criticality, the usual scaling arguments imply the scaling form $p_\ell(t, t_0) = t^{-\kappa/\nu_\parallel} \tilde{F}(t/t_0)$ with an unknown exponent κ. This exponent can be determined by keeping the difference $\Delta t = t - t_0$ fixed and taking t to infinity. In this limit the critical system approaches the absorbing state and therefore the probability $p_\ell(t, t_0)$ of finding *no* activity between t_0 and $t_0 + \Delta t$ tends to 1. This implies that $\kappa = 0$, i.e., we arrive at the scaling form

$$p_\ell(t, t_0) = \tilde{F}(t/t_0)\,, \tag{4.253}$$

where $\tilde{F}$ is a scaling function with the asymptotic behaviour

$$\begin{aligned} \tilde{F}(z) &\sim z^{-\Theta_\ell} && \text{for } z \gg 1, \\ \tilde{F}(z) &\to 1 && \text{for } z \to 1. \end{aligned} \tag{4.254}$$

In order to verify the scaling form (4.253) and its asymptotic behaviour (4.254) we simulated a (1+1)-dimensional contact process and measured the probability that a given site is not activated in the time interval between t_0 and t. The left graph in Fig. 4.33 shows the raw data for fixed t_0 as a function of t. The graph on the right side demonstrates the corresponding data collapse according to the scaling form (4.253), plotting $p_\ell(t, t_0)$ versus the quotient $t/t_0 - 1$. As can be seen, one obtains a convincing data collapse. However, the collapse is not entirely perfect. Possible reasons for the observed deviations are discussed in [230].

Problems

34. Show that the Janssen-de Dominicis functional of directed percolation (4.180) is invariant under the rapidity-reversal (4.181) at zero field.

35. Consider a single species A of particles moving diffusively on a lattice and undergoing pair annihilation $2A \to \emptyset$ with a rate λ which occurs when two particles meet each other. Using a stochastic Langevin equation for the coarse-grained space-time-dependent particle-density $\varrho(t, \boldsymbol{r})$, after averaging over the noise one writes down the following reaction-diffusion equation

$$\partial_t \varrho = D \Delta \varrho - \lambda \varrho^2 \tag{4.255}$$

where Δ is the spatial Laplacian and D, λ are positive constants. It is well-known that for spatially homogeneous initial conditions, the density decays algebraically as $\varrho(t) = \varrho_0/(1 + \varrho_0 \lambda t) \simeq (\lambda t)^{-1}$ for $t \to \infty$.

Derive (or bound), for *any* initial distribution of particles, the time-dependence of the average particle-density $\bar{\varrho}(t) := |\Omega|^{-1} \int_\Omega \mathrm{d}\boldsymbol{r}\, \varrho(t, \boldsymbol{r})$ in some spatial domain Ω of volume $|\Omega|$. Consider the cases of (i) Dirichlet boundary conditions $\varrho|_{\partial\Omega} = 0$ and (ii) Neumann boundary conditions $\boldsymbol{n} \cdot \boldsymbol{\nabla} \varrho|_{\partial\Omega} = 0$, where $\partial\Omega$ denotes the boundary of $\Omega \subset \mathbb{R}^d$ and $\boldsymbol{n}$ is a unit vector orthogonal to $\partial\Omega$. Show in particular that for large volumes $|\Omega| \to \infty$ and large times $t \to \infty$, one has $\bar{\varrho}(t) \sim (\lambda t)^{-1}$ for both boundary conditions.

A useful technical tool are the **Poincaré inequalities**. For Dirichlet and Neumann boundary conditions, they read (see [575] for an elementary proof)

$$\begin{aligned}
&\int_\Omega \mathrm{d}\boldsymbol{r}\, |\boldsymbol{\nabla} u|^2 \geq C^{-1} |\Omega|^{-2/d} \int_\Omega \mathrm{d}\boldsymbol{r}\, |u|^2 \quad ; \text{ Dirichlet } u|_{\partial\Omega} = 0 \qquad (4.256)\\
&\int_\Omega \mathrm{d}\boldsymbol{r}\, |\boldsymbol{\nabla} u|^2 \geq C^{-1} |\Omega|^{-2/d} \int_\Omega \mathrm{d}\boldsymbol{r}\, |u - \bar{u}|^2 \quad ; \text{ Neumann } \boldsymbol{n} \cdot \boldsymbol{\nabla} u|_{\partial\Omega} = 0
\end{aligned}$$

where C is some constant independent of $|\Omega|$ (but *a priori* dependent on the boundary conditions) and $\bar{u} = |\Omega|^{-1} \int_\Omega \mathrm{d}\boldsymbol{r}\, |u|$.

36. Consider the diffusion-coagulation model of a single species A with the rates $A + \emptyset \overset{D}{\leftrightarrow} \emptyset + A$ and $A + A \overset{\gamma}{\longrightarrow} A + \emptyset, \emptyset + A$ and consider the special case when $\gamma = D$ (what is the physical interpretation of this condition?) on a one-dimensional lattice with $L + 1$ sites which contains at least one particle. Consider the empty-interval probability $E_n(t)$ that n consecutive sites are empty. Show that the $E_n(t)$ satisfy a closed system of equations of motion [53]. In the continuum limit, $E_n(t) \to E(x, t)$ which satisfies

$$\frac{\partial E}{\partial t} = 2D \frac{\partial^2 E}{\partial x^2} + 2\beta \frac{\partial E}{\partial x} \tag{4.257}$$

together with the boundary conditions

material	y	reaction(s)	Reference
$C_{10}H_8$	$0.52 - 0.59$	$\bullet\bullet \to \begin{cases} \circ\circ \\ \bullet\circ \end{cases}$	[530]
P1VN/PMMA film	0.47(3)	$\bullet\bullet \to \begin{cases} \circ\circ \\ \bullet\circ \end{cases}$	[397]
TMMC	0.48(4)	$\bullet\bullet \to \bullet\circ$	[401]

Table 4.11 Measured decay exponent y of the mean exciton density $\bar{\varrho}(t) \sim t^{-y}$ on polymer chains. The error bar for TMMC comes from averaging over the results of [401] for different initial particle densities.

$$E(0,t) = 1 \ , \ E(\infty,t) = 0. \tag{4.258}$$

How does one obtain the mean particle-density $\bar{\varrho}(t)$?

37. Find from the empty-interval (4.257) together with the boundary condition (4.258), for the special case $\beta = 0$ of a pure diffusion-coagulation model, the long-time behaviour by making a scaling ansatz $E(x,t) = f(x t^{-1/z})$ where the dynamical exponent z is to be determined. Derive in particular the long-time behaviour of the mean particle-density and show that $\bar{\varrho}(t) \sim t^{-1/2}$.

Use the results of exercise 26 to show that the same conclusion also applies to the diffusion-coagulation-annihilation model, where one has in addition to the above also the reaction $A + A \xrightarrow{2\alpha} \emptyset + \emptyset$ such that $D = \alpha + \gamma$. Compare the exact result derived here with the expectation of mean-field theory as obtained in the previous exercise 35 and the experimental results listed in Table 4.11 on the kinetics of excitons moving on long polymeric chains. The experimental examples correspond to different branching ratios $\Gamma(\bullet\bullet \longrightarrow \circ\circ)/\Gamma(\bullet\bullet \longrightarrow \bullet\circ) = 2\alpha/\gamma$ of about 10% in the first two examples [530, 397] and essentially zero in the last case [401].

38. Use a random-walk argument to re-derive the decay law for the mean-particle density of the diffusion-coagulation process $2A \to A$ of a single species A [598]. Try to generalise to a two-species annihilation process $A + B \to \emptyset$, where initially both species A and B occur equally frequently and where both kinds of particles diffuse with the same diffusion rate $D_A = D_B = D$.

39. Consider the $1D$ **diffusion-coagulation-production** process, where particles of a single species A diffuse on a chain and undergo coagulation reactions $2A \to A$ and production $A\emptyset A \to 3A$. Analyse the case where the diffusion and coagulation rates are equal, by the empty-interval method [273].

40. Analyse the **triplet-annihilation process**, which is described by a system of particles of a single species A, such that single particles move diffusively and if *three* particles meet on neighbouring sites of the lattice, they may undergo the reaction $3A \to \emptyset$.

Within simple mean-field, use **Hölder's inequality**, for a domain $\Omega \subset \mathbb{R}^d$

$$\int_\Omega \mathrm{d}\boldsymbol{r}\, |u(\boldsymbol{r})v(\boldsymbol{r})| \leq \left(\int_\Omega \mathrm{d}\boldsymbol{r}\, |u(\boldsymbol{r})|^p\right)^{1/p} \left(\int_\Omega \mathrm{d}\boldsymbol{r}\, |v(\boldsymbol{r})|^q\right)^{1/q} \tag{4.259}$$

and with $\frac{1}{p} + \frac{1}{q} = 1$, to derive a (rough) upper bound on the mean particle-density $\bar{\varrho}(t) = \frac{1}{|\Omega|}\int_\Omega \mathrm{d}\boldsymbol{r}\ \varrho(t,\boldsymbol{r})$ and then study the system in the pair-approximation, especially for $D = 0$.

41. Consider the reaction-diffusion equation for a density $\varrho = \varrho(t,\boldsymbol{r})$

$$\partial_t \varrho = D\Delta\varrho + \tau\varrho - \varrho^2 \tag{4.260}$$

which arises either from simple mean-field in DP or, for $\tau = 0$, in a diffusion-coagulation process. Show that (4.260) is invariant under the scale-transformation

$$t \mapsto t' = \Lambda^z t \ , \ \boldsymbol{r} \mapsto \boldsymbol{r}' = \Lambda \boldsymbol{r} \ , \ \tau \mapsto \tau' = \Lambda^{-1/\nu_\perp}\tau \ , \ \varrho \mapsto \varrho' = \Lambda^{-\beta/\nu_\perp}\varrho \tag{4.261}$$

and derive the corresponding mean-field values of $z, 1/\nu_\perp$ and $\beta/\nu_\perp$.

42. Perform the same kind of scaling analysis as in exercise 41 for the reaction-diffusion process $3A \to \emptyset$, $A \to 2A$.

43. Perform a dimensional analysis of the field-theoretic action associated to directed percolation (or Reggeon field-theory). What do you find from a dimensional analysis directly on the Langevin equation?

44. Consider the contact process on the half-infinite line $r \geq 0$. If $a(t,r)$ denotes the mean particle-density at time t and position r, the mean-field kinetic equation (site approximation) can be written as

$$\dot{a} = Da'' + 4(p - 1/2)\, a - 2a^2 \tag{4.262}$$

In the steady state, $\dot{a} = 0$. We write for the steady-state density $a(r) = a_\infty \varphi(r/\xi_\perp)$, where $a_\infty = 2p - 1 \sim \xi_\perp^{-2}$ is the bulk density and $\xi_\perp = \sqrt{D/(p-1/2)}$ the spatial correlation length. Show that the mean-field profile is given by [125]

$$\varphi(y) = \frac{3}{2}\left(\frac{\sqrt{(2\varphi_0+1)/3} + \tanh(y)}{1 + \sqrt{(2\varphi_0+1)/3}\ \tanh(y)}\right)^2 - \frac{1}{2} \tag{4.263}$$

where $\varphi_0 = a(0)/a_\infty$ is related to the boundary density. Show that $\varphi(y) \to 1$ as $y \to \infty$ monotonously.

45. Use the field-theoretical formulation of absorbing phase transitions to show that the survival probability

$$P_{\text{sur}}(t) = \frac{\int \mathrm{d}\boldsymbol{r}\, \Upsilon(t,\boldsymbol{r})}{\int \mathrm{d}\boldsymbol{r}\, \varrho(t,\boldsymbol{r})}$$
$$= \lim_{t_1\to\infty} \frac{\int \mathrm{d}\boldsymbol{r}\, \langle \varrho(t+t_1,\boldsymbol{r})\widetilde{\varrho}(t_1,\boldsymbol{0})\rangle_{\text{empty}}}{\int \mathrm{d}\boldsymbol{r}\, \langle \varrho(t,\boldsymbol{r})\rangle_{\text{full}}}$$

where $\widetilde{\varrho}$ is the response field conjugate to the order parameter ϱ and 'empty' and 'full' refer to the initial state of the lattice.

Use this result to derive equations (4.96-4.98).

46. Consider the empty intervals of size ℓ of the Cantor set (see appendix H). Show that at each step of its iterative construction, empty intervals of a certain size ℓ are added and that their number $n(\ell) \sim \ell^{-d_{\text{f}}}$, where d_{f} is the fractal dimension of the Cantor set. Deduce that the probability $P(\ell)$ of empty intervals is

$$P(\ell) \sim \ell^{-1-d_{\text{f}}}$$

47. Relate the metric factors in (4.139) to those used habitually in this book, as in equations (4.41) and (4.80).

48. Derive the duality relation (4.182) between the stationary density and the survival probability for bond directed percolation [330].

Chapter 5
Universality Classes Different from Directed Percolation

As we have seen in the previous chapters, universality classes of continuous phase transitions are usually characterised by the dimensionality, the type of order parameters, and a set of certain symmetries. For example, in equilibrium statistical mechanics, the hallmark of an Ising transition is a discrete $\mathbb{Z}_2$-symmetry under spin reversal. Sometimes these symmetries are implemented as exact symmetries on the microscopic level. In many cases, however, they emerge only as asymptotic symmetries. A simple example is directed percolation, which is symmetric under rapidity-reversal (see Sect. 4.1.2) within the corresponding path integral formulation [496, 331]. Generally this symmetry is not present on the level of the microscopic dynamics, instead it emerges only asymptotically on a coarse-grained scale near criticality, where all irrelevant terms of the underlying field theory can be neglected. It is therefore not always possible to determine a system's universality class just by identifying the symmetries of its microscopic dynamics.

Fortunately, the DP universality class can be characterised by very few properties: According to the DP-conjecture by Janssen and Grassberger [326, 240] (see Sect. 3.2.2), systems with short-range interactions, exhibiting a continuous phase transition into a single absorbing state, belong generically to the DP universality class, provided that they are characterised by a one-component order parameter without additional symmetries and without unconventional features such as quenched disorder. Non-DP behaviour is expected to occur in systems where at least one of these requirements is not fulfilled. Therefore, it is interesting to search systematically for other universality classes of non-equilibrium phase transitions.

In this chapter, we shall consider several universality classes different from DP. First we consider absorbing phase transitions with additional symmetries, namely, the parity-conserving (PC) universality class, the voter universality class, and the universality class of compact directed percolation (CDP). Additional sections are devoted to dynamical percolation (DyP), the special case of tricritical directed percolation, and the Manna universality class. An open problem is the critical behaviour of the pair-contact process (PCPD) with diffusion and this remains a matter of controversial discussions at present. Another section deals with first-order transitions. At the end of this chapter, we consider crossover phenomena between different universality classes.

We also comment on attempted classifications of absorbing phase transitions [395, 203]. The interesting recent idea [203] to relate the dynamics to a classical Hamiltonian and study its topological properties might be an important ingredient for more comprehensive studies.

5.1 Parity-Conserving Universality Class

The **parity-conserving universality class** (PC), comprises phase transitions that occur in reaction-diffusion processes of the form

$$\begin{aligned} A &\to (n+1)A \\ 2A &\to \emptyset \end{aligned} \tag{5.1}$$

combined with single-particle diffusion, where the number of offspring n is assumed to be *even*.[1] Sometimes, models in this class are also referred to as **branching-annihilating random walks** with an even number of offspring. As an essential feature, these processes conserve the number of particles modulo 2. This is meant when one speaks, in the context of absorbing phase transitions, of systems with conserved *parity*, and should not be confused with the parity quantum number which arises under spatial reflections at the origin (or left-right exchanges in one dimension) and plays an important role in elementary particle physics. A particularly simple one-dimensional model in this class with $n = 2$ was proposed by Zhong and ben-Avraham [644]. The estimated critical exponents in $1D$

$$\beta = \beta' = 0.92(2)\,, \quad \nu_{\|} = 3.22(6)\,, \quad \nu_{\perp} = 1.83(3) \tag{5.2}$$

differ significantly from those of DP in one dimension, establishing PC transitions as an independent universality class.

In contrast to DP, the actual values of δ and Θ of parity-conserving transitions in seed simulations turn out to depend on the initial condition. If one starts with a *pair* of particles at the origin, one obtains as usual the exponents $\delta = \beta'/\nu_{\|} = 0.285(5)$ and the slip exponent $\Theta = 0.00(5)$. However, if the process starts with a *single* particle, it will never stop because of parity conservation, hence the survival probability is constant so the survival expo-

[1] With an odd number of offspring the transition turns out to belong to the DP universality class.

nent δ has to vanish, meaning that the usual relation $\delta = \beta'/\nu_\|$ no longer holds. Instead the slip exponent now takes the value $\Theta = 0.285(5)$, i.e., the roles of δ and Θ are exchanged. The theoretical reason for this exchange is not yet fully understood.

The relaxational properties in the subcritical phase differ significantly from the standard DP behaviour. While in DP, the particle density below the critical point decays exponentially as $\varrho(t) \sim e^{-t/\xi_\|}$, in PC models it decays *algebraically* as $t^{-1/2}$ since the asymptotic decay is governed by the annihilation process $2A \to \emptyset$.

The same conclusion of PC models being in a universality class different from DP is also reached by considering the surface critical behaviour. In 1+1 dimensions, and for an absorbing surface (ordinary transition), estimates $\beta_1/\nu_\perp = 0.73(1)$ and $0.720(2)$ were reported from Monte Carlo simulations [228] and the DMRG [40], respectively. For an active wall (extraordinary transition), however, $\beta_1/\nu_\perp = 1.11(1)$ is found in Monte Carlo simulations [228] and the DMRG gives $\beta_1/\nu_\perp = 1.10(1)$ [40].

A systematic field-theory for PC models can be found in [124, 114], confirming the existence of the annihilation fixed point in the inactive phase. However, the traditional perturbative field-theoretic treatment at criticality is extremely difficult as there are *two* critical dimensions: $d_c = 2$, above which mean-field theory applies, and $d_c' \approx 4/3$, where for $d > d_c'$ ($d < d_c'$) the branching process is relevant (irrelevant) at the annihilation fixed point. Therefore, the physically interesting spatial dimension $d = 1$ cannot be accessed by a controlled ϵ-expansion down from upper critical dimension $d_c = 2$. Here, the new method of the non-perturbative renormalisation group allows much easier access to the structure of the phase diagram and to genuinely non-gaussian fixed points [104, 105].

5.2 Voter Universality Class

Order-disorder transitions in models with a $\mathbb{Z}_2$-symmetry which are driven by *interfacial noise* belong to the so-called voter universality class [191]. As will be explained below, the voter class and the parity-conserving class are identical in one spatial dimension but differ in higher dimensions.

5.2.1 The Classical Voter Model

The classical **voter model** [422] is a simple lattice model which caricatures the formation of opinions before an election (schematically, this may also be

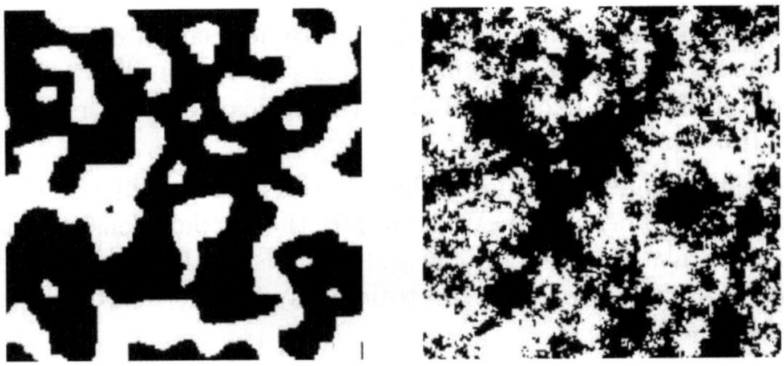

Glauber–Ising model at T=0 Classical voter model

Fig. 5.1 Coarsening of a random initial state in the Glauber-Ising model at zero temperature compared to the coarsening in the classical voter model with two opinions. Both snapshots were taken after 1000 Monte Carlo updates.

described in terms of the competing reactions $\circ\bullet \longrightarrow \circ\circ$ and $\circ\bullet \longrightarrow \bullet\bullet$). The lattice sites represents voters, each of them having a certain political opinion. The process evolves by random-sequential updates by randomly selecting a voter who then adopts the opinion of a randomly chosen nearest neighbour. More specifically, if $s_i \in \mathbb{N}$ denotes the 'opinion' at site i the update consists of the following steps:

1. Select a random site i.
2. Select one of its nearest neighbours j of i randomly.
3. Set $s_i := s_j$.
4. Increase time by $t = t + 1/\mathcal{N}$, where $\mathcal{N}$ is the total number of sites.

Obviously, the process reaches an absorbing state when all voters arrive at the same opinion. Starting with random initial conditions (see Fig. 5.1) the time-evolution of the voter model is characterised by coarsening domains. Note that the process is invariant under permutation of the opinions, in particular it has a $\mathbb{Z}_2$-symmetry in the case of two opinions. The classical voter model has no order parameter and no phase transition. The study of coarsening phenomena can be included into the studies of ageing phenomena, and will be taken up in Vol. 2. In Sect. 3.3.1 we discussed the relation with the DK automaton.

At this point, it is worthwhile comparing the classical voter model with two opinions to the Glauber-Ising model at zero temperature. Both models are $\mathbb{Z}_2$-symmetric and have two absorbing states. Starting with a random initial configuration, one observes in both cases a coarsening process forming ordered domains whose linear size grows as $L(t) \sim \sqrt{t}$. However, as can be seen in Fig. 5.1, in two spatial dimensions the form of the domains is clearly different. This exemplifies that symmetry alone does not always specify the asymptotic dynamical behaviour. In the Ising model at $T = 0$, domain growth is curvature-driven, leading to an effective surface tension of the domain walls. This is the reason why domains in the Ising model appear to be smooth. Contrarily, in the voter model domains are much more fuzzy and seem to coarsen only by fluctuation effects. In fact, in the voter model the density of domain walls decays only logarithmically as $1/\ln t$. This marginal behaviour is usually attributed to the exceptional character of its analytic properties [135, 556, 227] and may be interpreted physically as the *absence* of surface tension, see Vol. 2 for further information.

5.2.2 Voter-Type Phase Transitions

Let us now turn to the question of how one can generate an order-disorder phase transition in both Ising and voter models. For the Ising model, the answer is simple: One has to increase **temperature** until the long-range order is destroyed. As discussed before, temperature leads to spin flips inside ordered domains, forming oppositely oriented minority islands. For small temperatures, the influence of surface tension is still strong enough to eliminate these minority islands, stabilising the ordered phase. However, increasing T above a certain critical threshold T_c, this mechanism breaks down, leading to the well-known order-disorder phase transition in the Ising model. Thus, from the perspective of a dynamical process, the Ising transition results from competition between the surface tension of domain walls and bulk noise.

On the other hand, in the voter model, even very weak thermal bulk noise would immediately destroy global order. Therefore, thermally induced noises are too strong to allow for a phase transition at a finite T_c.[2] However, adding **interfacial noise** one observes a non-trivial continuous phase transition at a finite value of the noise amplitude. Unlike bulk noise, which flips spins everywhere inside the ordered domains, interfacial noise restricts spin flips to sites adjacent to domain walls. It turns out that transitions of this type constitute a new universality class, the so-called **voter universality class** [191]. Such transitions were first observed in the so-called non-equilibrium Ising model [473] (NEKIM) and in models with several absorbing states [285].

In one spatial dimension, **kinks** between domain walls may be interpreted as particles. Using this interpretation the one-dimensional classical voter model corresponds to diffusing particles which annihilate by the reaction $2A \to \emptyset$. On the other hand, interfacial noise at the boundaries splits domain walls locally into three domain walls. In the particle language, this process can be interpreted as offspring production $A \to 3A$ which conserves parity. For this reason the voter class and the parity-conserving class coincide in one spatial dimension. However, in higher dimensions they are expected to be dif-

---

[2] One could also argue that such a 'transition' would take place at zero temperature.

ferent. Roughly speaking, the parity-conserving class deals with the dynamics of zero-dimensional objects (particles), while in the voter class the dynamical objects of interest are (d-1)-dimensional manifolds (domain walls).

Recently, Al Hammal, Chaté, Dornic and Muñoz [265] introduced a Langevin equation describing voter transitions. It is given by

$$\frac{\partial}{\partial t}\varrho = (a\varrho - b\varrho^3)(1-\varrho^2) + D\nabla^2\varrho + \sigma\sqrt{1-\varrho^2}\,\eta\,, \tag{5.3}$$

where η is a Gaussian noise with constant amplitude. For $b > 0$, this equation is found to exhibit separate Ising and DP transitions [197], while for $b \le 0$, a genuine voter transition is observed. With these new results, the voter universality class is now on a much firmer basis than before.

Voter-type phase transitions can also be interpreted as non-equilibrium phase transitions into *several* symmetric absorbing states.[3] A model defined in the spirit of this interpretation is the **generalised Domany-Kinzel model** [285].This model has $n+1$ possible states per site: one active state A and n different inactive states $I_1, I_2, \ldots, I_n$. The conditional probabilities for the local updates are given by $(k, l = 1, \ldots, n;\ k \neq l)$

$$\begin{aligned} P[I_k\,|\,I_k, I_k] &= 1\,,\\ P[A|A,A] = 1 - n\,P[I_k|A,A] &= q\,,\\ P[A|I_k,A] = P[A|A,I_k] &= p\,,\\ P[I_k|I_k,A] = P[I_k|A,I_k] &= 1-p\,,\\ P[A|I_k,I_l] &= 1\,. \end{aligned} \tag{5.4}$$

For $n = 1$ the model defined above reduces to the original Domany-Kinzel model in Sect. 3.3.1, see page 73. For $n = 2$ one has two competing types of absorbing domains. The transition $I_1, I_2 \to A$ creates active sites between two inactive domains of different colours and thus it can be interpreted as interfacial noise. Varying p, one observes a non-equilibrium phase transition which belongs to the voter universality class.[4] For $n > 2$ one obtains a trivial behaviour, as discussed in [296].

---

[3] With two absorbing states such transitions are also said to belong to the *DP2 universality class*.

[4] For example, if one sets $p = q$ for $n = 2$, one has in one dimension, $p_c = 0.5673(5)$. The exponent estimates $\delta = 0.285(10)$, $z = 1.15(1)$, $\beta = 0.90(5)$ and $\Theta = 0.00(1)$ agree with those of the PC class [296].

5.3 Compact Directed Percolation

The term **compact directed percolation** (CDP) stands for a universality class of absorbing phase transitions, where the percolation clusters are compact objects [188, 208].

CDP can be regarded as a special limit of DP, where an additional $\mathbb{Z}_2$-symmetry emerges. To see this, let us recall the $(1+1)$-dimensional Domany-Kinzel (DK) automaton introduced in Sect. 3.3.1. The phase diagram of this model (see Fig. 3.8 and Fig. 3.9) comprises a critical line of second-order phase transitions, belonging to the DP-universality class. This critical line terminates at the upper point $p_1 = 1/2$ and $p_2 = 1$, where a different critical behaviour is observed. This special point is characterised by an additional particle-hole symmetry

$$\varrho \longleftrightarrow 1 - \varrho \,. \tag{5.5}$$

Along the upper line $p_2 = 1$, the DK model has two absorbing states, namely the empty and the fully occupied lattice. Here the dynamics is special in so far as particles can no longer disappear spontaneously in the interior of fully active domains. Therefore, the dynamics can be described in terms of diffusing and annihilating domain walls. The random walk of these domain walls is generally biased: For $p_1 > 1/2$ active domains preferentially grow while for $p_1 < 1/2$ they tend to shrink. Right at the CDP critical point, the random walk of the diffusing domain walls is unbiased and one recovers a simple annihilating random walk or, equivalently, the classical voter model in one dimension.

The situation is particularly simple if one starts with a single active site. In this case, one has only a single domain, bounded by a pair of random walkers which annihilate each other when they meet. This allows one to calculate the percolation probability exactly [208]

$$P_{\text{perc}}(p_1) \;=\; \begin{cases} 0 & ;\ \text{if } p_1 < 1/2\,, \\ (2\,p_1 - 1)/p_1^2 & ;\ \text{if } p_1 > 1/2\,, \end{cases} \tag{5.6}$$

yielding the percolation exponent $\beta' = 1$. Furthermore, the probability that the two walkers have not annihilated each other at time t scales as $t^{-1/2}$, yielding the survival exponent $\delta = 1/2$. In that way all exponents are related to the scaling behaviour of the random walkers. A complete list of critical exponents is given in Table 5.1.

According to these results, CDP stands for a transition with a continuously vanishing order parameter P_{perc} and algebraically diverging correlation lengths [208], indicating a second order phase transition. On the other hand, the stationary density of active sites ϱ_s is zero below $p_1 = 1/2$ and $\varrho_s = 1$ above $p_1 = 1/2$, exhibiting a discontinuous jump. This discontinuity at the critical point implies that the critical exponent β vanishes. Because of

d	β	$\nu_\perp$	$\nu_\parallel$	d_f	β'	δ	Θ	z
1	0	1	2	1	1	1/2	0	2
>2	0	1/2	1	2	1	1	0	2

Table 5.1 The critical exponents of the compact directed percolation (CDP) universality class [180]. At criticality, the $1D$ CDP is equivalent to the so-called voter model [422]. The crossover exponent from CDP to ordinary DP is given by $\phi = 2/d$ [331]. At the upper critical dimension $d_c = 2$ the scaling behaviour is affected by logarithmic corrections.

$$\beta \neq \beta' , \tag{5.7}$$

CDP has *four* instead of three independent critical exponents.

Phase transitions in the CDP universality class can be described by an effective Langevin equation [331]

$$\lambda^{-1}\,\partial_t \varrho = \tau \varrho (1-\varrho) + \nabla^2 \varrho + \eta \tag{5.8}$$

with noise correlations of the form

$$\langle \eta(t,\boldsymbol{r})\eta(t',\boldsymbol{r}')\rangle = \lambda^{-1}\kappa\,\varrho(t,\boldsymbol{r})\,[1-\varrho(t,\boldsymbol{r})]\;\delta(t-t')\,\delta(\boldsymbol{r}-\boldsymbol{r}') . \tag{5.9}$$

Note that the Langevin equation and the noise are constructed in such a way that they obey the particle-hole symmetry (5.5). Simple dimensional counting reveals that the noise is irrelevant for $d > 2$, i.e., the value of the upper critical dimension of CDP is $d_c = 2$. The corresponding field-theoretic response functional [331]

$$\mathcal{J}[\tilde{\varrho},\varrho] = \lambda \int \mathrm{d}^d\boldsymbol{r}\,\mathrm{d}t\,\tilde{\varrho}\left[\lambda^{-1}\partial_t\varrho - \tau\varrho(1-\varrho) - \nabla^2\varrho - \frac{\kappa}{2}\tilde{\varrho}\varrho(1-\varrho)\right] \tag{5.10}$$

is invariant under the transformation

$$\begin{aligned} \varrho(t,\boldsymbol{r}) &\longleftrightarrow 1 - \varrho(-t,\boldsymbol{r}) , \\ \tilde{\varrho}(t,\boldsymbol{r}) &\longleftrightarrow -\tilde{\varrho}(-t,\boldsymbol{r}) , \\ \tau &\longleftrightarrow -\tau , \end{aligned} \tag{5.11}$$

characterising the symmetry of the CDP universality class. For $\tau = 0$, one recovers the response functional of annihilating random walks and the corresponding relevant diagrams can be summed exactly (see e.g. [516, 415, 416, 331]). In Sect. 5.10.2, we shall consider the crossover from the CDP to the ordinary DP universality class in the example of the Domany-Kinzel automaton.

5.4 Tricritical Directed Percolation

A universality class of absorbing phase transitions which is different but directly related to DP is **tricritical directed percolation** (TDP). As for tricritical phenomena at equilibrium [411], where the tricritical points are usually the meeting points of lines of first- and second-order transitions,[5] the TDP emerges as a special point in the phase diagram of DP, where the coefficient of a non-linear term in the action vanishes such that the next-leading term takes over. To see this explicitly, let us consider the Langevin equation of DP with an additional cubic term

$$\partial_t \varrho = \tau \varrho - g\varrho^2 - c\varrho^3 + \nabla^2 \varrho + h + \eta\,, \tag{5.12}$$

$$\langle \eta(t, \boldsymbol{r})\eta(t', \boldsymbol{r}')\rangle = \kappa\, \varrho(t, \boldsymbol{r})\, \delta(t-t')\, \delta^d(\boldsymbol{r}-\boldsymbol{r}'). \tag{5.13}$$

Dynamical stability requires the coefficient c to be positive. Under generic conditions, i.e. when $g > 0$, this additional term is irrelevant under RG transformation and thus it merely generates corrections to the leading DP-like scaling behaviour. However, in a suitably parametrised phase diagram it is possible to follow a line of DP-like critical points such that g decreases and eventually vanishes. At this so-called **tricritical point** the additional term $-c\varrho^3$ becomes relevant.[6] Obviously, the presence of this term violates the rapidity reversal symmetry of DP, leading to a different type of universal critical behaviour. Following the transition line further, where g is negative, the transitions become of first order. Therefore, the tricritical point separates a line of continuous transitions from a line of first-order transitions.

The study of TDP started with the seminal work by Grassberger [240] in 1982. Later, a field-theoretical analysis was performed by Ohtsuki and Keyes [502, 501] (see also [331, 334]), where several tricritical exponents were estimated by means of an ϵ-expansion. A systematic numerical analysis of the TDP was carried out only recently in [436, 244].

There are three fundamental differences between ordinary and tricritical directed percolation. Firstly, as mentioned above, the response functional of TDP no longer satisfies the rapidity-reversal symmetry. This implies that TDP is characterised by *four* instead of three independent critical exponents. Secondly, the upper critical dimension is $d_c = 3$ for TDP while for DP, one has $d_c = 4$ [123, 489, 502, 501]. Thirdly, in contrast to DP, it is impossible to observe TDP transitions in one spatial dimension $d = 1$. This is due

---

[5] In a field-theoretical setting, tricritical points may be described in terms of a ϕ^6-theory, with the action

$$S[\phi] = \int \mathrm{d}\boldsymbol{r} \left[\frac{1}{2}(\nabla\phi)^2 + \frac{\mu^2}{2}\phi^2 + \frac{g}{4}\phi^4 + \frac{g'}{6}\phi^6 \right],$$

where $g' > 0$ for stability. The tricritical point corresponds to $\mu = g = 0$.

[6] Note that the bare parameter g is in general not in one-to-one correspondence with the corresponding physical parameters of lattice models.

to the circumstance that first-order transitions cannot occur in fluctuating one-dimensional systems where the surface tension does not depend on the domain size (see Sect. 5.9 for a detailed discussion). Thus in one-dimensional systems it is impossible to have a tricritical point that separates a first-order regime from a second-order regime.

5.4.1 Mean-Field Approximation of TDP

We start by recalling a simple but instructive mean-field calculation. Neglecting the noise term as well as spatial variations of the order parameter the mean-field equation for zero field

$$\partial_t \varrho = \tau\varrho - g\varrho^2 - c\varrho^3 \tag{5.14}$$

has two stationary solutions. One is the absorbing state $\varrho_s = 0$ which is stable for $\tau < 0$ and unstable for $\tau > 0$. The other solution represents a fluctuating state with the particle density

$$\varrho_s = -\frac{g}{2c} + \sqrt{\frac{\tau}{c} + \left(\frac{g}{2c}\right)^2}\,. \tag{5.15}$$

For $g > 0$, the second solution is real-valued and stable in the active phase $\tau > 0$ but is in general complex and unstable otherwise. Close to the transition, i.e. for small τ, (5.15) may be approximated by

$$\varrho_s = \frac{\tau}{g} - c\frac{\tau^2}{g^3} + O(\tau^3)\,. \tag{5.16}$$

Thus, to leading order the density scales linearly with $\beta_{MF} = 1$ as in ordinary DP while the additional term merely leads to a correction to scaling. The amplitude of this correction, compared to the leading term, diverges in the limit $g \to 0$, signalling the crossover to the tricritical behaviour. This crossover will be analysed in more detail in Sect. 5.10.

For negative g the solution (5.15) is stable for $\tau > -g^2/4c$. This means that the system can be in a stable active state even when the control parameter is negative. Therefore a new region emerges in the phase diagram (shown as a shaded region in Fig. 5.2), where the nontrivial active phase and the absorbing state are both thermodynamically stable. In other words, for $g < 0$ the absorbing state and the active phase *coexist* between the spinodal $\tau = -g^2/4c$ and the line $\tau = 0$, giving rise to a first-order phase transition.

The tricritical behaviour is observed for $g = 0$. In the active phase $\tau > 0$ the order parameter varies as

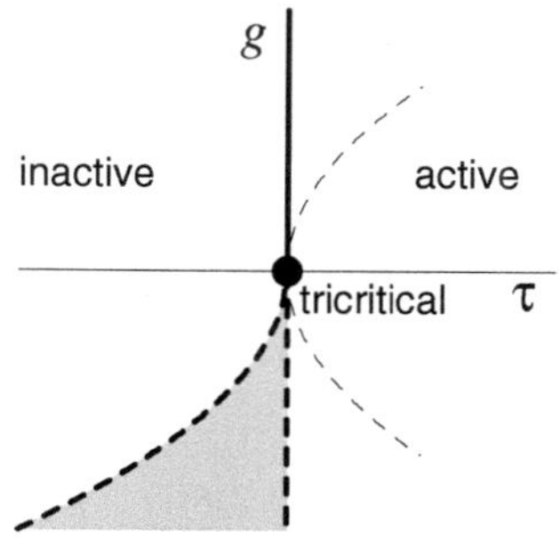

Fig. 5.2 The mean-field phase diagram of tricritical directed percolation (TDP) as a function of the coarse-grained variables τ and g [see (5.12)]. The thick lines represent the second-order phase transitions with $\beta_{\mathrm{DP}} = 1$. The bold circle at the origin marks the TDP transition point. In the shaded area the absorbing state ($\varrho = 0$) and the active phase ($\varrho > 0$) are found to coexist (see text). The thin dashed lines illustrate the crossover to the tricritical behaviour. Reproduced from [436] with kind permission of Springer Science and Business Media.

$$\varrho_{\mathrm{s}} = (\tau/c)^{\beta_{\mathrm{t}}} \tag{5.17}$$

with the tricritical order parameter exponent $\beta_{\mathrm{t}} = 1/2$. To determine the other mean-field tricritical exponents we consider how the stationary order parameter at criticality depends on the field

$$\varrho_{\mathrm{s}} = (h/c)^{\beta_{\mathrm{t}}/\sigma_{\mathrm{t}}}\,, \tag{5.18}$$

giving the exponent $\sigma_{\mathrm{t}} = 3/2$. Furthermore, the order parameter decays at the tricritical point as

$$\varrho = \left(\varrho_0^{-2} + 2ct\right)^{-1/2} \underset{t\to\infty}{\longrightarrow} (2ct)^{-\alpha_{\mathrm{t}}} \tag{5.19}$$

with $\alpha_{\mathrm{t}} = 1/2$. Finally, in the active phase the density relaxes towards its stationary value according to

$$\varrho(t;\tau) = \sqrt{\frac{\tau}{c}}\,\left[1 - c_0 \mathrm{e}^{-t/\xi_\|} + \mathrm{O}(\mathrm{e}^{-2t/\xi_\|})\right]. \tag{5.20}$$

Here, the constant c_0 contains information about the initial state and $\xi_\|$ denotes the temporal correlation length

$$\xi_\| = (2\tau)^{-\nu_{\|,\mathrm{t}}} \tag{5.21}$$

with $\nu_{\|,\mathrm{t}} = 1$. Incorporating spatial variations of the order parameter, the spatial correlation length $\xi_\perp$ can be derived via an Ornstein-Zernicke approach. The resulting correlation length exponent $\nu_{\perp,\mathrm{t}} = 1/2$ leads to the dynamical exponent $z_{\mathrm{t}} = \nu_{\|,\mathrm{t}}/\nu_{\perp,\mathrm{t}} = 2$. This completes the computation of the mean-field tricritical exponents, which are summarised in Table 5.2.

	$d=2$	$d>3$
β_t	0.100(4)	1/2
$\nu_{\perp,\mathrm{t}}$	0.547(3)	1/2
$\nu_{\parallel,\mathrm{t}}$	1.156(4)	1
σ_t	0.848(12)	3/2
γ'_t	0.894(10)	1/2
γ_t	0.748(11)	1
$\eta_{\perp,\mathrm{t}}$	0.366(16)	1
$d_{\mathrm{f},\mathrm{t}}$	1.817(8)	2
β'_t	1.408(10)	1
δ_t	1.218(7)	1
α_t	0.087(3)	1/2
Θ_t	−0.353(9)	0
z_t	2.110(6)	2
$\frac{\mathrm{X}_\mathrm{t}(+1,0)}{\mathrm{X}_\mathrm{t}(-1,0)}$ [436]	0.35(5)	1/2
U_t	0.84(4)	

Table 5.2 The critical exponents and various universal amplitude combinations of tricritical directed percolation (TDP). The crossover exponent from tricritical to ordinary DP is given in $2D$ by $\phi = 0.36(4)$ and $\phi_{\mathrm{MF}} = 1/2$. The values of the exponents are obtained from spreading simulations [244]. Note that the estimates of some exponents are the subject of controversy in the literature. The numbers in brackets give the estimated uncertainty in the last given digit(s).

Note that the tricritical exponent of the order parameter fluctuations is given by $\gamma'_{\perp,\mathrm{t}} = 1/2$ [502]. This value reflects a qualitative difference between the mean-field scaling behaviour of DP and TDP. In the latter case the fluctuations diverge at the transition point whereas they remain finite (with a discontinuous jump) in case of DP [480, 445].

5.4.2 Numerical Simulations of TDP

Lattice models for TDP can be obtained by modifying the dynamical rules of an ordinary DP process in a suitable manner. For example, tricritical behaviour may occur if higher-order reactions, such as the pair reaction [502, 501]

$$2A \longrightarrow 3A \tag{5.22}$$

are added to the reaction scheme of DP

$$A \longrightarrow 0\,, \quad A \longrightarrow 2A\,. \tag{5.23}$$

Examples are the tricritical contact process (TCP), which may be defined by the following reactions [436]

$$\bullet \xrightarrow{(1-q)(1-p)} \circ\,,\ \circ\bullet \xrightarrow{q(1-p)} \circ\circ\,,\ \circ\bullet \xrightarrow{(1-q)p} \bullet\bullet\,,\ \bullet\bullet \xrightarrow{(1-q)(1-p)} \circ\bullet\,,\ \circ\bullet\bullet \xrightarrow{q} \bullet\bullet\bullet$$

controlled by the parameters p and q such that for $q = 0$, one is back to the contact process. The tricritical point is at $q_t = 0.9055$, $p_t = 0.29931(3)$ [435]. Analogous modifications of the Domany-Kinzel automaton have also been studied [26, 244].

A numerical or experimental analysis of the scaling behaviour of tricritical systems poses particular challenges. First, an accurate determination of the tricritical point is pivotal since the numerical accuracy of universal quantities is ultimately limited by the uncertainty of the tricritical point.[7] Moreover, the parameters of the model do not always coincide with the scaling parameters of the Langevin equation, meaning that it matters from which direction the tricritical point is approached [547]. Despite these difficulties, numerical simulations of TDP have been successfully performed. The existing estimates of the tricritical exponents are listed in Table 5.2.

It is worth comparing the numerical estimates with the field-theoretical results in linear order of $\epsilon = d_c - d$ (with $d_c = 3$) [502, 501, 331]

$$\begin{aligned}
&\beta_t = \frac{1}{2} - \epsilon\,0.4580\ldots\,, \quad \beta_t' = 1 + O(\epsilon^2)\,, \quad z_t = 2 + \epsilon\,0.0086\ldots\,,\\
&\gamma_t = 1 + O(\epsilon^2)\,, \quad \gamma_t' = \frac{1}{2} + \epsilon\,0.4386\ldots\,, \quad \nu_{\perp,t} = \frac{1}{2} + \epsilon\,0.0075\ldots\,,\\
&\phi = \frac{1}{2} - \epsilon\,0.0121\ldots\,.
\end{aligned} \tag{5.24}$$

The last exponent ϕ describes the crossover from TDP to DP. The existing numerical values are still controversially discussed in the literature [436, 244] and the agreement between numerical and field-theoretical results is rather poor. Thus further numerical as well as analytical work, in particular an ϵ-expansion in two-loop order, would be desirable.

In addition to the critical exponents, universal scaling functions of the TDP universality class have been determined numerically [436]. For example, Fig. 5.3 shows the universal scaling form for the equation of state. Sufficiently close to the tricritical point ($g = 0$) the order parameter obeys the scaling form

$$\varrho(\tau_{\text{path}}, h) \simeq \lambda^{-\beta_t}\,\tilde{R}(\lambda a_{\text{path}} \tau_{\text{path}}, 0, a_h h \lambda^{\sigma_t})\,, \tag{5.25}$$

where τ_{path} describes the distance to the tricritical point along a certain path in the phase diagram. Different paths, often referred to as different scaling directions, lead to different corrections to scaling but do not affect the leading order of the scaling behaviour. This can be seen in Fig. 5.3, where the

---

[7] Since tricritical points are, in the RG sense, more unstable than ordinary critical points, one must be very close to them in order to see the distinctive tricritical behaviour, rather than the cross-over to ordinary criticality.

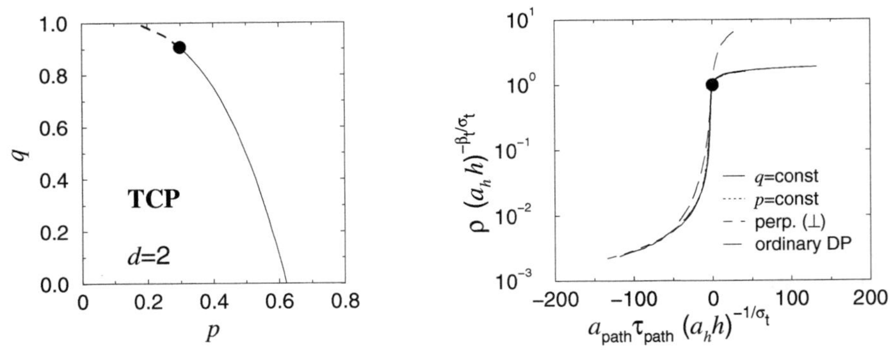

Fig. 5.3 Left: Phase diagram of the tricritical contact process (TCP). The parameter p controls spreading $A \to 2A$ while the parameter q controls the pair creation process $2A \to 3A$. The solid line marks continuous phase transitions belonging to the DP universality class while the dashed line stands for first-order phase transitions. Both lines are separated by a tricritical point marked by the bold circle. Right: Universal scaling function of the equation of state, measured along three different paths crossing the tricritical point in the p-q-plane: *i*) along q = const, *ii*) along p = const, and *iii*) perpendicular to the phase boundary. For all three paths at least four different curves are plotted corresponding to four different field values which all collapse onto a single curve. The circle marks the normalisation point $\tilde{R}(0,0,1) = 1$. The dashed line corresponds to the universal scaling function of ordinary DP, proving that both universality classes are different. Reproduced from [436] with kind permission of Springer Science and Business Media.

tricritical point is crossed along three different paths. All curves collapse onto a single one, namely, the universal scaling function of the equation of state of the TDP universality class. Furthermore, the tricritical scaling function differs significantly from the corresponding scaling function of DP, confirming that the two universality classes are indeed different.

For a detailed analysis of the tricritical scaling behaviour, including dynamical and steady-state universal scaling functions of the order parameter fluctuations and of the susceptibility as well as an investigation of universal finite size scaling functions, we refer the interested reader to [436, 244]. Beyond the tricritical scaling behaviour itself, systems exhibiting tricriticality provide us with an opportunity to study crossover effects as well as first-order transitions (cf. Sect. 5.9 and Sect. 5.10).

5.5 Dynamical Percolation

Dynamical percolation (DyP) is a universality class of phase transitions which differs significantly from directed percolation. In the language of epidemic spreading, dynamical percolation is most easily introduced as a generalisation of directed percolation including the effect of **immunisation**.

An often-studied model is the so-called **generalised epidemic process** (GEP) [111, 113, 327]. In this model a lattice site can be in three different states, namely, susceptible, infected, and immune. Initially all sites are susceptible and an infected seed is placed at the origin. The model then evolves in the same way as a contact process, the only difference being that a recovered site becomes immune. The immunisation is assumed to be perfect, i.e. a recovered site cannot be infected again.

The GEP has the following phenomenological properties. If the infection rate is small, activity will spread locally for a short time until it dies out, leaving a certain cluster of immune sites behind. This is the inactive phase of the system. On the other hand, if the infection rate is large, there is a finite probability that a front of active sites will spread concentrically over the whole system, constituting the phase of **annular growth**. Both phases are separated by a non-equilibrium phase transition which belongs to the universality class of dynamical percolation.

As first suggested by Cardy [111], the effect of immunisation can be implemented by adding a term on the right hand side of the DP Langevin equation (3.10)

$$\begin{aligned}\partial_t \varrho(t,\boldsymbol{r}) &= \tau\varrho(t,\boldsymbol{r}) - g\varrho(t,\boldsymbol{r})^2 + D\nabla^2\varrho(t,\boldsymbol{r}) + \eta(t,\boldsymbol{r}) \\ &\quad + \mu\varrho(t,\boldsymbol{r}) \exp\left(-w\int_0^t \mathrm{d}t'\,\varrho(t',\boldsymbol{r})\right) . \end{aligned} \tag{5.26}$$

In this term, the integral sums up the past activity at position $\boldsymbol{r}$ between the initial condition $t = 0$ and the actual time t. If this integrated activity is still small, the exponential function is essentially equal to 1, marking a non-immune site. However, when the integrated activity exceeds a certain threshold of the order $1/w$, the exponential function 'switches' to zero, representing an immunised site. Since the exponential function is coupled to the order parameter $\varrho(\boldsymbol{r},t)$, the effect of immunisation effectively modifies the spreading rate τ [245]. More specifically, non-immune sites are infected with rate $\tau + \mu$, while immune sites are infected with rate τ. This allows one to control the rates for the first and all subsequent infections of an indi-

exponent	$d=2$ [578, 481]	$d=3$ [481, 141]	$d=4$ [33, 141]	$d=5$ [33, 141]	$d\geq 6$ mean-field
β'	5/36	0.417	≈ 0.64	≈ 0.84	1
$\nu_\perp$	4/3	0.875	≈ 0.68	≈ 0.57	1/2
$\nu_\parallel$	1.506	1.169			1
δ	0.092	0.346(6)	0.595(8)	0.806(12)	1
Θ	0.586	0.488(7)	0.30(1)	0.134(10)	0
z	1.1295	1.375(5)	1.605(9)	1.815(10)	2

Table 5.3 Critical exponents of dynamical percolation. Empty entries indicate unknown estimates. The values for $\nu_\perp$ and for β' are those of isotropic percolation (where β' is commonly denoted by β) [578, 635, 636].

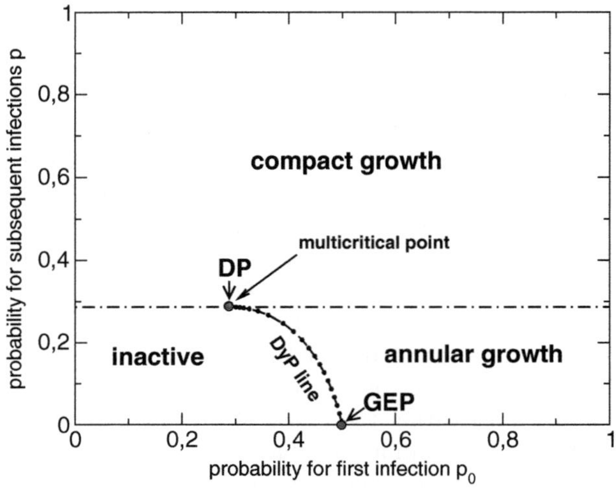

Fig. 5.4 Phase diagram of the general epidemic process (see text).

vidual separately. The additional term modifies the critical behaviour of the transition, leading to a universality class which is different from DP. This so-called GEP-class comprises all models which are defined in the spirit of this Langevin equation.

Using the Janssen-de Dominicis formalism by introducing a response field $\tilde{\varrho}(\boldsymbol{r},t)$ and integrating out the Gaussian noise, one is led to a field-theoretic action [111, 113]

$$\mathcal{J}[\tilde{\varrho},\varrho] = \int \mathrm{d}t\,\mathrm{d}\boldsymbol{r}\,\tilde{\varrho}\Big[\lambda^{-1}\partial_t\varrho - (\tau+\nabla^2)\varrho - \Big(\frac{\kappa}{2}\tilde{\varrho} - g\,\varrho\Big) + \frac{\mu}{2}\,\varrho\,\exp\Big(-\int_0^t \mathrm{d}t'\,\varrho(t',\boldsymbol{r})\Big)\Big] \qquad (5.27)$$

which is expected to be valid in arbitrary dimensions. By dimensional analysis one can see that the upper critical dimension of the GEP is $d_c = 6$. Moreover, this action is invariant under the replacement

$$\frac{\partial}{\partial t}\tilde{\varrho}(t,\boldsymbol{r}) \;\leftrightarrow\; \varrho(-t,\boldsymbol{r})\,. \qquad (5.28)$$

which differs from the rapidity-reversal symmetry of DP discussed in Sect. 4.1.2, see e.g. [496, 481, 331, 342] for further details.

Apart from the interpretation as an epidemic process with immunisation, dynamical percolation has the important properties that it produces *isotropic* (undirected) percolation clusters in d dimensions. More specifically, whenever the process terminates, it leaves behind a certain cluster of immune sites. This cluster can be shown to be an ordinary isotropic (undirected) percolation cluster [578]. In this sense the GEP has infinitely many absorbing states [482] and can be used as a dynamical prescription or as an algorithmic tool to grow ordinary percolation clusters.

The critical exponents of isotropic percolation are known (in two dimensions even exactly) and can be expressed in terms of two standard exponents β and $\nu \equiv \nu_\perp$. In dynamical percolation, the additional temporal degree of freedom induces another exponent, namely, the dynamical exponent z. Knowing β, ν, and z, the exponents for the survival probability $P(t) \sim t^{-\delta}$ and the average number of particles $N(t) \sim t^{\Theta}$ can be expressed as

$$\delta = \frac{\beta}{\nu_\parallel}, \qquad \Theta = \frac{d}{z} - \frac{2\beta}{\nu_\parallel} - 1, \tag{5.29}$$

where $\nu_\parallel = z\nu_\perp$. The exponents of dynamical percolation are summarised in Table 5.5.

The form of the action (5.27) suggests that dynamical percolation can also be observed in models with partial immunisation, where an immune site can be re-infected, although with a lower probability than susceptible sites. The phase diagram of a contact process with separate rates for first infections and re-infections [245, 482, 363] is shown in Fig. 5.4. Along the diagonal, where both rates coincide, the model reduces to the usual contact process with a DP transition while for perfect immunisation one recovers the GEP. Both transition points are connected by a curved transition line, which belongs to the DyP universality class.

More recently, Janssen and Stenull investigated dynamical percolation at the upper critical dimension $d_c = 6$ by field-theoretical methods, computing logarithmic corrections [340]. A tricritical version of DyP was investigated in [334].

5.6 Long-Range Interactions

So far we have considered spreading processes with *local* interactions, where activity spreads to nearest-neighbour sites. However, in many realistic situations one observes that the spreading agent is occasionally transported over long distances. For example, infections may be carried by insects or the spreading agent may be transported by a turbulent flow.

There are two important types of long-range interactions. One possibility is to consider models with a large but *finite* interaction range. In these models one expects a crossover between an effective mean-field behaviour on small scales, where diffusive mixing dominates, to an effective short-range behaviour on large scales. For example, such a crossover is observed in processes where the infection spreads within a given radius R (see Sect. 5.10.3 for further details).

Another possibility, on which we shall focus in the following, is to study situations in which the interaction range is in principle unlimited. For example, one could study a DP-like process with infections over long distances r which are randomly distributed according to a power law. In the literature such long-range moves are known as **Lévy flights.** More specifically, one assumes that the distribution $P(r)$ of the spreading distance r decays as

$$P_\ell(r) \sim \frac{1}{r^{d+\sigma_\ell}}\,, \tag{5.30}$$

controlled by an exponent σ_ℓ which may be regarded as an external parameter. Here the index 'ℓ' stands for 'long range'.

Similarly, one may also study an analogous long-range mechanism in the temporal direction. Such 'temporal' Lévy flights can be interpreted as **waiting times** Δt before an infected site can infect another site. As in the previous case, these waiting times are assumed to be algebraically distributed as

$$P(\Delta t) \sim \frac{1}{(\Delta t)^{1+\kappa_\ell}} \tag{5.31}$$

with a control exponent κ_ℓ. Generally, it turns out that such long-range interactions in systems with absorbing states may change the critical behaviour at the transition in a certain range of the control exponents. In the following we summarise some of the main results. For a more detailed account we refer the reader to a recent review [290].

5.6.1 DP with Spatial Lévy Flights

Let us first consider directed percolation with long-range spreading by *spatial* Lévy flights [335, 292]. On the level of the Langevin equation (4.175), long-range infections by Lévy flights can be implemented by replacing the term for short-range diffusion by a non-local integral expression

$$\begin{aligned}\lambda^{-1}\partial_t\varrho(t,\boldsymbol{r}) &= [(\tau - g\,\varrho(t,\boldsymbol{r})]\,\varrho(t,\boldsymbol{r}) + \eta(t,\boldsymbol{r}) \\ &\quad + D\int \mathrm{d}\boldsymbol{r}'\,P(|\boldsymbol{r}-\boldsymbol{r}'|)\,[\varrho(t,\boldsymbol{r}') - \varrho(t,\boldsymbol{r})]\end{aligned} \tag{5.32}$$

with the power-law kernel $P(r)$ in (5.30). Keeping the most relevant terms in a small-momentum expansion, this equation may also be written as [335]

$$\lambda^{-1}\partial_t\varrho(t,\boldsymbol{r}) = \left[(\tau - g\,\varrho(t,\boldsymbol{r}) + \nabla^2 + D_\ell\nabla^{\sigma_\ell}\right]\varrho(t,\boldsymbol{r}) + \eta(t,\boldsymbol{r})\,, \tag{5.33}$$

where ∇^{σ_ℓ} is a linear operator representing the action of the integral in (5.32). This operator is a so-called *fractional derivative* ∇^{σ_ℓ}, which is most easily defined through its action in momentum space

$$\nabla^{\sigma_\ell} \mathrm{e}^{\mathrm{i}\boldsymbol{k}\cdot\boldsymbol{r}} = -k^{\sigma_\ell} \mathrm{e}^{\mathrm{i}\boldsymbol{k}\cdot\boldsymbol{r}} . \tag{5.34}$$

The term for ordinary diffusion ∇^2 is retained in (5.33), reflecting the short-range component of the Lévy flights on a lattice. It is important to note that this term, even if it were not included initially, would be generated under renormalisation.

Neglecting spatial fluctuations as well as the noise in (5.33), we obtain the mean-field values of the critical exponents

$$\beta_{\mathrm{MF},\ell} = 1 , \qquad \nu_{\|,\mathrm{MF},\ell} = 1 . \tag{5.35}$$

which coincide with those of ordinary DP. Moreover, analysing (5.34) by an Ornstein-Zernicke-like approach one gets the correlation length exponent

$$\nu_{\perp,\mathrm{MF},\ell} = \begin{cases} 1/2 & \text{if} \quad \sigma_\ell \geq 2 \\ 1/\sigma_\ell & \text{if} \quad \sigma_\ell < 2 \end{cases} \tag{5.36}$$

which varies continuously with the control exponent σ_ℓ over a certain range. Finally, dimensional analysis of the noise term in (5.34) shows that the upper critical dimension is given by

$$d_\mathrm{c} = \begin{cases} 4 & \text{if} \quad \sigma_\ell > 2 \\ 2\sigma_\ell & \text{if} \quad \sigma_\ell < 2 . \end{cases} \tag{5.37}$$

Thus, for $\sigma_\ell < 2$ the Lévy flights are relevant, such that the exponent $\nu_{\perp,\mathrm{MF},\ell}$ as well as d_c vary continuously with σ_ℓ, while for $\sigma_\ell \geq 2$ the distribution decays sufficiently fast that the Lévy flights remain irrelevant and one is back to the usual nearest-neighbour (diffusive) motion of individual particles.

We note that within the mean-field regime, the boundary between the short-range ($\nu_\perp = 1/2$) and the long-range ($\nu_\perp = 1/\sigma_\ell$) scaling behaviour is exactly $\sigma_\ell^* = 2$. However, as a surprising result, one finds that fluctuation effects below the upper critical dimension shift this threshold to slightly higher values. The resulting phase structure is sketched in Fig. 5.5.

5.6.1.1 Spatial Lévy flights: Field-Theoretical Results

The starting point of a field-theoretic renormalisation group analysis is the dynamic functional [335]

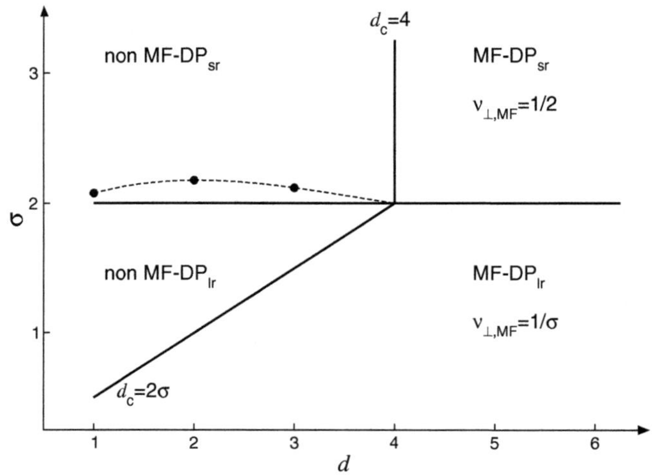

Fig. 5.5 Four different scaling regimes of directed percolation with long-range interactions by means of isotropic Lévy flights. The phase boundaries reflect the stability and instability of the corresponding renormalisation group fixed points. The indices 'lr' and 'sr' denote the long-range and the short-range fixed points, respectively. Along the phase boundaries, the stability of the fixed points is exchanged, leading to logarithmic corrections on top of the usual power-laws. The long-range (lr) interactions remain irrelevant for $\sigma_\ell > \sigma_\ell^*$. In the non-mean-field regime, one would naively expect that $\sigma_\ell^* = 2$ (solid line) but a more careful analysis reveals $\sigma_\ell^* = 2 - \eta_\perp + z$ (dashed line).

$$\mathcal{J}[\tilde{\varrho}, \varrho] = \lambda \int \mathrm{d}t\, \mathrm{d}\boldsymbol{r}\; \tilde{\varrho} \Big[\lambda^{-1} \partial_t \varrho - (\tau + \nabla^2 + D_\ell \nabla^{\sigma_\ell}) \varrho - \Big(\frac{\kappa}{2}\, \tilde{\varrho} - g\, \varrho \Big) \varrho \Big] . \tag{5.38}$$

Note that the additional long-range operator ∇^{σ_ℓ} appears in the *free* part of the action. Therefore, the structure of the Feynman diagrams in a loop expansion is exactly the same as in ordinary DP, the only difference being that the free propagator in the loop integrals has to be replaced by its generalised counterpart. Moreover, the functional still obeys the rapidity-reversal symmetry [335], although the critical exponents and scaling functions are generally different from those of short-range DP.

Within an ϵ-expansion to one-loop order, the critical exponents in the non-MF regime are given by [335, 292]

$$\beta_\ell = 1 - \frac{2\epsilon}{7\sigma_\ell} + \mathrm{O}(\epsilon^2) , \tag{5.39}$$

$$\nu_{\perp,\ell} = \frac{1}{\sigma_\ell} + \frac{2\epsilon}{7\sigma_\ell^2} + \mathrm{O}(\epsilon^2) , \tag{5.40}$$

d	1	2	3	≥ 4 (MF)
σ_ℓ^*	2.076660(21)	2.179(15)	2.118(17)	2

Table 5.4 The values of the borderline exponent σ_ℓ^*, separating the long-range scaling regime ($\sigma_\ell < \sigma_\ell^*$) from the short-range scaling regime ($\sigma_\ell > \sigma_\ell^*$). The values are determined by inserting the numerically known DP exponents into the scaling relation (5.48).

$$\nu_{\parallel,\ell} = 1 + \frac{\epsilon}{7\sigma_\ell} + \mathrm{O}(\epsilon^2)\,, \tag{5.41}$$

$$z_\ell = \sigma_\ell - \frac{\epsilon}{7} + \mathrm{O}(\epsilon^2)\,, \tag{5.42}$$

where $\epsilon = d - d_c$ denotes the distance from the upper critical dimension $d_c = 2\sigma_\ell$. Note that all exponents depend continuously on the external parameter σ_ℓ. Furthermore, the usual DP hyperscaling relation (4.91)

$$\frac{d}{z} = \Theta + 2\delta \tag{5.43}$$

holds for all values of σ_ℓ outside the mean-field regime below the upper critical dimension. Thus the dynamical exponents are given by

$$\Theta_\ell = \frac{\epsilon}{7\sigma_\ell} + \mathrm{O}(\epsilon^2)\,, \tag{5.44}$$

$$\delta_\ell = 1 - \frac{7\epsilon}{\sigma_\ell} + \mathrm{O}(\epsilon^2)\,. \tag{5.45}$$

At the upper critical dimension, the scaling behaviour is characterised by the mean-field exponents modified by logarithmic corrections. For example, along the line $d_c = 2\sigma_\ell$ the order parameter obeys

$$\varrho \sim \tau\,|\ln\tau|^{2/7} \tag{5.46}$$

for $\tau > 0$ and $h = 0$ [335].

As a remarkable result of the field-theoretic calculation one finds that the fractional operator ∇^{σ_ℓ} does not renormalise itself, instead it renormalises its short-range counterpart ∇^2 to all orders of perturbation theory. This implies an additional exact scaling relation of the form [292]

$$\nu_\perp \sigma_\ell = d\nu_{\perp,\ell} - 2\beta_\ell + \nu_{\parallel,\ell} \tag{5.47}$$

valid within the long-range scaling regime. Using the Fisher scaling law (B12) this relation can also be written as $\sigma_\ell = 2 - \eta_{\perp,\ell} + z_\ell$. Because of this scaling relation, DP with spatial Lévy flights is characterised by three exponents, e.g. β, $\nu_\parallel$, and σ_ℓ. Only two of them are independent since the latter plays the role of an external control parameter.

The real importance of the scaling relation (5.47), however, lies in a different aspect. Since renormalisation group analysis indicates that the critical exponents change continuously from the long-range to the short-range scaling regime [335] the threshold σ_ℓ^* between the long-range and the short-range regime can be computed by inserting the numerically known exponents of short-range DP into (5.47). This then leads to the expression

$$\sigma_\ell^* = 2 - \eta_\perp + z\,, \tag{5.48}$$

where z and $\eta_\perp$ denote the exponents of short-range DP. Remarkably, σ_ℓ^* is shifted away from the mean-field threshold $\sigma_\ell^* = 2$ in (5.36) towards higher values (see Table 5.4 and Fig. 5.5). Therefore, in an interacting field theory below its upper critical dimension, fluctuation effects may modify the boundary between the long-range and the short-range regime. Only recently, a similar shift of such a boundary was also observed in the ϕ^4-theory at equilibrium (see e.g. [449] and references therein), although in that case $\sigma_\ell^* = 2 - \eta$ is shifted to values smaller than 2.

5.6.2 DP with Temporal Long-Range Interactions

Finally, we briefly mention that long-range interactions can also be established in the temporal direction and may be investigated separately [537] or in combination with the previously studied spatial Lévy flights [3]. In the language of epidemic spreading, such temporal long-range interactions can be implemented as stochastically distributed waiting times Δt before an infected individual can infect other individuals. These 'incubation times' are assumed to be distributed algebraically according to (5.31), where κ_ℓ is an external parameter. In the functional $\mathcal{J}[\tilde{\varrho}, \varrho]$ such waiting times can be taken into account by adding a term with a fractional derivative in time. This fractional operator differs from the one in (5.33) in so far as it is directed forward in time respecting causality and therefore acts differently in Fourier space (see e.g. [290]).

A field-theoretic analysis (see [537, 3]) leads to the phase diagram shown in Fig. 5.6. It comprises seven different phases, namely, three mean-field phases where either long-range flights (MFL), incubation times (MFI), or both of them (MF) are relevant. Similarly, there are three different fluctuation-dominated regimes, while in the short range limit of large σ_ℓ and κ_ℓ one recovers ordinary DP. The mean-field phase MF is characterised by the exponents

$$\beta_{\mathrm{MF},\ell} = 1\,, \qquad \nu_{\perp,\mathrm{MF},\ell} = \frac{1}{\sigma_\ell}\,, \qquad \nu_{\parallel,\mathrm{MF},\ell} = \frac{1}{\kappa_\ell}\,, \tag{5.49}$$

which are valid above the upper critical dimension

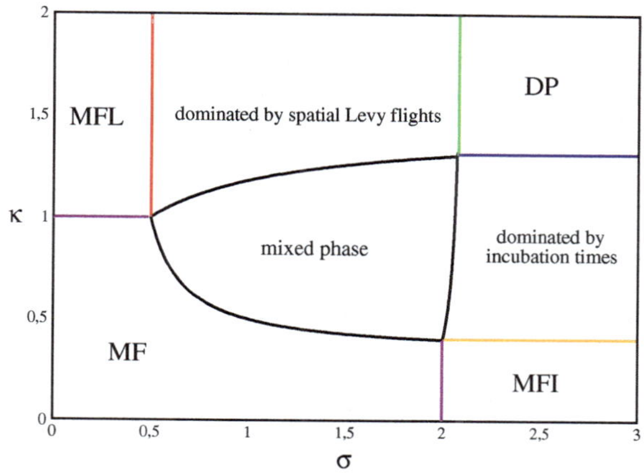

Fig. 5.6 Phase diagram for DP with spatio-temporal long-range interactions in one spatial dimension (see text).

$$d_c = \left(3 - \frac{1}{\kappa_\ell}\right) \sigma_\ell \,. \tag{5.50}$$

Below d_c in the mixed phase (LI), the exponents vary continuously with the external parameter σ_ℓ and κ_ℓ [3]:

$$\beta_\ell = 1 - \frac{\epsilon}{4\sigma_\ell} + O(\epsilon^2) \,, \tag{5.51}$$

$$\nu_{\perp,\ell} = \frac{1}{\sigma_\ell} + \frac{1\epsilon}{4\sigma_\ell^2} + O(\epsilon^2) \,, \tag{5.52}$$

$$\nu_{\parallel,\ell} = \frac{1}{\kappa_\ell} + \frac{\epsilon}{4\kappa_\ell\sigma_\ell} + O(\epsilon^2) \,. \tag{5.53}$$

Neither of the fractional derivatives renormalise themselves, implying two additional scaling laws

$$\sigma_\ell = 2 - \eta_{\perp,\ell} + z_\ell \,, \qquad \kappa_\ell z_\ell = 2 - \eta_{\perp,\ell} + z_\ell \tag{5.54}$$

and leading to the intuitively reasonable result [221]

$$z_\ell = \frac{\sigma_\ell}{\kappa_\ell} \,. \tag{5.55}$$

Apparently the combined effect of spatial and temporal long-range interactions predominates the scaling behaviour and the dynamical exponent is locked onto the ratio of σ_ℓ and κ_ℓ [3].

An interesting aspect of phase transitions with long-range interactions is the possibility to modify the upper critical dimension continuously. In numerical simulations this allows one to tune the upper critical dimension to a value close to the actual physical dimension, in which the simulation is carried out. In this way it is possible to verify the results of a field-theoretic ϵ-expansion quantitatively.

5.6.3 Other Models with Long-Range Interactions by Lévy Flights

In principle, any stochastic model involving short-range diffusion can be generalised to spatial long-range interactions by Lévy flights. For example, the simple annihilation process $2A \to \emptyset$ with long-range interactions was studied both numerically and by field-theoretical methods in [292]. Branching-annihilating random walks, normally belonging to the parity-conserving universality class, were generalised to long-range flights by Vernon and Howard [606]. Moreover, the classical voter model with long-range interactions was studied in [641]. Similarly, the general epidemic process (GEP) discussed in Sect. 5.5 can be endowed with infections by spatial Lévy flights, leading to a long-ranged version of dynamical percolation [335, 423].

Algebraically distributed waiting times can be implemented in any model, where the temporal derivative ∂_t in the evolution equation can be replaced by a directed fractional derivative $\partial_t^{\kappa_\ell}$. However, compared to spatial Lévy flights there are less studies since the physical motivation of algebraically distributed waiting times is not always clear. Besides epidemic spreading with incubation times, temporal long-range interactions may play a role in disordered systems [291]. Moreover, such waiting times can be used to describe algebraically distributed first-return probabilities in an effective manner [36]. Stochastic processes with waiting times are conceptually challenging as they are intrinsically non-Markovian.

5.6.4 Simulating Models with Long-Range Interactions

Let us finally outline how models with long-range interactions can be simulated numerically.

1. A Lévy flight from site i over distance r on an infinite one-dimensional lattice can be generated as follows. Let $z \in [0,1]$ be an ordinary random number and let $r = z^{-1/\sigma_\ell}$. As can be shown easily, $r \in [1,\infty]$ is distributed as $P(r) = \sigma_\ell r^{1+\sigma_\ell}$. In one spatial dimension the target site of the Lévy flight is $i \pm \lfloor r \rfloor$ with equal probability, where $\lfloor r \rfloor$ denotes truncation to an

integer. In higher dimensions, a Box-Müller-type algorithm can be used to determine a direction isotropically [423].

2. Systems with spatial long-range interactions are extremely sensitive to finite-size effects. For this reason one should prefer simulation methods where finite-size effects are virtually absent as, for example, seed simulations using a dynamically generated list of particle coordinates (see Sect. 4.3.6.2). On a finite lattice, past experience has shown that one should cut off the flight distance, instead the best results are obtained if one works with periodic boundary conditions. For example, on a one-dimensional lattice with L sites, the target site would be located at $(i \pm \lfloor r \rfloor) \bmod L$.
3. In seed simulations with spatial Lévy flights the measurement of the mean square spreading from the origin, defined as an arithmetic average $R^2(t) = \langle |\boldsymbol{x}(t)|^2 \rangle$ may not converge as the second moment of the flight distance diverges after only a single update. To circumvent this problem, one should instead study the *geometric average* $R^2(t) = \exp\left[\langle \ln(|\boldsymbol{x}(t)|^2)\rangle\right]$. This average is finite for all $\sigma_\ell > 0$ and yields consistent results in the short-range limit.
4. In order to simulate a system with waiting times, one has to implement an event queue for activations in the future. Moreover, in seed simulations the survival probability $P_{\rm sur}(t)$ is no longer defined and has to be replaced by the probability of finding at least one active site at time t.

5.7 Manna Universality Class

In this section, we discuss a class of absorbing phase transitions in which the number of particles is strictly conserved. This additional conservation law leads to another type of universal scaling behaviour different from DP. Analogous to the DP-conjecture by Janssen and Grassberger [326, 240] (cf. Sect. 3.2.2) it is believed that continuous transitions in stochastic models with an infinite number of absorbing states, where the order parameter is coupled to a non-diffusive conserved field, define a unique universality class [552], the so-called **Manna universality class**. As we shall see, this class is related to the concept of **self-organised criticality** (SOC) introduced in the study of avalanches of sandpiles (see e.g. [29, 30, 28, 601]) and is therefore of particular interest. Table 5.5 below collects estimates of critical exponents and amplitudes.

Compared to directed percolation, the Manna universality class is not yet well established. Beside field-theoretical approaches [181, 510, 602, 331], series expansions [583] and path integral representations [182] most quantitative results have been obtained by simulations. In particular, a systematic ϵ-expansion is still lacking. In the following we present several lattice models belonging to the Manna universality class and focus again on the universal scaling behaviour. At the end of this section, we shall discuss the relation to self-organised criticality.

A comment is worth making concerning the designation of the universality class. In the literature numerous names have been used, causing never-ending misunderstandings and confusion. However, the so-called **Manna model** (see below) has always been considered as a paradigmatic and particularly simple example of a model exhibiting an absorbing phase transition with a conserved non-diffusing field. Since universality classes are often named after the simplest model belonging to them, this class will be termed the *Manna universality class* throughout this book.

It is worth mentioning that the DP and the Manna universality class are characterised by the same value of the upper critical dimension $d_c = 4$. Furthermore, their mean-field scaling behaviour above the upper critical dimension is identical, while below d_c the classes are clearly different (see [433] for a detailed discussion). Such a scenario is quite familiar in statistical physics. For example, the Ising model, the XY-model as well as the Heisenberg model exhibit different scaling behaviours below d_c but share the same type of mean-field behaviour above d_c.

5.7.1 Manna Model

The simplest representative of the Manna universality class is the **Manna model**. Manna introduced this model as a stochastic cellular automaton in order to describe the dynamics of sandpiles in the context of self-organised criticality [460]. In this lattice model the sites of a lattice are associated with integer values representing e.g. local energies, number of sand-grains or particles (see Fig. 5.7). The Manna model is a bosonic lattice model, i.e., it allows for an unlimited number of particles per site $n = 0, 1, 2, \ldots, \infty$. Lattice sites are considered as inactive if the particle occupation is below a certain threshold $n < N_c$, e.g. $N_c = 2$ in all dimensions d or $N_c = d$ [442]. On the other hand, lattice sites with $n \geq N_c$ are considered as active. The model evolves in time by the dynamical rules that all active sites redistribute their particles among randomly chosen nearest neighbours, setting

$$n \longrightarrow 0 \qquad \text{for all sites with} \qquad n \geq N_c\,. \tag{5.56}$$

Apart from this original version of the Manna model [460], some authors have considered a modified version, where the occupation number of active lattice sites is not set to zero but reduced by randomly redistributing N_c particles:

$$n \longrightarrow n - N_c \qquad \text{for all sites with} \qquad n \geq N_c\,. \tag{5.57}$$

In the latter case the Manna model is analytically tractable if the active particles are distributed in a particular way [165]. In both versions the redistribution process conserves the number of particles.

The properties of the Manna model depend crucially on the boundary conditions. In the context of SOC, one uses open boundary conditions where particles can disappear, combined with a slow driving by continuous deposition of grains in the bulk. In this case, the system self-organises in a critical state where arbitrarily large sequences of redistribution events can be observed. The sizes of such avalanches are distributed algebraically, reflecting the critical properties of the Manna universality class.

As another possibility, one may consider periodic boundary conditions without external driving, so that the total number of particles N is conserved. In this case the model exhibits a continuous phase transition from an active into an inactive phase when the particle density $\varrho_n = N/L^d$, which now plays the role of a control parameter, is varied. The order parameter, which characterises this transition, is the density ϱ of lattice sites occupied by at least N_c particles [608]. Obviously a configuration is absorbing if all sites contain less than N_c particles. Thus the Manna model is characterised by an infinite number of absorbing configurations in the thermodynamic limit.

As in DP it is possible to apply an external field h which is conjugated to the order parameter. This field must be defined in a way that it does not violate the conservation law. A possible realisation was proposed in [431], where h is implemented as some kind of diffusion-like field, allowing particles in locally frozen configurations to move to nearest neighbours.

The driven-dissipative version (with open boundaries and slow external driving) of the Manna model was investigated intensely in the context of self-organised criticality (see e.g. [56, 165, 131, 429, 428, 174, 66]). Following the arguments by Ben-Hur and Biham [56] the universality class of a self-organised critical system is determined by the way in which active particles are distributed to the nearest neighbours (deterministic, stochastic, directed, undirected, etc.) and the symmetries of the system. For finite N_c the Manna model is characterised by a stochastic redistribution of particles. However, in the limit $N_c \to \infty$ the active particles are equally distributed to the nearest neighbours, i.e., the stochastic character of the process is lost. In fact, it can be shown that in this limit the Manna model exhibits a different type of scaling behaviour. The crossover between the two universality classes was studied in numerical simulations [428], confirming the conjecture of [56].

5.7.2 Conserved Threshold Transfer Process (CTTP)

A second lattice model belonging to the Manna universality class is the **conserved threshold transfer process** (CTTP) [552]. It is a modification of

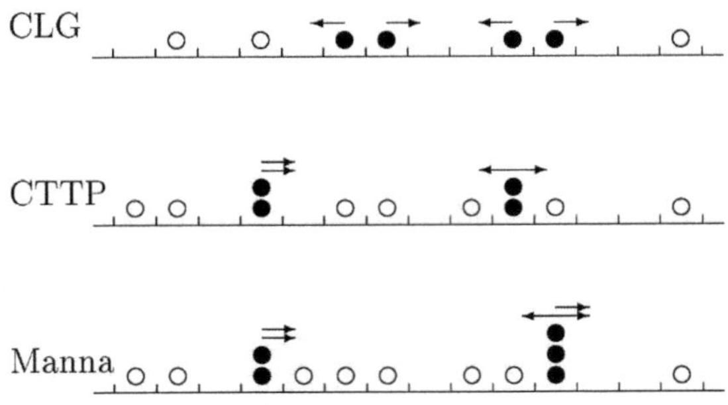

Fig. 5.7 Sketch of the dynamics of the Manna model, the CTTP, and the CLG in one dimension. Filled (hollow) circles mark active (inactive) particles. The arrows denote the possible directions in which the particles may be moved stochastically in the next update. In the CLG model particles move deterministically. In the case of the CTTP the redistribution can be either stochastic (left) or deterministic (right). Only the one-dimensional Manna model with $N_c = 2$ has a fully stochastic redistribution process. Reprinted with permission from [437]. Copyright (2003) by the American Physical Society.

the ordinary threshold transfer process discussed above in Sect. 3.3.4. Again lattice sites may be empty, occupied by one particle, or occupied by two particles. Empty and singly occupied lattice sites are considered as inactive, whereas doubly occupied lattice sites are considered as active. In the latter case the model evolves in that it tries to transfer both particles of a given active site to randomly chosen empty or singly occupied sites. In contrast to the threshold transfer process creation and annihilation of isolated particles is forbidden so that the number of particles is conserved. As in the Manna model the conjugated field can be implemented by a diffusion-like field [430]. Numerical simulations reveal that the CTTP exhibits the same scaling behaviour as the Manna model in $d \geq 2$ dimensions [432]. In one dimension a splitting of the universality class occurs [437] since in this case a significant amount of the relaxation events is deterministic. Thus in $d = 1$ dimensions the CTTP differs from the pure stochastic Manna model.

The CTTP can be considered as a modification of the Manna model in which the maximal number of particles per site is restricted. This restriction simplifies the analytical treatment significantly. For example, the CTTP can be analysed easily within a mean-field approximation [439] whereas the unrestricted configuration space of the Manna model leads to a system of coupled equations (see [433] and references therein). Denoting the density of unoccupied (empty) sites by ϱ_e, the density of single occupied sites by ϱ_i, and the density of double occupied sites by ϱ, the conditions for normalisation and particle conservation can be expressed as [433]

$$\varrho_e + \varrho_i + \varrho = 1\,, \qquad \varrho_n = \varrho_i + 2\varrho \tag{5.58}$$

leading to the mean-field equation for the CTTP:

$$\partial_t \varrho(\varrho_n, h) \;=\; \varrho\,(-1 + 2\varrho_n - 4\varrho + \varrho^2) + h(\varrho_n - 2\varrho)^2\,. \tag{5.59}$$

For zero conjugated field $h = 0$, the steady-state condition $\partial_t \varrho = 0$ yields the stationary order parameter as a function of the particle density [439, 179]

$$\varrho_s \;=\; 2 \;-\; \sqrt{5 - 2\varrho_n} \qquad \text{for} \quad \varrho_n \geq \varrho_{n,c} = 1/2\,. \tag{5.60}$$

The order parameter vanishes to leading order as $\varrho \;\propto\; \tau^\beta$ with $\tau = (\varrho_n - \varrho_{n,c})/\varrho_{n,c}$ and $\beta = 1$. For non-zero conjugated field the asymptotic behaviour of the equation of state is given by [439]

$$\varrho_s \;\simeq\; \frac{\tau}{8} \;+\; \sqrt{\frac{h}{16} + \left(\frac{\tau}{8}\right)^2}\,. \tag{5.61}$$

Thus the mean-field solution of the Manna universality class coincides with the mean-field behaviour of DP.

5.7.3 Conserved Lattice Gas (CLG) and Other Reaction-Diffusion Processes

A further lattice model belonging to the Manna universality class is the so-called **conserved lattice gas** (CLG) [552] which is a stochastic variant of a model introduced by Jensen [346]. In contrast to the Manna model the lattice sites of the CLG-model may be empty or occupied by a single particle. Motivated by experiments on flux flow in type-II superconductors one assumes a repulsive interaction [346], i.e., a particle is considered as active if at least one of its neighbouring sites is occupied as well. Otherwise, if all neighbouring sites are empty, the particle remains inactive. Active particles are moved in the next update step to one of their empty nearest neighbour sites selected at random. Obviously, the CLG conserves the number of particles. Moreover, all configurations with solitary particles are absorbing. As for the Manna model and the CTTP, the conjugated field is implemented by particle diffusion [431].

As in the Manna model, the CLG exhibits a continuous transition controlled by the particle density. In $d \geq 2$ dimensions this transition belongs to the Manna universality class, taking place at a critical threshold $\varrho_{n,c} < 1/2$. In one dimension, however, the phase transition is trivial and the critical density is $\varrho_{n,c} = 1/2$ (see [552] and references therein). This is due to the fact

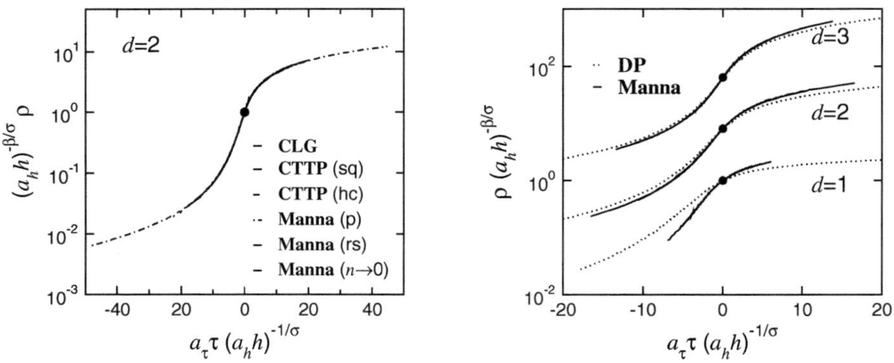

Fig. 5.8 The universal scaling function $\tilde{R}(x, 1)$ of the equation of state of the Manna universality class in various dimensions. The curves are plotted according to the scaling form (4.206). For $d = 2$ the CTTP data are obtained from simulations on a square (sq) and honeycomb (hc) lattice. In the case of the Manna model, data of the original model (5.56) using random sequential (rs) and parallel (p) updates as well as data of the modified model in the limit $n \to 0$ (5.57) are presented. As shown in the left panel, all curves collapse onto a single one. Moreover, the scaling functions depend on the dimension and differ from the corresponding scaling functions of DP (see right panel). Reprinted with permission from [433]. Copyright (2004) World Scientific Publishing Company.

that the CLG in one dimension is characterised by a deterministic dynamics, hence it does not belong to the Manna universality class.

Finally let us mention the reaction-diffusion process [400, 603, 510, 511]

$$A \longrightarrow B\,, \qquad A + B \longrightarrow 2\,A\,, \tag{5.62}$$

where A-particles and B-particles diffuse with diffusion rates D_A and D_B, respectively. Obviously, this process conserves the total number of particles $N = N_A + N_B$. By varying the particle density, the model exhibits a continuous phase transition. In the limit $D_B \to 0$ the order parameter ρ_A is coupled to a non-diffusive field, hence this transition belongs to the Manna universality class. Unfortunately, field-theoretic renormalisation group approaches for this model run into difficulties and are a matter of great controversy [496, 510, 602, 331].

5.7.4 Scaling Properties

In Fig. 5.8 we present the scaling form of the equation of state as an impressive exemplification of the universal scaling behaviour of the Manna class. Despite the different lattice structures and update schemes, the data of the Manna model, the CTTP and of the CLG collapse onto a single curve. This result

	$d=1$		$d=2$	$d=3$	Mean-field
	Manna	CTTP			
β	0.38(2)	0.382(19)	0.639(9)	0.840(12)	1
β'	0.32(5)	0.42(8)	0.624(29)	0.827(34)	1
$\nu_\perp$	1.35(9)	1.76(6)	0.799(14)	0.593(13)	1/2
$\nu_\parallel$	1.88(14)	2.45(11)	1.225(29)	1.081(27)	1
σ	2.710(40)	1.770(58)	2.229(32)	2.069(43)	2
γ'	0.550(40)	0.670(40)	0.367(19)	0.152(17)	0
γ	2.328(44)	1.388(40)	1.590(33)	1.229(45)	1
$\eta_\perp$	1.592(40)	1.619(28)	1.541(25)	1.744(29)	2
α	0.141(24)	0.14(2)	0.419(15)	0.745(17)	1
δ	0.170(25)	0.170(25)	0.510(20)	0.765(25)	1
Θ	0.35(3)	0.35(3)	0.310(30)	0.140(30)	0
z	1.393(37)	1.39(5)	1.533(24)	1.823(23)	2
$\tilde{D}(1,0)$			1.81(3)	1.91(8)	2
$\frac{\tilde{X}(+1,0)}{\tilde{X}(-1,0)}$			0.41(5)	0.80(5)	1
U	0.80(4)		0.69(3)	0.60(1)	1/2

Table 5.5 The critical exponents and various universal amplitude combinations of the Manna and the CTTP models, below the upper critical dimension $d_c = 4$. In $1D$, the Manna model and the CTTP are in different universality classes. The data of the exponents β, σ, $\nu_\perp$, and γ' are obtained from steady-state simulations [430, 437], whereas seed simulations were used to determine the values of α, δ, Θ, and z [437]. The exponents β' and $\nu_\parallel$ are derived via scaling laws. Note that the values of $\nu_\parallel$ are in good agreement with those of direct measurements of the order parameter persistence distribution [441].

is a strong manifestation of the robustness of the Manna universality class with respect to variations of the microscopic interactions.

The right panel of Fig. 5.8 compares the scaling functions $\tilde{R}(x,1)$ of the Manna and the DP class. As can be seen both classes are characterised by significantly different scaling functions. Evidently, these differences become less important with increasing dimension since both classes converge to the same type of mean-field behaviour at the upper critical dimension.

A complete analysis of the scaling behaviour of the Manna universality can be found in [433], including an investigation of universal scaling functions of the order parameter fluctuations and of the susceptibility as well as of finite size scaling functions. Numerical estimates of the critical exponents and of various universal amplitude combinations are listed in Table 5.5. As a field theoretical treatment is still lacking, it is not yet known how many of these exponents are independent. Numerical investigations indicate that the general hyperscaling law

$$\Theta + \frac{\beta}{\nu_\parallel} + \delta = \frac{d}{z} \tag{5.63}$$

is fulfilled. Furthermore, the values of $\beta/\nu_\parallel$ and δ agree within the error-bars. This suggests that $\beta = \beta'$ although the Manna universality class does not obey the rapidity-reversal symmetry. In addition, the Manna class does exhibit a scaling anomaly [552, 433], i.e., it is numerically observed that

$$\alpha \neq \frac{\beta}{\nu_{\|}} \tag{5.64}$$

So far no scaling law is known involving the order parameter decay exponent α (see also [541]). This indicates that the number of independent exponents of the Manna universality class is at least four or perhaps even five. To clarify this point further numerical as well as analytical work is desirable.

5.7.5 Relationship Between Absorbing Phase Transitions and Self-Organised Criticality (SOC)

Let us finally discuss the close relationship between absorbing phase transitions and self-organised criticality. The term SOC refers to driven-dissipative systems that naturally evolve towards a critical state, characterised by power-law distributions of relaxation events (see [29, 30, 601, 166] for introductory reviews). Such systems differ from ordinary critical phenomena in that they self-organise into a critical state without fine-tuning of an obvious control parameter to a critical value. In this critical state the systems jump among absorbing configurations via avalanche-like relaxation processes, triggered by a slow external driving. Scale-invariance manifests itself as an algebraic behaviour of certain distribution functions, e.g. for the size and duration of avalanches.

The paradigmatic realisations of SOC are sandpile models such as the **Bak-Tang-Wiesenfeld model** (BTW) [29] and the aforementioned Manna sandpile model [460]. The BTW model is analytically tractable because of its Abelian structure, allowing one to determine height distributions and the corresponding correlation functions [164, 457, 456, 532, 318]. However, the dynamical properties of the avalanches are not yet fully understood. Progress was achieved by decomposing the avalanches into a sequence of relaxation events termed *waves* [321, 533, 320] which can be mapped to spanning trees and loop-erased random walks [403]. Nevertheless, the behaviour of the avalanches is a complex phenomenon and the corresponding exponents are not known analytically (see [433] for a detailed discussion). Since reliable numerical estimates are difficult to obtain due to a lack of simple finite-size scaling, their interpretation is discussed controversially (see e.g. [459, 247, 442, 472, 594, 132, 471]). Contrarily, the Manna model was investigated successfully by numerical simulations (see for example [460, 56, 442, 133, 594, 429, 174, 66]) but no exact results are known so far.

The mechanism that drives the system into a critical state is also not completely understood. Phenomenologically, the critical stationary state is reached by a delicate balance of driving and dissipation in the slow driving limit, leading to a **separation of time scales** [253, 605, 610]. An important advance was achieved by mapping sandpile models to absorbing phase transitions, i.e., to a conventional non-equilibrium critical phenomenon. This mapping (see the next subsection) was first discussed by Tang and Bak within a mean-field approach [590] and elaborated further in a series of works, including [609, 181, 610, 607]. As discussed above, the Manna model exhibits an absorbing phase transition in closed ensembles (often called fixed-energy sandpiles [181]), where the particle density is strictly conserved.

5.7.6 Absorbing Phase Transitions and SOC: Mean-Field Approximation

In the following we show by a mean-field treatment how a system self-organises into a critical state when moving from a closed to an open ensemble with dissipation and external driving. In contrast to numerous other mean-field studies of SOC systems [590, 20, 167, 324, 231, 642, 319, 378, 605, 99, 610], which focus primarily on the determination of the avalanche exponents, we present here an instructive mean-field approach [433] for the CTTP which explains the attraction towards the critical state by the dynamics.

Contrariwise to the CTTP with global particle conservation (see (5.59)) we now consider a driven-dissipative modification. Firstly, we introduce an external driving by adding particles from outside with probability p at empty or single-occupied lattice sites. Obviously this driving breaks the global conservation law. Secondly we incorporate particle dissipation at rate ϵ which competes with the external driving. The corresponding rate equations lead to coupled differential equations of the form [433]

$$\partial_t \varrho = p\,(\varrho_{\mathrm{n}} - 2\,\varrho) \; + \; \varrho\,[-1 + 2(1-\epsilon)\varrho_{\mathrm{n}} - 4(1-\epsilon)\varrho + (1-\epsilon)^2\varrho^2] \tag{5.65}$$

$$\quad + \, h\,(\varrho_{\mathrm{n}} - 2\varrho)^2\,,$$

$$\partial_t \varrho_{\mathrm{n}} = p(1-\varrho) \; - \; 2\,\epsilon\,\varrho\,. \tag{5.66}$$

In the slow driving limit ($p \ll 1$) the steady-state solution can be expressed in terms of the drive-dissipation ratio $\kappa = p/\epsilon$:

$$\varrho_{\mathrm{s}} \; = \; \frac{\kappa}{2+\kappa}\,, \qquad \varrho_{\mathrm{n}} = \frac{2\,+\,6\,\kappa\,+\,2\,\kappa^2}{(2+\kappa)^2}\,. \tag{5.67}$$

It turns out that the parameter κ, which was phenomenologically introduced by Grinstein [253], is the appropriate parameter to describe the self-organisation to the critical point. Eliminating κ one recovers (5.60), describing the order parameter behaviour in the active phase of the closed systems.

In this way the control parameter in the active phase $\varrho_{\rm n} \in [\varrho_{\rm n,c}, 2]$ is mapped onto the interval of the drive-dissipative ratio $\kappa \in [0, \infty]$. In the weak driving limit $p \ll \epsilon$ ($\kappa \to 0$) we find to leading order [433]

$$\varrho_{\rm s} = \frac{\kappa}{2}, \qquad \varrho_{\rm n} = \varrho_{\rm n,c} + \kappa\,. \tag{5.68}$$

Therefore, as $\kappa \to 0$, the driven-dissipative system approaches the critical point of the corresponding absorbing phase transition ($\varrho_{\rm n} = \varrho_{\rm n,c}$, $\varrho_{\rm s} = 0$). The fundamental role of the Grinstein parameter κ is even valid beyond mean-field. In fact, simulations of the driven-dissipative CTTP in two dimensions indicate that the critical point is approached as $\varrho_{\rm s} = \kappa/2$ and $\varrho_{\rm n} = \varrho_{\rm n,c} + \kappa^{1/\beta}$ for $\kappa \to 0$ [433].

Let us now turn to the dynamical behaviour in the vicinity of the steady-state. Here (5.66) and (5.68) lead to

$$\partial_{pt}\varrho_{\rm n} = O(\kappa)\,, \qquad \partial_{\epsilon t}\varrho_{\rm n} = O(\kappa^2)\,. \tag{5.69}$$

In other words, the control parameter varies for $\kappa \to 0$ only on the perturbation scale $\tau_{\rm perp} = 1/p$, whereas $\varrho_{\rm n}$ can be considered as approximately constant on the dissipation scale $\tau_{\rm diss} = 1/\epsilon$. Furthermore, it is assumed that the system displays an avalanche-like response for infinitesimal driving ($p \to 0$), i.e., no further perturbations take place until an absorbing configuration ($\varrho = 0$) is reached. This extreme separation of time scales ($\kappa \to 0$) is necessary in order to identify individual relaxation events (avalanches). Therefore, one can justify characterisation of the scale-invariance via avalanche distribution functions [253].

Within this approximation, the dynamics of avalanches is determined by (5.65) with $p = 0$. At criticality ($\varrho_{\rm n} = \varrho_{\rm c}$) the order parameter decays for sufficiently small ϵ as [433]

$$\varrho(t) \simeq \frac{1}{\varrho_0^{-1} + 4\,(1-\epsilon)\,t} \qquad \text{for} \qquad \epsilon\, t \ll 1\,, \tag{5.70}$$

$$\varrho(t) \simeq \frac{\epsilon}{4(1-\epsilon)}\,{\rm e}^{-\epsilon\, t} \qquad \text{for} \qquad \epsilon\, t \gg 1\,. \tag{5.71}$$

Thus close to the critical point ($\kappa \to 0$) an avalanche, caused by the perturbation ϱ_0, decays algebraically before an exponential cutoff occurs. This cutoff is caused by the particle dissipation and takes place at $\tau_{\rm cutoff} = 1/\epsilon$. A pure power-law behaviour is obtained in the limit $\epsilon \to 0$ only. Therefore the driven-dissipative CTTP self-organises itself to the critical point and exhibits scale-invariant avalanches in the limit $\kappa \to 0$ and $\epsilon \to 0$ which is often termed the **SOC limit** [605, 610].

In conclusion, the key to understand SOC is the insight that the limit $\kappa \to 0$ tunes the control parameter to the boundary of the active phase, i.e., to the critical point. This explains why the *trivial* limit $\kappa \to 0$ in an open system corresponds to a fine-tuning of the control parameter to the critical point of an absorbing phase transition in a closed system. Remarkably, this limit occurs frequently in driven-dissipative systems in nature, i.e., many physical processes are characterised by a large separation of timescales that makes κ extremely small [28, 253]. For example, the motion of the tectonic plates that causes earthquakes is extremely slow compared to the time scale on which earthquakes proceed (see [253]). Thus, the dissipation and driving rate are separated by many orders of magnitude. Although the drive-dissipation rate remains finite, the system is very close to the critical point ($\kappa = 0$) and displays a power-law behaviour over several decades, namely, the celebrated *Gutenberg-Richter* law [257].

5.7.7 Relating Critical Exponents of SOC and Absorbing Phase Transitions

The avalanches of SOC systems are characterised by several quantities (see for example [56, 443]), e.g. the size s (number of elementary relaxation events), the area a (number of distinct relaxed sites), the time t (number of parallel updates until the configuration is stable), as well as the radius exponent r (radius of gyration). In the critical steady-state the corresponding probability distributions are found to decay algebraically

$$P_x \sim x^{-\tau_x} \tag{5.72}$$

with the so-called avalanche exponents τ_x where $x \in \{s, a, t, r\}$. Assuming that the four quantities scale relative to each other by power laws of the form $x \sim x'^{\gamma_{xx'}}$ one obtains the scaling relation

$$\gamma_{xx'} = \frac{\tau_{x'} - 1}{\tau_x - 1} \,. \tag{5.73}$$

For example, the exponent γ_{tr} is equal to the dynamical exponent z, the exponent γ_{ar} corresponds to the fractal dimension of the avalanches, and the exponent γ_{sa} indicates whether multiple relaxations of a lattice site are relevant ($\gamma_{sa} > 1$) or irrelevant ($\gamma_{sa} = 1$). This relationship [481, 608, 437], combined with certain hyperscaling laws [433], allows the avalanche exponents to be expressed in terms of the ordinary critical exponents of the corresponding absorbing phase transition:

$$\tau_r = 1 + z\,\delta = 1 + \frac{\beta'}{\nu_\perp}, \tag{5.74}$$

$$\tau_t = 1 + \delta = 1 + \frac{\beta'}{\nu_\parallel}, \tag{5.75}$$

$$\tau_a = 1 + \frac{z\,\delta}{d} = 1 + \frac{\beta'}{d\nu_\perp}, \tag{5.76}$$

$$\tau_s = 1 + \frac{\delta}{1+\Theta+\delta} = 1 + \frac{\beta'}{\nu_\parallel + \nu_\perp d - \beta}\,. \tag{5.77}$$

In this way, the critical state of SOC systems is related to the critical state of an ordinary second-order phase transition.

Eventually, we briefly comment about the relation of SOC to the famous $1/f$-noise, to be precise a $1/f^\alpha$-decay of the power-spectrum of a certain signal with a non-trivial exponent $\alpha < 2$. Originally, the concept of SOC aimed to explain the ubiquity of $1/f^\alpha$ [29]. But this idea is criticised in a series of works, including [348, 387, 347, 73], where a simple Lorentzian spectrum with a trivial $1/f^2$-decay is observed. This result has recently been questioned. According to [410], SOC leads rather generally to an $1/f^\alpha$-noise with the exponent $\alpha = \gamma_{st}$, calling for an re-investigation of numerical and experimental results.

5.8 Pair-Contact Process with Diffusion

A controversially discussed model is the so-called **pair-contact process with diffusion** (PCPD). This model generalises the ordinary pair-contact process (PCP) *without* diffusion (see Sect. 3.3.3). More specifically the PCPD is a stochastic process of diffusing particles A which interact according to the reaction-diffusion scheme

$$\begin{aligned} \text{fission:} &\qquad 2A \to 3A\,, \\ \text{annihilation:} &\qquad 2A \to \emptyset\,. \end{aligned}$$

The PCPD was originally suggested by Grassberger [240] and after 15 years rediscovered by Howard and Täuber [309], in the form of a (non-renormalisable) field-theory which does not take into account the 'fermionic' constraint that each site can be occupied by at most a single particle. Non-perturbative studies of lattice models with the 'fermionic' constraint were started in [126]. The model exhibits a continuous non-equilibrium phase transition caused by the competing character of fission and annihilation. At the time of writing, no consensus has been achieved on whether this transition belongs to one of the known universality classes or else represents a new kind of absorbing phase transition. A recent review article [274] discusses in detail

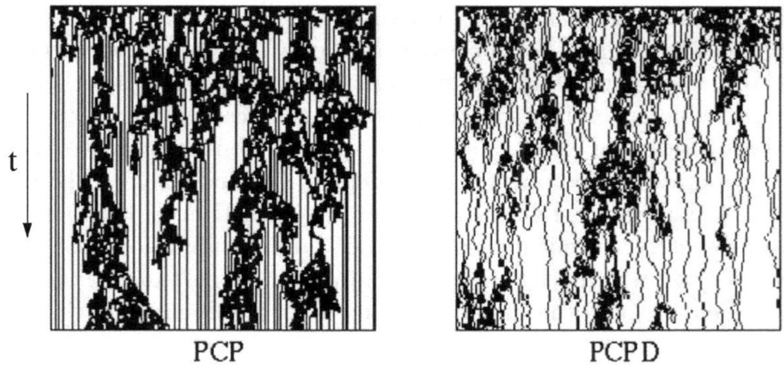

Fig. 5.9 One-dimensional pair-contact process starting with a fully occupied lattice at criticality. Left: In the pair-contact process without diffusion (PCP) solitary particles cannot diffuse, leading to frozen patterns of separated vertical lines. Right: In the PCPD, where individual particles are allowed to diffuse, offspring production can be restarted after long times when two diffusing particles meet, leading to a very different visual appearance of the process. Reprinted with permission from [272].

the ongoing debate about the PCPD. In the following we summarise (and slightly update) the main points.

As we have seen in Sect. 3.3.3, the pair-contact process *without* diffusion is a model with infinitely many absorbing states which nevertheless falls into the DP universality class. The pair-contact process *with* diffusion, however, has only two absorbing states, namely, the empty lattice and a homogeneous state with a single diffusing particle. Hence, introducing diffusion into the pair-contact process can be regarded as a *singular perturbation* which changes the structure of the absorbing phase. In fact, even the visual appearance changes significantly, as may be seen by comparing the left and right panels of Fig. 5.9.

In contrast to DP, the absorbing phase of the PCPD is characterised by an algebraic decay driven by particle annihilation. This has been taken by some as an indication that the PCPD might represent a universality class different from DP. This view is supported by the numerically confirmed observation that the upper critical dimension of the process is $d_c = 2$ instead of 4. The mostly studied form of the PCPD considers the evolution of a population of particles of a single species A, with the following reactions

$$\begin{cases} AA\emptyset \longrightarrow AAA & ;\ \text{with rate } (1-p)(1-D)/2 \\ \emptyset AA \longrightarrow AAA & ;\ \text{with rate } (1-p)(1-D)/2 \\ AA \longrightarrow \emptyset\emptyset & ;\ \text{with rate } p(1-D) \\ A\emptyset \longleftrightarrow \emptyset A & ;\ \text{with rate } D \end{cases} \tag{5.78}$$

There is a second-order phase transition, and a general agreement on its location $p_c(D)$ exists between different groups, see Fig. S.5 for the phase diagram in $1D$. However, even after intensive efforts, the numerical results for the critical exponents are unclear, so a surprising variety of possible scenarios have been proposed: Some authors believe that the PCPD should represent a novel universality class with a unique set of critical exponents, while others expect to find *two* different universality classes along the critical line, depending on the diffusion rate and/or the number of space dimensions. According to another viewpoint, the exponents might depend continuously on the diffusion constant. According to yet a third opinion, the PCPD in 1+1 dimensions can be interpreted as a cyclically coupled DP- and annihilation-process which crosses over to ordinary DP behaviour after very long time, in spite of the algebraic behaviour in the inactive phase. Table 5.6 documents the different conclusions reached by several authors by using different numerical methods.[8] See Table 4.4 for an illustrative example of the difficulties, encountered in the necessarily purely numerical studies, in reaching a clear and non ambiguous conclusion on the truly asymptotic scaling behaviour of the PCPD.

The broad spectrum of opinions is caused by numerical and conceptual difficulties. It turns out that various quantities, such as the particle or the pair densities, do not exhibit clean power laws, rather they are superseded by strong corrections to scaling which extend over the entire numerically accessible range. A reliable interpretation of the existing numerical results is made difficult by the lack of understanding of what these corrections to scaling in the PCPD really are. It is intriguing that the entire arsenal of available techniques should fail so utterly in attempts to clarify the scaling behaviour. Considerably more work will be needed in order to clarify the widely different and partially contradicting conclusions formulated so far on this apparently so simple-looking model.

5.9 First-Order Phase Transitions

So far, we have discussed continuous (second-order) phase transitions. In this section, we shall focus on discontinuous (first-order) phase transitions. Although theoretical physicists often prefer to study continuous phase transitions because of their universal properties, discontinuous transitions do occur frequently in nature and in general require less fine-tuning of control parameters. Already, but not only, for this reason, the study of discontinuous phase transitions under non-equilibrium conditions, especially with absorbing states, is in important field.

Here, we first address the question of whether first-order transitions can occur in low-dimensional non-equilibrium systems. In particular we shall study characteristic phenomena attributed to first-order transitions such as

[8] Concerning the critical behaviour near to a surface, $\beta_1/\nu_\perp = 0.72(1)$ was estimated for the ordinary transition and $\beta_1/\nu_\perp \simeq 1.11$ for the extraordinary transition, both in 1+1 dimensions and for $D \lesssim 0.5$ [40].

Authors	Ref.	D	δ	β	z	$\beta/\nu_\perp$
Carlon, Henkel and Schollwöck	[126]	0.1	-	-	1.87(3)	0.50(3)
		0.15	-	-	1.84(3)	0.49(3)
		0.2	-	-	1.83(3)	0.49(3)
		0.35	-	-	1.72(3)	0.47(3)
		0.5	-	-	1.70(3)	0.48(3)
		0.8	-	-	1.60(5)	0.51(3)
Hinrichsen	[288]	0.1	0.25	< 0.67	1.83(5)	0.50(3)
Ódor	[490, 496]	0.05	0.273(2)	0.57(2)	-	-
		0.1	0.275(4)	0.58(1)	-	-
		0.2	0.268(2)	0.58(1)	-	-
		0.5	0.21(1)	0.40(2)	-	-
		0.9	0.20(1)	0.39(2)	-	-
Park, Hinrichsen and Kim	[506]	*	0.236(10)	0.50(5)	1.80(2)	-
Park and Kim	[508]	*	0.241(5)	0.496(22)	1.80(10)	-
		*	0.242(5)	0.519(24)	1.78(5)	-
Dickman and de Menezes	[176]	0.1	0.249(5)	0.546(6)	2.04(4)	0.503(6)
		0.5	0.236(3)	0.468(2)	1.86(2)	0.430(2)
		0.85	0.234(5)	0.454(2)	1.77(2)	0.412(2)
Ódor	[492]	0.05	0.216(9)	0.411(10)	2.0(2)	0.53(7)
		0.1	0.206(7)	0.407(7)	1.95(1)	0.49(2)
		0.2	0.217(8)	0.402(8)	1.95(1)	0.46(3)
		0.5	0.206(7)	0.402(8)	1.84(1)	0.41(2)
		0.7	0.214(5)	0.39(1)	1.75(1)	0.38(2)
Kockelkoren, Chaté	[395]	*	0.200(5)	0.37(2)	1.70(5)	-
Barkema and Carlon	[40]	0.1	0.17	-	-	-
		0.2	0.17	-	1.70(1)	0.28(4)
		0.5	0.17(1)	-	-	0.27(4)
		0.9	0.17	-	1.61(3)	-
Noh and Park	[488]	0.1	0.27(4)	0.65(12)	1.8(2)	0.50(5)
Park and Park	[509]	*	0.20(1)	-	-	-
Hinrichsen	[289]	*	< 0.185	< 0.34	< 1.65	-
de Oliveira and Dickman	[151]	0.1	-	-	2.08(15)	0.505(10)
		0.5	-	-	2.04(5)	0.385(11)
		0.85	-	-	1.88(12)	0.386(5)
Kwon and Kim	[404]	*	-	-	1.61(1)	-
Smallenburg and Barkema	[574]	0.5	0.16(1)	0.28(2)	1.58	-

Table 5.6 Estimates for the critical exponents of the 1D PCPD, according to various authors. The diffusion rate D is as in (5.78) and with sequential update; a star in the D-column is used for PCPD models with a different microscopic definition.

phase coexistence, hysteresis cycles and nucleation effects. Several lattice models are known which exhibit a first-order absorbing phase transition. Besides the ZGB model [647, 646, 91] and models with several absorbing states [178, 27], tricritical systems give us an opportunity to study first-order transitions [436, 244].

5.9.1 Stabilisation by Elimination of Minority Islands

In equilibrium statistical mechanics, it is not uncommon that systems, which exhibit a first-order phase transition in high space dimensions d, may display a second-order transition below a certain upper critical dimension. The reason is that in low dimensions fluctuations become more pronounced, destabilising the ordered phase. In non-equilibrium statistical physics the situation is similar. Therefore, the question arises under which conditions first-order phase transitions can be observed in low-dimensional, especially one-dimensional models.

Systems with a first-order phase transition are usually characterised by at least two stable ordered phases. For example, in the subcritical regime $T < T_c$ of the two-dimensional Ising model there are two stable magnetised states. To ensure their stability, the Ising model provides a robust mechanism for the elimination of minority islands of the opposite phase generated by thermal fluctuations. In the kinetic Ising model this mechanism relies on the fact that the boundary of an island costs energy, leading to an effective surface tension. Attempting to minimise surface tension, the island feels an attractive 'force' and begins to shrink. This force decreases algebraically as $1/r$ with the typical radius r of the island so that thermal fluctuations of any size are safely eliminated.

Because of the $\mathbb{Z}_2$-symmetry under spin reversal, the ordered phases of the Ising model are both equally attracting. Thus, starting from a disordered state with zero magnetisation, one observes coarsening patterns of ordered domains. However, if an external field is applied, one type of minority island becomes unstable above a certain critical size. Since there is a finite probability of generating such islands by fluctuations, one of the two ordered phases eventually takes over, i.e., the system undergoes a first-order phase transition.

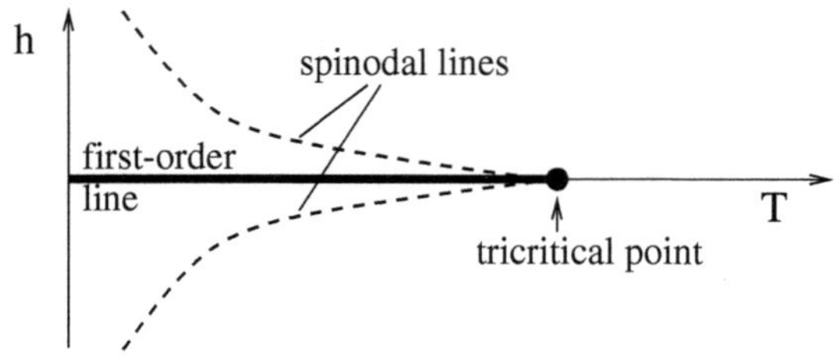

Fig. 5.10 Schematic phase diagram of Toom's north-east-centre voting model as a function of the temperature T and the external magnetic field h. In the coexistence region between the two spinodal lines the ordered phases are both thermodynamically stable.

Turning to non-equilibrium systems, there are even more efficient mechanisms for the elimination of minority islands. An interesting example is

Toom's two-dimensional north-east-centre **voting model** [597, 57]. In contrast to the ordinary Ising model with heat bath dynamics, where spins are oriented probabilistically according to the field of the neighbouring spins plus the external field h, a spin in Toom's model is oriented according to the field produced by the spin itself and its 'northern' and 'eastern' neighbours *only*. It can be shown that this type of spatially biased update does *not* obey detailed balance in the stationary state, hence the model is genuinely out of equilibrium.

Like the kinetic Ising model, Toom's model displays a discontinuous phase transition between two stable ordered phases below T_c. However, in contrast to the Ising model, where smooth domain walls perform an unbiased diffusive motion, the domain walls in Toom's model propagate with a velocity that depends on their orientation. Because of this anisotropy, minority islands quickly assume a triangular shape and begin to shrink at *constant* velocity. Thus, the effective 'force' by which an island shrinks is independent of its size, leading to much more stable phases as in the standard kinetic Ising model. In fact, the elimination mechanism is so robust that the ordered phases may even remain stable if an oppositely oriented external field h is applied. In order to flip the whole system, the intensity of this field has to exceed a certain threshold. The corresponding phase stability boundaries (also called spinodal lines) are sketched in Fig. 5.10. Between these lines, the two ordered states are both thermodynamically stable, i.e., they coexist in a whole region of the parameter space. Crossing the coexistence region by varying the external field, the magnetisation of the system follows in a hysteresis loop.

5.9.2 First-Order Transitions in One Spatial Dimension

According to a classical theorem by Landau, *first-order phase transitions are impossible in one-dimensional equilibrium systems.* In non-equilibrium statistical mechanics, however, several models have been shown to display a first-order transition even in one spatial dimension. For example, the (1+1)-dimensional Ziff-Gulari-Barshad model [647] for heterogeneous catalysis (cf. Sect. 3.3.5 and exercise 30) is known to exhibit a first-order phase transition which relies on the interplay of three different kinds of particles. Similarly, models for phase separation on a ring [213] use more than two different species of particles. Another example is the so-called *bridge model* [237] for bidirectional traffic on a single lane, where special boundary conditions induce a discontinuous transition in the currents. Even more subtle is the mechanism in the two-species model introduced by Oerding et al., where a first-order phase transition is induced by fluctuations [500]. Therefore, in attempting to comprehend the full range of first-order phase transition under non-equilibrium conditions, it would be interesting to seek the simplest (1+1)-dimensional model which exhibits a discontinuous transition. By 'simple' we mean that such a model should involve only one species of particle evolving by local dynamic rules without macroscopic currents, conservation laws, and unconventional symmetries. Moreover, the choice of the boundary conditions should be irrelevant.

The prototype of such a dynamic process is the one-dimensional Glauber-Ising model at zero temperature in a magnetic field, which is also known as compact directed percolation (see Sect. 5.3). The model evolves random-sequentially by randomly selecting a pair of adjacent sites. If the two spins are in opposite states, they are aligned with the probabilities p and $1-p$, respectively:

$$\uparrow\downarrow, \downarrow\uparrow \xrightarrow{p} \uparrow\uparrow; \qquad \uparrow\downarrow, \downarrow\uparrow \xrightarrow{1-p} \downarrow\downarrow . \tag{5.79}$$

As there are no fluctuations inside ordered domains, the two fully ordered configurations are absorbing.

Obviously, $p-1/2$ plays the role of an external field h. For $h \neq 0$, one of the two absorbing states is stable while the other one is unstable under small perturbations. Therefore, starting with random initial conditions, the system approaches one of the fully magnetised states in an exponentially short time since minority islands of the opposite phase shrink linearly with time. For $h = 0$, however, the two absorbing states of the Glauber model are only *marginally* stable against perturbations. For example, by flipping a single spin in a fully magnetised domain, a pair of kinks is created. These kinks perform an unbiased random walk until they annihilate one another. Thus, minority islands do not shrink by virtue of an attractive force, rather they are eliminated solely by the circumstance that random walks in one dimension are recurrent [543]. On the one hand the survival probability of such an island is known to decay as $1/\sqrt{t}$, on the other hand it reaches a typical size of the order $\sqrt{t}$. Therefore, averaging over many independent samples, the mean size of minority islands is asymptotically constant. This demonstrates that the ordered phase is only marginally stable against perturbations at $T=0$. In fact, raising the temperature above zero by introducing a small rate for spontaneous spin flips, the first-order transition in the Glauber-Ising model is lost.

5.9.3 Impossibility of Discontinuous Phase Transitions in Fluctuating One-Dimensional Systems

Is it generally possible to observe first-order transitions in *fluctuating* (1+1)-dimensional two-state systems with short-range interactions? Regarding this question it is instructive to study the (1+1)-dimensional triplet creation pro-

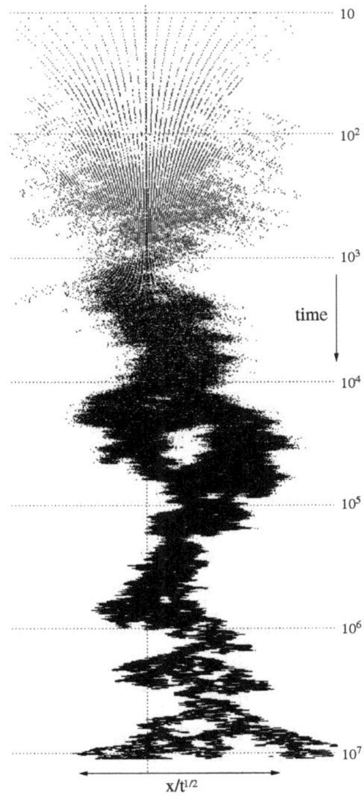

Fig. 5.11 Apparent first-order behaviour in the triplet creation process $3A \to 4A$, $A \to \emptyset$. The space coordinate x is rescaled by the mean cluster extension $t^{1/2}$ and plotted versus $\ln t$. Initially the domains appear to be compact. Only after a transient of 10^6 update steps the ordered phase disintegrates and one eventually observes DP-like branching.

cess introduced by Dickman and Tomé [178] which resembles the reaction scheme

$$3A \to 4A\,, \qquad A \to \emptyset \tag{5.80}$$

combined with single particle diffusion. In this model, the high-density phase is not strictly absorbing, rather islands of unoccupied sites are spontaneously created in the bulk so that one of the ordered states fluctuates. In numerical simulations, it was observed that above a certain tricritical point the second-order DP transition line splits up into two spinodal lines, where the transition becomes first order. Moreover, the order parameter seemed to follow a hysteresis loop when the parameter for offspring production was varied. However, refined simulations [286] revealed that the transition only *appears* to be a discontinuous one but actually crosses over to DP after a sufficiently long time. Therefore, it seems that the entire transition line in this model belongs to DP. The gradual disintegration of ordered domains can also be observed in a rescaled logarithmic space-time plot, as demonstrated in Fig. 5.11.

This example illustrates that is very difficult, if not impossible, to find a short-range model with fluctuating ordered states that displays a first-order

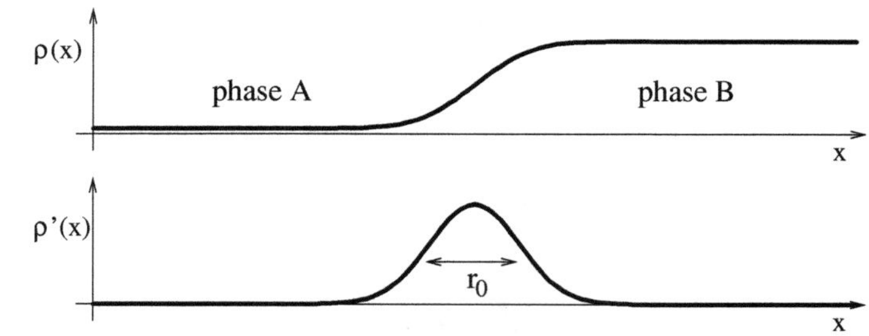

Fig. 5.12 Domain wall between two stable ordered phases A and B in one spatial dimension. In models with short-range interactions and no conservation laws the derivative of the profile is expected to decay exponentially.

transition in 1+1 dimensions. Obviously such a model would need a much more robust mechanism for the elimination of minority islands than in the Glauber model. A simple random walk of a pair of kinks is not sufficient, rather there has to be an attractive force which prevents small minority islands from growing. However, without long-range interactions (mediated by explicit interactions or conservation laws) a domain wall between two ordered domains should always have a finite width, i.e. the order parameter profile is localised and its derivative is expected to decay exponentially, as sketched in Fig. 5.12. If the ordered phase fluctuates there is always a finite probability of generating a minority island which is sufficiently large that its domain walls do not interact. As in the Glauber model such an island could reach any size and therefore disintegrate the ordered phase. Consequently, on scales much larger than this critical island size, the process is expected to cross over to DP.

These considerations suggest that numerical evidence for first-order transitions in one-dimensional models with fluctuating ordered states should always be taken with caution, unless one can clearly identify the mechanism that stabilises the ordered phase. Usually such a mechanism requires long-range interactions or conservation laws.

The impossibility of a discontinuous transition in one spatial dimension in models with short-range interactions explains the behaviour of the tricritical contact process discussed in Sect. 5.4. This model displays a first-order transition with an associated tricritical point only in two or more dimensions, whereas in one spatial dimension the transition always belongs to the DP universality class.

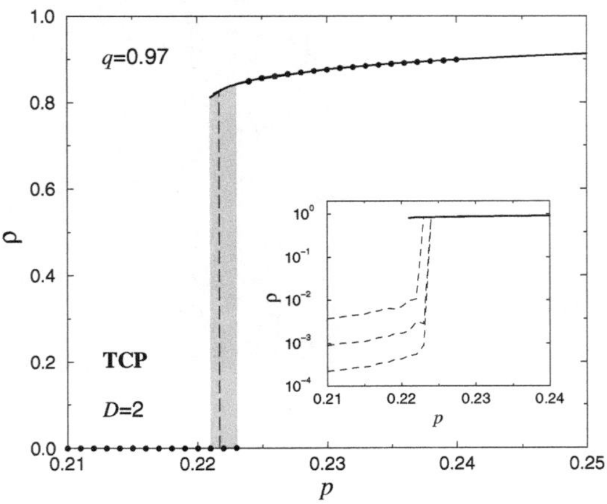

Fig. 5.13 Hysteresis loop of the order parameter in the first-order regime of the tricritical contact process (TCP). The solid line is obtained from steady-state measurements at zero field for various system sizes. During the simulations the control parameter p is slowly decreased until the absorbing state is reached. The circles correspond to data which are obtained from finite-field simulations and slowly increasing parameter p. In the shadowed area the active and the absorbing phase coexist. The dashed line marks the transition point p_c obtained from a stability analysis of separated phases (see text). The inset shows how the order parameter switches back for growing p. Here the three dashed lines correspond to three different field values (from $h = 4 \cdot 10^{-4}$ (top) to $h = 3 \cdot 10^{-5}$). As can be seen, there exists a well defined upper limit of the supercooled low density phase. Reproduced from [436] with kind permission of Springer Science and Business Media.

5.9.4 Phase Coexistence and Hysteresis Cycles

A general phenomenon associated with first-order transitions is the presence of hysteresis cycles when the transition is crossed. In equilibrium, hysteresis effects manifest themselves as supercooling and superheating.

Analogous effects occur in the case of first-order absorbing phase transitions [436]. This is shown in Fig. 5.13 for the first-order regime of the tricritical contact process in two dimensions (see Sect. 5.4). Decreasing the control parameter τ the order parameter jumps at a certain point p_o from a finite value to zero. On the other hand, the order parameter jumps back to a finite value at p_u on cooling, i.e., with increasing control parameter. Note that this part of the hysteresis can only be observed if a small external field is applied which allows the system to escape from the absorbing state (see inset of Fig. 5.13 and [436] for details). Thus the first-order regime of the tricritical contact process exhibits a finite hysteresis, where the two phases coexist between $p_o < p < p_u$.

Snapshots of the system within the supercooled and superheated state are shown in Fig. 5.14. Analogous to equilibrium, the dynamics of the discontin-

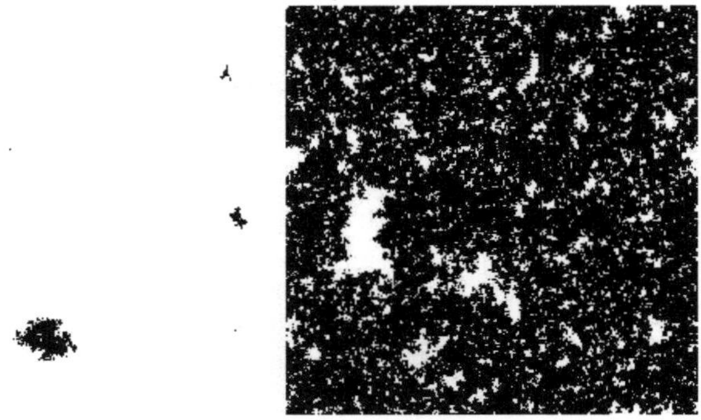

Fig. 5.14 Snapshots of the two coexisting phases in the first-order regime of the tricritical contact process. The left figure shows a typical low density configuration (supercooled phase), where small droplets are triggered by a weak external field. The right figure displays a typical high density configuration (superheated phase). Here, order parameter fluctuations lead to droplets of various sizes. In both cases the droplets are subcritical, i.e., they disappear after a certain lifetime. Reproduced from [436] with kind permission of Springer Science and Business Media.

uous transition is characterised by nucleation effects, i.e., the formation of critical seeds or droplets. The weak external field creates small fluctuating droplets of the minority phase within the majority phase. Subcritical droplets disappear after a certain lifetime whereas critical droplets grow continuously, creating larger and larger domains of the favoured phase. Examples of subcritical droplets can be seen in Fig. 5.14.

A crucial problem of non-equilibrium first-order transitions concerns the location of the transition point. In equilibrium, the transition point is related to a thermodynamical potential such as the free energy. At the critical temperature the free energy of both phases are equal. Unfortunately, this definition cannot be applied to non-equilibrium phase transition. Alternatively one can define the first-order transition point in terms of the drift velocity of a moving interface between both phases. In equilibrium the interface velocity is zero whereas it is non-zero if one of the phases is favoured by the dynamics.

This definition allows one to estimate the critical point as follows. First, the system is prepared in a state with a stripe of the active phase bounded by two interfaces [91, 436, 244]. Depending on the control parameter, the active phase grows or shrinks, i.e., the system reaches after a transient either the absorbing phase or a steady-state of a homogeneous non-zero particle density (see Fig. 5.15). Typically one obtains a parameter range where both phases are equally favoured by the dynamics. As the system size and the simulation time are increased, the size of this parameter range shrinks, leading to a well-defined value of the transition point [436].

In the case of the TCP the estimate of the critical point p_c indicates that the hysteresis is asymmetric and quite narrow (see Fig. 5.13). Finally we mention that the interface behaves like generic fluctuating non-equilibrium sur-

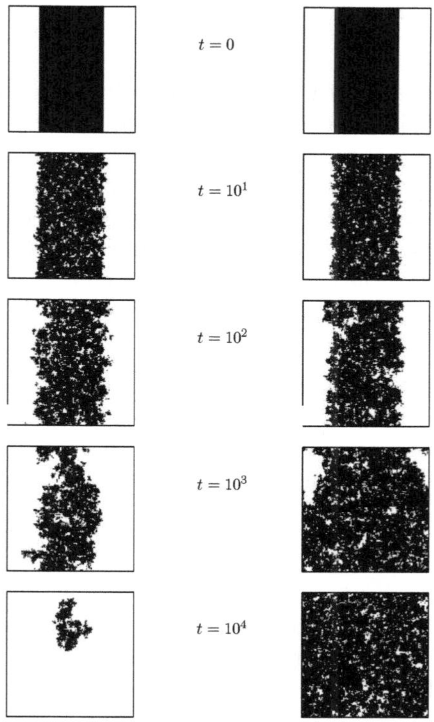

Fig. 5.15 Estimating the critical point of a first-order phase transition by analysing the interface motion. The figure shows snapshots of the tricritical contact process with periodic boundary conditions close to the first-order transition. Starting with a stripe of particles the dynamics is attracted either by the empty lattice (left, $p < p_c$) or by a steady-state of a homogeneous non-zero particle density (right, $p > p_c$). Reproduced from [436] with kind permission of Springer Science and Business Media.

faces at first-order transition points, i.e., the width of the interface increases for sufficiently small time as $t^{1/3}$ [376, 244].

In summary, discontinuous non-equilibrium transitions share many properties with their equilibrium counterparts. For example, one can observe hysteresis loops and metastability effects such as supercooling and superheating. As in equilibrium, first-order transitions do not occur in fluctuating one-dimensional systems. Nevertheless first-order transitions under non-equilibrium conditions may differ from discontinuous equilibrium transitions. For example, in Fig. 5.15 an initially planar interface between the two phases roughens as $t^{1/3}$ at the transition point [376, 244], while under equilibrium conditions one expects the width to grow only as $t^{1/4}$.

5.10 Crossover Phenomena

Crossovers, i.e., passages from one universality class to another, are well-known from equilibrium phase transitions. In the topological language of renormalisation group theory crossover phenomena are understood in terms of competing fixed points (see e.g. [520]). Suppose that more than one fixed point are embedded into the critical surface and suppose that one of the two fixed points is characterised by a certain number of irrelevant scaling fields and one relevant scaling field. If this relevant scaling field is weak, a trajectory will approach that fixed point until it is eventually driven away to another fixed point.

In the framework of phenomenological scaling theory, crossover effects can be described by including the relevant scaling field as an additional parameter in the corresponding scaling forms. For example, if this parameter was denoted by κ, the equation of state would read

$$\varrho_{\mathrm{s}}(\tau, h, \kappa) \simeq \lambda^{-\beta}\, \tilde{r}(\lambda\tau, \lambda^{\sigma}h, \lambda^{\phi}\kappa)\,, \tag{5.81}$$

where τ denotes the distance from criticality. The exponent ϕ which is associated with the scaling field κ is the so-called **crossover exponent**. A negative exponent ϕ, meaning that the scaling field is irrelevant, gives rise to corrections to the leading scaling behaviour [620] (cf. Sect. 2.3.4). These non-universal corrections are often termed **confluent singularities**.

The situation is different for a relevant scaling field, i.e., $\phi > 0$. Setting $\lambda\tau = 1$ the above scaling form reads for $\tau > 0$ and $h = 0$

$$\varrho_{\mathrm{s}}(\tau, \kappa) \;\simeq\; \tau^{\beta}\, \tilde{r}(1, 0, \kappa\tau^{-\phi})\,. \tag{5.82}$$

For small $\kappa \ll \tau^{\phi}$ one observes again corrections to ordinary scaling. However, approaching the transition point, $\tau \to 0$, the scaling argument diverges, leading to a different scaling behaviour. Thus, (5.81) describes a crossover between two different universality classes controlled by the scaling field κ. Usually, the point at which

$$\mathrm{O}(\kappa|\tau|^{-\phi}) = 1 \tag{5.83}$$

is considered as the point where the crossover takes place [546].

Classical examples of crossover phenomena are ferromagnetic systems exhibiting weak uniaxial spin anisotropy (see for e.g. [6]). Similar crossover phenomena are also known from non-equilibrium systems, including crossover effects in self-organised critical systems [196, 588, 428], absorbing phase transitions [432, 434, 436], as well as depinning transitions [553]. Here we shall consider three different examples which are related to the models considered so far. First, the crossover from ordinary DP to tricritical DP is reviewed. Second, we study the crossover from DP to compact DP in the one-dimensional

Domany-Kinzel automaton. Third, we discuss systems with variable interaction ranges which cross over to mean-field behaviour when the interaction range is increased.

5.10.1 Crossover from DP to TDP

The starting point of the analysis is the mean-field approximation of tricritical directed percolation in Sect. 5.4. In this approximation the steady-state of the order parameter is given, at zero-field $h = 0$, by

$$\varrho_{\mathrm{s}}(\tau, g) \;=\; -\frac{g}{2c} + \sqrt{\frac{\tau}{c} + \left(\frac{g}{2c}\right)^2}\,. \tag{5.84}$$

Obviously, the order parameter behaviour can be described in terms of a generalised homogeneous function, i.e.,

$$\varrho_{\mathrm{s}}(\tau, \kappa) = \lambda^{-\beta_{\mathrm{t}}}\,\tilde{r}(\tau\lambda, 0, g\lambda^{\phi}) \tag{5.85}$$

with the crossover scaling field $\kappa = g$ and the exponents $\beta_{\mathrm{t}} = 1/2$ and $\phi = 1/2$. Omitting the metric factor c, the crossover scaling function reads

$$\tilde{r}(x, 0, y) = -\frac{y}{2} + \sqrt{x + \left(\frac{y}{2}\right)^2}\,. \tag{5.86}$$

The scaling function $\tilde{r}(x, 0, 1)$ is of particular interest since its asymptotic behaviour

$$\tilde{r}(x, 0, 1) \;\sim\; \begin{cases} x^{\beta_{\mathrm{t}}} & \text{with } \beta_{\mathrm{t}} = 1/2 \text{ for } x \to \infty \\ x^{\beta_{\mathrm{DP}}} & \text{with } \beta_{\mathrm{DP}} = 1 \;\; \text{for } x \to 0 \end{cases} \tag{5.87}$$

reflects the crossover between both universality classes.

Beyond mean-field theory, an analogous scaling behaviour can be observed, e.g. for the tricritical contact process (TCP). As discussed in Sect. 5.4, the microscopic model parameters p and q of the TCP are not identical to the scaling fields τ and g. Instead there is a certain window of scaling, where the coarse-grained variables τ and g can be replaced by $p - p_{\mathrm{c}}(q)$ and $q_{\mathrm{t}} - q$, respectively [436]. In this case the scaling form at zero field $h = 0$ is given by

$$\varrho_{\mathrm{s}}(p, q) \simeq \lambda^{-\beta_{\mathrm{t}}}\,\tilde{r}_q(\delta p\lambda, 0, \delta q\lambda^{\phi})\,, \tag{5.88}$$

with $\delta p = p - p_{\mathrm{c}}(q)$ and $\delta q = q_{\mathrm{t}} - q$. As indicated by the index q this scaling form is valid only if the critical point $p_{\mathrm{c}}(q)$ is approached along paths with constant q. Moreover, the simulations have to be performed in a way that the distance to the phase boundary is smaller than the distance to the tricritical point, i.e., $\delta p \ll \delta q$.

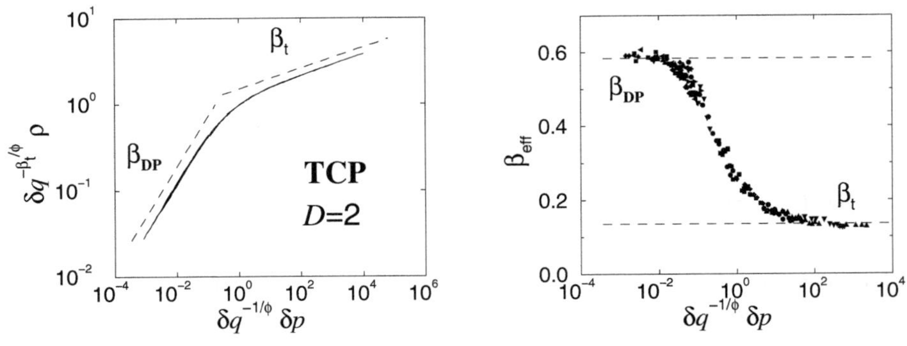

Fig. 5.16 The crossover scaling function of the order parameter at zero field (left). The dashed lines indicate the asymptotic power laws of ordinary and tricritical DP. The right figure shows the associated effective exponent $\beta_{\rm eff}$. Both asymptotic scaling regimes ($\beta_{\rm t}$ and $\beta_{\rm DP}$) as well as the crossover regime are clearly recovered. Reproduced from [436] with kind permission of Springer Science and Business Media.

If both conditions are fulfilled, then different curves corresponding to different values of q collapse onto a single curve if the rescaled order parameter $\varrho\, \delta q^{-\beta_{\rm t}/\phi}$ is plotted as a function of the rescaled control parameter $\delta p\, \delta q^{-1/\phi}$. Such a data collapse is shown in the left panel of Fig. 5.16. As can be seen, both asymptotic behaviours, tricritical as well as ordinary directed percolation, are recovered. However, as the entire crossover region spans several decades it is usually difficult to identify small but systematic differences between different curves. It is therefore instructive to scrutinise the quality of the representation by plotting the so-called effective exponent [548, 448, 110, 432, 434]

$$\beta_{\rm eff} = \frac{\partial}{\partial \ln x} \ln \tilde{r}_{_q}(x,0,1)\,. \tag{5.89}$$

The effective exponent is shown in the right panel of Fig. 5.16. The data collapse of $\beta_{\rm eff}$ over more than six decades reflects the crossover between the universality class of TDP and DP. A detailed discussion of this crossover scenario, including an analysis of the order parameter fluctuations and of the dynamical scaling behaviour, is presented in [436] and [244], respectively.

5.10.2 Crossover from DP to CDP

In Sect. 3.3.1 we have seen that the phase diagram of the one-dimensional Domany-Kinzel (DK) automaton comprises two different universal scaling behaviours: A critical line of second-order DP-like phase transitions and its upper terminal point which belongs to the universality class of compact DP

(CDP). Moving along the critical line it is possible to examine the crossover from the DP to the CDP scaling behaviour. For the sake of concreteness, we focus here on the order parameter decay $\varrho(t)$ at various critical points along the DP-line. As pointed out in Sect. 4.3.4 the duality-symmetry factor μ can be used to parametrise the critical line, or more precisely, to parameterise the distance to the terminal point along the critical line. Approaching the terminal point the parameter μ vanishes, reflecting the violation of the rapidity-reversal for CDP (see Fig. 4.16). On a phenomenological level the crossover can be understood as a growing mean size of active islands when approaching the terminal point (see [287]).

Similar to (5.81) this crossover is described by the scaling form

$$\varrho \simeq \lambda^{-\beta_{\text{CDP}}}\, \tilde{r}(t\lambda^{-\nu_{\parallel,\text{CDP}}}; \tau\lambda, \kappa\lambda^{\phi}) . \tag{5.90}$$

The crossover exponent $\phi = 2/d$ is known exactly from a field theoretical treatment [331]. Sufficiently close to the terminal point the phase transition line is given by $\kappa = (1-p_2) \propto (p_1 - 1/2)^{\phi}$. As can be seen from Fig. 4.16, the parameter μ^2 vanishes linearly with $p_1 - 1/2$, leading to $\kappa \propto \mu^{2\phi}$. Setting $\kappa\lambda^{\phi} = 1$, the order parameter is described along the critical line ($\tau = 0$) by

$$\varrho \simeq \tilde{r}(t\mu^{2\nu_{\parallel,\text{CDP}}}; 0, 1) , \tag{5.91}$$

where $\beta_{\text{CDP}} = 0$ is already recognised. The asymptotic scaling behaviour of the scaling function $\tilde{r}$ is given by

$$\tilde{r}(x; 0, 1) \sim \begin{cases} x^{-\alpha_{\text{DP}}} & \text{for } x \to \infty , \\ \text{const} & \text{for } x \to 0 . \end{cases} \tag{5.92}$$

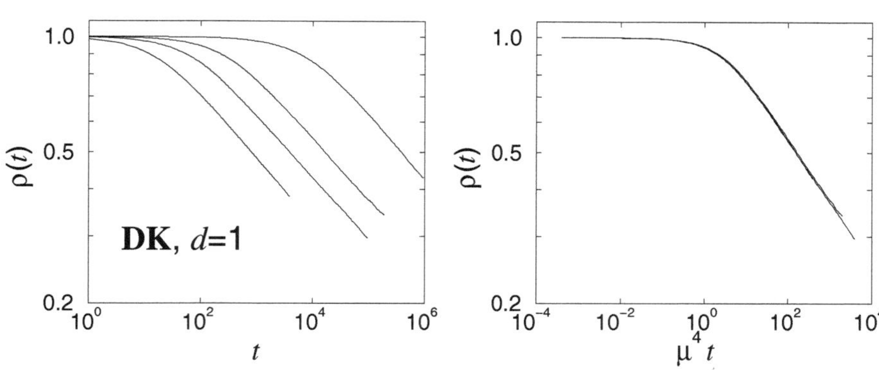

Fig. 5.17 The order parameter decay along the critical line of the one-dimensional Domany-Kinzel (DK) automaton. Unscaled data is shown in the left figure ($p_1 = 0.54865$, $p_1 = 0.52469$, $p_1 = 0.512425$, $p_1 = 0.5024969$, from left to right). A data collapse is obtained by rescaling the order parameter according to (5.91) with $\nu_{\parallel,\text{CDP}} = 2$ and $\phi = 2/d = 2$ [331] (see right figure). Both asymptotic scaling regimes ($\alpha_{\text{CDP}} = 0$ and α_{DP}) are recovered. Reprinted with permission from [435].

Numerical simulations confirm this crossover scaling form. Corresponding numerical data are shown in Fig. 5.17. A crossover between both scaling regimes ($\alpha_{\mathrm{CDP}} = 0$ and α_{DP}) takes place at

$$\mathrm{O}(\kappa^{\nu_{\|,\mathrm{CDP}}/\phi} t) = \mathrm{O}(\mu^4 t) = 1\,. \tag{5.93}$$

Note that the full crossover again spans several decades. Alternatively, the crossover from the DP to the CDP universality class can be described via the scaling functions of the survival probability P_{sur} or in terms of the scaling function of the number of active sites N_{a}, respectively.

5.10.3 Crossover to Mean-Field Scaling Behaviour

The critical behaviour of a system exhibiting a second order phase transition with non-mean-field scaling behaviour is strongly affected by the range of interactions R. The longer the range of interactions, the stronger will be the reduction of the critical fluctuations, and in the limit of infinite interaction the system will be characterised by mean-field scaling. According to the **Ginzburg criterion** [235, 19], mean-field-like behaviour occurs even for finite interaction ranges sufficiently far away from the critical point. If one approaches the transition point in such a situation, a crossover from mean-field to a different type of scaling behaviour is observed.

From the renormalisation group point of view, the crossover reflects the instability of the mean-field fixed point associated with the mean-field scaling. This crossover is described by a corresponding exponent ϕ which is known exactly due to the Ginzburg criterion. This crossover scenario is considered as a paradigm of crossover phenomena and it was intensively investigated in equilibrium (including [21, 52, 477, 450, 451, 448, 513, 514, 110, 389]) as well as in non-equilibrium systems [432, 434].

Here, we discuss several systems belonging to the Manna universality class. Changing the range of interactions the full crossover from the non-mean-field to mean-field critical behaviour can be observed. In the standard versions of the CLG model, the CTTP, and the Manna model (see Sect. 5.7), particles of active sites are moved to nearest neighbours only, i.e., the interaction range is $R = 1$. It is straightforward to generalise these models to a finite interaction range [432] by moving particles not only to nearest neighbours but to sites randomly selected within a radius R. It is important that the simulations are performed in the so-called critical crossover limit, i.e., the associated correlation length has to be sufficiently large in the whole crossover region [451, 513]. For any finite interaction range R the phase transition is characterised by a non-mean-field scaling behaviour, whereas mean-field scaling occurs for infinite interactions. For example, the order parameter ϱ of the two-dimensional

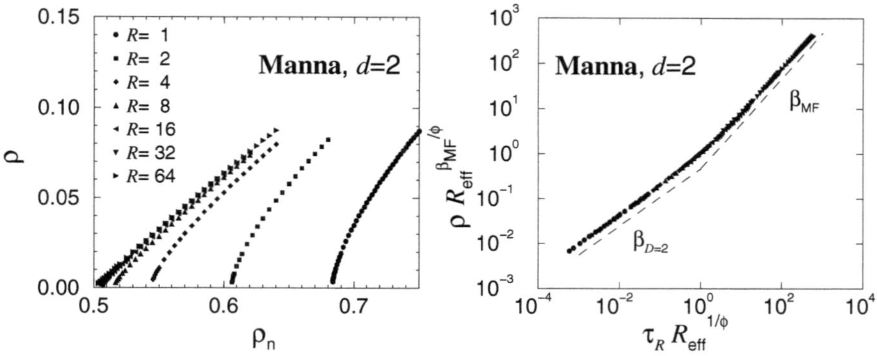

Fig. 5.18 The order parameter (left) of the two-dimensional Manna model for various values of the interaction range R. With increasing range the critical density tends to the mean-field value $\varrho_{\mathrm{n,c,R}} = 1/2$. As can be seen, the power-law behaviour of ϱ changes with increasing R. The data is obtained from simulations of various system sizes $L \leq 2048$. The right figure shows the corresponding scaling plot. The curves are rescaled according to (5.95). The scaling variable $\tau_{\mathrm{R}} = \varrho_{\mathrm{n}} - \varrho_{\mathrm{n,c,R}}$ denotes the distance to the critical point. The dashed lines correspond to the power-law behaviour of the two-dimensional model and to the mean-field scaling behaviour, respectively.

Manna model is plotted in Fig. 5.18 for various interaction ranges at zero field. As can be seen, the transition point $\varrho_{\mathrm{n,c,R}}$ depends on the range of interactions and the power-law behaviour of ϱ changes with increasing R.

As mentioned above, the crossover scaling function has to incorporate the range of interactions as a third scaling field. But the parameter R is not an appropriate quantity to describe the scaling behaviour since it describes the maximum range of interactions. Furthermore, the actual strength of the interactions may depend on the given lattice structure. Thus instead of R the effective interaction range [477]

$$R_{\mathrm{eff}}^2 = \frac{1}{z_n} \sum_{i \neq j} |\boldsymbol{r}_i - \boldsymbol{r}_j|^2, \qquad \text{for} \qquad |\boldsymbol{r}_i - \boldsymbol{r}_j| \leq R \tag{5.94}$$

is often used for the scaling analysis. Here, z_n denotes the number of lattice sites within radius R. The values of the effective interaction ranges are listed for two- and three dimensional simple cubic lattices e.g. in Tables 6 and Table 7 of [433]. Using the effective range of interactions the equation of state obeys the scaling form ($\kappa = R_{\mathrm{eff}}^{-1}$)

$$\varrho_{\mathrm{s}}(\tau, h, R_{\mathrm{eff}}) \simeq \lambda^{-\beta_{\mathrm{MF}}}\, \tilde{r}(\tau_R \lambda, h\lambda^{\sigma_{\mathrm{MF}}}, R_{\mathrm{eff}}^{-1}\lambda^{\phi})\,. \tag{5.95}$$

Here, the distance to the critical point $\tau = \varrho_{\mathrm{n}} - \varrho_{\mathrm{n,c,R}}$ depends on the interaction range R. At zero-field, both scaling regimes are recovered

$$\tilde{r}(x,0,1) \sim \begin{cases} x^{\beta_{\mathrm{MF}}} & \text{for } x \to \infty \\ x^{\beta_d} & \text{for } x \to 0\,. \end{cases} \tag{5.96}$$

The crossover exponent ϕ is exactly known due to the Ginzburg criterion [235, 19]. It states that the mean-field picture is self-consistent in the active phase as long as the fluctuations within a correlation volume are small compared to the order parameter itself

$$\Delta_{\varrho} \ll \xi^d \, \varrho_{\mathrm{s}}^2 . \tag{5.97}$$

In the steady-state, ξ corresponds to the spatial correlation length that diverges at the critical point according to $\xi_{\perp} \propto \tau_{\mathrm{R}} R_{\mathrm{eff}}^{-\nu_{\perp,\mathrm{MF}}}$. Thus, for zero-field the mean-field theory applies if

$$1 \ll R_{\mathrm{eff}}^d \, \tau_{\mathrm{R}}^{\gamma'_{\mathrm{MF}}+2\beta_{\mathrm{MF}}-\nu_{\perp,\mathrm{MF}}d} = \left(R_{\mathrm{eff}} \tau^{(4-d)/2d}\right)^d , \tag{5.98}$$

where we have used the power-laws

$$\varrho_{\mathrm{s}} \sim \tau_{\mathrm{R}}^{\beta_{\mathrm{MF}}} , \qquad \Delta_{\varrho} \sim \tau_{\mathrm{R}}^{-\gamma'_{\mathrm{MF}}} \tag{5.99}$$

as well as the mean-field values of the Manna universality class $\beta_{\mathrm{MF}} = 1$, $\nu_{\perp,\mathrm{MF}} = 1/2$, and $\gamma'_{\mathrm{MF}} = 0$. On the other hand, mean-field scaling behaviour occurs as long as $1 \ll R_{\mathrm{eff}} \, \tau_{\mathrm{R}}^{\phi}$. Consequently, the crossover exponent is given by

$$\phi = \frac{4-d}{2d} . \tag{5.100}$$

According to the scaling form (5.95), data obtained from simulations at zero field and various interaction ranges have to collapse onto a single curve by plotting $\varrho_{\mathrm{s}} \, R_{\mathrm{eff}}^{\beta_{\mathrm{MF}}/\phi}$ as a function of $\tau_{\mathrm{R}} \, R_{\mathrm{eff}}^{1/\phi}$. The corresponding plot of the two-dimensional Manna model is shown in Fig. 5.18. A fairly good collapse is observed over the entire crossover which spans roughly six decades. The asymptotic power-law behaviour of the two-dimensional Manna class as well as of the mean-field scaling behaviour are convincingly recovered. A corresponding scaling plot of three-dimensional models can be found in [433].

In order to obtain universal crossover scaling functions metric factors have to be taken into account [432]:

$$\varrho_{\mathrm{s}}(\tau, h, R_{\mathrm{eff}}) \simeq \lambda^{-\beta_{\mathrm{MF}}} \, \tilde{R}(a_\tau \tau_R \lambda, a_h h \lambda^{\sigma_{\mathrm{MF}}}, a_R R_{\mathrm{eff}}^{-1} \lambda^{\phi}) . \tag{5.101}$$

Once the non-universal metric factors a_τ, a_h, and a_R are chosen in a specified way, the universal scaling function $\tilde{R}(x,y,z)$ is the same for all systems within the same universality class. Since the mean-field scaling function should be recovered for $R \to \infty$, i.e.,

$$\tilde{R}(x,y,\infty) = \tilde{R}_{\mathrm{MF}}(x,y) , \tag{5.102}$$

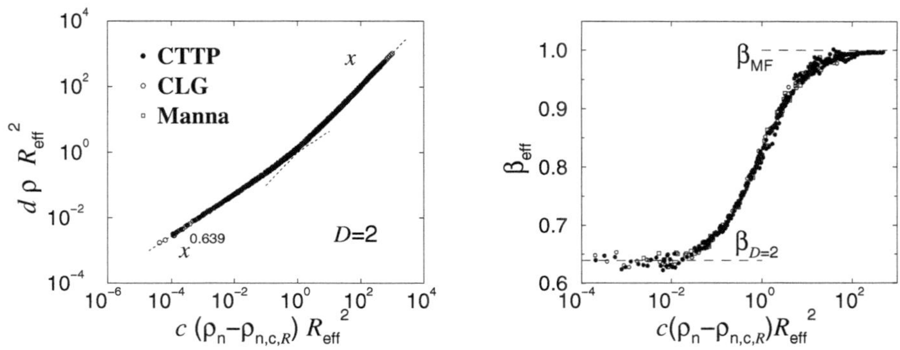

Fig. 5.19 The universal crossover scaling function of the order parameter at zero field for various two-dimensional models of the Manna universality class (left). The metric factors are given by $c = a_\tau a_R^2$ and $d = a_R^2$. The dashed lines correspond to the asymptotic behaviour of the two-dimensional system ($\beta_{d=2} = 0.639$) and of the mean-field behaviour ($\beta_{\mathrm{MF}} = 1$). The right panel shows the corresponding effective exponent β_{eff}. Reprinted with permission from [433]. Copyright (2004) World Scientific Publishing Company..

it is convenient to require

$$\tilde{R}(1,0,\infty) = \tilde{R}_{\mathrm{MF}}(1,0) = 1\,, \qquad \tilde{R}(0,1,\infty) = \tilde{R}_{\mathrm{MF}}(0,1) = 1\,. \tag{5.103}$$

In this way, a_τ, a_h, and a_R can be expressed in terms of the metric factors and critical parameters of the ordinary model ($R = 1$) and of the mean-field limit ($R \to \infty$). A straightforward calculation yields [432]

$$a_\tau = \frac{a_{\tau,R\to\infty}}{\rho_{c,R\to\infty}}\,, \qquad a_h = a_{h,R\to\infty}\,, \qquad a_R = \left(\frac{\rho_{c,R=1}}{a_{\tau,R=1}}\,\frac{a_{\tau,R\to\infty}}{\rho_{c,R\to\infty}}\right)^{\phi\beta_d/(\beta_{\mathrm{MF}}-\beta_d)}. \tag{5.104}$$

In this case the asymptotic scaling behaviour of the universal function $\tilde{R}$ equals

$$\tilde{R}(x,0,1) \simeq \begin{cases} x^{\beta_{\mathrm{MF}}} & \text{for } x \to \infty \\ x^{\beta_d} & \text{for } x \to 0\,. \end{cases} \tag{5.105}$$

According to the universal scaling form (5.101), the rescaled order parameter is plotted in Fig. 5.19 for three different two-dimensional systems. An excellent data-collapse is observed over the entire range of the crossover. The quality of the data-collapse is confirmed by the analysis of the associated effective exponent

$$\beta_{\mathrm{eff}} = \frac{\partial}{\partial \ln x}\ln\tilde{R}(x,0,1)\,. \tag{5.106}$$

Thus, the full crossover region can be described in terms of universal scaling functions. This is a remarkable result since the range where the universal scaling behaviour applies is usually restricted to a small vicinity around the critical point. For a detailed discussion of this issue we refer the interested reader to [432, 433, 434] and references therein. An analysis of universal order parameter fluctuations and the order parameter susceptibility can be found in the same works.

5.11 Quenched Disorder

One of the possible reasons why DP is so difficult to realise experimentally (see Sect. 3.4) is **quenched disorder**. For example, the local density of open channels in a porous rock will vary because of inhomogeneities of the material. It is therefore important to investigate how quenched disorder affects the critical properties of a spreading process. It turns out that even small disorder can affect or even destroy the critical behaviour of DP.

In the DP Langevin equation (3.10), the control parameter τ plays the role of the percolation probability. Quenched disorder may be introduced by random variations of τ, i.e., by adding another noise field χ

$$\tau \to \tau + \chi(t, \boldsymbol{r})\,. \tag{5.107}$$

Thus, the resulting Langevin equation reads

$$\partial_t \varrho(t,\boldsymbol{r}) \;=\; \tau\varrho(t,\boldsymbol{r}) - g\varrho(t,\boldsymbol{r})^2 + D\nabla^2\varrho(t,\boldsymbol{r}) + \eta(t,\boldsymbol{r}) + \varrho(t,\boldsymbol{r})\chi(t,\boldsymbol{r})\,. \tag{5.108}$$

The noise χ is *quenched* in the sense that quantities like the particle density are averaged over many independent realisations of the intrinsic noise η while the disorder field χ is kept fixed. In the following we distinguish two different types of quenched disorder χ, namely *spatially* quenched disorder $\chi_s(\boldsymbol{r})$ and *temporally* quenched disorder $\chi_t(t)$.

5.11.1 Temporally Quenched Disorder

Temporally quenched disorder is defined by the correlations

$$\overline{\chi_t(t)\chi_t(t')} = \gamma\,\delta(t-t')\,. \tag{5.109}$$

In this case the additional term scales as a *relevant* perturbation $\varrho\chi_t \mapsto \lambda^{-\nu_\parallel/2-\beta}\varrho\chi_t$. Therefore, we expect the critical behaviour and the associated critical exponents to change entirely. The influence of spatio-temporally quenched randomness was investigated in detail in [351]. Employing series expansion techniques it was demonstrated that the three exponents $\beta, \nu_\perp, \nu_\parallel$ vary continuously with the disorder strength. Thus the transition no longer belongs to the DP universality class.

5.11.2 Spatially Quenched Disorder

For spatially quenched disorder, the disorder field $\chi = \chi_s(\boldsymbol{r})$ is defined through the correlations

$$\overline{\chi_s(\boldsymbol{r})\chi_s(\boldsymbol{r}')} = \gamma\,\delta^d(\boldsymbol{r}-\boldsymbol{r}')\,, \tag{5.110}$$

where the bar denotes the average over independent realisations of the disorder field (in contrast to averages $\langle\ldots\rangle$ over the intrinsic noise η). The parameter γ is an amplitude which controls the intensity of disorder. In order to find out whether this type of noise affects the critical behaviour of DP, let us again consider the properties of the Langevin equation under the scaling transformation (4.9) and (5.110). Checking the scaling behaviour of the additional term $\varrho\chi_s$ in (5.108) at the critical dimension, we observe that it scales as

$$\varrho\chi_s \mapsto \lambda^{-\beta-d_c\nu_\perp/2}\varrho\chi_s\,, \tag{5.111}$$

i.e., spatially quenched disorder is a *marginal* perturbation. Therefore, it may seriously affect the critical behaviour at the transition. Janssen [329] showed by a field-theoretic analysis that the stable fixed point is shifted to an unphysical region, leading to runaway solutions of the flow equations in the physical region of interest. Therefore, spatially quenched disorder is expected to crucially disturb the critical behaviour of DP. The findings are in agreement with earlier numerical results by Moreira and Dickman [479] who reported non-universal logarithmic behaviour instead of power laws. Later Cafiero et al. [98] showed that DP with spatially quenched randomness can be mapped onto a non-Markovian spreading process with memory.

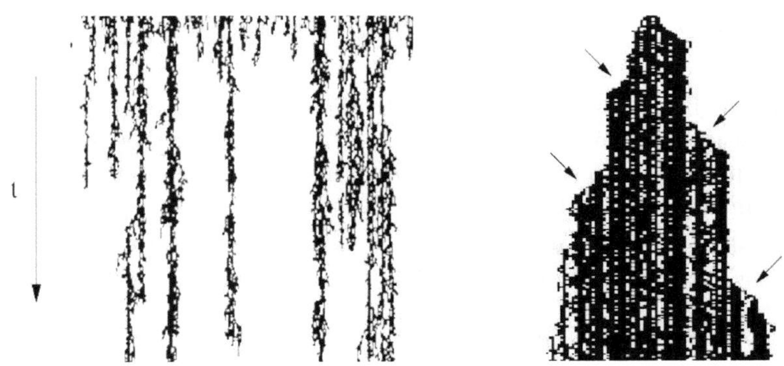

Fig. 5.20 (1+1)-dimensional DP with spatially quenched disorder. Left: in the glassy phase the disorder forces active sites to percolate in 'channels' where the local percolation probability is high. Right: Supercritical disordered DP process starting from a single seed, leading to avalanches (marked by the arrows) where the spreading agent overcomes a local barrier. Figure reprinted with permission from [287].

From a more physical point of view, spatially quenched disorder in (1+1)-dimensional systems was studied by Webman et al. [619]. It turns out that even very weak randomness drastically modifies the phase diagram. Instead of a single critical point one obtains a whole phase of very slow glassy-like dynamics. The glassy phase is characterised by non-universal exponents which depend on the percolation probability and the disorder amplitude. For example, in an absorbing (1+1)-dimensional DP process without quenched disorder the boundaries of a cluster propagate at constant velocity v. However, in the glassy phase v decays *algebraically* with time. The corresponding exponent turns out to vary continuously with the mean percolation probability. The power-law behaviour is due to 'blockages' at certain sites where the local percolation probability is small (see Fig. 5.20). Similarly, in the active edge of the glassy phase, the spreading agent becomes localised at sites with high percolation probability. In $d > 1$, however, numerical simulations indicate that a glassy phase does not exist.

More recently, Hooybergs et al. [303, 304] applied the Hamiltonian formalism to the contact process with spatially quenched disorder. Using a strong-disorder renormalisation group technique developed by Ma-Dasgupta-Hu [452] they demonstrated that the transition for sufficiently strong disorder is controlled by an infinite-randomness fixed point with a non-algebraic behaviour. Vojta [613] showed that the sharp transition is in fact destroyed due to the fact that strongly coupled spatial regions can undergo local phase transitions independent from the bulk system. In subsequent studies of the diluted contact process [614, 615, 140] a Griffiths phase and a multicritical point with unconventional scaling behaviour could be identified.

5.12 Attempts of Classification

The reader might have become impressed by the immense variety of absorbing phase transitions we have mentioned in this volume. It may therefore not be entirely inappropriate to summarise some recent attempts to put some order into these many universality classes.

1. Kockelkoren and Chaté [395, 128] suggested that the universality class of absorbing phase transitions in reaction-diffusion processes of the form

$$mA \to (m+k)A\ , \quad nA \to (n-l)A\ , \tag{5.112}$$

is determined by the orders m, n of creation and removal, while the numbers k and l determine additional symmetries such as parity conservation. In terms of m, n, they propose the general classification scheme shown in Table 5.7. In addition, they investigate the role of parity-conservation in more detail and find that such a symmetry does *not* alter the universality class whenever

$m\backslash n$	1	2	3	4
1	DP	DP/PC	DP	DP
2	DP	PCPD	PCPD	PCPD
3	DP	DP	TCPD	TCPD
4	DP	DP	DP	?

Table 5.7 Classification scheme of $1D$ absorbing phase transitions described by (5.112), according to [395, 128].

every sector includes an absorbing state. Their conjecture is in agreement with the observation, based on $1D$ numerical simulations, that an additional parity-conservation does not change the critical behaviour of the PCPD [506] and evidence of a PCPD transition in the diffusive $2A \to 3A$, $4A \to \emptyset$ system, where $m = 2, n = 4$ [494].[9]

2. A very interesting approach has been proposed by Elgart and Kamenev [203]. They consider a *Hamiltonian* action

$$\mathcal{J}[p,q] = \int \mathrm{d}t\, \mathrm{d}\boldsymbol{r}\ \left[p\partial_t q + D\boldsymbol{\nabla} p \cdot \boldsymbol{\nabla} q - H_R(p,q)\right] , \tag{5.113}$$

where $q = q(t,\boldsymbol{r})$ represents the order parameter such that $\varrho(t,\boldsymbol{r}) = \langle q(t,\boldsymbol{r})\rangle$, $p = p(t,\boldsymbol{r})$ is the canonically conjugate momentum and the reaction Hamiltonian $H_R(p,q)$ describes the reactions allowed in a given model. The Hamiltonian is determined from the (imaginary-time) Schrödinger equation $\partial_t F = H_R(p,q)F$ satisfied by the generating function

$$F(\{p\},t) = \sum_{\{n\}} p_1^{n_1} \cdots p_{\mathcal{N}}^{n_{\mathcal{N}}} P(n_1,\ldots,n_{\mathcal{N}};t) \tag{5.114}$$

and with $q_i = \frac{\partial}{\partial p_i}$, see exercise 52 for an explicit example. The time-dependence follows from the associated variational equations (classical equations of motion in a physicist's terminology)

$$\partial_t q = \partial_p H_R(p.q) \ \ , \ \ \partial_t p = -\partial_q H_R(p.q) \,. \tag{5.115}$$

Because of the conservation law $\frac{\mathrm{d}}{\mathrm{d}t} H_R(p,q) = 0$, absorbing phase transitions are characterised by the **zero-energy condition**

$$H_R(p,q) = 0 \tag{5.116}$$

The corresponding zero-energy curves acts as separatrices which decompose the phase space in disjoint sectors.

According to Elgart and Kamenev [203], close to phase transitions the topology of the phase space should be determined by the form of $H_R(p,q)$

[9] In two spatial dimensions, a re-entrant phase diagram is seen [493].

model	reactions	$H_R(p,q)$	d_c	symmetries	
kCPD	$kA \longrightarrow (k+1)A$ $(k+1)A \longrightarrow kA$ $kA \longrightarrow \emptyset$	$(m+up-vq)pq^k$	$4/k$	–	
kPC	$kA \longrightarrow (k+2)A$ $(k+1)A \longrightarrow A$	$(up^{k-1} - vh_k(p)q)$ $\times p(p^2-1)q^k$	$2/k$	$p \mapsto -p$ $q \mapsto -q$	k even h_k even
kPC	$kA \longrightarrow (k+2)A$ $(k+1)A \longrightarrow A$	$(up^{k} - vh_{k+1}(p)q)$ $\times p(p^2-1)q^k$	$2/k$	$p \mapsto -p$ $q \mapsto -q$	k odd h_{k+1} odd
kR	$kA \leftrightarrow (k+1)A$	$(p-1)(u-vq)p^kq^k$	$2/k$	$p \mapsto \frac{v}{u}q$ $q \mapsto \frac{u}{v}p$ $t \mapsto -t$	
kRPC	$kA \leftrightarrow (k+2)A$	$(p^2-1)(u-vq)p^kq^k$	$2/(k+1)$	i) $p \mapsto -p$, $q \mapsto -q$ ii) $p \mapsto \frac{v}{u}q$, $q \mapsto \frac{u}{v}p$ $t \mapsto -t$	

Table 5.8 List of the four series of models ($k \in \mathbb{N}$) with absorbing phase transitions and a scalar order parameter as identified in [203]. For a typical set of reactions, the reaction Hamiltonian is given in normal form and m, u, v are control parameters. For the model 1CPD=DP, one can arrange for $u = v$ and then one has, for $m = 0$, the rapidity-reversal symmetry $p \mapsto -q$, $q \mapsto -p$, $t \mapsto -t$.

only. By suitable shifts of variables, they identify, for non-disordered systems described by a single scalar order parameter, four infinite series of models with absorbing phase transitions and we list their main results in Table 5.8. One has two series of intrinsically irreversible models and two series of reversible models. The four series are essentially distinguished by their global symmetry properties, although it may happen that for certain members of a series, a richer symmetry arises. For example, this is the case for directed percolation (=1CPD), which is rapidity-reversal-invariant. From the upper critical dimensions, the following five universality classes with an upper critical dimension $d_c > 1$ are read off: DP = 1CPD, PCPD=2CPD, TCPD=3CPD, PC=1PC and 1R. There are four marginal classes with $d_c = 1$, namely 4CPD, 2PC, 2R and 1RPC. All other models are fully described by simple mean-field for the physically interesting dimensions $d \geq 1$.

This approach does have some predictive power: for example, if one adds to the 2PC process the reaction $2A \longrightarrow \emptyset$ and considers the process

$$2A \longrightarrow 4A \;\;,\;\; 3A \longrightarrow A \;\;,\;\; 2A \longrightarrow \emptyset\,, \tag{5.117}$$

the reduced Hamiltonian $H_R(p,q) = (m+u(p^2-1)-vpq)(p^2-1)q^2$, although it does obey the characteristic symmetry of the PC universality class, leads indeed to the same phase-space topology as the model 2CPD=PCPD [203], in agreement with $1D$ numerical simulations [506, 498]. Similarly, the kPC processes with competition in the k-particle channel belong to the kCPD universality class [203]. These result emphasise that the PC universality class, in spite of its usual name, may be better characterised in terms of phase-space topology rather than by the conservation of parity.

These thought-provoking studies [395, 203] deserve a few comments: First, both approaches are essentially based on some symmetry arguments but are also capable of going beyond too rigid an application of symmetry requirements, as seems to be required by the results of numerical simulations. Second, the PCPD is treated as an independent universality class in both schemes. In view of the repeated assertions that the PCPD should at truly long times cross over to directed percolation [40, 289, 574], it remains to be seen whether this is a really non-trivial prediction or rather a failure of the proposed classification schemes. Third, the mere facts that (i) the two classifications disagree with each other and (ii) some of the classes discussed in this book do not appear to be represented suggest that both might capture some important aspects of a larger and still unknown structure. For example, both schemes are based on models where the strict 'fermionic' constraint of at most one particle per lattice site is relaxed and allow, in principle, for an arbitrary number of particles per site, leading to 'bosonic' variants of the models discussed in this volume. However, as we shall show explicitly in Vol. 2, the *bosonic* variants of the contact process and the pair-contact process (with single-particle diffusion) are exactly solvable on the critical line [305, 504, 46], but their critical behaviour is completely different from the one of their 'fermionic' counterparts which we have described in this volume. Furthermore, the construction of $H_R(p, q)$ and the analysis of the resulting phase-space topology completely neglects possible spatial variations. While this already suppresses phenomena related to the difference between diffusive transport and Lévy flights, the *bosonic* contact and pair-contact processes may show a different kind of phase transition, related to the eventual *collapse* of all particles on a single site (but where the total number of particles is conserved on *average*), quite analogous to a (real-space) Bose-Einstein condensation, and with a tricritical point in the bosonic pair-contact process [504, 46, 48]. Fourth, classification attempts such as those discussed here are based directly on the microscopic reaction schemes. However, as pointed out earlier in this chapter, it is not always possible to determine a universality class by identifying the symmetries of a system's microscopic dynamics. Rather, universality classes are fixed by (hidden) asymptotic symmetries which emerge on a coarse-grained scale close to a critical point. Future attempts for the classification of absorbing phase transitions will have to take into account that these are ultimately described by renormalised field-theories. Since these field-theories are still unknown for many models, the goal of classification appears to be remote at present.

5.13 Some Open Questions

Progress in science can sometimes be appreciated by returning to questions asked some time ago. In 1997, Grassberger [243] raised some intriguing questions in relationship to directed percolation and we close this volume with a short discussion of them.

1. *"Designing and performing [...] an experiment is the outstanding problem in this field"*. The long-standing and *"anomalous"* [243] absence of any reliable experimental evidence for the DP universality class might begin to change [589], as described in Sect. 3.4.
2. *"Verify by renormalisation-group methods whether systems with fluctuating passive states are in the DP universality class"*. This still seems to be an open problem.
3. *"Relevance of frozen randomness"*. We reviewed some results on quenched randomness in Sect. 5.11. Adding either temporally or spatially correlated quenched randomness may change the universality class of the transition.
4. *"The relationship of self-organised criticality (SOC) with DP"*. We discussed in Sect. 5.7 the relationship of SOC with absorbing phase transitions, in the context of the Manna model and the CTTP.
5. *"It is frustrating that the DP model [...] cannot be solved exactly in the lowest dimensions"*. This still stands unsolved and may well turn out to be an unsolvable problem.

All in all, absorbing phase transitions have provoked researchers to apply tremendous ingenuity to their precise study and they have served as an important stimulus in the development of many new theoretical tools. In this wider perspective, it seems appropriate to terminate this volume with the following quotation [243]:

> *"Since DP in its epidemic interpretation shows one of the most basic non-equilibrium phase transitions, similar only to the Ising model in equilibrium statistical mechanics, it is certain to provide the stimulus for further studies."*
> P. Grassberger, *Directed percolation: results & open problems* (1997)

Problems

49. Consider the Langevin equation of long-range interacting directed percolation (5.32). Consider small spatial deviations from the homogeneous steady-state order parameter $\varrho_{\mathrm{s}}(\tau)$ by introducing $\delta\varrho(t,\boldsymbol{r}) = \varrho(t,\boldsymbol{r}) - \varrho_{\mathrm{s}}(\tau)$. Derive within an linear approximation (Ornstein-Zernicke-like approach) in $\delta\varrho(t,\boldsymbol{r})$ the spatial correlation length $\xi_{\perp}(\tau)$. Derive the correlation exponent $\nu_{\perp}$ as a function of the interaction parameter σ_{ℓ}.

50. Consider the pair-contact process with diffusion in one spatial dimension. Find the phase diagram in the site and pair approximations.

51. Consider a reaction-diffusion system, consisting of two species A, B of particles, with the allowed reactions

$$2A \stackrel{\lambda}{\to} \emptyset \;,\; 2B \stackrel{\lambda}{\to} \emptyset \;,\; AB \stackrel{\delta}{\to} \emptyset$$

together with diffusion of the single particles. What kind of behaviour would you expect, depending on the ratio λ/δ ?

Analyse the system with the site-approximation. Show that for the ratio $\rho = b/a$, where a and b denote the mean densities of As and Bs, respectively

$$\frac{\mathrm{d}\rho}{\mathrm{d}\tau} = (\lambda - \delta)\frac{\rho(1-\rho)}{\lambda + \delta\rho} \tag{5.118}$$

with the fictitious time $\tau = -\ln a$, in order to analyse the long-time behaviour [587].

52. Reconsider the simple model of a population of $n(t)$ particles with procreation rate λ and death rate 1, studied in exercises 1 and 2. Consider the generating function $F(p,t) := \sum_{n=0}^{\infty} p^n P_n(t)$ and derive the reaction Hamiltonian $H_R(p,q)$ from the Schrödinger equation $\partial_t F = H_R F$. Analyse the zero-energy solutions and the flow, according to the variational equations.

53. An often-studied, albeit very simple, model of population dynamics is the **Volterra-Lotka** model, which at the level of simple mean-field is described by the normalised equation

$$\dot{p} = p - xp \;,\; \dot{x} = -x + \beta xp \tag{5.119}$$

where p represents a plant/prey population, x represents an animal/predator and β is a rate constant. Show that this model predicts oscillations by finding a conserved quantity.

More modern approaches are sceptical about the applicability of the law of mass-action implicit in (5.119) to population dynamics. A recent proposal is [571]

$$\dot{p} = p - p^{1-\lambda}x^{\lambda} \;,\; \dot{x} = -x + \beta^{1-\lambda}p^{1-\lambda}x^{\lambda} \tag{5.120}$$

with $0 \leq \lambda \leq 1$. Analyse these mean-field equations and compare with (5.119).

Appendices

A. Equilibrium Models

We recall briefly the definition and some basic properties of the most important equilibrium models. For two-dimensional models, a lot of information has been found either by exact solutions [50] or by methods based on conformal invariance [270]. We refer to the sources quoted for more detailed information.

A.1 Potts Model

The q**-state Potts model** is defined by the Hamiltonian

$$\mathcal{H} = -J \sum_{(i,j)} \delta(\sigma_i, \sigma_j), \qquad \delta(a,b) = \begin{cases} 1 \; ; & \text{if } a = b \\ 0 \; ; & \text{if } a \neq b \end{cases} \tag{A1}$$

where the local variables σ_i can take the values $\sigma_i = 0, 1, \ldots q-1$, J is the coupling constant, and the sum runs over the nearest neighbours of a hypercubic (or any other) lattice. The global symmetry of the Potts model is the permutation group $\mathbb{S}_q$ of q elements. For $q = 2$ one recovers the Ising model with T replaced by $2T$ because of $\delta(\sigma, \sigma') = (1 + \sigma\sigma')/2$. The case $q = 1$ corresponds to the universality class of **isotropic percolation**.

In two dimensions, the Potts-q model[1] is self-dual and its critical point is given by $1/T_c = \ln(1 + \sqrt{q})$. The phase-transition is continuous (second order) if $q \leq 4$ and discontinuous (first order) if $q > 4$. The values of the critical exponents are listed in Table A1. In three dimensions, the transition

[1] For brevity, we shall sometimes refer to the q-states Potts model as the *Potts-q model*. Similar conventions will apply to other spin models with a discrete global symmetry.

remains continuous if $q = 2$ but is of first order already for $q = 3$. A very detailed review can be found in [635, 636].

We also include, besides the usual bulk critical exponents, *surface exponents* such as $\eta_{\|}$ which describes the decay of the two-point correlation function close to a plane free surface, and the local magnetisation exponent β_1. Here and throughout we assume that the surface does *not* order before the bulk (i.e. we have the case without an external field located at the surface) and are therefore dealing with the **ordinary transition.**

Although there is no obvious up-down symmetry, an order parameter is readily defined by

$$\langle M(t)\rangle = \frac{1}{N_s}\frac{1}{(q-1)L^d}\sum_{n=1}^{N_s}\sum_{i=1}^{L^d}\left[q\delta(\sigma_{i,[n]}(t),1)-1\right] , \tag{A2}$$

where $\sigma_{i,[n]}(t)$ denotes the ith spin of sample number n at the Monte Carlo sweep t; N_s is the number of Monte Carlo samples and $\mathcal{N} = L^d$ is the total number of lattice sites. This is repeated N_B times for the final estimates of the order parameter as a function of t. See appendix G for simulational methods.

A.2 Clock Model

The **clock model** is a variant of the Potts model and is defined by the Hamiltonian

$$\mathcal{H} = -J\sum_{(i,j)}\cos(\vartheta_i-\vartheta_j) \ , \ \vartheta_i = \frac{2\pi}{p}n_i \tag{A3}$$

where $n_i = 0, 1, \ldots, p-1$. Here the global symmetry is the cyclic group $\mathbb{Z}_q$, which is a true subgroup of $\mathbb{S}_q$. For $p = 2$ and $p = 3$ one recovers the Ising and Potts-3 models, respectively, whereas in the limit $p \to \infty$ one retrieves the XY model. In two dimensions, for $p \leq 4$, there is a single equilibrium phase-transition with conventional power-law behaviour. On the other hand, for $p > 4$, there exist two distinct transitions at temperatures T_1 and $T_2 > T_1$.

q	α	ν	β	γ	δ	η	$\eta_{\|}$	β_1
1	−2/3	4/3	5/36	43/18	91/5	5/24	2/3	8/9
2	0(log)	1	1/8	7/4	15	1/4	1	1/2
3	1/3	5/6	1/9	13/9	14	4/15	4/3	5/9
4	2/3	2/3	1/12	7/6	15	1/4	2	2/3

Table A1 Some equilibrium bulk and surface critical exponents (ordinary transition) of the q-state Potts model in two dimensions.

The system is paramagnetic for $T > T_2$ and ferromagnetic for $T < T_1$ whereas one has a Kosterlitz-Thouless phase with a non-local order parameter in between. Both phase transitions at T_1 and at T_2 are of Kosterlitz-Thouless type with exponentially diverging correlation lengths, susceptibilities etc.

A.3 Turban Model

In this model, also referred to in the literature as the **multispin Ising model**, one studies the combined effect of the standard nearest-neighbour interactions and a higher-order term in one direction [600, 599]. The Hamiltonian is

$$\mathcal{H}^{(m)} = -\sum_{(i,j)} \left(J_y s_{i,j} s_{i+1,j} + J_x \prod_{\ell=0}^{m-1} s_{i,j+\ell} \right), \tag{A4}$$

where $s_{i,j} = \pm 1$ is an Ising spin on the site (i,j) of a square lattice. The model is self-dual with a critical line given by $\sinh(2J_x/T)\sinh(2J_y/T) = 1$ [600, 599, 152, 517]. A symmetry analysis suggests that the model has the same equilibrium properties as the q-state Potts model, where $q = 2^{m-1}$. In particular, for $m = 3$ one recovers the four-state Potts model [152]. On the other hand, for $m \geq 4$, one expects a first-order transition. This conjecture has been well confirmed numerically, see [572, 524, 398] and references therein.

A.4 Baxter-Wu Model

This model is named after its exact solution and is defined in terms of Ising spins $s_i = \pm 1$ on a *triangular* lattice with the Hamiltonian

$$\mathcal{H} = -J \sum_{(i,j,k)} s_i s_j s_k \,, \tag{A5}$$

where the sum involves the three spins on each triangle of the lattice. Since the triangular lattice can be decomposed into three sublattices, $\mathcal{H}$ is invariant under reversal of all spin belonging to two of the sublattices and the ground state is hence four-fold degenerate. The model is self-dual and undergoes a second-order phase-transition at $1/T_c = \frac{1}{2}\ln(1+\sqrt{2})$. The exact solution gives the same equilibrium critical exponents (e.g. $\alpha = \nu = 2/3$ and $\eta = 1/4$) as for the Potts-4 model, see [50].

T/J	η	η_{SW}	$\eta_{\parallel}$	$\eta_{\parallel,SW}$
0.1	0.017(1)	0.016	0.032(1)	0.032
0.2	0.036(3)	0.032	0.069(1)	0.064
0.3	0.052(5)	0.048	0.103(2)	0.095
0.4	0.074(6)	0.064	0.143(4)	0.127
0.5	0.100(8)		0.186(5)	
0.6	0.122(13)		0.231(8)	
0.7	0.154(14)		0.298(12)	
0.8	0.188(22)		0.374(17)	
0.893	0.250(28)		0.542(22)	

Table A2 Equilibrium bulk exponent η and surface exponent $\eta_{\parallel}$ in the low-temperature phase of the $2D$ XY model as a function of temperature T. The corresponding results from the spin-wave approximation (SW) are also listed. After [58].

A.5 Blume-Capel Model

This model is defined in terms of spin variables $S_i = -1, 0, 1$ and the Hamiltonian

$$\mathcal{H} = -J \sum_{(i,j)} S_i S_j + D \sum_i S_i^2 , \tag{A6}$$

where in addition to the exchange coupling J a crystal field D is introduced. The model shows a line of phase transitions which are of second order if D/J is small (and which is in the Ising equilibrium universality class) but which become of first order for sufficiently large values of D. The meeting point of first- and second-order transitions is a **tricritical point** with a completely different critical behaviour [411] which occurs at $D/J \simeq 1.9655$ and $T_c/J = 0.610$ [138].

A.6 XY Model

The XY model (or *planar rotator* model) is defined by the Hamiltonian

$$\mathcal{H} = -J \sum_{(i,j)} \cos(\vartheta_i - \vartheta_j) = -J \sum_{(i,j)} \boldsymbol{S}_i \cdot \boldsymbol{S}_j , \tag{A7}$$

where $\vartheta_i \in [0, 2\pi]$ or, equivalently, $\boldsymbol{S}_i \in \mathbb{R}^2$ is a two-dimensional unit vector. In more than two dimensions, the model undergoes a conventional second-order phase-transition between a paramagnetic and a ferromagnetic phase. In two dimensions, however, the Mermin-Wagner theorem asserts that a spontaneous magnetisation is impossible. Rather, the whole low-temperature phase in two dimensions remains critical in that the magnetic correlation functions decay algebraically as $C(\boldsymbol{r}) \sim |\boldsymbol{r}|^{-\eta(T)}$ with a temperature-dependent exponent $\eta(T)$. Using conformal-invariance techniques, precise numerical estimates for the bulk exponent $\eta(T)$ and also the surface exponent $\eta_{\|}(T)$ (ordinary transition) were found [58], see Table A2. The transition occurs at $T_{\rm KT}/J \simeq 0.893$ [386] and is of **Kosterlitz-Thouless** type, meaning that the correlation length, susceptibility and other quantities diverge according to a stretched exponential, viz. $\xi(T) \sim \exp(b/(T-T_{\rm KT})^{1/2})$. At $T = T_{\rm KT}$, the spin-spin correlator decays as $C(\boldsymbol{r}) = \langle \boldsymbol{S}(\boldsymbol{r}) \cdot \boldsymbol{S}(\boldsymbol{0})\rangle \sim |\boldsymbol{r}|^{-1/4} \ln(|\boldsymbol{r}|)^{1/8}$ [478], which implies $\eta(T_{\rm KT}) = \frac{1}{4}$.

A.7 O(n) Model

These systems are defined in terms of unit spin vectors $\boldsymbol{S}_i \in \mathbb{R}^n$ and $|\boldsymbol{S}_i| = 1$ and the Hamiltonian

$$\mathcal{H} = -J \sum_{(i,j)} \boldsymbol{S}_i \cdot \boldsymbol{S}_j \,. \tag{A8}$$

For $n = 1$ and $n = 2$ one recovers the Ising and the XY model, respectively, while the case $n = 3$ defines the Heisenberg universality class. In $d > 2$ dimensions, O(n) models have a conventional second-order ferromagnetic phase-transition while for $d = 2$ dimensions, the Mermin-Wagner theorem excludes a phase-transition if $n > 2$. Equilibrium critical exponents have been determined by several different methods, see [648, 515]. In Table A3 conservative estimates for values of critical exponents, as well as for $T_c(n)$, are listed for some values of n, but the reader should be aware that a lot of effort continues to be invested in further improving the exponent values. We refer to the very extensive reviews of Pelissetto and Vicari [515] and of Barmartz et al. [41] for details, in particular for universal amplitude combinations and for the comparison with experimental results. Re-summed variational perturbation theory with lists of exponents for $0 \leq n \leq 28$ is presented in [393]. A recent review for surface exponents is [184].

In the limit $n \to \infty$, the O(n)-model reduces to the exactly solvable **spherical model.** Rather than through infinite-dimensional unit spin vectors, one may also define the spherical model in terms of real spin variables $S_i \in \mathbb{R}$ attached to each site i of a d-dimensional lattice Λ with $\mathcal{N}$ sites and subject to the **spherical constraint** $\sum_{i\in\Lambda} S_i^2 = \mathcal{N}$. The Hamiltonian is usually chosen to describe the habitual nearest-neighbour interactions $\mathcal{H} = -J\sum_{(i,j)} S_i S_j$ [61] but it is one of the attractive features of the model that it may be solved for considerably more general interactions and that one may even include external fields. A classical review of many of the equilibrium properties is [366], for more recent discussions see [50, 80]. Calculations are considerably simplified if the spherical constraint is only imposed on average [419, 420]. The equilibrium *bulk* critical behaviour of the spherical model is the same as

n	T_c/J	α	β	γ	ν	η	$\eta_{\parallel}$	β_1
1	4.5115279(6)	0.110(1)	0.3265(3)	1.2375(5)	0.6301(4)	0.0364(5)	1.528	0.796
2	2.20183(1)	−0.0146(8)	0.3485(3)	1.3177(3)	0.6715(3)	0.0380(4)	1.422	0.810
3	1.44299(1)	−0.134(2)	0.3689(3)	1.3960(9)	0.7112(5)	0.0375(5)	1.338	0.824
4	1.068535(9)	−0.247(6)	0.388(3)	1.471(4)	0.749(2)	0.0365(10)		
5	0.8559(3)	−0.298	0.396	1.506	0.766	0.034		
10	0.41187(1)	−0.61(2)	0.44(1)	1.721(14)	0.871(7)	0.025(20)		
∞		−1	$\frac{1}{2}$	2	1	0	1	1
SM	3.956776...	−1	$\frac{1}{2}$	2	1	0	2	$\frac{3}{2}$

Table A3 Critical temperatures and equilibrium critical exponents of the $3D$ O(n) model. The bulk exponents for $1 \leq n \leq 5$ are from [515], the estimates of $T_c(n)$ for $n = 1, 2, 3, 10$ and the bulk exponents for $n = 10$ are from [623], $T_c(4)$ is from [32] and $T_c(5)$ is from [313]. The estimates of surface exponents $\eta_{\parallel}$, β_1 at the ordinary transition are from [68, 184]. SM denotes the $3D$ spherical model.

in the $n \to \infty$ limit of the O(n) model [577] but we warn the reader that this no longer holds true for the *surface* critical behaviour [68, 183].

A.8 Double Exchange Model

This model plays a role in the study of perovskite manganites and is defined in terms of classical spin vectors $\boldsymbol{S}_i \in \mathbb{R}^3$ by the Hamiltonian

$$\mathcal{H} = -J \sum_{(i,j)} \sqrt{1 + \boldsymbol{S}_i \cdot \boldsymbol{S}_j} \tag{A9}$$

with nearest-neighbour interactions on a simple hypercubic lattice. It undergoes in $3D$ a second-order phase transition at $T_c/J = 0.74515$. Some $3D$ equilibrium exponents are $\nu = 0.68(2)$ and $\beta = 0.356(6)$, see [215] and references therein. It is believed that the equilibrium phase transition is in the universality class of the $3D$ Heisenberg model.

A.9 Frustrated Spin Models

A different kind of very rich behaviour is found in spin systems with a non-random competition such that there is a macroscopic number of equilibrium states [421, 637]. The best-known example is probably the **triangular Ising antiferromagnet**. On the square lattice, one may consider the **fully frustrated Ising/Potts model** which is defined by the Hamiltonian

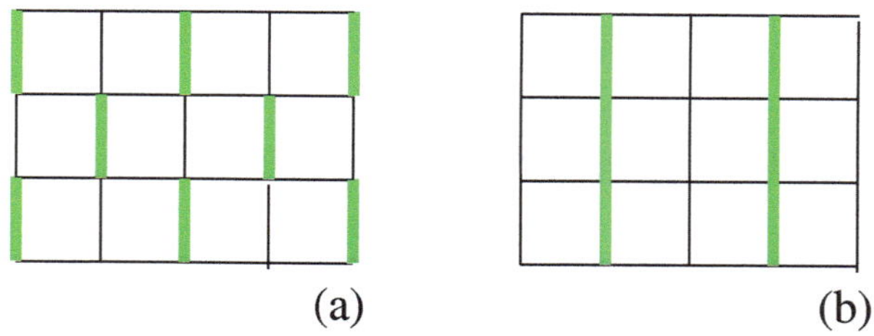

Fig. A1 Patterns of coupling constants for (a) checkerboard and (b) domino-tiled modulations of the square lattice. The antiferromagnetic bonds are indicated by the thick grey lines.

$$\mathcal{H} = -J \sum_{i,j} \left[\delta(\sigma_{i,j}, \sigma_{i+1,j}) + (-1)^{f(i,j)} \delta(\sigma_{i,j}, \sigma_{i,j+1}) \right] \tag{A10}$$

where the frustration variable $f(i, j)$ is chosen such that on each plaquette of the square lattice one bond is antiferromagnetic while the other three are ferromagnetic. The choices $f(i, j) = i + j$ and $f(i, j) = i$, respectively, led to the checkerboard and domino-tile pattern of frustration, see Fig. A1.

The triangular Ising antiferromagnet [306, 616, 617] and the fully frustrated Ising model (both with checkerboard and domino-tile modulation) [611] are disordered for any $T > 0$ but the zero-temperature state, which has a finite entropy, is critical. From an exact mapping onto the Baxter model, the equilibrium exponent $\eta = 1/2$ [582, 222] and the surface exponent $\eta_{\|} = 1$ if $J_s/J = \frac{1}{2}$ and $\eta_{\|} = 2$ if $J_s/J > \frac{1}{2}$ [256] can be found, where J_s is the coupling constant at the surface. The finite-size scaling of the correlation length and of the free energy are in agreement with conformal invariance, and a central charge $c = 1$ is found [256].

The fully frustrated three-states Potts model with checkerboard modulation is disordered for any $T > 0$ but critical at $T = 0$. The exponent $\nu = 1.139$ was estimated from transfer matrix calculations [226]. On the other hand, for a domino-tile modulation, one has a second-order phase-transition between a paramagnetic and a ferromagnetic state at $T_c/J = 0.365(1)$, while the exponent ν may be compatible with the pure Potts-3 result $\nu = \frac{5}{6}$ [225].

Equilibrium properties of the fully frustrated XY model are reviewed in detail in [268].

A.10 Hilhorst-van Leeuwen Model

This model is a variant of the $2D$ Ising model on a semi-infinite square lattice. The Hamiltonian is [284]

$$\mathcal{H} = -\sum_{r_\parallel \in \mathbb{Z}} \sum_{r_\perp \in \mathbb{N}} \left[J_1 \sigma_{r_\parallel, r_\perp} \sigma_{r_\parallel+1, r_\perp} + J_2(r_\perp) \sigma_{r_\parallel, r_\perp} \sigma_{r_\parallel, r_\perp+1} \right] \tag{A11}$$

and the coupling perpendicular to the surface varies as follows [71]

$$J_2(r_\perp) = J_2(\infty) + \frac{T_c \sinh(2J_2(\infty)/T_c)}{4} \frac{A}{r_\perp^\omega} . \tag{A12}$$

While the bulk critical behaviour is the same as in the usual $2D$ Ising model, the surface critical behaviour depends on the form of $J_2(r_\perp)$, see [316] for a review. For $\omega > 1$ the modified couplings are irrelevant, while for $\omega < 1$ they are relevant. The case $\omega = 1$ is marginal. Then the surface exponent of the order parameter depends continuously on the amplitude A. If one takes $J_1 = J_2(\infty)$ and $A < 1$, this reads simply

$$x_1 = \frac{\beta_1}{\nu} = \frac{1}{2}(1 - A) \tag{A13}$$

and one recovers the usual semi-infinite $2D$ Ising model for $A = 0$.

B. Scaling Laws for Absorbing Phase Transitions

In general, absorbing phase transitions are characterised by four independent exponents, for example β, β', $\nu_\perp$, and $\nu_\parallel$ (see [287, 496, 433] and Table 4.1 for the definitions). All other exponents can be expressed in terms of these exponents. The spreading exponents δ, Θ, and the dynamical exponent z are given by

$$\delta = \frac{\beta'}{\nu_\parallel}, \qquad \Theta = \frac{d}{z} - \frac{\beta}{\nu_\parallel} - \frac{\beta'}{\nu_\parallel}, \qquad z = \frac{\nu_\parallel}{\nu_\perp}. \tag{B1}$$

The steady-state critical exponents γ, γ' and σ are related to the standard exponents (β, β', $\nu_\perp$, $\nu_\parallel$) via

$$\gamma = \sigma - \beta, \qquad \sigma = d\nu_\perp + \nu_\parallel - \beta', \qquad \gamma' = d\nu_\perp - 2\beta. \tag{B2}$$

Furthermore, the fractal dimension of the spreading clusters is given by

$$d_f = z(\Theta + \delta) = d - z\delta = d - \frac{\beta}{\nu_\perp}. \tag{B3}$$

(where possible contributions from intermittency, as they may arise in DP [269, 277], are neglected). Simple dimensional analysis offers a convenient way to derive these scaling laws. Therefore, we remind ourselves that the various quantities of interest enter the scaling forms in combinations with their scaling powers, e.g.,

$$\begin{aligned}&\lambda\tau, \quad \lambda^{\beta}\varrho, \quad \lambda^{\sigma}h, \quad \lambda^{-\gamma}\chi_{\text{a}}, \quad \lambda^{-\gamma'}\Delta_{\varrho}, \quad \lambda^{-\nu_\perp}\boldsymbol{x}, \quad \lambda^{-\nu_\perp}L,\\ &\lambda^{-\nu_\perp}R, \quad \lambda^{-\nu_\parallel}t, \quad \lambda^{d\nu_\perp-\nu_\parallel\delta}\varrho_0, \quad \lambda^{-\Theta\nu_\parallel}N_{\text{a}}, \quad \lambda^{\delta\nu_\parallel}P_{\text{sur}}, \quad \ldots .\end{aligned} \tag{B4}$$

In the language of real-space renormalisation, the rescaling is usually related to the transformation $\boldsymbol{x} \mapsto b\boldsymbol{x}$, corresponding to $\lambda = b^{1/\nu_\perp}$. Enforcing scale-invariance, the definition of e.g. the susceptibility leads to

$$\chi = \frac{\partial\, \varrho_{\text{s}}}{\partial\, h} \quad \Longrightarrow \quad \gamma = \sigma - \beta\,. \tag{B5}$$

Similarly, the hyperscaling law of the fluctuation exponent is obtained from

$$\Delta_\varrho = L^d\,(\langle\varrho_{\text{s}}^2\rangle - \langle\varrho_{\text{s}}\rangle^2) \quad \Longrightarrow \quad \gamma' = d\nu_\perp - 2\beta\,. \tag{B6}$$

Taking into account that an initial homogeneous particle density ϱ_0 may be represented by the external (conjugated) field we find

$$h = \varrho_0\,\delta(t) \quad \Longrightarrow \quad \sigma = d\nu_\perp + \nu_\parallel - \nu_\parallel\delta\,. \tag{B7}$$

Considering activity spreading, the average number of active sites (usually averaged over all runs) scales as

$$N_{\text{a}} = \text{const}\,\varrho\,P_{\text{sur}}\,R^d \quad \Longrightarrow \quad \nu_\parallel\Theta = d\nu_\perp - \beta - \nu_\parallel\delta\,. \tag{B8}$$

Finally, the connected correlation function of active sites

$$\Gamma(\boldsymbol{r}_i,\boldsymbol{r}_j;\tau,h) \;=\; \Big\langle \big(\varrho(\boldsymbol{r}_i) \;-\; \langle\varrho(\boldsymbol{r}_i)\rangle\big)\,\big(\varrho(\boldsymbol{r}_j) \;-\; \langle\varrho(\boldsymbol{r}_j)\rangle\big)\Big\rangle \tag{B9}$$

is expected to decay for translational invariant systems ($\boldsymbol{r} = \boldsymbol{r}_i - \boldsymbol{r}_j$) as

$$\Gamma(\boldsymbol{r},\boldsymbol{0};0,0) \;\sim\; r^{-d+2-\eta_\perp}\,. \tag{B10}$$

The correlation function is related to the order parameter fluctuations

$$\Delta_\varrho(\tau,h) \;=\; \sum_{\boldsymbol{r}}\,\Gamma(\boldsymbol{r},\boldsymbol{0};\tau,h)\,, \tag{B11}$$

yielding the scaling power of the correlation function 2β. Thus, the correlation function exponent $\eta_\perp$ obeys the scaling law

$$(2\;-\;\eta_\perp)\,\nu_\perp \;=\; d\,\nu_\perp\;-\;2\,\beta \;=\; \gamma' \tag{B12}$$

which corresponds to the Fisher scaling law of equilibrium phase transitions.

C. Diagonalisation of Time-Evolution Operators

In this appendix, we demonstrate for the technically interested reader how the Liouville operator of a reaction-diffusion model with random-sequential updates such as the contact process can be constructed using a symbolic computer language such as *Mathematica*[2] and then briefly comment on sparse-matrix methods for obtaining low-lying eigenvalues of Liouville operators.

Starting point is the 2-site interaction matrix ℓ_{CP} which in case of the contact process is given by (3.39):

```
lcp = {{0,-1,-1,0}, {0,1+λ,0,-1}, {0,0,1+λ,-1}, {0,-λ,-λ,2}};
```

In order to construct a chain of L sites, we need to carry out tensor products. The following function performs the tensor product of two arbitrary (not necessarily quadratic) matrices in the standard basis:

```
CircleTimes[a_,b_]:=
       Transpose[Flatten[Transpose[Flatten[Transpose[
       Outer[Times,a,b],{1,3,2,4}],1],{3,1,2}],1]];
```

The function `CircleTimes` can be accessed conveniently using the infix notation ⊗ by typing ESC `c * ` ESC. With the n-site identity matrix

```
Id[n_]:=IdentityMatrix[2^n];
```

the interaction matrix acting at sites i and $i+1$ is just given by

```
Id[i-1]⊗ lcp ⊗ Id[L-i-1]
```

However, periodic boundary conditions require us to handle the interaction between site L and site 1 separately. To this end one first defines a set of basis matrices e_{ij} through

```
e[i_,j_]:=Table[If[i1==i && j1==j,1,0],{i1,1,2},{j1,1,2}]
```

as well as the function

```
PlaceAtSite[matrix_,j_,L_]:=Module[{k,l},  If[j<L,
      Id[j-1]⊗matrix⊗Id[L-j-1],
      Sum[matrix[[Range[2*k-1,2*k],Range[2*l-1,2*l]]]
          ⊗Id[L-2]⊗e[k,l], {k,1,2},{l,1,2}]]];
```

To construct a chain with periodic boundary conditions we have to perform the sum in (3.38), i.e.:

---

[2] Version 5.0 or higher. *Mathematica* is a registered trademark of Wolfram Research Inc., see `http://www.wolfram.com`

```
PeriodicChain[matrix_,L_]:=
      Sum[PlaceAtSite[matrix,j,L],{j,1,L}];
```

Using these functions the Liouville operator of a (1+1)-dimensional contact process with 3 sites and periodic boundary conditions is obtained by typing

```
PeriodicChain[lcp,3]
```

rendering the output

$$\begin{pmatrix} 0 & -2 & -2 & 0 & -2 & 0 & 0 & 0 \\ 0 & 2\lambda+2 & 0 & -2 & 0 & -2 & 0 & 0 \\ 0 & 0 & 2\lambda+2 & -2 & 0 & 0 & -2 & 0 \\ 0 & -\lambda & -\lambda & 2\lambda+4 & 0 & 0 & 0 & -2 \\ 0 & 0 & 0 & 0 & 2\lambda+2 & -2 & -2 & 0 \\ 0 & -\lambda & 0 & 0 & -\lambda & 2\lambda+4 & 0 & -2 \\ 0 & 0 & -\lambda & 0 & -\lambda & 0 & 2\lambda+4 & -2 \\ 0 & 0 & 0 & -2\lambda & 0 & -2\lambda & -2\lambda & 6 \end{pmatrix}.$$

Now we can compute the first four eigenvalues of the Liouville operator for a contact process with 10 sites at $\lambda = \lambda_c$ by typing

```
λ=3.297847;
ev=Eigenvalues[PeriodicChain[lcp,10]];
Take[Sort[ev],4]
```

After approximately 20 seconds *Mathematica* has diagonalised this 1024×1024 matrix, printing

```
{0., 0.0585524, 1.39426, 1.49716}
```

The first value vanishes since the absorbing state is stationary. The second value corresponds to the non-trivial eigenvalue μ_1 whose scaling properties are analysed in Sect. 4.3.1.

For larger system sizes, sparse-matrix techniques will become more efficient. A workhorse for a long time has been the *Lanczos algorithm* [136]. In the context of absorbing phase transitions, the Liouvillian/quantum Hamiltonian matrices to be diagonalised are in general non-symmetric and also, because detailed balance is not valid in general, cannot be brought to a symmetric form by a known similarity transformation. For real non-symmetric matrices the **Arnoldi algorithm** can be an useful alternative, which brings the $n \times n$ Liouvillian matrix $\mathcal{L}$ to an upper Hessenberg form. We indicate the main loop:

$$\begin{array}{rcl} w & = & \mathcal{L}v_j \\ h_{ij} & = & v_i \cdot w \qquad ; \; i = 1, \ldots, j \\ w' & = & w - \sum_{i=1}^{j} h_{ij} v_i \\ h_{j+1,j} & = & (w' \cdot w')^{1/2} \\ v_{j+1} & = & w'/h_{j+1,j} \end{array} \tag{C1}$$

where v_i, w, w' are n-component vectors. A normalised starting vector v_1, which should not coincide with one of the eigenvectors of H, must be given. The numbers $h_{i,j}$ form an upper Hessenberg matrix. Rapidly converging estimates for the lowest eigenvalues can be found by diagonalising the Hessenberg submatrix obtained after a moderate number m of iterations.

In practise, the v_i should *always* be re-orthogonalised, since the Hessenberg matrix is more sensitive to rounding errors than the tridiagonal matrix encountered with the Lanczos algorithm. Extended numerical precision (quadruple) may be required.

D. Langevin Equations and Path Integrals

Instead of describing a stochastic process by a Langevin equation it is often more convenient to use a path integral representation. In particular, correlation functions as well as response functions can be expressed conveniently in terms of path integrals [325, 49, 190]. Furthermore, the path integrals can be treated by well established methods of normalised perturbation theory such as the ϵ-expansion. Here, we sketch the equivalence of both approaches, considering stochastic processes as described by the **Langevin equation**

$$\lambda^{-1}\partial_t \varrho(t, \boldsymbol{r}) = F[\varrho(t, \boldsymbol{r})] + \eta(t, \boldsymbol{r}), \tag{D1}$$

with the noise correlator

$$\langle \eta(t, \boldsymbol{r}) \eta(t', \boldsymbol{r}') \rangle = \lambda^{-1} \kappa N[\varrho(t, \boldsymbol{r})] \, \delta^d(\boldsymbol{r} - \boldsymbol{r}') \, \delta(t - t'). \tag{D2}$$

Here, $F[\varrho(t, \boldsymbol{r})]$ denotes the deterministic part of the Langevin equation. For example, the Langevin equation of DP is obtained for

$$F[\varrho(t, \boldsymbol{r})] = \left(\tau - g\,\varrho(t, \boldsymbol{r}) + \nabla^2\right)\varrho(t, \boldsymbol{r}) + h, \tag{D3}$$

$$N[\varrho(t, \boldsymbol{r})] = \varrho(t, \boldsymbol{r}). \tag{D4}$$

Similarly, for a relaxing ferromagnet without any conservation laws but coupled to a bath of temperature T one has

$$F[\varrho(t, \boldsymbol{r})] = -\frac{\delta \mathcal{H}[\varrho]}{\delta \varrho} + h \quad , \quad \kappa N[\varrho(t, \boldsymbol{r})] = 2T \tag{D5}$$

where $\mathcal{H}$ is the equilibrium Ginzburg-Landau functional. Note that both $F[\varrho(t, \boldsymbol{r})]$ and $N[\varrho(t, \boldsymbol{r})]$ may include in general differential operators. For example, $N[\varrho(t, \boldsymbol{r})]$ involves a Laplacian operator in the case of conserved fields.

The partition sum of the stochastic process is defined as the functional integral over all realisations of the field $\varrho(t, \boldsymbol{r})$ and the noise $\eta(t, \boldsymbol{r})$ which satisfy the above Langevin equation, i.e.,

$$Z = \int \mathcal{D}\eta\, P(\eta) \int \mathcal{D}\varrho\, \delta\Big(\lambda^{-1}\partial_t \varrho(t, \boldsymbol{r}) - F[\varrho(t, \boldsymbol{r})] - \eta(t, \boldsymbol{r})\Big), \tag{D6}$$

where the noise field is distributed according to a Gaussian, i.e.

$$P(\eta) \sim \exp\left(-\int \mathrm{d}t\, \mathrm{d}\boldsymbol{r}\, \frac{\eta^2}{2\kappa N[\varrho]}\right). \tag{D7}$$

In order to eliminate the integration over the noise η, auxiliary fields $\widetilde{\varrho}(t, \boldsymbol{r})$ are introduced [463, 325], making use of the identity

$$\delta(x) = \frac{1}{2\pi} \int_{\mathbb{R}} \mathrm{d}\widetilde{\varrho}\, \exp\left(\mathrm{i}\widetilde{\varrho}x\right). \tag{D8}$$

Applying this to the functional δ-function in (D6) one finds after an appropriate Wick rotation within the complex plane

$$Z \sim \int \mathcal{D}\varrho \mathcal{D}\widetilde{\varrho}\, \mathrm{e}^{-\int \mathrm{d}t\, \mathrm{d}\boldsymbol{r}\, \widetilde{\varrho}\left(\lambda^{-1}\partial_t \varrho - F[\varrho]\right)} \int \mathcal{D}\eta\, P(\eta)\, \mathrm{e}^{-\int \mathrm{d}t\, \mathrm{d}^d\boldsymbol{r}\, \widetilde{\varrho}\eta}. \tag{D9}$$

Performing the integration over η by completing the square yields

$$Z \sim \int \mathcal{D}\varrho \mathcal{D}\widetilde{\varrho}\, \mathrm{e}^{-\mathcal{J}[\varrho, \widetilde{\varrho}]} \tag{D10}$$

with the so-called Janssen-De Dominicis functional

$$\mathcal{J}[\varrho, \widetilde{\varrho}] = \int \mathrm{d}t\, \mathrm{d}\boldsymbol{r}\, \widetilde{\varrho}\left(\lambda^{-1}\partial_t \varrho - F[\varrho] - \frac{\kappa}{2} N[\varrho]\widetilde{\varrho}\right). \tag{D11}$$

This allows one to derive correlation functions from the path integral, e.g.

$$\langle \varrho(t, \boldsymbol{r})\varrho(t', \boldsymbol{r}')\rangle \sim \int \mathcal{D}\varrho \mathcal{D}\widetilde{\varrho}\, \varrho(t, \boldsymbol{r})\varrho(t'\boldsymbol{r}')\, \mathrm{e}^{-\mathcal{J}[\varrho, \widetilde{\varrho}]}. \tag{D12}$$

The auxiliary field $\widetilde{\varrho}$ has the following physical meaning. Applying the external conjugated field h the linear response of the order parameter is given by

$$\left.\frac{\delta\langle\varrho\rangle}{\delta h}\right|_{h\to 0} = \langle \varrho\widetilde{\varrho}\rangle. \tag{D13}$$

Therefore, the auxiliary field is often referred to as the **response field**. Note that the two-point correlation function (D12) and the response function $\langle\varrho\widetilde{\varrho}\rangle$ are different, reflecting the violation of the fluctuation-dissipation theorem in non-equilibrium systems.

Non-stationary initial conditions can be included within the present formalism as follows [338, 328]. The paradigmatic example is a distribution of initial states with mean order parameter $\langle \varrho(0, \boldsymbol{r}) \rangle = m_0(\boldsymbol{r})$ and short-ranged Gaussian fluctuations

$$\Big\langle \left[\varrho(0,\boldsymbol{r}) - m_0(\boldsymbol{r})\right]\left[\varrho(0,\boldsymbol{r}') - m_0(\boldsymbol{r}')\right]\Big\rangle = \tau_0^{-1}\delta^d(\boldsymbol{r}-\boldsymbol{r}') . \tag{D14}$$

The corresponding contribution to the action will be a Gaussian of the form

$$\mathcal{J}_0[\varrho] = \int \mathrm{d}\boldsymbol{r}\, \frac{\tau_0}{2}\left[\varrho(0,\boldsymbol{r}) - m_0(\boldsymbol{r})\right]^2 . \tag{D15}$$

Since the canonical dimension of the parameter τ_0 is 2, the only fixed-point value which leads to a normalisable distribution is $\tau_0 = \infty$. Since corrections coming from a finite value of τ_0 will be irrelevant, one can set from the outset $\tau_0 = \infty$, unless in situations where terms vanish in this limit. Using the short-time relation $\varrho(0, \boldsymbol{r}) = \tau_0^{-1}\widetilde{\varrho}(0, \boldsymbol{r})$ [328], the Janssen-de Dominicis functional now turns into

$$\mathcal{J}_{\mathrm{eff}}[\varrho, \widetilde{\varrho}] = \mathcal{J}[\varrho, \widetilde{\varrho}] + \int \mathrm{d}\boldsymbol{r}\left(\frac{1}{2\tau_0}\widetilde{\varrho}^2(0,\boldsymbol{r}) - m_0\widetilde{\varrho}(0,\boldsymbol{r})\right) . \tag{D16}$$

E. Mean-Field Approximations

Mean-field theories are useful in order to obtain a first survey of the possible critical behaviour of a statistical system. Here we describe a technique for a particle-reaction model with the following random-sequential updates:

$$\begin{aligned} A\emptyset,\ \emptyset A &\to AA && \text{with rate } \sigma \\ AA &\to \emptyset\emptyset && \text{with rate } \lambda \\ A\emptyset &\leftrightarrow \emptyset A && \text{with rate } D\ , \end{aligned} \tag{E1}$$

where σ, λ and D are the control parameters. This model exhibits a non-equilibrium phase transition which belongs to the DP universality class.

All mean-field treatments must to some extent neglect correlations. A systematic way of constructing a sequence of improving mean-field theories was devised by ben-Avraham and Köhler [55] on the basis of earlier work [258, 169]. In their so-called (n, m)-approximation one considers clusters of n sites, assuming that clusters with more than n sites factorise in terms of n-site probabilities with an overlap of m sites between adjacent clusters. More specifically, if $\mathfrak{A}, \mathfrak{B}, \mathfrak{C}$ are such overlapping clusters of a one-dimensional system, and if $P(\mathfrak{A}|\mathfrak{B}) = P(\mathfrak{A})/P(\mathfrak{B})$ is the conditional probability to find $\mathfrak{A}$ for given $\mathfrak{B}$, one uses the approximation

reaction	$\Delta N_\bullet$	rate
$\bullet\circ \longrightarrow \bullet\bullet$	$+1$	$\frac{1}{2}\sigma\,\varrho(1-\varrho)$
$\circ\bullet \longrightarrow \bullet\bullet$	$+1$	$\frac{1}{2}\sigma\,\varrho(1-\varrho)$
$\bullet\bullet \longrightarrow \circ\circ$	-2	$\lambda\,\varrho^2$

Table E1 Rates in the site approximation for the process (E1).

$$P(\mathfrak{ABC}) \approx P(\mathfrak{AB})P(\mathfrak{BC}|\mathfrak{B}), \tag{E2}$$

meaning that correlations up to n sites are taken into account. It is assumed throughout that the cluster probabilities are spatially translation-invariant. For example, consider a five-site cluster in the state $ABCDE$. In the $(3,1)$- and $(3,2)$-approximations, respectively, the probability $P(ABCDE)$ of this cluster is expressed in terms of mutually overlapping 3-site probabilities as

$$P^{(3,1)}(ABCDE) \approx P(ABC)P(CDE|C) = \frac{P(ABC)\,P(CDE)}{P(C)} \tag{E3}$$

$$\begin{aligned} P^{(3,2)}(ABCDE) &\approx P(ABC)\,P(BCDE|BC) \\ &\approx P(ABC)\frac{P(BCD)\,P(CDE|CD)}{P(BC)} \\ &= \frac{P(ABC)\,P(BCD)\,P(CDE)}{P(BC)\,P(CD)}\,. \end{aligned} \tag{E4}$$

It has been argued that the $(n, n-1)$-approximations are qualitatively the most reliable [55] and this conclusion has been generally accepted. The simplest of these schemes is the $(1,0)$-approximation, also called the **site approximation** or **simple mean-field** approximation. Similarly, the $(2,1)$-approximation is called the **pair approximation** while the $(3,2)$- and $(4,3)$-approximations are referred to as the **triplet approximation** and **quartet approximation**, respectively.

In what follows we give examples of these approximation schemes. For simplicity we consider the one-dimensional case, denoting an occupied site by $\bullet$ and an empty site by $\circ$. For example, a three-site cluster $AA\emptyset$ will then be denoted as $\bullet\bullet\circ$ and has the probability $P(AA\emptyset) = P_{\bullet\bullet\circ}$.

E.1 Simple Mean-Field/Site Approximation

The simplest mean-field approach considers only the single-site probabilities. Assuming that the density $\varrho(t) = P_\bullet(t)$ is translational invariant the corresponding rate equation is easily derived. In Table E1 we summarise those reactions from (E1) which change the mean particle-density, together with the change $\Delta N_\bullet$ in the number of occupied sites $N_\bullet$ by each of these reactions. Adding all these contributions one easily finds

$$\frac{\mathrm{d}\varrho(t)}{\mathrm{d}t} = \sigma\,\varrho(t)\Big(1-\varrho(t)\Big) - 2\lambda\,\varrho(t)^2 = \sigma\varrho(t) - (\sigma+2\lambda)\varrho(t)^2 \tag{E5}$$

which has of course the same structure as the simple mean-field equation (3.6).

E.2 Pair-Approximation

Treating our model in (E1) in the pair-approximation, we assume again that the probabilities $P(AB)$ are translation-invariant and furthermore that the system is left/right-invariant, viz. $P_{\bullet\circ}(t) = P_{\circ\bullet}(t)$. Then, because of $P_\bullet = P_{\bullet\circ} + P_{\bullet\bullet}$ and $P_{\bullet\bullet} + 2P_{\bullet\circ} + P_{\circ\circ} = 1$, there are two independent variables which may be chosen as the particle-density $\varrho(t) = P_\bullet(t)$ and the pair-density $u(t) = P_{\bullet\bullet}(t)$.

The equations of motion of the pair-approximation are now found in a straightforward way. Introducing the shortcuts $v = P_{\bullet\circ} = \varrho - u$ and $w = P_{\circ\circ} = 1 - 2\varrho + u$, the reactions changing $N_\bullet$ and $N_{\bullet\bullet}$ are listed with their rates in Table E2, grouped into three separate groups corresponding to the three reactions. In the last group, we only need to take into account those reactions $A \to 2A$ which modify the particle-configuration on a given site. Furthermore, there are three kinds of symmetry factors. For diffusion, an extra factor 2 comes from the inverse reaction. For pair-annihilation, parity-symmetric configurations need only be calculated explicitly once. Finally, for particle-creation, the two reactions of (E1) contribute equally. Collecting these contributions one obtains

$$\dot{\varrho}(t) = -2\lambda\, u(t) + \sigma\Big(\varrho(t) - u(t)\Big) \tag{E6}$$

$$\dot{u}(t) = -2D\frac{(\varrho(t)-u(t))(u(t)-\varrho(t)^2)}{\varrho(t)(1-\varrho(t))} - 2\lambda\, u(t)\frac{2u(t)+\varrho(t)}{\varrho(t)} + \sigma\Big(\varrho(t) - u(t)\Big)\,. \tag{E7}$$

The pair-approximation is the lowest-order cluster approximation which allows the effects of diffusion to be treated explicitly. In particular, in the $D \to 1$ limit, the site-approximation (E5) is recovered for $\varrho(t)$, while the pair density becomes $u(t) = \varrho(t)^2$ as if the two sites were uncorrelated.

reaction	$\Delta N_{\bullet}$	$\Delta N_{\bullet\bullet}$	rate	symmetry factor
● ● ○○ → ● ○ ●●	0	− 1	$D\,uvw/\varrho(1-\varrho)$	×2
○ ● ○● → ○ ○ ●●	0	+1	$D\,v^3/\varrho(1-\varrho)$	×2
● ● ●● → ● ○ ○●	− 2	− 3	$\lambda\,u^3/\varrho^2$	
● ● ●○ → ● ○ ○○	− 2	− 2	$\lambda\,u^2 v/\varrho^2$	×2
○ ● ●○ → ○ ○ ○○	− 2	− 1	$\lambda\,uv^2/\varrho^2$	
● ○ ● → ● ● ●	+1	+2	$\frac{1}{2}\sigma\,v^2/(1-\varrho)$	×2
● ○ ○ → ● ● ○	+1	+1	$\frac{1}{2}\sigma vw/(1-\varrho)$	×2

Table E2 Rates in the pair-approximation for the process (E1).

E.3 The 'Hop-Away' Mean-Field Approximation

The long-time behaviour of the simple model (E1) has been analysed in great detail by renormalisation-group methods. From the standard perturbative renormalisation-group [124, 114] it was concluded that simple mean-field was qualitatively correct in $d > 2$ dimensions, but for $d \leq 2$ a minimal branching rate $\sigma_c > 0$ is needed for an active state to exist. On the other hand, recent results from the non-perturbative renormalisation group [107, 104] show that for all dimensions $d > 2$ there is a finite value λ_c such that

(i) for $\lambda < \lambda_c$ there is an active state for all $\sigma > 0$, and
(ii) for $\lambda > \lambda_c$ an active state only exists if $\sigma > \sigma_c(\lambda)$.

This result has been qualitatively confirmed in numerical simulations [495]. In order to get a better qualitative understanding of these results, Canet and Hilhorst [108] have recently proposed a new type of mean-field theory which we shall now briefly present.

Rather than writing a continuum kinetic equation, Canet and Hilhorst [108] modify the system, allowing multiple occupancy per site. The stochastic rules are as follows.

1. Each particle is subject to the on-site creation reaction $A \to 2A$ at rate σ.
2. Each pair of particles on the same site is subject to the on-site annihilation reaction $2A \to 0$ at rate λ.
3. Each particle may *hop away* from its site at a rate D to another (not necessarily neighbouring) site which is empty (of some abstract lattice that need not be specified).

While the first two rules are merely the translation of (E1) into a particle language, the third rule introduces an important modification which makes the problem exactly solvable. As the third rule ignores spatial correlations, the so defined model is called the **hop-away mean-field model**. One may hope that for $D \to 0$ this model might provide a good qualitative agreement with the behaviour of the original system.

The solution of the hop-away mean-field model proceeds in two steps: First, one considers the evolution on a single site. If $P(n,t)$ is the probability that a specific site contains exactly n particles at time t, it satisfies the master equation

$$\begin{aligned}\frac{\mathrm{d}}{\mathrm{d}t}P(n,t) = \sigma(n-1)P(n-1,t) - \sigma n P(n,t)\\ +\tfrac{1}{2}\lambda(n+1)(n+2)P(n+2,t) - \tfrac{1}{2}\lambda n(n-1)P(n,t)\\ +D(n+1)P(n+1,t) - DnP(n,t)\end{aligned} \tag{E8}$$

for $n = 0, 1, 2, \ldots$ and with the convention that $P(-1,t) = 0$. Defining the vector $\boldsymbol{P}(t) = (P(0,t), P(1,t), \ldots)$ this may be rewritten in a matrix form $\mathrm{d}\boldsymbol{P}/\mathrm{d}t = \mathfrak{M}\boldsymbol{P}$, where

$$\mathfrak{M} = \begin{pmatrix} 0 & D & \lambda & 0 & 0 & \cdots \\ 0 & -D-\sigma & 2D & 3\lambda & 0 & \cdots \\ 0 & \sigma & -2D-2\sigma-\lambda & 3D & 6\lambda & \cdots \\ 0 & 0 & 2\sigma & -3D-3\sigma-3\lambda & 4D & \cdots \\ 0 & 0 & 0 & 3\sigma & -4D-4\sigma-6\lambda & \cdots \\ \vdots & \vdots & \vdots & \vdots & \vdots & \ddots \end{pmatrix}. \tag{E9}$$

Writing $G_{n,n_0}(t-t_0)$ for the solution with the initial condition $P(n,t_0) = \delta_{n,n_0}$ the analysis of Canet and Hilhorst is based on the assumption that, for $\sigma > 0$, $G_{n,n'}(t) \sim \exp(\mu_1 t)$ for $t \to \infty$, where $\mu_1 = \mu_1(\sigma, \lambda, D)$ is the smallest non-vanishing eigenvalue of $\mathfrak{M}$.

Second, the coupling between sites is treated as follows. There are $S_n(t)$ sites with occupation number n and the total particle number is $N(t) = \sum_{n=1}^{\infty} nS_n(t)$. Some sites are already occupied initially while others become occupied only during the evolution of the system. Hence $\langle S_n(t)\rangle = \langle S_n(t)\rangle^{(\mathrm{ini})} + \langle S_n(t)\rangle^{(\mathrm{evo})}$ where the average is taken over the initial distribution and over the temporal evolution. For the initially occupied sites one has

$$\langle S_n(t)\rangle^{(\mathrm{ini})} = \sum_{n'=1}^{\infty} G_{n,n'}(t) S_{n'}(0)\,. \tag{E10}$$

Since new occupied sites are created at time t' with a rate $D\langle N(t')\rangle$, one also has

$$\langle S_n(t)\rangle^{(\mathrm{evo})} = D\int_0^{\infty} \mathrm{d}t'\; G_{n,1}(t-t')\langle N(t')\rangle\,. \tag{E11}$$

Adding these two contributions and assuming the initial condition $S_n(0) = N(0)\delta_{n,1}$ one finally obtains, with $H(t) = \sum_{n=1}^{\infty} nG_{n,1}(t)$

$$\langle N(t)\rangle = N(0)H(t) + D\int_0^{\infty} \mathrm{d}t'\; H(t-t')\langle N(t')\rangle \tag{E12}$$

which can be solved by a Laplace-transformation.

With the assumptions made above on $G_{n,n'}(t)$, this leads to $\langle N(t)\rangle \simeq N(0)\exp(-(\mu_1-D)t)$ and this will only vanish for large times if $\mu_1-D>0$. In the special case $D=0$, the single-site master equation describes the evolution of each site independently and $\mu_1=\mu_1(\sigma,\lambda,0)>0$ for $\sigma>0$. In addition, one expects that $\lim_{\sigma\to\infty}\mu_1(\sigma,\lambda,0)=\infty$ since then the metastable state present in the system decays infinitely slowly. If furthermore μ_1 is continuous in D, the condition $\mu_1-D>0$ will indeed be satisfied for D small enough, i.e. at least up to a threshold $D\le D_c(\sigma,\lambda)$. Conversely, this means that for all ratios σ/λ there exists a $D_c(\sigma/\lambda)$ such that for

$$\frac{\lambda}{D} > \frac{1}{D_c(\sigma/\lambda)} \tag{E13}$$

the stationary state is absorbing [108], in qualitative agreement with the results of the non-perturbative renormalisation group. See [108] for further details, in particular quantitative comparisons.

F. Finite-Size Scaling Techniques

F.1 Sequences of Finite-Size Estimates

We illustrate the use of finite-size scaling techniques for the determination of critical points and critical exponents [37, 201, 270]. Since in transfer matrix/Liouvillian/quantum Hamiltonian calculations the size of the lattices within reach is often quite small, finite-size methods are particularly useful in this context. In Fig. F1a, the finite-size behaviour of the lowest energy gap $\Gamma_L(\tau)=E_1(\tau)-E_0(\tau)$ of the quantum Hamiltonian [264]

$$H=-\sum_{\ell=1}^{L}\sigma_\ell^z-(1+\tau)\sum_{\ell=1}^{L}\sigma_\ell^x\sigma_{\ell+1}^x \tag{F1}$$

defined on a periodic chain with L sites is shown,[3] where the $\sigma_\ell^{x,z}$ are Pauli matrices attached to the site ℓ. For L large enough, the data converge to the limit $\Gamma_\infty(\tau)=2|\tau|$ for $\tau<0$, but a direct estimate of the critical value τ_c is difficult. Because of the expected finite-size scaling behaviour $\Gamma_L\sim L^{-1}$, estimates for pseudo-critical points $\tau_c(L)$ may be found by looking for

[3] In view of the recent fashion in nano-physics, this quantum system may be studied in its own right, but we remind the reader that its phase transition is in the same universality class as the classical $2D$ Ising model, see e.g. [270]. The ground state energy E_0 of H corresponds to the classical free energy and the lowest energy gap $\Gamma=E_1-E_0\sim\xi^{-1}$ is related to the inverse magnetic correlation length. Finally, τ corresponds to the control parameter of the $2D$ Ising model.

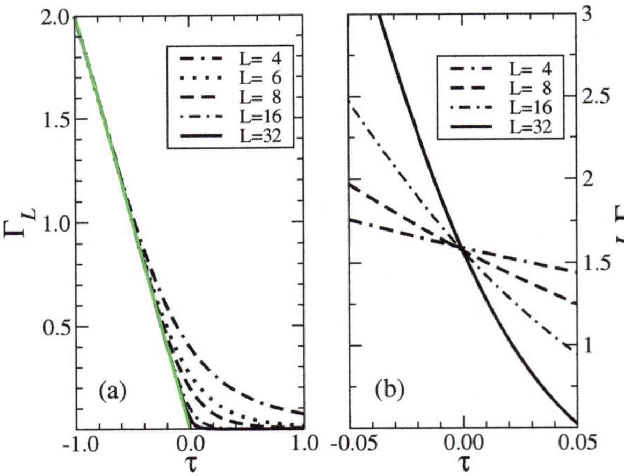

Fig. F1 (a) Finite-size behaviour of the smallest energy gap of the Ising quantum chain (F1) for $L = [4, 6, 8, 16, 32]$ from top to bottom. The grey line gives the limit $\Gamma_\infty(\tau) = 2|\tau|$. (b) Estimating the pseudo-critical point $\tau_c(L)$ from intersections of $L\Gamma_L(\tau)$.

intersections of $L\Gamma_L(\tau)$ for different values of L, as illustrated in Fig. F1b, and from which it can be seen that very likely $\tau_c(L) \approx 0$ ($\tau_c = 0$ can be verified by an exact analytic calculation [264]).

In principle, the same kind of technique may also be used for systems with non-equilibrium steady-states, since then the lowest gap should display the finite-size scaling $\Gamma_L \sim L^{-1/z}$ in the critical region, see Fig. 4.14 for an illustration in the contact process. However, there are cases when the critical line in the phase diagram of a non-equilibrium model separates a non-critical ordered phase from a *critical* disordered phase. Known examples are the branching-annihilating random walk with an even number of offspring or the pair-contact process with diffusion. In these cases, the analysis of finite-size data needs some modifications.

Consider, following [126], the finite-size scaling of the lowest gap Γ of the time evolution operator. As usual, we begin with the finite-size scaling form

$$\Gamma_L(\tau) = L^{-z} f\left((\tau - \tau_c)L^{1/\nu_\perp}\right) , \tag{F2}$$

where τ is the control variable which measures the distance to the critical point τ_c and f is assumed to be continuously differentiable. One expects the asymptotic behaviour

$$\Gamma_L(\tau) \sim \begin{cases} e^{\sigma L} & ; \text{ if } \tau < \tau_c \\ L^{-2} & ; \text{ if } \tau > \tau_c \end{cases} \tag{F3}$$

where $\sigma = \sigma(\tau)$ is a constant. In the usual case, one would have instead $\Gamma_L(\tau) \sim \Gamma_\infty$ for $\tau > \tau_c$. From (F3), one finds for the scaling function

$$f(\mathfrak{z}) \sim \begin{cases} \exp(-A|\mathfrak{z}|^{\nu_\perp}) & ; \text{ if } \mathfrak{z} \to -\infty \\ \mathfrak{z}^{(z-2)\nu_\perp} & ; \text{ if } \mathfrak{z} \to +\infty \end{cases} \tag{F4}$$

where A is a positive constant. Therefore, since $f(\mathfrak{z}) > 0$, it follows that for $z < 2$ the scaling function f must have a maximum at some finite value $\mathfrak{z}_{\max}$. Next, consider the logarithmic derivative

$$Y_L(\tau) := \frac{\ln[\Gamma(\tau; L+1)/\Gamma(\tau; L-1)]}{\ln[(L+1)/(L-1)]} = -z + \frac{\ln[f(\mathfrak{z}_+)/f(\mathfrak{z}_-)]}{\ln[(L+1)/(L-1)]} \tag{F5}$$

where $\mathfrak{z}_\pm = (\tau - \tau_c)(L \pm 1)^{1/\nu_\perp}$.

The standard method [37, 201] of finding estimates for both the critical point τ_c and the exponent z generalises the procedure for equilibrium systems, as illustrated in Fig. F1, by now considering two lattices of sizes L and L' and looking for the intersection of the two curves

$$Y_L(\tau^*) = Y_{L'}(\tau^*) = -z_{L,L'}\,, \tag{F6}$$

where the intersection point $\tau^* = \tau_{L,L'}$ defines a sequence of estimates converging towards τ_c as $L, L' \to \infty$ and similarly $z_{L,L'}$ converges towards z. A correlation length exponent can be found by forming first

$$Z_L(\tau) := \frac{\ln\left[\partial\Gamma_{L+1}/\partial\tau\right] - \ln\left[\partial\Gamma_L/\partial\tau\right]}{\ln(L+1) - \ln L} \tag{F7}$$

and since

$$\frac{1}{2}\left(Z_{L+1} + Z_L\right) = \zeta = \frac{1}{\nu_\perp} - z \tag{F8}$$

estimates for $\nu_\perp$ can be obtained when the derivatives are taken at $\tau = \tau_{c,L}$ which solves (F6)). In practice, finite-size corrections are often minimised by choosing L and L' as close as possible, i.e. $L' = L \pm 1$.

We illustrate how this may work, using as an example the quantum Hamiltonian/Liouvillian [542, 272]

$$\mathcal{L} = -\frac{1}{2}\sum_{j=1}^{L}\left[\tau\sigma_j^z + \sigma_j^x\sigma_{j+1}^x + \sigma_j^y\sigma_{j+1}^y - \sigma_j^z\sigma_{j+1}^z \right. \\ \left. +\mathrm{i}\left((1-\sigma_j^z)\sigma_{j+1}^y + \sigma_j^y(1-\sigma_{j+1}^z)\right)\right] \tag{F9}$$

which can be derived from a transfer matrix in $(1+1)D$ Reggeon field-theory and hence is in the universality class of directed percolation [123]. Although $\mathcal{L}$ is complex and non-hermitean, because of the global quasi-symmetry $U\mathcal{L}U^+ = \mathcal{L}^+$ with $U = \prod_{\ell=1}^{L}\sigma_\ell^z$, the eigenvalues of $\mathcal{L}$ are either real or occur in complex conjugate pairs. For $\tau \geq 2$, the ground-state energy is $E_0 = -\frac{1}{2}(\tau - 1)L$.

sizes	τ_c	z	ζ
2,3,4	2.61575	1.47580	-0.76573
3,4,5	2.62261	1.45710	-0.72164
4,5,6	2.62043	1.46444	-0.70633
5,6,7	2.61773	1.47542	-0.69918
6,7,8	2.61553	1.48589	-0.69518
7,8,9	2.61384	1.49508	-0.69265
8,9,10	2.61255	1.50298	-0.69089
9,10,11	2.61155	1.50974	-0.68958
10,11,12	2.61078	1.51556	-0.68854
11,12,13	2.61016	1.52059	-0.68769
12,13,14	2.60966	1.52497	-0.68696
13,14,15	2.60925	1.52881	-0.68633
14,15,16	2.60892	1.53219	-0.68577
∞	2.60640(7)	1.5807(10)	-0.6728(25)

Table F1 Finite-size estimates for the critical point τ_c and the exponents z and ζ, obtained from the quantum Hamiltonian (F9) of $(1+1)D$ Reggeon field-theory. The line labelled ∞ gives the infinite-lattice estimated obtained from the BST algorithm.

In Table F1, we list finite-size estimates [272] for the critical point τ_c, and the exponents z and ζ, which were obtained by calculating the intersections of $Y_{L,L'}(\tau)$ and Z_L for triplets of three lattices. These estimates appear to form convergent sequences, whose limit may be estimated by using extrapolation methods such as the BST algorithm, to be described below. We already notice here that the estimate for the critical point τ_c is in excellent agreement with an estimate $\tau_c = 2.60628(4)$ [542] which comes from a series-expansion study of (F9). Also, the resulting exponent estimates $z = 1.5807(10)$ and $\nu_\perp = 1.101(3)$ are in good agreement with the more precise values known from other methods, as listed in Table 4.3 on page 159.

A similar application of the method to the $(1+1)D$ contact process gives, after extrapolation of the finite-size data with the BST algorithm, the estimate $\lambda_c = 3.29792(4)$ for the critical point and the exponents $z = 1.58077(2)$, $\nu_\perp = 1.09681(2)$ and $\delta = 0.162(2)$ [150].

However, it may happen that the logarithmic derivatives $Y_L(\tau)$ do not intersect. *A priori*, there is no guarantee that they should do so, since the intersection of the curves $Y_L(\tau)$ for different values of L depends on the structure of the *corrections* to the leading finite-size scaling behaviour. A known example for non-intersection is furnished by the $1D$ PCPD with free boundary-conditions, where the curves $Y_L(p)$ do not intersect for different values of L. We illustrate this in Fig. F2.

It is easy to see from (F5) that in the scaling limit $\tau \to \tau_c$ and $L \to \infty$ such that $\mathfrak{z} = (\tau - \tau_c)L^{1/\nu_\perp}$ is kept fixed, one has

$$\lim \frac{\mathrm{d}Y_L}{\mathrm{d}\tau} \simeq \begin{cases} L^{1/\nu_\perp} A(2-\nu_\perp)(-\mathfrak{z})^{\nu_\perp - 1} & ; \text{ if } \mathfrak{z} \to -\infty \\ L^{1/\nu_\perp}(z-2)\mathfrak{z}^{-1} & ; \text{ if } \mathfrak{z} \to +\infty \end{cases} \tag{F10}$$

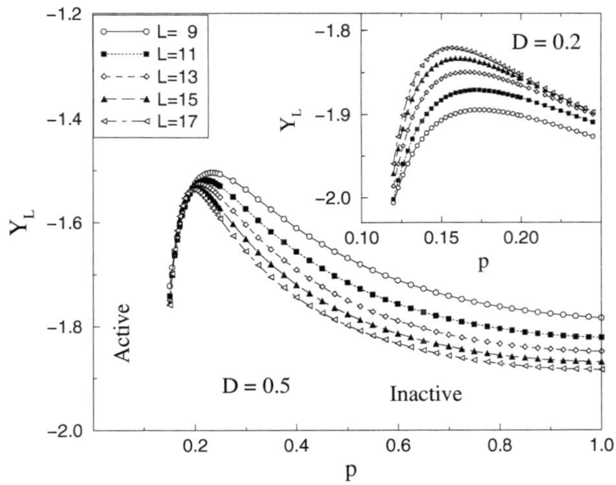

Fig. F2 Finite-size behaviour of the logarithmic derivative $Y_L(p)$ in the one-dimensional PCPD as a function of p and for several lattice sizes L. The main plot shows $D = 0.5$ and the inset shows $D = 0.2$. Reprinted with permission from [126]. Copyright (2001) by the American Physical Society.

Provided that $z < 2$ and $\nu_\perp < 2$, from Rolle's theorem there *must* exist a finite $\mathfrak{z}^*$ such that $\mathrm{d}Y_L/\mathrm{d}\tau|_{\mathfrak{z}=\mathfrak{z}^*} = 0$. However, since

$$Y_L(\mathfrak{z}^*) = -z + \frac{1}{\nu_\perp}\frac{\mathfrak{z}^* f'(\mathfrak{z}^*)}{f(\mathfrak{z}^*)} \tag{F11}$$

that maximum value of Y_L *cannot* be used to estimate the dynamical exponent z. This is the main difference with respect to the more usual situation. Rather, one has to form first a sequence of estimates τ_L of the critical point τ_c from the above extremal criterion which should converge according to $\tau_L \simeq \tau_c + \mathfrak{z}^* L^{-1/\nu_\perp}$. Having found τ_c, estimates of z can finally be obtained from the extrapolation $z = \lim_{L\to\infty} Y_L(\tau_c)$.

F.2 Sequence Extrapolation

Having a found a sequence of finite-size estimates for either critical points or exponents, the limit for $L \to \infty$ of such a sequence must be estimated and we outlined how such an extrapolation might be attempted. Ideally, "*... the method should be able to represent the underlying singular structure of the function under investigation*" [259]. But since in practice it is never really clear whether the assumed asymptotic behaviour is sufficiently well realised by the data available, *there is nothing like a straightforward and foolproof* *extrapolation method* and the use of *any* extrapolation technique does require a considerable amount of care and judgement. This means that necessarily the final estimates and in particular their error estimates are to some extent subjective. Formally, if f_N are the elements of a convergent sequence with $f = \lim_{N\to\infty} f_N$, one considers

$$\rho := \lim_{N\to\infty} \frac{f_{N+1} - f}{f_N - f}. \tag{F12}$$

in order to define two important types of convergence of sequences [259, 624].

1. The case of **linear convergence** is defined by $0 < |\rho| < 1$. As an example, take for $N \to \infty$
$$f_N = f + a\exp(-bN) + \ldots, \tag{F13}$$
where a and $b > 0$ are constants.
2. The case of **logarithmic convergence** is defined by $\rho = 1$. Examples of this are, for $N \to \infty$
$$f_N = f + a_1 N^{-\omega_1} + a_2 N^{-\omega_2} + \ldots \tag{F14}$$
$$f_N = f + a\exp(-bN^{\omega}) + \ldots \ ; \ \omega < 1, \tag{F15}$$
where a_1, a_2, a and $b, \omega_1, \omega_2, \omega > 0$ are constants.

Finite-size sequences obtained from observables close to a critical point are in general logarithmically convergent. In practise, it has turned out that the **BST algorithm** (originally devised as a method for numerical integration [96], before having been slightly adapted for applications to critical phenomena [280]) is probably the best-performing generic method known at present, in the context of this particular kind of application. It is defined as follows [96, 280].

Consider a sequence h_N $(N = 0, 1, 2, \ldots)$ converging to zero as $N \to \infty$. Form a table of extrapolants

$$\begin{array}{ccc} T_0^{(0)} & & \\ & T_1^{(0)} & \\ T_0^{(1)} & & T_2^{(0)} \\ & T_1^{(1)} & \\ T_0^{(2)} & & \ddots \end{array} \tag{F16}$$

where

$$T_{-1}^{(N)} = 0 \tag{F17}$$
$$T_0^{(N)} = T(h_N) = f_N \tag{F18}$$

$$T_m^{(N)} = T_{m-1}^{(N+1)} + \left(T_{m-1}^{(N+1)} - T_{m-1}^{(N)}\right) \times \left[\left(\frac{h_N}{h_{N+m}}\right)^{\omega}\left(1 - \frac{T_{m-1}^{(N+1)} - T_{m-1}^{(N)}}{T_{m-1}^{(N+1)} - T_{m-2}^{(N+1)}}\right) - 1\right]^{-1} \tag{F19}$$

with ω as a *free parameter.* This algorithm arises from approximating the function $T(h)$ by a sequence of rational functions in the variable h^{ω}. One expects that $T_1^{(N)}$ converges faster than $T_0^{(N)}$ to the desired limit $f = T(0)$ and so on. One tries to choose ω such as to minimise simultaneously the differences $T_m^{(N+1)} - T_m^{(N)}$.

For detailed reviews on the extrapolation methods by sequence-transformation, see [259, 624]. In the context of applications to finite-size scaling, it has been shown [38] that the so-called VBS algorithm performs better than its many alternatives. On the other hand, the VBS algorithm is outperformed by the BST method [280], see also [270].

As a practical example, consider the determination of the critical point of the quantum Hamiltonian/Liouvillian (F9). The finite-size estimates for the pseudo-critical point τ_c, see Table F1, for $L = 8, 9, \ldots 15$ are listed in the first column of Table F2. The following columns show the construction of the extrapolation table, with an assumed value $\omega = 1.94$. If that value were indeed the final one, one might read off the estimate $\tau_c \simeq 2.60641$. Of course, the effects of varying ω must be thoroughly investigated and in general, the precision of the final estimate increases with the length of the sequence available.

2.6138388							
	2.6075368						
2.6125490		2.6063534					
	2.6071762		2.6064220				
2.6115541		2.6063865		2.6064067			
	2.6069493		2.6064116		2.6064131		
2.6107756		2.6063975		2.6064099		2.6064068	
	2.6068001		2.6064107		2.6064121		2.6064107
2.6101573		2.6064029		2.6064029		2.6064103	
	2.6066987		2.6064029		2.6064109		
2.6096593		2.6064029		2.6064029			
	2.6066275		2.6064321				
2.6092529		2.6064141					
	2.6065788						
2.6089176							

Table F2 Finite-lattice extrapolation with the BST algorithm, $\omega = 1.94$.

G. Numerical Methods

We recall some background knowledge on Monte Carlo simulation and detailed balance, before describing some methods for the numerical computation of response functions. For further information, see [323, 485, 405, 59].

G.1 Simulational Techniques

Classical spin systems do not have a natural dynamics. In order to describe the relaxational behaviour of such systems, it is common to create an artificial dynamics. The most simple (and most usual) way to do this is through a **Markov process**. If one denotes by $\{\sigma\}$ a configuration of spins and by $P(\{\sigma\};t)$ the probability to find this configuration at time t, the change of this probability over a short time interval δt is given by a **master equation**

$$\frac{P(\{\sigma\};t+\delta t)-P(\{\sigma\};t)}{\delta t}=\sum_{\{\tau\}}\Big[w(\tau\to\sigma)P(\{\tau\};t)-w(\sigma\to\tau)P(\{\sigma\};t)\Big] \tag{G1}$$

where $w(\tau \to \sigma)$ is the (time-independent) transition rate from the configuration $\{\tau\}$ to the configuration $\{\sigma\}$. Clearly, the change of the probability $P(\{\sigma\};t)$ only depends on that same probability and not on the configurational probabilities of earlier times. This is the defining property of a Markov process. From a physical point of view it is not trivial at all that a description of the dynamics of a given classical system in terms of a Markov process should be adequate.

In many instances, Markov processes are used in order to generate an importance sampling for the calculation of averages at equilibrium, where implicitly an ergodicity hypothesis is used in order to be able to relate averages of a thermodynamic ensemble to a time-average over the Markov dynamics. In this context, the condition of **detailed balance** is important, which involves the stationary probability distribution

$$P_\infty(\{\sigma\}) := \lim_{t\to\infty} P(\{\sigma\};t) \tag{G2}$$

and reads

$$w(\tau\to\sigma)P_\infty(\{\tau\})-w(\sigma\to\tau)P_\infty(\{\sigma\})=0 \tag{G3}$$

If $P_\infty(\{\sigma\}) \neq 0$ for all configurations $\{\sigma\}$, one may define an equilibrium Hamiltonian $\mathcal{H}$ through $P_\infty(\{\sigma\}) = Z^{-1}\exp(-\mathcal{H}[\{\sigma\}]/T)$ where T is the temperature (implicitly contained in the rates w) and Z the canonical partition function. It can then be shown that for Markov processes with detailed balance, for $t \to \infty$ one always has convergence to this unique equilibrium state.

A simple argument can be given through a Boltzmann H-theorem. The derivation given here follows [484]. First, since $w(\sigma \to \sigma) = 1 - \sum_{\{\tau\} \neq \{\sigma\}} w(\sigma \to \tau)$ is the probability that the system remains in the state $\{\sigma\}$ during one update, one has, for all configurations $\{\tau\}$, the normalisation

$$\sum_{\{\sigma\}} w(\tau \to \sigma) = 1. \tag{G4}$$

Consider now the time-dependent quantity

$$G(t) := \sum_{\{\tau\}} \frac{[P(\{\tau\};t) - P_{\text{eq}}(\{\tau\})]^2}{P_{\text{eq}}(\{\tau\})} = \sum_{\{\tau\}} \frac{P(\{\tau\};t)^2}{P_{\text{eq}}(\{\tau\})} - 1 \tag{G5}$$

and also define $\Delta G := G(t + \delta t) - G(t)$. Then

$$\begin{aligned}
\Delta G &= \sum_{\{\tau\}} \left(\frac{P(\{\tau\};t+\delta t)^2}{P_{\text{eq}}(\{\tau\})} - \frac{P(\{\tau\};t)^2}{P_{\text{eq}}(\{\tau\})} \right) \\
&= \sum_{\{\tau\}} \left(\sum_{\{\sigma,\rho\}} w(\sigma \to \tau) w(\rho \to \tau) \frac{P(\{\sigma\};t)P(\{\rho\};t)}{P_{\text{eq}}(\{\tau\})} - \frac{P(\{\tau\};t)^2}{P_{\text{eq}}(\{\tau\})} \right) \\
&= \sum_{\{\tau,\sigma,\rho\}} w(\tau \to \sigma) w(\tau \to \rho) P_{\text{eq}}(\{\tau\}) \frac{P(\{\sigma\};t)}{P_{\text{eq}}(\{\sigma\})} \frac{P(\{\rho\};t)}{P_{\text{eq}}(\{\rho\})} \\
&\quad - \sum_{\{\tau,\sigma\}} w(\sigma \to \tau) \frac{P(\{\sigma\};t)^2}{P_{\text{eq}}(\{\sigma\})}
\end{aligned}$$

where in the second line the master equation (G1) was used twice. In the third line detailed balance (G3) was used twice and in the last term the normalisation condition (G4) was used and the summation indices were exchanged. The second term in the last line is rewritten in a symmetrised way as follows

$$\begin{aligned}
&\sum_{\{\tau,\sigma\}} w(\tau \to \sigma) P_{\text{eq}}(\{\tau\}) \left(\frac{P(\{\tau\};t)}{P_{\text{eq}}(\{\tau\})} \right)^2 \cdot \overbrace{\sum_{\{\rho\}} w(\tau \to \rho)}^{=1} \\
&= \frac{1}{2} \sum_{\{\tau,\sigma,\rho\}} w(\tau \to \sigma) w(\tau \to \rho) P_{\text{eq}}(\{\tau\}) \left[\left(\frac{P(\{\sigma\};t)}{P_{\text{eq}}(\{\sigma\})} \right)^2 + \left(\frac{P(\{\rho\};t)}{P_{\text{eq}}(\{\rho\})} \right)^2 \right]
\end{aligned}$$

and upon inserting this into ΔG one finally arrives at

$$\Delta G = -\frac{1}{2} \sum_{\{\tau,\sigma,\rho\}} w(\tau \to \sigma) w(\tau \to \rho) P_{\text{eq}}(\{\tau\}) \left[\frac{P(\{\sigma\};t)}{P_{\text{eq}}(\{\sigma\})} - \frac{P(\{\rho\};t)}{P_{\text{eq}}(\{\rho\})} \right]^2 \leq 0$$

Consequently, $G(t)$ will always decrease unless all states $\{\sigma\}$ which one can reach by a single move from the state $\{\tau\}$ satisfy $P(\{\sigma\};t) \sim P_{\text{eq}}(\{\sigma\})$. This calculation only depends on (i) detailed balance and (ii) that $w(\sigma \to \sigma) \neq 0$ for all configurations $\{\sigma\}$. In order for the argument to be applicable, one finally needs (iii) that the system is ergodic, or in other words, that the configuration space cannot be decomposed into independent components. Under these conditions, *the master equation has a unique stationary solution which is given by the equilibrium distribution* $P_{\text{eq}}(\{\sigma\})$ *and for any initial state, the evolution according to the master equation will lead to that equilibrium solution* [484].

We now list a few popular choices for the transition rates which give rise to several stochastic algorithms.[4] They all satisfy ergodicity and detailed balance. We begin with several algorithms where configurations are modified by changing a single spin-variable at a time. These correspond to dynamics with a non-conserved order parameter which is often also referred to as **Glauber dynamics**.

Metropolis Algorithm

This well-known algorithm was introduced in 1953 by Metropolis et al. and is given by the rates

$$w(\tau \to \sigma) = \begin{cases} \exp(-(\mathcal{H}[\{\sigma\}] - \mathcal{H}[\{\tau\}])/T) & ; \text{ if } \mathcal{H}[\{\sigma\}] - \mathcal{H}[\{\tau\}] > 0 \\ 1 & ; \text{ otherwise} \end{cases} \tag{G6}$$

Glauber Algorithm

This is a specific algorithm which should be carefully distinguished from the Glauber dynamics defined above, of which it is merely a special case. The rates are

$$w(\tau \to \sigma) = \frac{1}{2} \left(1 + \tanh \left[\frac{1}{2T} (\mathcal{H}[\{\sigma\}] - \mathcal{H}[\{\tau\}]) \right] \right) \tag{G7}$$

[4] In the simulation of equilibrium systems, the relaxation towards the equilibrium state can often be increased considerably by using so-called *cluster algorithms*. Since these imply non-local update schemes, they are beyond the scope of this book, see e.g. [323].

Heatbath Algorithm

While in the two algorithms discussed so far, the change of the state depends on the current state $\{\sigma\}$, the flips of the heatbath algorithm are independent of the current state. For illustration, consider a spin system where the local variable σ_i at the site i has q possible values (e.g. the q-states Potts model). In order to go from the old configuration $\{\tau\}$ to the new configuration $\{\sigma\}$, one selects a single site $\boldsymbol{i}$ and sets the spin variable σ_i to an assigned value n (which may be the same value as before). The transition rate is

$$w(\tau \to \sigma) = \frac{\exp(-\mathcal{H}_n/T)}{\sum_{m=1}^{q} \exp(-\mathcal{H}_m/T)} \tag{G8}$$

where $\mathcal{H}_n = \mathcal{H}[\{\sigma\}]$ is the energy with the spin variable σ_i set to n. For the Ising model, the Glauber and the heatbath algorithms are equivalent.

Between these algorithms, the Glauber algorithm and the heatbath algorithms are considerably more efficient than the Metropolis algorithm, although they all give a dynamical exponent $z = 2$ when applied to phase-ordering kinetics. When selecting an algorithm for, say, the study of phase-ordering kinetics, it is sensible to choose an algorithm which moves slowly through phase space (measured in terms of updates required), in order to obtain a time range as large as possible for measurements of time-dependent quantities. Test runs easily show (e.g. [426]) that according to this criterion the Metropolis algorithm is inferior to both the Glauber and the heatbath algorithm.

Another important aspect is the choice of the update scheme. While a naïve sequential update scheme is likely to introduce unphysical correlations, reasonable choices are either (i) *random* update, where an additional random number is drawn to select the next site to be visited and (ii) *checkerboard* update, where the hypercubic lattice is divided into two sublattices such that all the neighbours of each site of the first sublattice are on the second sublattice and conversely. Now, first all sites on the first sublattice are updated, and only then all sites on the second sublattice, which can be efficiently implemented e.g. for the heatbath algorithm.

In comparing the efficiency of these two schemes, two considerations are important. First, the extra random number needed for the random update considerably slows down the calculation (a factor 8 between the two schemes was reported for the Ising/Potts model [426]). On the other hand, the passage through the phase space is somewhat faster for the checkerboard update. In order to understand by how much, observe that for the checkerboard update, each spin is updated exactly once for a passage over the entire lattice (a **sweep**). For a random update, the probability of a given spin having *not* been visited in one try is $p_{\rm not} = 1 - \mathcal{N}^{-1}$. During a sweep, one makes $\mathcal{N}$ tries, hence an arbitrary spin has been visited

$$p_{\text{visit}} = 1 - \left(1 - \mathcal{N}^{-1}\right)^{\mathcal{N}} \stackrel{\mathcal{N}\to\infty}{\simeq} 1 - e^{-1} \approx 0.63 \tag{G9}$$

which means that the checkerboard update passes about 1.6 times faster through the phase space than the random update [426, 427]. Hence checkerboard update is more efficient than random update if the generation of the extra random number slows execution by at least a factor 1.6.

In contrast to Glauber dynamics, whose elementary moves modify a single spin, **Kawasaki dynamics** select a *pair* σ_i, σ_j of distinct spins ($i \neq j$) and then perform a spin exchange $\sigma_i \mapsto \sigma_i' = \sigma_j, \sigma_j \mapsto \sigma_j' = \sigma_i$. This may then be combined with a Metropolis, Glauber or heatbath algorithm. We quote here the transition rate for the heatbath algorithm of the Ising model with Kawasaki dynamics [236]

$$w_{ij}(\tau \to \sigma) = \frac{\exp[\sigma_i(h_i - h_j + h_{ij}^{\mathrm{W}})/T]}{2\cosh[(h_i - h_j + h_{ij}^{\mathrm{W}})/T]} \prod_{k\neq i,j} \delta_{\tau_k,\sigma_k} \tag{G10}$$

and the Weiss field is given by

$$h_{ij}^{\mathrm{W}} = \sum_{k\in\partial i\backslash j} J_{ik}\sigma_k - \sum_{\ell\in\partial i\backslash j} J_{j\ell}\sigma_\ell \tag{G11}$$

and $\partial i \backslash j$ is the set of neighbours of the site i, with the site j excluded.

G.2 Computation of Response Functions

Obtaining a precise estimate of response functions is notoriously difficult because of the functional derivative in their definition. As a rule, this will lead to very noisy data, considerably more then ever obtained for correlations.

G.2.1 Finite External Magnetic Field

We first discuss the method proposed by Barrat [44] around this difficulty. We begin with the case of a non-conserved order parameter. Barrat's proposal is to consider integrated response functions and apply a magnetic field, according to some time protocol. Since for ferromagnetic systems the application of even rather small spatially homogeneous magnetic fields rapidly leads to a relaxation of the unique thermodynamic state, one rather chooses a spatially disordered magnetic field $h_{\boldsymbol{i}} = \pm\mathfrak{h}$ on the sites $\boldsymbol{i}$ of the lattice Λ with $|\lambda| = \mathcal{N}$ sites. Then the thermoremanent magnetisation is for the Ising model with spin variables $\sigma_{\boldsymbol{i}} = \pm 1$ given by [44]

$$TM_{\mathrm{TRM}}(t,s) = \frac{1}{\mathcal{N}} \sum_{\boldsymbol{i} \in \Lambda} \overline{\langle \sigma_{\boldsymbol{i}}(t) h_{\boldsymbol{i}}(s) \rangle} \tag{G12}$$

where $\langle . \rangle$ denotes the thermal average and $\overline{\cdot}$ the average over the disordered magnetic field. The corresponding integrated response function is then $\chi_{\mathrm{TRM}}(t,s) = M_{\mathrm{TRM}}(t,s)/\mathfrak{h} = \int_0^s \mathrm{d}u\, R(t,u)$.

For the q-states Potts model, this may be generalised as follows [427]

$$TM_{\mathrm{TRM}}(t,s) = \frac{1}{q-1} \left(\frac{q}{\mathcal{N}} \sum_{\boldsymbol{i} \in \Lambda} \overline{\left\langle \delta_{\sigma_i(t), h_i(s)} \right\rangle} - 1 \right) \tag{G13}$$

Here $\sigma_{\boldsymbol{i}} = 1, \ldots q$ is a Potts spin and the field $h_{\boldsymbol{i}}$ takes the values $1, \ldots, q$ randomly and with an amplitude $\mathfrak{h}$ in the classical Potts Hamiltonian $\mathcal{H}$.

While this method works rather well, it has the obvious disadvantages that (i) only integrated response functions can be found and (ii) it contains explicitly an amplitude $\mathfrak{h}$ which must be carefully chosen (large enough for a tolerable signal/noise ratio and small enough to remain in the linear response regime). In practice, it may be necessary to repeat the simulation several times, with different values of $\mathfrak{h}$. Statements found in the literature such as " *... a magnetic field* $\mathfrak{h} = 0.05$ *was used ...* " are at best statements about the results of such a detailed study (in a certain time regime of a specific model) and cannot be taken over to a different model or even to a different realisation of the same universality class.

Obviously, the same idea can also be applied to systems with a conserved order parameter.

For the measurement of the surface thermoremanent magnetisation of an Ising model, defined on a slab with two free surfaces of length L, one may use [47]

$$TM_{\mathrm{TRM},1}(t,s) = \frac{1}{2L} \sum_{\boldsymbol{i} \in \partial\Lambda} \overline{\langle \sigma_{\boldsymbol{i}}(t) h_{\boldsymbol{i}}(s) \rangle} \tag{G14}$$

where $\partial\Lambda$ denotes the total boundary of the lattice Λ.

G.2.2 Infinitesimal Magnetic Field

We next describe the method of Ricci-Tersenghi [545] which allows one to find integrated response functions, with an infinitesimal field. Consider the integrated response

$$\chi_{\boldsymbol{j},\boldsymbol{k}}(t; t_1, t_2) := \int_{t_1}^{t_2} \mathrm{d}u\, R(t,u; \boldsymbol{j} - \boldsymbol{k}) \tag{G15}$$

where $R(t,u;\boldsymbol{j}-\boldsymbol{k}) = \delta\langle\phi(t,\boldsymbol{j})\rangle/\delta h(s;\boldsymbol{k})|_{h=0}$ is the usual space-time response function of the order parameter and for the simplicity of notation, we assumed spatial translation-invariance although that is not necessary for what follows. The integrated responses usually considered are recovered as special cases: for $t_1 = s, t_2 = t$ we have $\chi_{\rm ZFC}(t,s) = \chi_{\boldsymbol{r},\boldsymbol{r}}(t;s,t)$, for $t_1 = 0, t_2 = s$ we have $\chi_{\rm TRM}(t,s) = \chi_{\boldsymbol{r},\boldsymbol{r}}(t;0,s)$, finally, for $t_1 = s/2, t_2 = s$ we obtain $\chi_{\rm Int}(t,s) = \chi_{\boldsymbol{r},\boldsymbol{r}}(t;s/2,s)$.

To be specific, we restrict the discussion from now on to the Ising model, but generalisations are obvious. We begin with the case of a non-conserved order parameter. Using the heatbath algorithm, the transition rate between the configurations $\{\tau\}$ and $\{\sigma\}$ through a spinflip at site i reads

$$w_i(\tau \to \sigma) = \frac{\exp[\sigma_i(h_i + h_i^{\rm W})/T]}{2\cosh[(h_i + h_i^{\rm W})/T]} \prod_{j\neq i} \delta_{\tau_j,\sigma_j} \tag{G16}$$

where h_i is the external magnetic field at site i and h_i^W is the Weiss field which describes the interactions of the spin variable σ_i with its neighbours. For a spin Hamiltonian $\mathcal{H} = \sum_{i,j} J_{ij}\sigma_i\sigma_j$ with binary interactions, the Weiss field is given by $h_i^{\rm W} = \sum_{j\neq i} J_{ij}\sigma_j$. Since the average magnetisation at site j is given by

$$\langle\sigma_j(t)\rangle = \sum_{\{\sigma(t')\}} \left[\sigma_j(t) \prod_{t'=1}^{t} w_{I(t')}(\sigma(t'-1) \to \sigma(t'))\right] \tag{G17}$$

where $I(t)$ is the site index of the spin to be updated at time t, a straightforward calculation now gives the susceptibility [545]

$$\begin{aligned} T\chi_{\boldsymbol{j},\boldsymbol{k}}(t;t_1,t_2) &= \langle\sigma_{\boldsymbol{j}}(t)\Delta\sigma_{\boldsymbol{k}}(t_1,t_2)\rangle \\ \Delta\sigma_{\boldsymbol{k}}(t_1,t_2) &= \sum_{u=t_1+1}^{t_2} \delta_{I(u),\boldsymbol{k}} \left(\sigma_{\boldsymbol{k}}(u) - \tanh\left(h_{I(u)}^{\rm W}(u)/T\right)\right) \end{aligned} \tag{G18}$$

Here, the time is discretised in units of Monte Carlo steps.

Following [545], we point out that $\Delta\sigma_k$ is a centred random variable with variance $\langle\Delta\sigma_k^2\rangle \sim t_2 - t_1$. This means that the calculation of the long-time behaviour of χ using this method requires a computational effort which increases with time (in particular, if quantities such as $\chi_{\rm TRM}, \chi_{\rm ZFC}$ or $\chi_{\rm Int}$ are required). On the other hand, since no longer a small field has to be introduced by hand, one is automatically in the linear response regime. Furthermore, a single simulation can be used to produce data for correlators and responses. The method was tested in several ageing systems, with and without disorder [545].

A very similar method works for the case of a conserved order parameter. It is enough to quote the result [236]

$$
\begin{aligned}
T\chi_{\boldsymbol{j},\boldsymbol{k}}(t;t_1,t_2) &= \langle \sigma_{\boldsymbol{j}}(t) \Delta\sigma_{\boldsymbol{k}}(t_1,t_2)\rangle \\
\Delta\sigma_{\boldsymbol{k}}(t_1,t_2) &= \sum_{u=t_1+1}^{t_2} \left\{ \delta_{I(u),\boldsymbol{k}} \left(\sigma_{\boldsymbol{k}}(u) - \tanh\left(h^{\mathrm{W}}_{I(u)J(u)}(u)/T \right)\right) \right. \\
&\quad \left. + \delta_{J(u),\boldsymbol{k}} \left(\sigma_{\boldsymbol{k}}(u) - \tanh\left(h^{\mathrm{W}}_{I(u)J(u)}(u)/T \right)\right) \right\}
\end{aligned} \tag{G19}
$$

and where the Weiss field h_{ij}^{W} was given in (G11).

G.2.3 Direct Calculation of the Response Function in the Ising Model

Chatelain [129] has shown how to calculate the response function directly from a certain correlation function, without having to apply a field at all. For an Ising model with Glauber dynamics, he obtains

$$
R(t,s;\boldsymbol{j}-\boldsymbol{i}) = \frac{1}{T}\left(1+\frac{\partial}{\partial t}+\frac{\partial}{\partial s}\right)\left\langle \sigma_{\boldsymbol{j}}(t)\left(\sigma_{\boldsymbol{i}}(s) - \tanh\left(h^{\mathrm{W}}_{I(s)}/T\right)\right)\right\rangle \delta_{I(s),\boldsymbol{i}} \tag{G20}
$$

In the derivation of this result, a slight modification of the usual Glauber dynamics with a *sequential* update scheme was used which makes the transition rates time-dependent but which should reduce to the usual one in the large-size limit $\mathcal{N} \to \infty$. Also Chatelain's method starts from a discrete-time master equation. It can be shown that from this form of the response function, one does recover the correct equilibrium limit and the fluctuation-dissipation theorem [129].

For practical applications, the same kind of remarks as for the method of [545] apply. In addition, the response function is intrinsically more noisy than the integrated responses so that a larger computational effort is required to obtain data which can be analysed. On the other hand, in many cases the functional form of the response function is considerably more simple than for the integrated responses.

It is possible to obtain also non-linear responses from the consideration of higher derivatives. We quote two examples [129]

$$
\begin{aligned}
R_{\boldsymbol{j};\boldsymbol{i}\boldsymbol{i}'}(t;s,s') &:= \left.\frac{\delta^2\langle\sigma_{\boldsymbol{k}}(t)\rangle}{\delta h(s,\boldsymbol{i})\delta h(s',\boldsymbol{i}')}\right|_{h=0} \\
&= \frac{1}{T}\left(1+\frac{\partial}{\partial t}+\frac{\partial}{\partial s}\right)\left(1+\frac{\partial}{\partial t}+\frac{\partial}{\partial s}+\frac{\partial}{\partial s'}\right) \\
&\quad \times \left\langle \sigma_{\boldsymbol{j}}(t)\left(\sigma_{\boldsymbol{i}}(s) - \tanh\Big(\frac{h^{\mathrm{W}}_{I(s)}}{T}\Big)\right)\left(\sigma_{\boldsymbol{i}}(s') - \tanh\Big(\frac{h^{\mathrm{W}}_{I(s)}}{T}\Big)\right)\right\rangle \\
&\quad \times \delta_{I(s),\boldsymbol{i}}\delta_{I(s'),\boldsymbol{i}'}
\end{aligned} \tag{G21}
$$

$$R^{(2)}_{\boldsymbol{j};\boldsymbol{i}}(t;s) := \left.\frac{\delta^2\langle\sigma_{\boldsymbol{k}}(t)\rangle}{\delta h(s,\boldsymbol{i})^2}\right|_{h=0} \tag{G22}$$
$$= -\frac{2}{T}\left(1+\frac{\partial}{\partial t}+\frac{\partial}{\partial s}\right)$$
$$\times\left\langle \sigma_{\boldsymbol{j}}(t)\tanh\left(\frac{h^{\mathrm{W}}_{I(s)}}{T}\right)\left(\sigma_{\boldsymbol{i}}(s')-\tanh\left(\frac{h^{\mathrm{W}}_{I(s)}}{T}\right)\right)\right\rangle\delta_{I(s),\boldsymbol{i}}$$

Extensions to other spin models are easily implemented if one replaces $\tanh(h^{\mathrm{W}}_{I(s)}/T)$ by the appropriate expression [129].

G.2.4 Alternative Method to Compute the Response Function of the Ising Model

Finally, we present an alternative algorithm [291] for the Ising model with standard heat bath dynamics. The main idea is to process perturbations at *all* sites simultaneously – a trick which allows one to obtain reasonable statistical averages within short time. A similar algorithm was already used for the contact process (see [289]). For the Ising model with heat bath dynamics the algorithm is somewhat more complicated, as will be explained in the following. However, it is conceptually simpler than the approaches described before in so far as it measures the response function directly.

Let us consider an Ising model with heat-bath dynamics which is prepared in a disordered state, where all spins independently take the values ± 1 with equal probability. Starting at time $t = 0$, the system evolves at finite temperature $\beta = 1/k_B T$ by random-sequential updates according to the following dynamical rules:

- Select a random site i.
- Calculate the local field $h_i = \sum_j \sigma_j$, where j runs over the $2D$ nearest neighbours of site i, respecting the chosen boundary conditions.
- Generate a random number $z \in (0,1)$ and set

$$\sigma_i^{(\mathrm{new})} := \begin{cases} +1 & \text{if } z < \frac{1}{2}\left(1+\tanh(\beta h_i)\right) \\ -1 & \text{otherwise} \end{cases} . \tag{G23}$$

The third step of this update procedure can be reformulated equivalently by defining a threshold field

$$h_i^{(c)} := \beta^{-1}\mathrm{atanh}(2z-1) , \tag{G24}$$

where $z \in (0,1)$ is a random number, and setting

$$\sigma_i^{(\text{new})} := \begin{cases} +1 & \text{if } h_i > h_i^{(c)} \\ -1 & \text{otherwise} \end{cases} . \tag{G25}$$

In order to measure the autoresponse function $R(t, s)$ directly we would have to apply a weak external field $0 < h_i^{(\text{ext})} \ll 1$ at time s locally at a randomly chosen site i and to study how much the ensemble average of the local magnetisation $\langle \sigma_j(t) \rangle$ at a different site j changes at some later time $t > s$. However, such a measurement is hard to perform since a localised weak field leads only to occasional spin flips in the future so that the average response to such a *single* spin flip is hard to quantify. This is probably the reason why most researchers avoid a direct measurement of the autoresponse function and prefer to study the integrated response. In the kinetic Ising model, however, there are two ways to optimise a direct measurement, namely, by (i) reweighting the probability of spin flips and (ii) simultaneous tagging of all spins in the system.

(i) Reweighting spin flip probabilities

As a first optimisation we study the response of an infinitely strong external perturbation instead of a weak one and extract the limit $h_j^{(\text{ext})} \to 0$ by an appropriate reweighting procedure. Reweighting is a standard technique in Monte-Carlo simulations to enhance their efficiency (see e.g. [216, 323]). The starting point is the observation that in heat bath dynamics as described above a weak perturbation by a positive localised external field $h_j^{(\text{ext})} \ll 1$ changes the probability of getting a positively oriented spin $\sigma_j^{(\text{new})} = +1$ to lowest order as

$$\frac{1}{2}\left(1 + \tanh\left[\beta(h_j + h_j^{(\text{ext})})\right]\right) = \frac{1}{2}\left(1 + \tanh\left[\beta h_j\right]\right) + \left(\frac{\beta}{2}\cosh^{-2}[\beta h_j]\right) h_j^{(\text{ext})} + \mathrm{O}([h_j^{(\text{ext})}]^2) . \tag{G26}$$

Therefore, we may equivalently turn up the spin σ_j with 100% probability and compensate this intervention by weighting the resulting response in the ensemble average by the corresponding first-order coefficient. Thus, the reweighting scheme consists of the following steps:

1. Select random site i at time s.
2. Define a weight $w_i = \frac{1}{2}\beta \cosh^{-2}(\beta h_i)$, where $h_i = \sum_j \sigma_j$ is the local field caused by the nearest neighbours of site i at time s, and store them in the memory.
3. Set $\sigma_i := +1$.
4. Measure how much this modification increases the magnetisation at site j at some later time $t > s$.

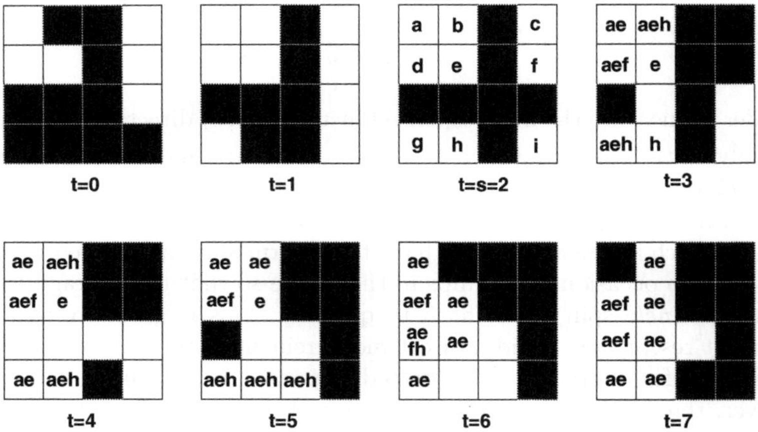

Fig. G1 Computation of the response function for a small system with 4×4 sites and periodic boundary conditions without quenched disorder. The figure shows eight snapshots at $t = 0, 1, \ldots, 7$. The sequence starts with a random initial state. Throughout the whole simulation the system evolves as an ordinary unperturbed Ising model with heat bath dynamics at $T = T_c$, where black boxes represent positive spin variables $\sigma_i = 1$ while white or grey boxes represent negative spin variables $\sigma_i = -1$. At time $t = s = 2$ all negatively oriented sites (white boxes) are tagged by individual labels $a, b, \ldots, i$ and the corresponding weights are recorded ($w_a \simeq 0.024482$, $w_b = w_c = w_d = w_g \simeq 0.110172$, $w_e = w_f = w_h = w_i \simeq 0.220343$). Subsequently the simulation tracks how the application of an external local field $h > 0$ would have changed the configuration. For example, if an external field would have turned up the spin in the left upper corner at time $t = s = 2$ (marked by label a) it would have changed the configuration at $t = 8$ in such a way that all sites marked by label a (in addition to those marked by black boxes) are positive as well. Note that each site may carry several labels, requiring a dynamically generated set structure in the code. At any time $t > s$ the autoresponse function is the sum over all sites which carry their original label assigned at $t = s$ (indicated by grey boxes), multiplied by the corresponding weight factor.

5. Multiply the measured response by the weight factor w_i and then perform the average over many independent realisations.

(ii) Simultaneous tagging of all possible spin flips caused by an external field

Even with the reweighting procedure described above the measurement of the autoresponse function is still difficult to perform since the response to a perturbation at a *single* site is quite small and fluctuates strongly. However, in the special case of the Ising model with heat bath dynamics it is possible to study the response to a perturbation of *all* spins simultaneously. The reason is that for a given set of random numbers used in the simulation a positive external field can only *increase* the magnetisation in the future, turning spins up but never turning spins down. Consequently, the cluster of positively oriented spins generated in a system *without* perturbation by an external field is always a subset of the actual cluster that would have been generated if the perturbation was applied. Thus, for each site i, where the field may have generated a spin flip, it is possible to trace and mark the cluster of all spins in the future that would have been turned up by this perturbation.

The other special property of the Ising model with heat bath dynamics, which is exploited by the algorithm, is the circumstance that these clusters do not interact. More specifically, if the external field had generated two spin flips, the resulting cluster would just be the union of the respective individual clusters. This allows one to tag all spins at time s with non-interfering labels and to identify the percolation cluster for each of the labels (see Fig. G1). The response of a particular site j is then proportional to the number of its labels (re-weighted as described above). This trick accelerates simulations by a factor of the order N, where N is the total number of sites.

The tagging algorithm can be implemented as follows. In addition to the local spin variables σ_i we attach to each site i a dynamically generated set S_i of labels. For example, in C++ such a dynamical set is provided by the class `set<...> S[N]` of the standard template library STL. Initially all these sets are empty ($S_i = \emptyset$) and the simulation evolves as usual. At time s, however, each site i is tagged with an individual label Λ_i and the weight w_i is recorded as described above. Therefore, at time s each set S_i has exactly one element, namely, $S_i = \{\Lambda_i\}$. During the subsequent simulation the spin variables σ_i evolve according to the usual heat bath dynamics as if there was no perturbation. However, in each update step the following additional steps are carried out

- When updating spin i, first clear the corresponding set by setting $S_i = \emptyset$ and let $U_i = \bigcup_j S_j$ be the union of all tags carried by the nearest neighbours.
- For all labels $\Lambda \in U_i$ count the number n_Λ of nearest neighbours which are oriented in the negative direction $\sigma_j = -1$ *and* carry the tag Λ. Obviously, if an external field had flipped the spin corresponding to the label Λ at time s it would have increased the local field h_j in the present update by $2n_\Lambda$.
- If $h_j < h_j^{(c)}$ and $h_j + 2n_\Lambda > h_j^{(c)}$ add label Λ to set S_i.

The autoresponse function $R(t, s)$ is then proportional to the weighted sum over all sets tagged by their original label, averaged over many independent realisations:

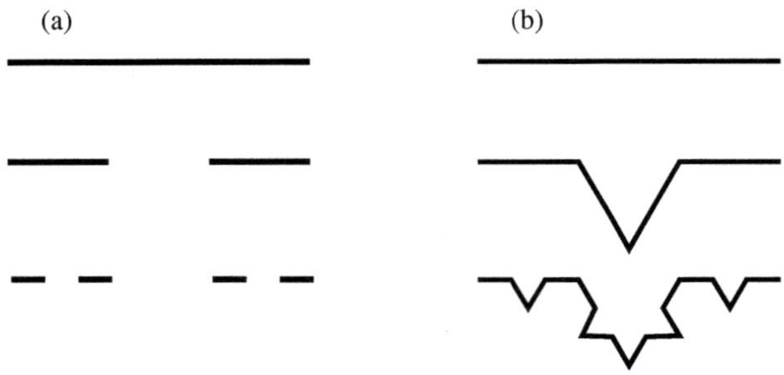

Fig. H1 First three steps, from top to bottom, in the construction of (a) the Cantor set and (b) the Koch curve.

$$R(t,s) \propto \left\langle \frac{1}{N} \sum_j w_j \delta_{\Lambda_j \in S_j} \right\rangle \tag{G27}$$

H. Fractal Dimensions

To make this volume a little more self-contained, we briefly collect the definitions of **fractal dimensions**, following [584]. For more than a century, mathematicians have studied so-called *fractal* ensembles which do not fit into standard continuum Euclidean geometry. Figure H1 shows two classic examples. The **Cantor set** is constructed, starting from the real interval $I = [0,1]$, by taking out the middle third of each line, for each iteration step, see Fig. H1a. It is not hard to see that the Cantor set has measure zero in the sense that it can be covered by intervals whose total length can be made arbitrarily small and yet, because of Cantor's famous *diagonal argument*, it is not countable. Similarly, Fig. H1b illustrates the construction of the **Koch curve**, again starting from $I = [0,1]$. For each step, one replaces the middle third of each straight line by an equilateral triangle. Since the length of the curve increases by a factor $\frac{4}{3}$ in each step, the total curve has infinite length, yet it can be embedded into a finite volume.

These and other non-standard ensembles may be characterised in terms of a non-integer, *fractal* dimension. Some possible definitions of a fractal dimension follow [214, 584]:

1. Consider a *self-similar* ensemble $E \subset \mathbb{R}^d$. By self-similarity, it is meant that E is made out of m copies of itself, but of a size reduced by a factor r. The fractal **similarity dimension** is then

$$d_{\text{self}} := \frac{\ln m}{\ln r} \tag{H1}$$

For example, both the Cantor set and the Koch curve are self-similar and

$$d_{\text{self}} = \begin{cases} \dfrac{\ln 2}{\ln 3} \approx 0.63\ldots & ; \text{ Cantor set} \\ \dfrac{\ln 4}{\ln 3} \approx 1.26\ldots & ; \text{ Koch curve} \end{cases} \tag{H2}$$

2. Consider an ensemble $E \subset \mathbb{R}^d$ which need not be self-similar. Cover the set with a set of hypercubic boxes, of linear size ε, and then count the number $N(\varepsilon)$ of boxes needed to cover E. The fractal **box dimension** is then
$$d_{\text{box}} := -\lim_{\varepsilon\to 0} \frac{\ln N(\varepsilon)}{\ln \varepsilon} \tag{H3}$$
For the Cantor set and the Koch curve, one has
$$d_{\text{box}} = \begin{cases} \dfrac{\ln 2}{\ln 3} \approx 0.63\ldots & ; \text{ Cantor set} \\ \dfrac{\ln 4}{\ln 3} \approx 1.26\ldots & ; \text{ Koch curve} \end{cases} \tag{H4}$$
Sometimes, d_{box} is referred to as **capacity**.
3. A method to find fractal dimensions which has become standard proceeds as follows [248]. Consider a possibly fractal ensemble $E \subset \mathbb{R}^d$ and a set of points $x_i \in E$, with $i = 1, \ldots, \mathcal{N}$. Set
$$N_{x_i}(\varepsilon) := \{\text{number of points } x \in E \text{ such that } |x_i - x| < \varepsilon\}.$$
Average over many points $x_i \in E$ to obtain the correlation $C(\varepsilon) := \lim_{\mathcal{N}\to\infty} \mathcal{N}^{-1} \sum_{i=1}^{\mathcal{N}} N_{x_i}(\varepsilon)$. Then the fractal **correlation dimension** is
$$d_{\text{corr}} := \lim_{\varepsilon\to 0} \frac{\ln C(\varepsilon)}{\ln \varepsilon} \tag{H5}$$
In general, one has $d_{\text{corr}} \leq d_{\text{box}}$, altough the numerical values are usually close to each other.

   As an example, consider the **logistic equation** $x_{n+1} = rx_n(1 - x_n)$, $n \in \mathbb{N}$ and with $r = r_\infty = 3.5699\ldots$ where, after an infinite series of period-doublings, the transition to chaos occurs. The fractal dimensions of the attractor are $d_{\text{corr}} = 0.500(5)$ [248] and $d_{\text{box}} \simeq 0.538$ [239].
4. If one considers a covering of the ensemble E by sets of variable size, one arrives at the fractal *Hausdorff dimension* d_{Haus}. While it is nicely invariant under a large class of geometric transformations, its actual computation is difficult. One has $d_{\text{Haus}} \leq d_{\text{box}}$. We refer to [214, 579] for further information.

For fractal sets, the probability $P(\ell)$ of empty intervals of size ℓ scales as

$$P(\ell) \sim \ell^{-1-d_{\text{f}}} \tag{H6}$$

where d_f is one of the fractal dimensions. Exercise 46 illustrates this for the Cantor set.

Solutions

Problems of Chapter 1

1 The only non-vanishing rates are

$$\begin{aligned} w(n \to n+1) &= \lambda n \quad &; \text{ if procreation} \\ w(n \to n-1) &= n \quad &; \text{ if death} \end{aligned}$$

These are irreversible reactions and (non-trivial) detailed balance is not possible. Considering the change in $P_n(t)$, one obtains the following **master equation**

$$\dot{P}_n(t) = (n+1)P_{n+1}(t) + \lambda(n-1)P_{n-1}(t) - n(1+\lambda)P_n(t)$$

which is valid for all $n \in \mathbb{N}$. Infinite recurrences of this kind are often solved by introducing a **generating function**

$$F(x,t) := \sum_{n=0}^{\infty} x^n P_n(t).$$

Using the master equation, it follows

$$\frac{\partial F}{\partial t} = \sum_{n=0}^{\infty} x^n \dot{P}_n(t) = (1-x)(1-\lambda x)\frac{\partial F}{\partial x} \tag{S.1}$$

and $F(x,0)$ is given by the initial condition. The partial differential equation (S.1) may be solved by the method of characteristics [373], but in practise, the following two simple rules are often sufficient.

Lemma: *(i) Let $f = f(x,y)$ be a solution of*

$$a\frac{\partial f}{\partial x} + b\frac{\partial f}{\partial y} = 0 \tag{S.2}$$

where a, b are functions depending on x and y. Then $F(f(x,y))$ is also a solution of (S.2), where F is an arbitrary differentiable function.
(ii) Consider the differential equation

$$a(x)\frac{\partial f}{\partial x} + b(y)\frac{\partial f}{\partial y} = 0$$

A solution of this equation is (provided $a(x) \neq 0$ and $b(y) \neq 0$)

$$f(x,y) = \int^x \frac{\mathrm{d}x'}{a(x')} - \int^y \frac{\mathrm{d}y'}{b(y')}$$

Proof: Both statements are verified by direct computation. □

Applying this to (S.1), for $\lambda \neq 1$ a special solution is

$$\int^t \mathrm{d}t \,+ \int^x \frac{\mathrm{d}x'}{(1-x)(1-\lambda x)} = t + \frac{1}{1-\lambda}\ln\frac{1-\lambda x}{1-x}.$$

Hence the general solution of (S.1) can be written in the form

$$F(x,t) = f\left(e^{(1-\lambda)t}\frac{1-\lambda x}{1-x}\right)$$

and the function f has to be found from the initial condition. If one has initially a single particle, $F(x,0) = x$, which leads to $f\left(\frac{1-\lambda x}{1-x}\right) = x$. To solve such a functional equation, let $y := \frac{1-\lambda x}{1-x}$. Solving this for x, we obtain[5] $f(y) = x(y) = \frac{1-y}{\lambda-y}$. Hence the generating function is

$$F(x,t) = \frac{(1-\lambda x)e^{(1-\lambda)t} - (1-x)}{(1-\lambda x)e^{(1-\lambda)t} - \lambda(1-x)}$$

and from this $P_n(t) = n!\,\frac{\partial^n}{\partial x^n}F(x,t)\big|_{x=0}$. On the other hand, for an initial Poisson distribution $P_n(0) = \frac{\mu^n}{n!}e^{-\mu}$, one has $F(x,0) = e^{\mu(x-1)}$. In the same way as before, we find $f(y) = \exp\frac{\mu(\lambda-1)}{y-\lambda}$, hence

$$F(x,t) = \exp\frac{\mu(\lambda-1)(1-x)}{\left(e^{(1-\lambda)t}-1\right) + x\left(e^{(1-\lambda)t}-\lambda\right)}.$$

The unpopulated state $n = 0$ is an **absorbing state**, since it can be reached but once there, one cannot leave it. Therefore, $P_0(t) = F(0,t)$ is the *extinction probability* until time t and $P(t) := 1 - P_0(t)$ is the *survival*

[5] As a check, note that one must have $F(1,t) = \sum_{n=0}^{\infty} P_n(t) = 1$, hence $f(\infty) \overset{!}{=} 1$.

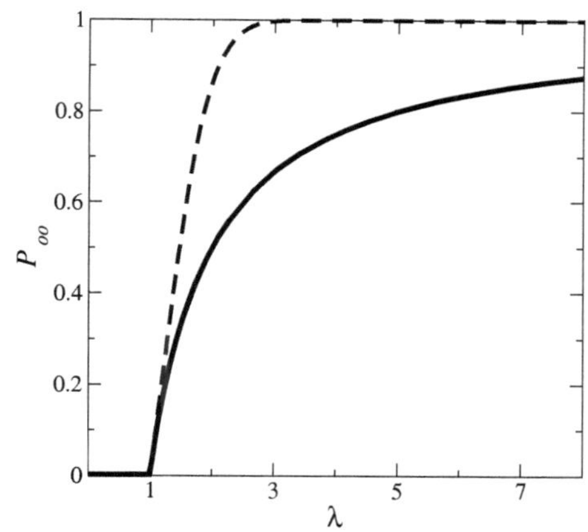

Fig. S.1 Survival probability P_∞ as a function of the rate λ. Full line: a single initial particle. Dashed line: initial Poisson distribution with $\mu = \lambda$.

probability until time t. We are interested in the (ultimate) **survival probability** $P_\infty := \lim_{t\to\infty} P(t)$. For the case of a single initial particle, we have $P(t) = (\lambda - 1)/(\lambda - e^{(1-\lambda)t})$ and

$$P_\infty = \begin{cases} 0 & ; \text{ if } \lambda < 1 \\ 1 - \lambda^{-1} & ; \text{ if } \lambda > 1 \end{cases} .$$

For an initial Poisson distribution, $P(t) = 1 - \exp\left[\mu(\lambda - 1)/(e^{(1-\lambda)t} - 1)\right]$ and

$$P_\infty = \begin{cases} 0 & ; \text{ if } \lambda < 1 \\ 1 - e^{-\mu(\lambda-1)} & ; \text{ if } \lambda > 1 \end{cases} .$$

As can be seen from Fig. S.1, there are two phases. For $\lambda < 1$, the population will certainly become extinct. For $\lambda > 1$, there is a non-vanishing probability for indefinite survival. Hence P_∞ serves as an **order parameter** to distinguish these phases.[6] At the **critical point** $\lambda_c = 1$ of this example of an **absorbing phase transition**, one finds, for a single initial particle, in a similar way $F(x,t) = (1+(t-1)(1-x))/(1+t(1-x))$, hence $P(t) = (1+t)^{-1}$ and $P_\infty = 0$. For an initial Poisson distribution, $F(x,t) = \exp\left(-\frac{\mu(1-x)}{t(1-x)+1}\right)$ and $P(t) = 1 - e^{-\mu/(t+1)} \simeq \frac{\mu}{t+1}$ for $t \to \infty$. The approach towards the steady state is, for large times

$$P(t) - P_\infty \sim \begin{cases} e^{-t/\tau_{\text{rel}}} \quad , \; \tau_{\text{rel}} = \frac{1}{|1-\lambda|} & ; \text{ if } \lambda \neq 1 \\ 1/t & ; \text{ if } \lambda = 1 \end{cases} .$$

The long-time behaviour close to the critical point $\lambda_c = 1$ is independent of the initial conditions, but we also see that away from criticality, the behaviour

[6] In contrast to equilibrium phase transitions, we have continuous phase transitions in a formally zero-dimensional system, without any long-range interaction. Still, P_∞ is singular at $\lambda = \lambda_c = 1$.

of $P(t)$ and P_∞ does depend on the initial state. In particular, we observe the typical distinction between an exponential approach away from criticality and an algebraic approach at criticality, which in turn is reflected in the divergence of the relaxation time $\tau_{\rm rel}$ as $\lambda \to 1$.

2 One has

$$\langle n\rangle(t) = \sum_{n=0}^{\infty} nP_n(t) \ , \ \ \langle n^2\rangle(t) = \sum_{n=0}^{\infty} n^2 P_n(t)$$

Using the explicit form of the generating function found in exercise 1, these may be obtained from $\langle n\rangle(t) = x\frac{\partial F}{\partial x}\big|_{x=1}$ and $\langle n^2\rangle(t) = x\frac{\partial}{\partial x}\left(x\frac{\partial F}{\partial x}\right)\big|_{x=1}$, but carrying out these calculations explicitly is cumbersome. Therefore, we present here an alternative and derive equations of motion for the averages from the master equation. This gives

$$\begin{aligned}
\frac{\mathrm{d}}{\mathrm{d}t}\langle n\rangle &= \sum_{n=0}^{\infty} n\dot{P}_n = (\lambda-1)\sum_{n=0}^{\infty} nP_n = (\lambda-1)\langle n\rangle \\
\frac{\mathrm{d}}{\mathrm{d}t}\langle n^2\rangle &= \sum_{n=0}^{\infty} n^2\dot{P}_n = 2(\lambda-1)\sum_{n=0}^{\infty} n^2 P_n + (\lambda+1)\sum_{n=0}^{\infty} nP_n \\
&= 2(\lambda-1)\langle n^2\rangle + (\lambda+1)\langle n\rangle
\end{aligned}$$

with the initial conditions $\langle n\rangle(0) = n_0$ and $\langle n^2\rangle(0) = n_0^2$. Therefore, $\langle n\rangle(t) = n_0 e^{(\lambda-1)t}$. For $\lambda < 1$, the population dies out rapidly, while for $\lambda > 1$ it will grow beyond any bound (reminding us of Malthusian scenarios). In particular, at criticality the mean density $\langle n\rangle(t) = n_0$ is *constant*! In order to understand how this may be compatible with a survival probability $P_\infty = 0$, see Fig. S.1, consider the *fluctuations* around this result. We have for $\lambda \neq 1$

$$\langle n^2\rangle(t) = \left(n_0^2 + n_0\frac{\lambda+1}{\lambda-1}\right)e^{2(\lambda-1)t} - n_0\frac{\lambda+1}{\lambda-1}e^{(\lambda-1)t}$$

and for $\lambda = 1$

$$\langle n^2\rangle(t) = 2n_0 t + n_0^2$$

At criticality, the variance $\sigma^2(t) = 2n_0 t$ grows unboundedly with time such that an extremely unlikely fluctuation will drive the system to extinction.

In Vol. 2 of this book, we shall meet this model again as the zero-dimensional case of the *bosonic contact process* [305].

Problems of Chapter 2

3 On a lattice Λ with $|\Lambda| = \mathcal{N}$ sites, the Ising model Hamiltonian is $\mathcal{H} = -J\sum_{(n,n')\in\Lambda\otimes\Lambda}\sigma_n\sigma_{n'} - h\sum_{n\in\Lambda}\sigma_n$, where the first sum is over pairs of nearest neighbours. For a fixed magnetisation $M = \sum_{n\in\Lambda} \sigma_n$, the partition function is, with $K = J/T$ and $\delta_{n,0} = \frac{1}{2\pi}\int_{-\pi}^{\pi} \mathrm{d}\alpha\, e^{\mathrm{i}\alpha n}$

$$\begin{aligned}\tilde{Z}(T,M) &= \sum_{\{\sigma\}} e^{-\mathcal{H}/T} \delta_{\sum_n \sigma_n, M}\\ &= \frac{1}{2\pi}\int_{-\pi}^{\pi} \mathrm{d}\alpha\, e^{-\mathrm{i}\alpha M} \sum_{\{\sigma\}} \exp\left[K \sum_{(n,n')} \sigma_n \sigma_{n'} + \mathrm{i}\alpha \sum_n \sigma_n\right].\end{aligned}$$

Writing $M = \mathcal{N}m$, $F(T,M) = \mathcal{N}f(T,m)$ and $G(T,h) = \mathcal{N}g(T,h)$, the relationship between f and g is given by

$$\exp\left[-\frac{\mathcal{N}}{T} f(T,m)\right] = \frac{1}{2\pi}\int_{-\pi}^{\pi} \mathrm{d}\alpha\, \exp\left[-\mathcal{N}\left(\frac{1}{T} g(T, \mathrm{i}\alpha T) + \mathrm{i}\alpha m\right)\right]. \tag{S.3}$$

For $\mathcal{N}$ large, integrals of this kind can be estimated from the saddle-point method, which in this simple case amounts to the identity

$$\int_{\mathbb{R}} \mathrm{d}x\, e^{-\lambda f(x)} \simeq e^{-\lambda f(x_0)} \sqrt{\frac{2\pi}{\lambda f''(x_0)}} \left(1 + \mathrm{O}(\lambda^{-2})\right) \quad ; \quad f'(x_0) = 0$$

Applied to (S.3), the extremal condition $\partial_\alpha\left(T^{-1} g + \mathrm{i}\alpha m\right) \stackrel{!}{=} 0$ leads to the standard thermodynamic relation $m = -\partial g/\partial h$, as it should be. The estimate of the integral in (S.3) gives

$$\exp\left[-\frac{\mathcal{N}}{T} f(T,m)\right] = \exp\left[-\frac{\mathcal{N}}{T}(g + hm) + \frac{1}{2}\ln\left(\frac{2\pi}{\mathcal{N}T\chi}\right)\right]\left(1 + \mathrm{O}(\mathcal{N}^{-2})\right)$$

where $\chi = \partial m/\partial h$ is the susceptibility. Hence

$$f(T,m) = g(T,h) + hm + \mathrm{O}\left(\frac{\ln \mathcal{N}}{\mathcal{N}}\right)$$

reproduces in the limit $\mathcal{N} \to \infty$ the standard equilibrium thermodynamics, where the two thermodynamic potentials $f = f(T,m)$ and $g = g(T,h)$ are related by a **Legendre transformation**.

For illustration, one may calculate the mean energy $\langle E\rangle = Z^{-1}\partial Z/\partial(1/T)$. Writing $\langle E\rangle = \mathcal{N}\varepsilon$, from the above it is easily seen that $\varepsilon|_m = \varepsilon|_h + \mathrm{O}(\mathcal{N}^{-1})$. For an explicit check in the $1D$ Ising model, one may calculate the two partition functions on a periodic chain of $\mathcal{N} = 2N$ sites, for $h = 0$ and $M = 0$, respectively. The result is

$$\begin{aligned}Z(T,0) &= (2\cosh K)^{2N} + (2\sinh K)^{2N} && ; \text{ if } h = 0\\ \tilde{Z}(T,0) &= (2\sinh 2K)^{N}\left[P_N(\coth 2K) - P_{N-1}(\coth 2K)\right] && ; \text{ if } M = 0\end{aligned} \tag{S.4}$$

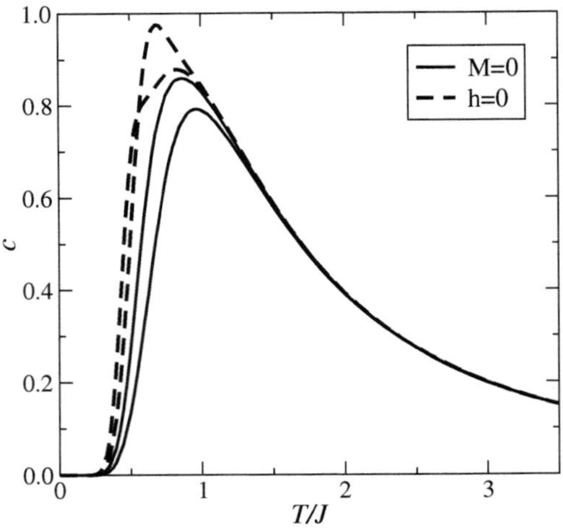

Fig. S.2 Specific heat of the Ising model on a periodic chain. The full lines give the case $M = 0$ for 30 sites (lower curve) and 60 sites (upper curve). The dashed lines give the case $h = 0$ for 30 sites (upper curve) and 60 sites (lower curve).

where P_n is the nth Legendre polynomial [2].

In Fig. S.2 the specific heats, calculated from either $\tilde{Z}(T, 0)$ (fixed magnetisation – full curves) or $Z(T, 0)$ (fixed field – dashed curves) are shown for two periodic lattices of different size. We see that for fixed magnetisation, the specific heat is lower than for a fixed magnetic field (as it should be for an equilibrium system) and that for $N \to \infty$, the two sets of curves converge towards each other. This convergence is very rapid for sufficiently large temperatures but becomes notably more slow for more small values of T and evolves in opposite directions for $M = 0$ fixed and $h = 0$ fixed.

While the calculation of $Z(T, 0)$ can be found in many textbooks [640], the analogous calculation for $\tilde{Z}(T, M)$ contains some technical complications which we now outline. Using the identity $\delta_{n,0} = \frac{1}{2\pi} \int_{-\pi}^{\pi} \mathrm{d}\alpha \, e^{\mathrm{i}\alpha n}$, one has for a periodic chain with $2N$ sites

$$\begin{aligned}
\tilde{Z}(T, M) &= \sum_{\{s\}} e^{-\mathcal{H}/T} \, \delta_{\sum_i s_i, M} \\
&= \frac{1}{2\pi} \int_{-\pi}^{\pi} \mathrm{d}\alpha \, e^{\mathrm{i}\alpha M} \underbrace{\sum_{\{s\}} \exp\left[K \sum_n s_n s_{n+1} + \mathrm{i}\alpha \sum_n s_n \right]}_{Z(T, \mathrm{i}\alpha T)}
\end{aligned}$$

where $K = J/T$ and we need the canonical partition function of the Ising chain in an *imaginary* external field. From the explicit expression of $Z(T, h)$ [640] and using the binomial theorem, we have

$$\tilde{Z}(T, M) = \frac{1}{\pi} \int_{-\pi}^{\pi} \mathrm{d}\alpha \, e^{\mathrm{i}\alpha M} \sum_{n=0}^{N} \binom{2N}{2n} e^{2nK} \left(e^{-2K} - e^{2K} \sin^2 \alpha \right)^{N-n} \cos^{2n} \alpha$$

$$= \frac{e^{2KN}}{\pi} \sum_{n=0}^{N} \binom{2N}{2n} e^{-4Kn} \int_{-\pi}^{\pi} \mathrm{d}\alpha\, e^{\mathrm{i}\alpha M} \cos^{2(N-n)} \alpha \left(1 - e^{4K} \sin^2 \alpha\right)^n$$

$$= \frac{e^{2KN}}{\pi} \sum_{n=0}^{N} \binom{2N}{2n} \sum_{j=0}^{n} \binom{n}{j} \left(\frac{1-x}{x}\right)^{n-j} \int_{-\pi}^{\pi} \mathrm{d}\alpha\, \cos(\alpha M) \cos^{2(N-n+j)} \alpha$$

where we have set $x := e^{4K}$. Next, we use the identity

$$\int_{-\pi}^{\pi} \mathrm{d}\alpha\, \cos(\alpha M) \cos^{2B} \alpha = \frac{\pi}{2^{2B}} \left[\binom{2B}{B + M/2} + \binom{2B}{B - M/2} \right]$$

and find

$$\tilde{Z}(T, M) = 2e^{2KN} \sum_{j=0}^{N} \binom{2(N-j)}{N-j-M/2} 4^{j-N} \left(\frac{1-x}{x}\right)^j \sum_{n=0}^{N} \binom{2N}{2n} \binom{n}{j}$$

Using the identity [538, eq. (4.2.5.79)]

$$\sum_{k=m}^{N} \binom{2N}{2k} \binom{k}{m} = 2^{2N-2m-1} \frac{N}{N-m} \binom{2N-m-1}{m}$$

and using identities between binomial coefficients, we finally have

$$\tilde{Z}(T, M) = 2Ne^{2KN} \sum_{\ell=0}^{N} \frac{1}{2N-\ell} \binom{2N-\ell}{N+M/2} \binom{N+M/2}{\ell} \left(\frac{1-x}{x}\right)^{\ell} \tag{S.5}$$

The case $M = 0$ of vanishing magnetisation is now found from the identities, where P_n is the nth Legendre polynomial [2]

$$\sum_{k=0}^{n} \binom{n}{k} \binom{n+k}{n} x^k = P_n(1+2x)$$

$$\int_1^Y \mathrm{d}y\, (y-1)^{N-1} P_N(y) = \frac{1}{2N} (Y-1)^N \left[P_N(Y) + P_{N-1}(Y)\right].$$

The first of these is [538, eq. (4.2.7.12)] and the second one follows from [539, eq. (1.14.6.4)], [540, eq. (7.3.1.141)] and [2, eq. (22.7.19)]. Then (S.4) follows.

The difference in the analytical effort required between the cases $h = 0$ and $M = 0$ is remarkable. We finally point out that in the often-invoked applications of the equilibrium Ising model to binary alloys or liquids, in full rigour, the case with M fixed is required.

4 The Hamiltonian is $\mathcal{H} = -J \sum_i \sigma_i \sigma_{i+1} - \frac{H}{2} \sum_i (\sigma_i + \sigma_{i+1}) - TC$ and the transfer matrix may be written as

$$\mathfrak{T} = \begin{pmatrix} e^{J/T+H/T+C} & e^{-J/T+C} \\ e^{-J/T+C} & e^{J/T-H/T+C} \end{pmatrix}$$

and hence its square is

$$\mathfrak{T}_2 = \mathfrak{T}^2 = \begin{pmatrix} e^{2J/T+2H/T+2C} + e^{-2J/T+2C} & e^{H/T+2C} + e^{-H/T+2C} \\ e^{H/T+2C} + e^{-H/T+2C} & e^{2J/T-2H/T+2C} + e^{-2J/T+2C} \end{pmatrix}.$$

Since the partition function $Z = \mathrm{tr}\ \mathfrak{T}^L = \mathrm{tr}\ {\mathfrak{T}_2}^{L/2}$ is invariant under the summing over every second spin, one has the following renormalisation-group relations between the couplings, written here using the new variables $x = e^{-4J/T}$, $y = e^{-2H/T}$ and $z = e^{-4C}$

$$x' = \frac{x(1+y)^2}{(x+y)(1+xy)} \;,\; y' = \frac{y(x+y)}{1+xy} \;,\; z' = \frac{z^2xy^2(1+y)^2}{(x+y)(1+xy)}.$$

The third of these does not influence the behaviour of x and y and will be discarded from now on. The fixed points are (i) $(x^*, y^*) = (1, y^*)$ with $0 \le y^* < 1$ which corresponds to infinite temperature and describes a paramagnetic state, (ii) $(x^*, y^*) = (0,0)$ which corresponds to zero temperature and an infinite magnetic field and (iii) $(x^*, y^*) = (1,0)$ which is a critical point at zero temperature and zero field. Linearisation around this fixed point gives $x' \simeq 4x$, $(y-1)' \simeq 2(y-1)$. Since one has performed a change of length scale by a factor two, from the above eigenvalues the critical exponents $y_\mathcal{T} = 2$ and $y_h = 1$ are read off. We leave it to the reader to derive from this the scaling of the singular part of the free energy, see (2.144), and to compare this with the exact solution [640].

5 The scaling relations follow from recalling the definitions of surface and excess magnetisation and a direct application of the scaling form (2.96).

6 The comparison of the derivatives gives explicitly, first for the susceptibility

$$A_1 A_2^2 |\tau|^{-\gamma} W_2^{\pm}(A_2 h|\tau|^{-\beta-\gamma}) = D_1 \xi_0^{2-\eta} |\tau|^{-(2-\eta)\nu} \tilde{X}^{\pm}(D_2 h|\tau|^{-\beta-\gamma})$$

which returns the well-known scaling relation $\gamma = (2-\eta)\nu$, and for the magnetisation density

$$A_1^2 A_2^2 |\tau|^{2\beta} W_1^{\pm\,2}(A_2 h|\tau|^{-\beta-\gamma}) = D_1 \xi_0^{2-\eta-d} |\tau|^{-(2-\eta-d)\nu} \tilde{Z}^{\pm}(D_2 h|\tau|^{-\beta-\gamma})$$

which implies the hyperscaling relation $2\beta = (d-2+\eta)\nu$ and (2.111) follows.

Very similar arguments can be raised for equilibrium systems with spatially strongly anisotropic interactions, as they may arise in systems with axial *competing interactions* between nearest neighbours and next-nearest neighbours [279]. An example for equilibrium critical points whose anisotropies lead to different correlation length exponents $\nu_\perp \neq \nu_\parallel$ in different spatial directions are so-called *Lifchitz points* to which we shall return in Vol. 2 of this book.

7 With the initial condition $v(0) = v_0$, the solution of (2.176) is for $F = 0$

$$v(t) = v_0 e^{-\gamma t} + e^{-\gamma t} \int_0^t \mathrm{d}t'\, e^{\gamma t'} \eta(t')$$

Squaring and averaging leads for large times to $\langle v^2(\infty)\rangle = B/\gamma$ and the Einstein relation follows from the stated equipartition condition, valid at equilibrium. Next, for the calculation of the equilibrium response function, recall that $\langle v(t)\rangle = \int_{\mathbb{R}} \mathrm{d}t'\, R(t-t')F(t')$ is a convolution. Hence, introducing the Fourier transform $\widehat{f}(\omega) = \int_{\mathbb{R}} \mathrm{d}t\, e^{\mathrm{i}\omega t} f(t)$, one has $\langle \widehat{v}(\omega)\rangle = \widehat{R}(\omega)\widehat{F}(\omega)$. Averaging over the noise in (2.176) then gives

$$\widehat{R}(\omega) = \frac{1}{m}\frac{1}{\gamma - \mathrm{i}\omega}$$

For $F = 0$, the correlation function $C(t) = C(-t) = (k_{\mathrm{B}}T/m)e^{-\gamma|t|}$ is symmetric in time (check it). Its Fourier transform is $\widehat{C}(\omega) = (2k_{\mathrm{B}}T/m)\gamma/(\gamma^2 + \omega^2)$ and comparison with $\widehat{R}(\omega)$ gives the *fluctuation-dissipation theorem*

$$\Im \widehat{R}(\omega) = \frac{\omega}{2\gamma k_{\mathrm{B}}T}\widehat{C}(\omega)$$

In order to see the equivalence of this with the form (2.125), one notes that

$$2\Im \widehat{R}(\omega) = \frac{1}{\mathrm{i}} \int_0^\infty \mathrm{d}t\, \left(e^{\mathrm{i}\omega t} - e^{-\mathrm{i}\omega t}\right) R(t)$$

where the lower integration limit comes from the *causality* condition $R(t - t') = 0$ for $t < t'$. In the second term, one now changes $t \mapsto -t$ and defines the *analytic continuation* $R(-t) \mapsto R(t)$ for $t > 0$. On the other hand, rewriting $\omega\widehat{C}(\omega) = \mathrm{i}\int_{\mathbb{R}} \mathrm{d}t\, e^{\mathrm{i}\omega t}\partial_t C(t)$ and comparing with the above form of the FDT leads back to (2.125), if one further scales $\gamma k_{\mathrm{B}} \stackrel{!}{=} 1$.

Lastly, since $x(t) = x(0) + \int_0^t \mathrm{d}t'\, v(t')$ (arrange for $x(0) = 0$) one has for the diffusion constant

$$D = \lim_{t\to\infty} \frac{\langle x^2(t)\rangle}{2t} = \lim_{t\to\infty} \frac{1}{t} \int_0^t \mathrm{d}t' \int_0^{t'} \mathrm{d}\tau\, \langle v(\tau)v(0)\rangle = \int_0^\infty \mathrm{d}\tau\, \langle v(\tau)v(0)\rangle = \widehat{C}(0)$$

where the limit $t \to \infty$ of the double integral was estimated by using l'Hôpital's rule. From this Einstein's expression of the diffusion constant is recovered.

8 The full Hamiltonian is $\mathcal{H} = \mathcal{H}_0 + \int \mathrm{d}t\, h(t)\phi(t)$ whereas the average order-parameter is given by a functional integral

$$\langle \phi(t)\rangle = \frac{1}{Z} \int \mathcal{D}\phi\, \phi(t) e^{-\mathcal{H}/T}$$

where Z is the partition function. Physically, this is justified if the relaxation time of the system is small compared to the time-scale on which $h(t)$ is varied. The announced result follows by straightforward functional differentiation and is equivalent to the FDT (2.121).

9 The proof of the commutator identity is straightforward via induction and will not be detailed here. Then, on one hand,

$$e^{\mathrm{i}t\mathcal{L}_0}A = \sum_{\ell=0}^{\infty}\frac{(\mathrm{i}t)^{\ell}}{\ell!}\mathcal{L}_0^{\ell}A = \sum_{\ell=0}^{\infty}\frac{1}{\ell!}\left(\frac{\mathrm{i}t}{\hbar}\right)^{\ell}\sum_{k=0}^{\ell}\binom{\ell}{k}(-1)^{\ell-k}H^k A H^{\ell-k}$$

where the iterated commutator $(\hbar\mathcal{L}_0)^{\ell}A$ was used. On the other hand,

$$e^{\mathrm{i}Ht/\hbar}A\,e^{-\mathrm{i}Ht/\hbar} = \sum_{k=0}^{\infty}\sum_{m=0}^{\infty}\frac{(-1)^m}{k!m!}\left(\frac{\mathrm{i}t}{\hbar}\right)^{k+m} H^k A H^m$$

which after a change of variables $\ell = k+m$ reduces to the above expression.

10 The verification of the hermiticity of $\mathcal{L}$ with respect to the scalar product on $\mathcal{V}$ amounts to the straightforward check that $S(\mathcal{L}A,B) = S(A,\mathcal{L}B)$. Clearly, for any pair of hermitean operators $\mathcal{A},\mathcal{B} : \mathcal{V}\to\mathcal{V}$ one has $\left(\mathcal{A}\mathcal{B}\right)^{\dagger} = \mathcal{B}\mathcal{A}$ for the natural definition of the adjoint operator. Then the assertion follows by expanding the exponential and term-wise application of (2.122).

11 Starting from (2.123), recognise that for $t\to\infty$ and local operators A and B, one should have, for widely separated time, the asymptotic relation $\langle A(-\mathrm{i}\hbar\beta)B(t)\rangle_0 \simeq \langle A(-\mathrm{i}\hbar\beta)\rangle_0\langle B(t)\rangle_0 = \langle A(0)\rangle_0\langle B(0)\rangle_0$ and the last step follows from time-translation-invariance. Since $\beta\leq 1/T$, the limit $\hbar\to 0$ can now be taken and this leads to the assertion.

12 Straightforward consequence of (2.129).

13 Close to equilibrium, the dynamical scaling form (2.180) of the correlation function gives, via the fluctuation-dissipation theorem, the instantaneous (and spatially averaged) response function

$$R(t;\tau) = -\frac{1}{T}\frac{\partial}{\partial t}\int_{\mathbb{R}^d}\mathrm{d}\boldsymbol{r}\; C(t,\boldsymbol{r};\tau) = |\tau|^{-\gamma+\nu z}\,\mathfrak{R}(t|\tau|^{\nu z})$$

where $\mathfrak{R}$ is a scaling function. The real and imaginary parts of the alternating susceptibility are now obtained from (see exercise 12)

$$\begin{aligned}\chi'(\omega,\tau) &= \int_0^{\infty}\mathrm{d}t'\;\cos(\omega t')R(t';\tau)\\ &\simeq |\tau|^{-\gamma}\int_0^{\infty}\mathrm{d}v\,\mathfrak{R}(v)\bigl(1+\mathrm{O}(\omega^2)\bigr)\sim|\tau|^{-\gamma}\end{aligned}$$

$$\chi''(\omega,\tau) = \int_0^\infty \mathrm{d}t' \, \sin(\omega t') R(t';\tau)$$
$$\simeq \omega|\tau|^{-\gamma-\nu z} \int_0^\infty \mathrm{d}v \, v \mathfrak{R}(v) \big(1 + \mathrm{O}(\omega^2)\big) \sim \omega|\tau|^{-\gamma-\nu z}.$$

See [207, 206] for confirmations of these, in the context of non-equilibrium thermodynamics (and within a mean-field approximation).

14 The single-particle generators, when acting on scalar fields with scaling dimension $x = \Delta = \overline{\Delta}$, are given by $X_{\boldsymbol{\alpha}} = \boldsymbol{\alpha} \cdot \boldsymbol{\nabla} + \frac{x}{d} \boldsymbol{\nabla} \cdot \boldsymbol{\alpha}$, where $\boldsymbol{\alpha} = \boldsymbol{\alpha}(\boldsymbol{r})$ is either a translation (generators $\ell_{-1}, \bar{\ell}_{-1}$), a rotation ($\mathrm{i}(\ell_0 - \bar{\ell}_0)$), a dilatation ($\ell_0 + \bar{\ell}_0$) or a special conformal transformation eq. (2.147) ($\ell_1, \bar{\ell}_1$). From these, n-particle generators are readily written down.

15 Because of the factorised structure, it is enough to check that the z-dependent and the $\bar{z}$-dependent factors satisfy separately the projective conformal Ward identities. Since for scalar fields, one may use rotation-invariance to bring the three points into the plane, and furthermore $\Delta_i = \overline{\Delta}_i$, the extension to $d \geq 2$ is immediate and we have for the two- and three-point functions

$$\langle \phi_1(\boldsymbol{r}_1)\phi_2(\boldsymbol{r}_2) \rangle = C_{12} \, |\boldsymbol{r}_1 - \boldsymbol{r}_2|^{-2x_1} \, \delta_{x_1,x_2}$$

$$\langle \phi_1(\boldsymbol{r}_1)\phi_2(\boldsymbol{r}_2)\phi_3(\boldsymbol{r}_3) \rangle =$$
$$C_{123} \, |\boldsymbol{r}_1 - \boldsymbol{r}_2|^{-(x_1+x_2-x_3)} \, |\boldsymbol{r}_2 - \boldsymbol{r}_3|^{-(x_2+x_3-x_1)} \, |\boldsymbol{r}_3 - \boldsymbol{r}_1|^{-(x_3+x_1-x_2)}$$

16 If one defines $\hat{P} := P_0 \cdot \mathbf{1} + \hat{K}$ such that $P_0 \geq \max_a \sum_b K_{ab}$, then $\hat{P}$ is positive definite. Then, where $\boldsymbol{\sigma} = (\sigma_1, \ldots, \sigma_{\mathcal{N}})$

$$Z = \mathrm{e}^{-\mathcal{N} P_0/2} \, 2^{-\mathcal{N}} \sum_{\{\sigma\}} \exp\left(\frac{1}{2} \boldsymbol{\sigma} \hat{P} \boldsymbol{\sigma} \right).$$

Under a change of variables $\boldsymbol{y} = \boldsymbol{s} + \hat{Q}^{-1}\boldsymbol{\sigma}$, one has $\boldsymbol{y}\hat{Q}\boldsymbol{y} = \boldsymbol{s}\hat{Q}\boldsymbol{s} + 2\boldsymbol{s} \cdot \boldsymbol{\sigma} + \boldsymbol{\sigma}\hat{Q}^{-1}\boldsymbol{\sigma}$ and with the choice $\hat{Q} = \hat{P}^{-1} = (P_0 \cdot 1 + \hat{K})^{-1}$, it follows

$$(2\pi)^{\mathcal{N}/2} (\det \hat{P})^{1/2} = \exp\left(-\frac{1}{2} \boldsymbol{\sigma} \hat{P} \boldsymbol{\sigma} \right) \int_{-\infty}^{\infty} \mathrm{d}^{\mathcal{N}} \boldsymbol{s} \, \exp\left(-\frac{1}{2} \boldsymbol{s} \hat{Q} \boldsymbol{s} - \sum_{a=1}^{\mathcal{N}} s_a \sigma_a \right)$$

and consequently

$$Z = \mathrm{e}^{-f_0} 2^{-\mathcal{N}} \sum_{\{\sigma\}} \int_{-\infty}^{\infty} \mathrm{d}^{\mathcal{N}} \boldsymbol{s} \, \exp\left[-{\sum_{(a,b)}}' Q_{ab} s_a s_b - \frac{1}{2} \sum_a Q_{aa} s_a^2 - \sum_a s_a \sigma_a \right],$$

where $a \neq b$ in the sum and $\mathrm{e}^{-f_0} = \mathrm{e}^{-\mathcal{N}P_0/2}\,(2\pi)^{-\mathcal{N}/2}(\det \hat{P})^{-1/2}$ is a smooth background term. Carrying out the sum, one has

$$Z = \mathrm{e}^{-f_0} \int_{-\infty}^{\infty} \mathrm{d}^{\mathcal{N}} \boldsymbol{s} \, \exp\left[-\sum_{(a,b)}{}' Q_{ab} s_a s_b - \sum_a \left(\frac{1}{2} Q_{aa} s_a^2 - \ln\cosh s_a \right) \right] \quad \text{(S.6)}$$

The first term in (S.6) describes the interactions between the continuous spin variables $\boldsymbol{s}$, related to the original interaction matrix $\hat{K}$ via $-P_0^2\hat{Q} = -P_0 \cdot 1 + K - P_0^{-1}\hat{K}^2 + P_0^{-2}\hat{K}^3 + \cdots$. The second term in (S.6) is a weight factor $w(s_a) = \frac{1}{2}Q_{aa}s_a^2 - \ln\cosh s_a = \frac{1}{2}(Q_{aa}-1)s_a^2 + \frac{1}{12}s_a^4 + \cdots$. If $\hat{K}$ describes nearest-neighbour interactions, the higher powers of $\hat{K}$ are suppressed with powers of P_0. In a formal continuum limit $s_a \mapsto \phi(\boldsymbol{r})$, the interactions described by $\hat{K}$ will create a leading term $\sim (\partial_\mu\phi)^2$. At least close to the upper critical dimension $d_c = 4$, the reader should show by standard power-counting arguments that all other terms are irrelevant. Hence the partition function becomes formally a functional integral $\mathcal{Z} \sim \int \mathcal{D}\phi\ e^{-S[\phi]}$ with the Landau-Ginsburg action $S[\phi] = \int \mathrm{d}\boldsymbol{r} \left[\frac{1}{2}(\nabla\phi)^2 + \frac{m^2}{2}\phi^2 + \frac{g}{4!}\phi^4 \right]$.

17 The Lagrangian density is $\mathcal{L} = \frac{1}{2}(\partial_\mu\phi)^2$ where $\partial_\mu = \partial_{\boldsymbol{r}_\mu}$, $\mu = 1, \ldots, d$. The classical equations of motion are $\partial_\mu\partial^\mu\phi = 0$. In Euclidean space, the canonical energy-momentum tensor is given by [648, 168]

$$\widetilde{T}_{\mu\nu} = \partial_\mu\phi \frac{\partial\mathcal{L}}{\partial\left(\partial_\nu\phi\right)} - \delta_{\mu\nu}\mathcal{L} = \partial_\mu\phi\partial_\nu\phi - \frac{1}{2}\delta_{\mu\nu}\left(\partial_\sigma\phi\right)^2$$

While this is symmetric $\widetilde{T}_{\mu\nu} = \widetilde{T}_{\nu\mu}$ and conserved $\partial^\mu\widetilde{T}_{\mu\nu} = 0$ (the equations of motion must be used), $\widetilde{T}$ is in general *not* traceless, since

$$\widetilde{T}^\mu_\mu = \left(1 - \frac{d}{2}\right)\left(\partial_\mu\phi\right)^2 = \left(1 - \frac{d}{2}\right)\left[\partial^\mu\left(\phi\partial_\mu\phi\right) - \phi\partial_\mu\partial^\mu\phi\right]$$

where the last term vanishes because of the equations of motion but the first one is a divergence and non-vanishing if $d \neq 2$. Such terms were neglected in the derivation of the canonical energy-momentum tensor. An improved energy-momentum tensor, which is symmetric, conserved and traceless, reads

$$T_{\mu\nu} = \partial_\mu\phi\partial_\nu\phi - \frac{1}{2}\delta_{\mu\nu}\left(\partial_\sigma\phi\right)^2 + \frac{1}{4}\frac{d-2}{d-1}\left(\delta_{\mu\nu}\partial_\sigma\partial^\sigma - \partial_\mu\partial_\nu\right)\phi^2$$

18 We write $\boldsymbol{r} = (r_\parallel, r_\perp) = (u, v)$ or $z = u + \mathrm{i}v$ such that the boundary is at $v = 0$ and the system is defined for $v \geq 0$ only. From the projective conformal transformations, only the *diagonal subalgebra* $\langle \ell_{-1}+\bar{\ell}_{-1}, \ell_0+\bar{\ell}_0, \ell_1+\bar{\ell}_1 \rangle$ remains. Imposing conformal invariance leads to the following projective Ward identities for the two-point function $G = \langle \phi_1(u_1, v_1)\phi_2(u_2, v_2)\rangle$ of two quasiprimary scaling operators $\phi_{1,2}$ with scaling dimensions $x_{1,2}$:

$$\sum_{i=1}^{2}\left[\frac{\partial}{\partial u_i}\right]G = 0$$

$$\sum_{i=1}^{2}\left[u_i\frac{\partial}{\partial u_i} + v_i\frac{\partial}{\partial v_i} + x_i\right]G = 0$$

$$\sum_{i=1}^{2}\left[\left(u_i^2 - v_i^2\right)\frac{\partial}{\partial u_i} + 2u_iv_i\frac{\partial}{\partial v_i} + 2x_iu_i\right]G = 0.$$

Because of translation-invariance parallel to the surface, one has

$$G = G(u; v_1, v_2),$$

with $u = u_1 - u_2$ and the two other identities become

$$u\frac{\partial G}{\partial u} + v_1\frac{\partial G}{\partial v_1} + v_2\frac{\partial G}{\partial v_2} + (x_1 + x_2)G = 0$$

$$\left(v_2^2 - v_1^2\right)\frac{\partial G}{\partial u} + uv_1\frac{\partial G}{\partial v_1} - uv_2\frac{\partial G}{\partial v_2} + u(x_1 - x_2)G = 0.$$

Using the methods of characteristics [373] or the method outlined in exercise 1, the general solution

$$G(\boldsymbol{r}_1, \boldsymbol{r}_2) = v_1^{-x_1}\, v_2^{-x_2}\, \Phi_{12}\left(\frac{(u_1 - u_2)^2 + v_1^2 + v_2^2}{v_1v_2}\right)$$

is readily found, where Φ_{12} remains an undetermined scaling function. In contrast to the bulk result (2.159), there is no constraint on the scaling dimensions x_1 and x_2 [270].

19 The logarithmic transformation $w = \frac{L}{\pi}\ln z$ maps the upper half-plane onto the strip with free boundary conditions. The remainder of the argument is based on the definition of the surface critical exponents.

Problems of Chapter 3

20 The solution of the mean-field equation of motion (3.47)

$$\lambda^{-1}\partial_t\varrho(t) = \tau\varrho(t) - g\varrho(t)^2,$$

is found by direct integration and reads

$$\varrho(t) = \frac{\tau}{g + (\tau\varrho_0 - g)\exp\left(-\lambda\tau t\right)}$$

for $\tau \neq 0$. The initial order-parameter at $t = 0$ is denoted by ϱ_0. Asymptotically ($t \gg 1/\lambda|\tau|$) the order-parameter behaves as

$$\varrho(t)\big|_{\tau<0} \sim -\tau \left(g - \frac{\tau}{\varrho_0}\right)^{-1} \mathrm{e}^{\lambda\tau t}$$

$$\varrho(t)\big|_{\tau>0} \sim \frac{\tau}{g} + \frac{\tau}{g^2}\left(g - \frac{\tau}{\varrho_0}\right) \mathrm{e}^{-\lambda\tau t}.$$

Apart from criticality ($\tau \neq 0$), the steady-state solutions $\varrho = \tau/g$ and $\varrho = 0$ are approached exponentially, independent of the initial value ϱ_0. The associated correlation (relaxation) time

$$\xi_{\parallel} = (\lambda|\tau|)^{-1}. \tag{S.7}$$

diverges at the critical point. At the critical point, the order-parameter decays as

$$\varrho(t) = \frac{1}{\varrho_0^{-1} + \lambda g t} \underset{t\to\infty}{\longrightarrow} \frac{1}{\lambda g t}$$

21 The steady-state order-parameter is given by

$$\varrho_{\mathrm{s}}(\tau, h) = \frac{\tau}{2g} + \sqrt{\frac{h}{g} + \left(\frac{h}{2g}\right)^2}.$$

Small deviations of the order-parameter $\delta\varrho(\boldsymbol{r}, t) = \varrho(\boldsymbol{r}, t) - \varrho_{\mathrm{s}}(\tau, h)$ obey the equation of motion

$$\lambda^{-1}\partial_t\delta\varrho(\boldsymbol{r}, t) = [\tau - 2g\varrho_{\mathrm{s}}(\tau, h)]\ \delta\varrho(\boldsymbol{r}, t) + \nabla^2\delta\varrho(\boldsymbol{r}, t) + \mathrm{O}(\delta\varrho^2). \tag{S.8}$$

In Fourier space, small deviations with vector $\boldsymbol{k}$ decay as

$$\lambda^{-1}\partial_t\, \delta\varrho(\boldsymbol{k}, t) = -\left[k^2 + \xi_{\perp}^{-2}\right]\ \delta\varrho(\boldsymbol{k}, t) \tag{S.9}$$

with the characteristic length scale

$$\xi_{\perp} = \left(4gh + \tau^2\right)^{-1/4}$$

leading to the mean-field exponent $\nu_{\perp,\mathrm{MF}} = 1/2$. The temporal correlation length $\xi_{\parallel}$ is obtained by the solution of (S.9)

$$\delta\varrho(\boldsymbol{k}, t) \propto \mathrm{e}^{-t/\xi_{\parallel}}$$

with

$$\xi_{\parallel}(\boldsymbol{k}) = \frac{1}{\lambda(k^2 + \xi_{\perp}^{-2})}\bigg|_{\boldsymbol{k}=\boldsymbol{0}} = \frac{\xi_{\perp}^2}{\lambda}, \tag{S.10}$$

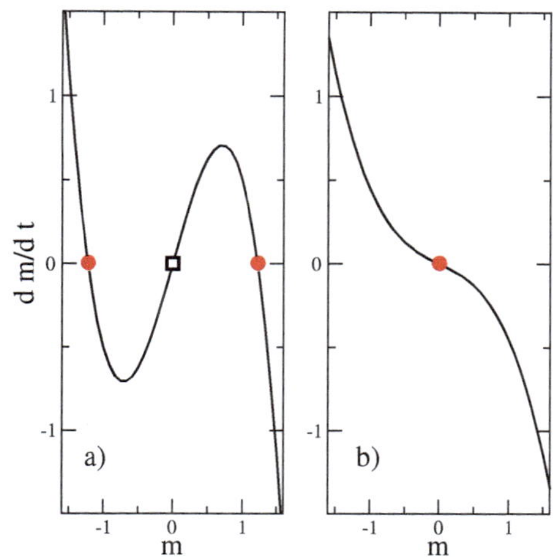

Fig. S.3 Vector field $(m(t), \dot{m}(t))$ for the relaxation of a simple magnet with a non-conserved order-parameter for (a) ordered phase with $\lambda^2 > 0$ ($T < T_c$) and (b) disordered phase with $\lambda^2 < 0$ ($T > T_c$). The fixed points m_∞ are shown, where filled points are stable and open points are unstable.

yielding $z_{\rm MF} = 2$ and $\nu_{\|,\rm MF} = 1$, respectively. Equation (S.10) shows that different modes of the order parameter decay on different correlation times. In the case of the homogeneous order parameter ($\boldsymbol{k} = \boldsymbol{0}$) we recover (S.7).

22 The stationary solutions are $m_\infty = 0, \pm\sqrt{3}\,\lambda$. In order to analyse the stability of these, consider the vector field $(m(t), \dot{m}(t))$ associated to the mean-field equation of motion, illustrated in Fig. S.3. Letting $\delta m(t) = m(t) - m_\infty$, a fixed point of the equation $\frac{\mathrm{d}m}{\mathrm{d}t} = f(m)$ is *stable*, if $f'(m_\infty) < 0$ and is *unstable*, if $f'(m_\infty) > 0$. Clearly, for $\lambda^2 > 0$ the fixed points $m_\infty = \pm\sqrt{3}\,\lambda$ are stable, and the sign of m_0 decides to which fixed point the system evolves. On the other hand, for $\lambda^2 < 0$ the only stable fixed point is $m_\infty = 0$. Unless $m_0 = 0$, there is for large times an exponential relaxation $m(t) - m_\infty \sim e^{-t/\tau_{\rm rel}}$, with the finite relaxation time

$$\frac{1}{\tau_{\rm rel}} = \begin{cases} 3|\lambda^2| & ; \text{ if } \lambda^2 < 0 \\ 6\lambda^2 & ; \text{ if } \lambda^2 > 0 \end{cases}$$

23 (i) Using the techniques and notations of appendix E, the mean particle-density is $\varrho(t) = P_{\bullet}(t)$. On a one-dimensional lattice, one has from the site-approximation (simple mean-field)

$$\dot{\varrho} = \underbrace{-\,\varrho}_{\text{free decay}} \quad \underbrace{+\,\lambda\varrho^2(1-\varrho)}_{\text{reaction }\bullet\circ\bullet\longrightarrow\bullet\bullet\bullet} \quad \underbrace{+\,\frac{\lambda}{2}\varrho(1-\varrho)^2\cdot 2}_{\text{2 reactions }\bullet\circ\circ\longrightarrow\bullet\bullet\circ}$$
$$= (\lambda-1)\varrho - \lambda\varrho^2$$

into which the diffusion constant D does not enter explicitly. The stationary solutions are $\varrho_\infty = 0, 1-\lambda^{-1}$. As in exercise 22, by plotting the vector field $(\dot{\varrho}, \varrho)$ one easily sees that the absorbing fixed point $\varrho_\infty = 0$ is stable for $\lambda < 1$

while the active fixed point $\varrho_\infty = 1 - \lambda^{-1}$ is stable for $\lambda > 1$. Therefore the critical point (or the **bifurcation**) occurs at $\lambda_c = 1$. Explicit integration of the simple mean-field equation gives

$$\varrho(t) = \begin{cases} \varrho_\infty \left(1 - e^{-\varrho_\infty \lambda (t-t_0)}\right)^{-1} & ; \text{ if } \lambda > 1 \\ \dfrac{\varrho_0}{1 + \varrho_0 t} & ; \text{ if } \lambda = 1 \\ \varrho_1 e^{-(1-\lambda)t} & ; \text{ if } \lambda = 1 \end{cases}$$

where $t_0, \varrho_0, \varrho_1$ are constants. The long-time behaviour is

$$\varrho(t) \simeq \begin{cases} \varrho_\infty (1 + e^{-t/\tau}) & , \ 1/\tau = \lambda - 1 & ; \text{ if } \lambda > 1 \\ t^{-1} & & ; \text{ if } \lambda = 1 \\ \varrho_1 e^{-t/\tau'} & , \ 1/\tau' = 1 - \lambda & ; \text{ if } \lambda < 1 \end{cases}$$

and the reader may identify the mean-field exponents from these expressions. (ii) Working with the mean particle-density $\varrho(t) = P_\bullet(t)$ and the mean pair-density $u(t) = P_{\bullet\bullet}(t)$, and using the abbreviations $v = \varrho - u = P_{\bullet\circ} = P_{\circ\bullet}$ and $w = 1 - 2\varrho + u = P_{\circ\circ}$, the relevant reactions and their rates of the pair-approximation are listed in the following table, including the symmetry factors. Reactions which change neither $N_\bullet$ nor $N_{\bullet\bullet}$ can be discarded.

event	rate	$\Delta N_\bullet$	$\Delta N_{\bullet\bullet}$	
● ○ ● → ● ● ●	$\lambda v^2/(1-\varrho)$	1	2	
● ○ ○ → ● ● ○	$\frac{1}{2}\lambda v w/(1-\varrho)$	1	1	×2
● ● ● → ● ○ ●	u^2/ϱ	−1	−2	
● ● ○ → ● ○ ○	uv/ϱ	−1	−1	×2
○ ● ○ → ○ ○ ○	v^2/ϱ	−1	0	
● ● ○○ → ● ○ ●○	$Duvw/(\varrho(1-\varrho))$	0	−1	×2
○ ● ○● → ○ ○ ●●	$Dv^3/(\varrho(1-\varrho))$	0	1	×2

Adding the various contributions, one obtains the equations of motion of the contact process in the pair-approximation

$$\begin{aligned} \dot{\varrho} &= \lambda(\varrho - u) - \varrho \\ \dot{u} &= \lambda \frac{(\varrho - u)(1-u)}{1-\varrho} - 2u - 2D\frac{(\varrho - u)(u - \varrho^2)}{\varrho(1-\varrho)} \end{aligned}$$

We observe the following. (a) For $D = 0$, the stationary solution is $\varrho_\infty = (\lambda-2)/(\lambda-1)$ and $u_\infty = (\lambda-2)/\lambda$, hence $\lambda_c = 2$ in the pair-approximation. This is already closer to the simulational result $\lambda_c = 3.29785\ldots$ than the site-approximation. (b) In the other extreme limit $D \to \infty$, a stationary solution $\dot{u} = 0$ is only possible if either $u = \varrho$, corresponding to the trivial solution, or else $u = \varrho^2$. Then $\dot{\varrho} = \lambda(\varrho - \varrho^2) - \varrho$ and one is back to the site-approximation. (c) In the intermediate case, the stationary densities are related, viz. $u_\infty = \frac{\lambda-1}{\lambda}\varrho_\infty$ which further leads to

$$\varrho_\infty = \frac{(\lambda-1)(1+2D/\lambda)-1}{(\lambda-1)(1+2D/\lambda)} \le 1$$

The positive zero of $\varrho_\infty(\lambda)$ gives the D-dependent critical point

$$\lambda_c(D) = 1 - D + \sqrt{1+D^2} \simeq 1 + \frac{1}{2D} + \ldots$$

As in the site-approximation, the approach to the steady-state is exponential if $\lambda \neq \lambda_c(D)$ and $\varrho(t) \sim u(t) \sim t^{-1}$ at the critical point. This may be easily checked by solving the above equations numerically.

24 The rate equation for the mean particle-density is

$$\dot{\varrho} = \lambda\varrho^2(1-\varrho) - \varrho$$

with the stationary solutions $\varrho_\infty = 0, \frac{1}{2}(1 \pm \sqrt{1-4/\lambda})$. The last two solutions only exist for $\lambda \le 4$. Since $\varrho_\infty = 0$ is always stable, a state with a density $\varrho(t) \gtrsim \frac{1}{2}$ jumps to zero when λ is increased beyond the **spinodal line** $\lambda_{\rm sp} = 4$. The reader should draw the vector field himself in order to analyse the stability more fully. We point out that in this simple model, the spinodal line $\lambda_{\rm sp}$ coincides with the critical point which, however, is not true in general (exercise 30 below gives an example).

25 Since we only have nearest-neighbour interactions, the Liouvillian may be decomposed as $\mathcal{L} = \sum_n \mathcal{L}_{n,n+1}$ in terms of two-site operators $\mathcal{L}_{n,n+1}$. In a natural basis, this may be written as

$$\mathcal{L}_{n,n+1} = \begin{pmatrix} 0 & & & \\ & D_R & -D_L & \\ & -D_R & D_L & \\ & & & 0 \end{pmatrix}$$

Alternatively, it is useful to re-express this in terms of the unit matrices

$$E^{00} = \begin{pmatrix} 1 & 0 \\ 0 & 0 \end{pmatrix}, \quad E^{01} = \begin{pmatrix} 0 & 1 \\ 0 & 0 \end{pmatrix}, \quad E^{10} = \begin{pmatrix} 0 & 0 \\ 1 & 0 \end{pmatrix}, \quad E^{11} = \begin{pmatrix} 0 & 0 \\ 0 & 1 \end{pmatrix}$$

and one has in a tensor-product representation

$$\mathcal{L}_{n,n+1} = -D_R\left(E^{01}\otimes E^{10} - E^{11}\otimes E^{00}\right) - D_L\left(E^{10}\otimes E^{01} - E^{00}\otimes E^{11}\right)$$

Here, we use the correspondence $|\emptyset\rangle = \begin{pmatrix} 1 \\ 0 \end{pmatrix}$ and $|A\rangle = \begin{pmatrix} 0 \\ 1 \end{pmatrix}$. Using the relationships with the Pauli matrices $\sigma^{x,y,z}$, namely $E^{01} = \sigma^+$, $E^{10} = \sigma^-$, $E^{00} = \frac{1}{2}(1+\sigma^z)$ and $E^{11} = \frac{1}{2}(1-\sigma^z)$, where $\sigma^\pm := \frac{1}{2}(\sigma^x \pm \mathrm{i}\sigma^y)$, the Liouvillian becomes, for periodic boundary conditions

$$\mathcal{L} = -\sum_{n=1}^{L} \left[D_R \sigma_n^+ \sigma_{n+1}^- + D_L \sigma_n^- \sigma_{n+1}^+ + \frac{D_R + D_L}{4} \left(\sigma_n^z \sigma_{n+1}^z - 1 \right) \right. \\ \left. + \frac{D_R - D_L}{4} \left(\sigma_n^z - \sigma_{n+1}^z \right) \right] \tag{S.11}$$

In order to see the relationship with the Heisenberg model, consider the similarity transformation generated by

$$U := \exp \left(\pi g \sum_{j=1}^{L} j \sigma_j^z \right) \quad , \quad U \sigma_j^{\pm} U^{-1} = \exp(\pm 2\pi g j) \sigma_j^{\pm}$$

Next, we introduce a parameter q by setting

$$\frac{D_R}{D_R + D_L} = \frac{q}{q + q^{-1}} \quad , \quad \frac{D_L}{D_R + D_L} = \frac{q^{-1}}{q + q^{-1}}$$

and if one further lets $q = e^{2\pi g}$, one has

$$\begin{aligned} \mathcal{L}' &= U \mathcal{L} U^{-1} \\ &= -\sqrt{D_R D_L} \sum_{n=1}^{L} \left[\sigma_n^+ \sigma_{n+1}^- + \sigma_n^- \sigma_{n+1}^+ + \frac{q + q^{-1}}{4} \left(\sigma_n^z \sigma_{n+1}^z - 1 \right) \right. \\ &\quad \left. + \frac{q - q^{-1}}{4} \left(\sigma_n^z - \sigma_{n+1}^z \right) \right] \\ &= -\frac{D}{2} \sum_{n=1}^{L} \left[\sigma_n^x \sigma_{n+1}^x + \sigma_n^y \sigma_{n+1}^y + \Delta \left(\sigma_n^z \sigma_{n+1}^z - 1 \right) + \frac{q - q^{-1}}{2} \left(\sigma_n^z - \sigma_{n+1}^z \right) \right] \end{aligned}$$

This is indeed the familiar XXZ Heisenberg Hamiltonian with anisotropy $\Delta = \frac{1}{2}(q + q^{-1})$ and global diffusion constant $D = \sqrt{D_R D_L}$. From the above similarity transformation one has the boundary conditions $\sigma_{L+1}^{\pm} = q^{\mp L} \sigma_1^{\pm}$ and $\sigma_{L+1}^z = \sigma_1^z$ [281] and the surface field generated by the last term vanishes.

In the case of biased diffusion ($q \neq 1$), the choice of the boundary conditions is important for the critical behaviour [402] and can be worked out from the low-energy spectrum of $\mathcal{L}$, for example using the Bethe ansatz, see [260, 281]. Indeed, for a periodic ring and any value of q, the low-lying spectrum becomes continuous in the $L \to \infty$ limit, which physically corresponds to particles freely diffusing along the ring, with a mean velocity $\sim \ln q$. On the other hand, for free boundary conditions (open chain), there is a finite gap between the ground state and the lowest excited state even in the $L \to \infty$ limit. Physically, this corresponds to the particles 'piling up' at one end of the open chain, with a density-profile depending on q.

Finally, note that the unbiased case $q = 1$ is also referred to as the **symmetric exclusion process**, whereas the case $q \neq 1$ is often called the **asym-**

metric exclusion process. In these models [565], one considers more general boundary conditions, such that particles may enter or leave at both ends. See [147] for a recent Bethe ansatz treatment of the asymmetric exclusion process with the most general open boundary conditions.

26 Because of the nearest-neighbour interactions, the Liouvillian is completely characterised by its two-body form, where we have also set $D = 1$

$$\mathcal{L}_{n,n+1} = \begin{pmatrix} 0 & -\delta & -\delta & -2\alpha \\ 0 & 1+\delta & -1 & -\gamma \\ 0 & -1 & 1+\delta & -\gamma \\ 0 & 0 & 0 & 2(\alpha+\gamma) \end{pmatrix}. \tag{S.12}$$

Consider the following similarity transformation

$$\widetilde{\mathcal{L}} = \mathcal{B}\mathcal{L}\mathcal{B}^{-1} \ , \ \mathcal{B} = \bigotimes_{j=1}^{L} B_j, \tag{S.13}$$

where B_j acts only on the site j. If F is a physical observable, its average transforms as follows

$$\langle F\rangle(t) = \langle s|\, Fe^{-\mathcal{L}t}\, |P_0\rangle = \langle s|\, \widetilde{F}e^{-\widetilde{\mathcal{L}}t} \left|\widetilde{P_0}\right\rangle$$

with the transformed observable $\widetilde{F} = F\mathcal{B}^{-1}$ and initial state $\left|\widetilde{P_0}\right\rangle = \mathcal{B}\,|P_0\rangle$. The form of the single-site matrix B is found by considering a single-site state, which must be of the form $|\varrho\rangle = \begin{pmatrix} 1-\varrho \\ \varrho \end{pmatrix}$, with $0 \leq \varrho \leq 1$. Since also in the transformed state $B\,|\varrho\rangle$, the probabilities must sum to unity, it follows $\forall \varrho$: $(b_{11} + b_{12})(1-\varrho) + (b_{21} + b_{22})\varrho = 1$, hence

$$B = \begin{pmatrix} b_1 & 1-b_2 \\ 1-b_1 & b_2 \end{pmatrix}$$

and consequently $\langle s|\, B = \langle s|$, as it should be. In what follows, consider an uncorrelated initial state $|P_0\rangle = \otimes_{j=1}^{L} \begin{pmatrix} 1-\varrho_j \\ \varrho_j \end{pmatrix}$ with $0 \leq \varrho_j \leq 1$ for all sites $j = 1, \ldots, L$.

A similarity transformation $\mathcal{S} \mapsto \widetilde{\mathcal{S}}$ between two systems, each characterised by its Liouvillian $\mathcal{L}$ and its densities $\{\varrho_j\}_j$ is called a **stochastic similarity transformation** (SST) if (i) (S.13) holds true with a non-singular $\mathcal{B}$ and (ii) $\left|\widetilde{P}\right\rangle = \mathcal{B}\,|P\rangle$ is again a probability distribution for each probability distribution $|P\rangle$. The particle-density a_j at the site j is represented by the matrix $\begin{pmatrix} 0 & 0 \\ 0 & 1 \end{pmatrix}_j$. Then, for the Liouvillian (S.12), explicit matrix multiplications show that:

1. for $\delta > 2\alpha + \gamma$ there is an SST such that $\widetilde{D} = D$, and the correlators transform as

$$C_n(t, \{r\}; \alpha, \gamma, \delta, \varrho) = \left(\frac{\delta - \alpha - \gamma}{\delta - 2\alpha - \gamma}\right)^n C_n(t, \{r\}; 0, \alpha + \gamma, \delta, \frac{\delta - 2\alpha - \gamma}{\delta - \alpha - \gamma}\varrho)$$

such that the transformed annihilation rate $\widetilde{\alpha} = 0$.

2. on the other hand, for $\delta < 2\alpha + \gamma$, one has

$$C_n(t, \{r\}; \alpha, \gamma, \delta, \varrho) = \left(\frac{\delta - \alpha - \gamma}{\delta - 2\alpha - \gamma}\right)^n C_n(t, \{r\}; \alpha + \gamma, 0, \delta, \frac{2\alpha + \gamma - \delta}{2\alpha + 2\gamma - \delta}\varrho)$$

and now the transformed coagulation rate $\widetilde{\gamma} = 0$.

These results imply that the universal aspects of the dynamics are independent of the branching ratio α/γ between annihilation and coagulation reactions. Since the densities between these processes are also transformed, the precise relationship between coagulation and annihilation configuration must be specified. For the map between pure annihilation ($\gamma = 0$) and pure coagulation ($\alpha = 0$) is has been shown [92] that a configuration of the pure annihilation process is turned into a configuration of pure coagulation by randomly eliminating half of the particles, but *without* stretching the space.

See [276, 275, 565] for systematic discussions of SSTs for two-state models and their relationship to $1D$ systems treatable with free-fermion methods.

27 Recall first the construction of the field-theory, following [186, 187]. The creation and destruction of particles is conveniently described in a Fock space formalism, where the state vector becomes

$$|P(t)\rangle = \sum_{\{\sigma\}} p(n_1, n_2, \ldots; t) \left(a^\dagger{}_1\right)^{n_1} \left(a^\dagger{}_2\right)^{n_2} \cdots |0\rangle$$

where $a^\dagger{}_i$ creates a particle at site i and $p(n_1, n_2, \ldots; t)$ is the probability of having n_1 particles at site 1, n_2 particles at site 2 and so on at time t. In the case when only diffusion and annihilation occur, the Hamiltonian is

$$\begin{aligned} H &= D\sum_{(i,j)} (a^\dagger{}_i - a^\dagger{}_j)(a_i - a_j) - 2\alpha \sum_i \left(a_i^2 - a^{\dagger 2}_i a_i^2\right) \\ &\to \int \mathrm{d}\boldsymbol{r} \left[D(\nabla a^\dagger)(\nabla a) - 2\alpha\left(a^2 - a^{\dagger 2} a^2\right)\right], \end{aligned}$$

where a formal continuum limit is taken. It is common to perform a shift $a^* = 1 + \widetilde{a}$ and the non-equilibrium action becomes,

$$\begin{aligned} \mathcal{J}[\widetilde{a}, a] &= \int \mathrm{d}t\, \mathrm{d}\boldsymbol{r}\, \widetilde{a}\left(\partial_t - D\nabla^2\right) a \\ &\quad + \int \mathrm{d}t\, \mathrm{d}\boldsymbol{r}\, \mathrm{d}\boldsymbol{r}'\, V(|\boldsymbol{r} - \boldsymbol{r}'|) \left[\widetilde{a}(\boldsymbol{r})\widetilde{a}(\boldsymbol{r}') + \lambda \widetilde{a}(\boldsymbol{r})\right] a(\boldsymbol{r}) a(\boldsymbol{r}') \end{aligned}$$

such that for $\lambda = 1$ only coagulation and for $\lambda = 2$ only annihilation reactions occur. Following [31], consider the potentials $V_{\alpha,\gamma}$ for pure annihilation and coagulation, respectively, to be of the form $V_\alpha(\boldsymbol{r}) = \alpha V(\boldsymbol{r})$ and $V_\gamma(\boldsymbol{r}) = \gamma V(\boldsymbol{r})$. Then $\lambda = \frac{2\alpha+\gamma}{\alpha+\gamma}$. Thus, through the transformation $a \to \lambda a, \widetilde{a} \to \lambda^{-1}\widetilde{a}$, the coagulation-annihilation model can be reduced to pure coagulation and the results of exercise 26 are recovered.

The field-theoretical renormalisation-group treatment of the more general process $kA \to \emptyset$ with diffusion of single particles gives for the long-time behaviour of the mean particle density $\varrho = \langle a \rangle$ [414]

$$\bar{\varrho}(t) \sim \begin{cases} (Dt)^{-d/2} & ; \text{ if } d < d_{\rm c} \\ \left(\ln t/(Dt)\right)^{1/(k-1)} & ; \text{ if } d = d_{\rm c} \\ (Dt)^{1/(k-1)} & ; \text{ if } d > d_{\rm c} \end{cases}$$

with $d_{\rm c} = 2/(k-1)$. The amplitudes are universal. They have been calculated to second order in $\epsilon = d_{\rm c} - d$ [414] and rigorously estimated for $k = 2$ [78, 79]. See [544] for a field-theoretical treatment with an additional creation $\emptyset \to A$.

28 The Liouville operator is $\mathcal{L}_{\alpha\beta} = -w_{\alpha\beta}(1 - \delta_{\alpha\beta}) + \delta_{\alpha\beta}\sum_\gamma w_{\gamma\alpha}$ and by hypothesis, the transition rates $w_{\alpha\beta} = w_{\beta\to\alpha}$ satisfy detailed balance. If one chooses a diagonal transformation matrix $\mathcal{V}_{\alpha\beta} = v_\alpha \delta_{\alpha\beta}$, then the transformed matrix

$$\mathcal{M}_{\alpha\beta} := \left(\mathcal{V}\mathcal{L}\mathcal{V}^{-1}\right)_{\alpha\beta} = -\frac{v_\alpha}{v_\beta} w_{\alpha\beta}(1 - \delta_{\alpha\beta}) + \delta_{\alpha\beta} \sum_\gamma w_{\gamma\alpha}.$$

This is symmetric for the choice $v_\alpha = \left(P_\alpha^{\rm eq}\right)^{-1/2}$.

29 It is useful to rewrite the rates (3.30) of the Domany-Kinzel automaton in terms of binary reaction rates of particles of a single species A. Denoting the empty and occupied state of a single site by $\circ$ and $\bullet$, respectively, one has

$$\overset{\circ\ \circ}{\bullet} = 0 \ ; \quad \overset{\bullet\ \circ}{\bullet} = p_1 \ ; \quad \overset{\bullet\ \bullet}{\bullet} = p_2$$

In Table S.1 we collect several notations commonly used in the literature for the rates of reaction-diffusion processes involving only two-site interactions. Here, we shall restrict ourselves furthermore to the case with left-right symmetry, that is, with the notation of [275, 271], $D_R = D_L = D$, $\beta_R = \beta_L = \beta$, $\gamma_R = \gamma_L = \gamma$, $\delta_R = \delta_L = \delta$ and so on. Since the evolution of a site only depends on the state of its neighbours, one may write schematically (where the upper line indicates the state of the pair at the initial time and the lower line the one at the final time)

diffusion to the left	○● → ●○	D_L	a_{32}	w_{32}	$w_{1,1}(1,0)$	Γ_{10}^{01}
diffusion to the right	●○ → ○●	D_R	a_{23}	w_{23}	$w_{1,1}(0,1)$	Γ_{01}^{10}
pair annihilation	●● → ○○	2α	a_{14}	w_{14}	$w_{1,1}(0,0)$	Γ_{00}^{11}
coagulation to the right	●● → ○●	γ_R	a_{24}	w_{24}	$w_{1,0}(0,1)$	Γ_{01}^{11}
coagulation to the left	●● → ●○	γ_L	a_{34}	w_{34}	$w_{0,1}(1,0)$	Γ_{10}^{11}
death at the left	●○ → ○○	δ_L	a_{13}	w_{13}	$w_{1,0}(0,0)$	Γ_{00}^{10}
death at the right	○● → ○○	δ_R	a_{12}	w_{12}	$w_{0,1}(0,0)$	Γ_{00}^{01}
decoagulation to the left	○● → ●●	β_L	a_{42}	w_{42}	$w_{1,0}(1,1)$	Γ_{11}^{01}
decoagulation to the right	●○ → ●●	β_R	a_{43}	w_{43}	$w_{0,1}(1,1)$	Γ_{11}^{10}
birth at the right	○○ → ○●	ν_R	a_{21}	w_{21}	$w_{0,1}(0,1)$	Γ_{01}^{00}
birth at the left	○○ → ●○	ν_L	a_{31}	w_{31}	$w_{1,0}(1,0)$	Γ_{10}^{00}
pair creation	○○ → ●●	2σ	a_{41}	w_{41}	$w_{1,1}(1,1)$	Γ_{11}^{00}
Rates defined after reference		[275]	[564]	[565]	[13]	[519]

Table S.1 Two-sites reaction-diffusion processes of a single species and their rates as denoted by various authors. After [271].

$$\begin{array}{cc}\bullet&\circ\\\bullet&\end{array} = \begin{array}{cc}\bullet&\circ\\\bullet&\circ\end{array} + \begin{array}{cc}\bullet&\circ\\\bullet&\bullet\end{array} = \begin{array}{cc}\bullet&\circ\\\circ&\bullet\end{array} + \begin{array}{cc}\bullet&\circ\\\bullet&\bullet\end{array}$$

by summing over the possible states of the right or left nearest neighbours. This implies $p_1 = (1 - D - \beta - \delta) + \beta = D + \beta$. Similar relations hold for the other rates and one obtains

$$p_1 = \frac{1}{2}(1 + \beta - \delta) \ , \ p_2 = 1 - 2\alpha - \gamma$$

and the diffusion rate is constrained to be $2D = 1 - \beta - \delta$ (and furthermore $\nu_R = \nu_L = \sigma = 0$). Therefore, on the level of the site-approximation, the mean density $a(t)$ satisfies

$$\dot{a} = (2\beta - \delta)a - 2(\alpha + \gamma + \beta)a^2$$

Upon identification of parameters, this is the same structure as for the mean-field theory of directed percolation, see exercise 23. On the other hand, the above identification suggests that for many choices of reaction rates in typical reaction-diffusion systems, there is a phase-transition in the steady-state which might be in the same universality class as directed percolation.

30 One must distinguish between an attempted deposition of a CO radical on a single site (probability y) and of an O_2 molecule on two sites (probability $1-y$). If one first considers the restriction of the ZGB model to a $1D$ lattice, the CO adsorption is controlled by the state of the *two* neighbours of the selected site and the O_2 adsorption is also controlled by the *two* sites left and right of the two selected sites. The mean-field rates are then as follows.

$1D$ event	rate	ΔN_O	ΔN_{CO}	
$O_2 + 2\underline{\emptyset} \to \underline{2\ O}$	$(1-y)\varrho_\emptyset^2(1-\varrho_{CO})^2$	2	0	
$O_2 + 2\underline{\emptyset} \to \underline{O + \emptyset}\ + CO_2$	$(1-y)\varrho_\emptyset^2(1-\varrho_{CO})\varrho_{CO}$	1	-1	$\times 2$
$O_2 + 2\underline{\emptyset} \to \underline{2\emptyset}\ + 2CO_2$	$(1-y)\varrho_\emptyset^2\varrho_{CO}^2$	0	-2	
$CO + \underline{\emptyset} \to \underline{CO}$	$y\varrho_\emptyset(1-\varrho_O)^2$	0	1	
$CO + \underline{\emptyset} \to \underline{\emptyset}\ + CO_2$	$y\varrho_\emptyset(1-(1-\varrho_O)^2)$	-1	0	

Here the underlined states represent the adsorbing surface, $\varrho_\emptyset$, ϱ_{CO}, ϱ_O are the concentrations of vacant sites, of CO and of O on the surface, and $\Delta N_O, \Delta N_{CO}$ give the change in the number of O or CO units on the surface after one time step. The last column gives an eventual symmetry factor. Writing $a := \varrho_O$ and $b := \varrho_{CO}$, hence $\varrho_\emptyset = 1 - a - b$, one has the $1D$ equations of motion

$$
\begin{aligned}
\dot{a} &= 2(1-y)(1-a-b)^2(1-b) - y(1-a-b)(1-(1-a)^2) \\
\dot{b} &= y(1-a-b)(1-a)^2 - 2(1-y)(1-a-b)^2 b
\end{aligned}
\tag{S.14}
$$

On the other hand, in the full $2D$ model, the state of the *four* neighbours of the site selected for CO adsorption and the *six* neighbours of the pair of sites selected for O_2 adsorption must be considered. The rates are

$2D$ event	rate	ΔN_O	ΔN_{CO}	
$O_2 + 2\underline{\emptyset} \to \underline{2\ O}$	$(1-y)\varrho_\emptyset^2(1-\varrho_{CO})^6$	2	0	
$O_2 + 2\underline{\emptyset} \to \underline{O + \emptyset} + CO_2$	$(1-y)\varrho_\emptyset^2(1-\varrho_{CO})^3(1-(1-\varrho_{CO})^3)$	1	-1	$\times 2$
$O_2 + 2\underline{\emptyset} \to \underline{2\emptyset}\ + 2CO_2$	$(1-y)\varrho_\emptyset^2(1-(1-\varrho_{CO})^3)^2$	0	-2	
$CO + \underline{\emptyset} \to \underline{CO}$	$y\varrho_\emptyset(1-\varrho_O)^4$	0	1	
$CO + \underline{\emptyset} \to \underline{\emptyset}\ + CO_2$	$y\varrho_\emptyset(1-(1-\varrho_O)^4)$	-1	0	

which finally yields the following $2D$ equations of motion

$$
\begin{aligned}
\dot{a} &= 2(1-y)(1-a-b)^2(1-b)^3 - y(1-a-b)(1-(1-a)^4) \\
\dot{b} &= y(1-a-b)(1-a)^4 - 2(1-y)(1-a-b)^2(1-(1-b)^3)
\end{aligned}
$$

We shall concentrate here on the $1D$ case, since its analysis is more simple and the results are qualitatively similar to the $2D$ case. Consider the steady-state, where $\dot{a} = \dot{b} = 0$. Taking the difference of (S.14), one has the stationary density of the empty sites $Z := \varrho_{\emptyset,\text{st}} = \frac{1}{2}y/(1-y)$. Then the other equation gives in the steady-state the condition $a^2 - a + Z = 0$, from which the stationary densities follow

$$
a = \frac{1}{2} \pm \frac{1}{2}\sqrt{1-4Z} \ , \ b = \frac{1}{2} + Z \mp \sqrt{1-4Z} .
$$

These are real-valued for $Z \le \frac{1}{4}$, hence $Z_{\text{sp}} = \frac{1}{4}$ is the **spinodal line**. Therefore $y \le y_{\text{sp}} = 2Z_{\text{sp}}/(1+2Z_{\text{sp}}) = \frac{1}{3}$ and one expects a *first-order* transition at some $y_c \le \frac{1}{3}$. In order to find y_c, one must consider the (linear) *stability* of the possible stationary solutions, namely (i) $\varrho_{\emptyset,\text{st}} = 0$ or equivalently $a = 1 - b$ and (ii) $a = \frac{1}{2}(1 - \sqrt{1-4Z})$ and (iii) $a = \frac{1}{2}(1 + \sqrt{1-4Z})$. The eigenvalues of the Jacobian of (S.14) of the first solution are 0 and $y(1-b^2)$, hence this one is stable, if $b \geq 1/\sqrt{2}$. For the second one, a straightforward numerical calculation gives stability for $y < y_c \simeq 0.27838\ldots$ and we note that $a(y_c) \simeq 0.2610\ldots$ and $b(y_c) \simeq 0.4539\ldots$. The third solution is never stable. Therefore, there is a first-order transition at $y = y_c$ where the order-parameter $\varrho_{\emptyset,\text{st}}$ jumps from $\simeq 0.285$ to 0. The spinodal line is at $y_{\text{sp}} = \frac{1}{3}$.

For the $2D$ case, one finds similarly that, again, $Z = \varrho_{\emptyset,\text{st}} = \frac{1}{2}y/(1-y)$ and has the further condition $(b+Z)^4 + (1-b)^3 - 1 = 0$. The spinodal line is now at $Z_{\text{sp}} \simeq 0.638986\ldots$, hence $y_{\text{sp}} \simeq 0.561013\ldots$, whereas the first-order transition occurs at $y_c \simeq 0.4787\ldots$. This mean-field estimate is not too far from the simulational estimate $y_c \simeq 0.525$.

31 (i) Free boundary conditions. Obviously, $N_a(1) = 2$ and $N_a(2) = 3$. Fix the leftmost site. If it is occupied, its neighbour must be empty in order to obtain a steady-state and it remains to consider an open chain of $L-2$ sites. Otherwise, if the leftmost site is empty, consider the remaining open chain of $L-1$ sites. Hence we have $N_a(L) = N_a(L-2) + N_a(L-1)$ which is the recursion relation of the Fibonacci numbers F_L. Therefore,

$$N_a(L) = F_{L+1} = \frac{g_+^{L+2} - g_-^{L+2}}{g_+ - g_-} \simeq 1.17 g_+^L$$

where $g_\pm = (1 \pm \sqrt{5})/2$ are the golden numbers.

(ii) Periodic boundary conditions. Fix one of the sites. If that site is occupied, both its left and right nearest neighbours must be empty and we are left with an *open* chain of $L-3$ sites. But if the site we have fixed is empty, we are left with an open chain of $L-1$ sites. Therefore

$$N_a^{(\text{per})}(L) = N_a(L-3) + N_a(L-1) = g_+^L + g_-^L \simeq g_+^L .$$

In general, recursions of the form encountered here may be analysed by considering the associated generating function $\mathcal{G}(z) := \sum_{n=0}^{\infty} N_a(L) z^L$. The recursion relation then implies a linear equation for $\mathcal{G}(z)$, which for the case at hand gives, for free boundary conditions $\mathcal{G}(z) = z(2+z)/(1-z-z^2)$. The coefficients of rational generating functions such as $\mathcal{G}(z)$ may now be readily found by applying the following

Lemma: *Let $P(z)$ be an entire function and $Q(z)$ be polynomials of order q such that $Q(z) = q_0(1-z\rho_1)\cdots(1-z\rho_q)$, where the ρ_i are pairwise distinct for $i = 1, \ldots, q$ and q_0 is a constant. Then*

$$f(z) = \frac{P(z)}{Q(z)} = \sum_{n=0}^{\infty} f_n z^n \ , \ \ f_n = -\sum_{j=1}^{q} \frac{\rho_j^{n+1} P(1/\rho_j)}{Q'(1/\rho_j)} \tag{S.15}$$

In [238, 274], this formula was proven under the additional conditions that $P(z)$ is a polynomial of order p and that $p < q$.

Proof: Step 1: we first prove the assertion for $P(z) = 1$, that is, we show that

$$\frac{1}{Q(z)} = -\sum_{n=0}^{\infty}\left[\sum_{j=1}^{q}\frac{\rho_j^{n+1}}{Q'(1/\rho_j)}\right] z^n \tag{S.16}$$

For $P(z) = 1$ we have, using Cauchy's formula

$$f_n = \frac{1}{n!}f^{(n)}(0) = \frac{1}{2\pi \mathrm{i}}\oint_{\mathcal{C}} \mathrm{d}w\,\frac{f(w)}{w^{n+1}} = \frac{1}{2\pi \mathrm{i}}\oint_{\mathcal{C}'} \mathrm{d}u\,\frac{u^n}{u\,Q(1/u)}$$

where we have set $u = 1/w$. The contour $\mathcal{C}$ is a circle around the origin with a radius smaller than the convergence radius of $f(z)$ and $\mathcal{C}'$ encloses all the points ρ_j, $j = 1, \ldots, q$. The only singularities of $u^{-1}f(1/u)$ are the simple poles located at $u = \rho_j$. If we concentrate on the singularity at $u = \rho_1$, we have $uQ(1/u) = (u-\rho_1)q_0\prod_{j=2}^{q}(1-\rho_j/u)$ such that the product is regular at $u = \rho_1$. On the other hand, the derivative $Q'(1/\rho_1) = -\rho_1 q_0 \prod_{j=2}^{q}(1-\rho_j/\rho_1)$. Therefore, close to $u \simeq \rho_1$, we have

$$uQ(1/u) \simeq -(u-\rho_1)\frac{Q'(1/\rho_1)}{\rho_1}$$

From the residue theorem and summing over all simple poles, (S.16) follows.
Step 2: Let now $P(z) = \sum_{n=0}^{\infty} p_n z^n$. Then, using (S.16)

$$\begin{aligned}
f(z) = \frac{P(z)}{Q(z)} &= \sum_{n=0}^{\infty}\sum_{m=0}^{\infty}\left[-\sum_{j=1}^{q}\frac{\rho_j^{m+1}}{Q'(1/\rho_j)}\right] p_n z^{n+m} \\
&= \sum_{N=0}^{\infty}\left[-\sum_{j=1}^{q}\frac{\rho_j^{N+1}}{Q'(1/\rho_j)}\left(\sum_{n=0}^{\infty}\rho_j^{-n}p_n\right)\right] z^N \\
&= \sum_{N=0}^{\infty}\left[-\sum_{j=1}^{q}\frac{\rho_j^{N+1}P(1/\rho_j)}{Q'(1/\rho_j)}\right] z^N
\end{aligned}$$

and the proof of (S.15) is complete. □

We leave it to the reader to verify that the Fibonacci numbers $N_a(L) = F_{L+1}$ can be recovered.

32 Analogously to the pair-contact process, one has for free boundary conditions the recursion $N_a(L) = N_a(L-1)+N_a(L-2)+N_a(L-3)$, together with the initial conditions $N_a(1) = 2$, $N_a(2) = 4$ and $N_a(3) = 7$. The generating function is

$$\mathcal{G}(z) = \sum_{n=0}^{\infty} N_a(L)z^L = \frac{z(2+2z+z^2)}{1-z-z^2-z^3}$$

From the determination of its simple poles and using the expansion lemma (S.15) of the previous exercise, the first part of the assertion follows. For periodic boundary conditions, we have

$$N_a^{(\mathrm{per})}(L) = N_a(L-1) + N_a(L-3) + 2N_a(L-4)$$

from which the second part is established.

33 Since the strong embedding constraint is broken if some of the bonds between nearest-neighbour occupied sites of the lattice animal are replaced by contacts, one must ask when that is possible. For directed lattice animals, replacing a bond by a contact is admissible if the animal contains closed loops and each loop contains exactly one 'final' site, such that only one of the two bonds leading to this final site can be replaced by a contact. Consider a directed animal with c loops. Let $g(c,k)$ be the number of ways one may replace exactly k of the bonds by contacts. Clearly, $g(c,k)=0$ if $k>c$ and $g(c,0)=1$ and $g(c,1)=2c$. For $k\geq 2$, we have the recursion

$$g(c,k) = 2g(c-1,k-1) + g(c-1,k)$$

and find the generating function, for all $c \geq 0$

$$f(c;\tau) := \sum_{k=0}^{c} g(c,k)\tau^k = \left(1+2\tau\right)^c .$$

On the square lattice, one has $2n = b+k+s$, where b is the number of bonds, k the number of contacts and s the solvent contact, that is the number of links between an occupied site of the animal with its empty nearest neighbours. In the strongly embedded case, one has $k = 0$. Then the Euler relation $c = b-n+1$ may be rewritten as $c = n-s+1$ and the directed-animal generating function becomes

$$\begin{aligned}
\mathcal{Z} &= \sum_{\{A\}} a_{n,k,s}\, x^n y^b \tau^k \\
&= \sum_{\{A_{\mathrm{str}}\}}{}' \, a_{n,0,s}\, x^n y^s \left(1+2\tau\right)^c \\
&= (1+2\tau) \sum_{\{A_{\mathrm{str}}\}}{}' \, a_{n,0,s} \left[x(1+2\tau)\right]^n \left(\frac{y}{1+2\tau}\right)^s
\end{aligned}$$

where A is any directed animal and A_{str} are the strongly embedded directed animals. Up to a redefinition of the fugacity y and a constant prefactor, Dhar's model [163] is recovered.

Problems of Chapter 4

34 The Janssen-de Dominicis functional of directed percolation at zero field is given by

$$\mathcal{J}[\tilde{n}, n] = \lambda \int \mathrm{d}t\, \mathrm{d}\boldsymbol{r}\; \tilde{n} \left[\lambda^{-1}\partial_t n - (\tau + \nabla^2)n - \sqrt{\frac{g\kappa}{2}}\,(\tilde{n} - n)n\right] . \quad \text{(S.17)}$$

In order to show that this functional is invariant under the **rapidity-reversal**

$$\tilde{n}(t, \boldsymbol{r}) \longleftrightarrow -n(-t, \boldsymbol{r}) \quad \text{(S.18)}$$

we consider the terms separately:

$$\begin{array}{llll}
\tilde{n}_t\partial_t n_t & \underset{t\to -t}{\longrightarrow} -\tilde{n}_{-t}\partial_t n_{-t} & \underset{\mathrm{RR}}{\longrightarrow} -n_t\partial_t\tilde{n}_t & \underset{\mathrm{PI}}{\longrightarrow} \tilde{n}_t\partial_t n_t \\
-\tilde{n}_t r n_t & \underset{t\to -t}{\longrightarrow} -\tilde{n}_{-t} r n_{-t} & \underset{\mathrm{RR}}{\longrightarrow} -\tilde{n}_t r n_t & \\
-\tilde{n}_t\nabla^2 n_t & \underset{t\to -t}{\longrightarrow} -\tilde{n}_{-t}\nabla^2 n_{-t} & \underset{\mathrm{RR}}{\longrightarrow} -n_t\nabla^2\tilde{n}_t & \underset{\mathrm{PI}^2}{\longrightarrow} \tilde{n}_t\nabla^2 n_t \\
-\tilde{n}_t g(\tilde{n}_t - n_t)n_t & \underset{t\to -t}{\longrightarrow} -\tilde{n}_{-t} g(\tilde{n}_{-t} - n_{-t})n_{-t} & \underset{\mathrm{RR}}{\longrightarrow} -\tilde{n}_t g(\tilde{n}_t - n_t)n_t &
\end{array}$$

where RR denotes a transformation according to the rapidity reversal and PI indicates a partial integration.

35 We begin with a domain Ω of finite volume $|\Omega| < \infty$ and shall eventually consider the limit $|\Omega| \to \infty$ at the end. A useful auxiliary quantity is $A(t) := |\Omega|^{-1} \int_\Omega \mathrm{d}\boldsymbol{r}\; \varrho(t, \boldsymbol{r})^2$. Integrating (4.255), one has with Gauß' theorem

$$\partial_t\bar{\varrho} = -\lambda A + \underbrace{\frac{1}{|\Omega|}\int_{\partial\Omega} \mathrm{d}\boldsymbol{\sigma}\cdot\boldsymbol{\nabla}\varrho}_{\leq 0\ \text{, since particles escape}} \leq -\lambda A.$$

Applying the Cauchy-Schwarz inequality, it follows that $\partial_t\bar{\varrho}(t) \leq -\lambda\bar{\varrho}(t)^2$, hence $\bar{\varrho}(t) \leq \varrho_0/(1 + \varrho_0\lambda t) \leq 1/(\lambda t)$ and independently of the boundary condition [53]. However, this upper bound may be too loose in certain cases.

Sharper bounds may be obtained by estimating $A(t)$.[7] From (4.255), it follows $\frac{1}{2}\partial_t\varrho^2 = D\varrho\Delta\varrho - \lambda\varrho^3$. Integrating, we have for the case of Dirichlet boundary conditions

$$\begin{aligned}
\partial_t A(t) &= \underbrace{\frac{2D}{|\Omega|}\int_{\partial\Omega} \mathrm{d}\boldsymbol{\sigma}\cdot(\varrho\boldsymbol{\nabla}\varrho)}_{=0} - \frac{2D}{|\Omega|}\int_\Omega \mathrm{d}\boldsymbol{r}\;(\boldsymbol{\nabla}\varrho)^2 - \frac{2\lambda}{|\Omega|}\int_\Omega \mathrm{d}\boldsymbol{r}\;\varrho^3 \\
&\leq -\frac{1}{\tau}A(t) - 2\lambda A(t)^{3/2}
\end{aligned}$$

[7] From now on we implicitly assume that $\varrho(t, \boldsymbol{r}) \geq 0$, which can be justified rigorously [575], but which is also physically reasonable, since $\varrho = 0$ is an absorbing steady-state.

with $\tau^{-1} = 2DC^{-1}|\Omega|^{-2/d}$ and Poincaré's and Hölder's inequalities have been used in the second and third term, respectively. This readily gives, where A_0 is a constant

$$A(t) \leq \frac{A_0 e^{-t/\tau}}{[1 + 2\lambda\tau A_0^{1/2}(1 - e^{-t/(2\tau)})]^{1/2}}.$$

Now, a sharper upper bound on ϱ is given by $\bar{\varrho}(t) \leq A(t)^{1/2}$ and a *lower* bound is, since the above results imply $\bar{\varrho}(t) \to 0$ for $t \to \infty$, $\bar{\varrho}(t) \geq \lambda \int_t^\infty \mathrm{d}t'\, A(t')$. Define the *finite-size scaling limit*

$$t \to \infty \;\;,\;\; \tau \to \infty \;\;,\;\; \text{such that } \mathfrak{z} := t/(2\tau) \text{ is kept fixed.}$$

Then

$$\frac{1}{e^{\mathfrak{z}} - 1} \geq 2\lambda\tau\bar{\varrho}(t) \geq \frac{1}{e^{\mathfrak{z}} - 1} + \ln\left(1 - e^{-\mathfrak{z}}\right) \tag{S.19}$$

which is universal in the sense that the bounds on the scaling function do *not* depend on the unknown constant C.

For systems evolving in such a large domain Ω that $\mathfrak{z} \ll 1$, both upper and lower bounds in (S.19) are $\simeq \mathfrak{z}^{-1}$ and consequently the algebraic decay $\bar{\varrho}(t) \simeq (\lambda t)^{-1}$ becomes exact. On the other hand, for finite volumes and large times such that $\mathfrak{z} \gg 1$, we have $e^{-\mathfrak{z}} \geq 2\lambda\tau\bar{\varrho}(t) \geq \frac{1}{2}e^{-2\mathfrak{z}}$ leading to an exponential decay. Since $|\Omega| \sim L^d$, where L is the typical linear size of the domain Ω, the relaxation time $\tau \sim L^2$, and the dynamical exponent $z = 2$, which is natural for this kind of mean-field treatment.

In the case of Neumann boundary conditions, we have $\partial_t\bar{\varrho} = -\lambda A$, from which Poincaré's and Hölder's inequalities imply

$$\partial_t A(t) \leq -\frac{1}{\tau}(A - \bar{\varrho}^2) - 2\lambda A^{3/2} \tag{S.20}$$

Because of

$$\begin{aligned}
\partial_t(A - \varrho^2) &\leq -\frac{1}{\tau}(A - \bar{\varrho}^2) - 2\lambda A^{3/2} - 2\bar{\varrho}\partial_t\bar{\varrho} \\
&= -\frac{1}{\tau}(A - \bar{\varrho}^2) - 2\lambda A^{3/2} + 2\lambda\bar{\varrho}A \\
&\leq -\frac{1}{\tau}(A - \bar{\varrho}^2)
\end{aligned}$$

the variance $\alpha(t) := A(t) - \bar{\varrho}(t)^2$ satisfies $0 \leq \alpha(t) \leq \alpha_0 e^{-t/\tau}$, is exponentially small and will merely make a negligible contribution to (S.20). Therefore, for times $t \gg \tau$, one has again, from identical upper and lower bounds, $\bar{\varrho}(t) \simeq (\lambda t)^{-1}$ and this time independently of whether the limit $|\Omega| \to \infty$ was taken or not.

In conclusion, in the limit $|\Omega| \to \infty$ the average density has the same behaviour as found in the case of spatially homogeneous systems, for both Dirichlet and Neumann boundary conditions. Remarkably, this rather slow *algebraic* decay of $\bar{\varrho}(t)$ remains valid for Neumann boundary conditions for finite values of $|\Omega|$, whereas for Dirichlet boundary conditions the particle density decays exponentially if $|\Omega| < \infty$ and t is large enough.

36 One writes $E_n(t) = P(\underbrace{\circ\cdots\circ}_{n}; t) = P([n]; t)$ for the probability of finding n consecutive empty sites. Since the average particle-density is $\langle\varrho\rangle(t) = \varrho_\bullet P(\bullet; t) + \varrho_\circ P(\circ; t) = P(\bullet; t) = 1 - P(\circ; t) = 1 - E_1(t)$. Similarly, since $P(\bullet\underbrace{\circ\cdots\circ}_{n}; t) + P(\circ\underbrace{\circ\cdots\circ}_{n}; t) = P(\underbrace{\circ\cdots\circ}_{n}; t)$, one has $P(\bullet[n]; t) = E_n(t) - E_{n+1}(t)$ and from $P(\bullet\bullet[n-1]; t) + P(\bullet\circ[n-1]; t) + P(\circ\bullet[n-1]; t) + P(\circ\circ[n-1]; t) = P([n-1]; t)$ it follows that $P(\bullet\bullet[n-1]; t) + P(\circ\bullet[n-1]; t) = E_{n-1}(t) - E_n(t)$.

For deriving an equation of motion for $E_n(t)$, notice that a change of the configuration $[n] = \underbrace{\circ\cdots\circ}_{n}$ can only arise through the boundaries. Therefore, for $n = 2, \ldots, L$, where the number of sites of the chain is $L + 1$, one has

$$\begin{aligned}
\dot{E}_n(t) &= 2\gamma P(\bullet\bullet[n-1]; t) + 2DP(\circ\bullet[n-1]; t) - 2(\beta + D)P(\bullet[n]; t) \\
&\overset{\gamma=D}{=} 2D(E_{n-1}(t) - E_n(t)) - 2(\beta + D)(E_n(t) - E_{n+1}(t)) \\
&= 2D(E_{n-1}(t) + E_{n+1}(t) - 2E_n(t)) + 2\beta(E_{n+1}(t) - E_n(t))
\end{aligned} \quad \text{(S.21)}$$

For the last remaining case $n = 1$, one has

$$\dot{E}_1(t) = 2\gamma P(\bullet\bullet; t) - 2\beta P(\bullet\circ; t)$$

Since $P(\bullet\circ) = E_1 - E_2$ and $P(\bullet\bullet) = 1 - 2E_1 + E_2$, it follows that if one imposes the boundary condition $E_0(t) = 1$ and for $\gamma = D$, (S.21) also holds true for $n = 1$. A further boundary condition is obtained as follows: if one has $L+1$ sites and at least a single particle (which can never disappear) then the largest possible hole can have at most L sites and $E_{L+1}(t) = 0$.

Physically, the condition $D = \gamma$ means that if two particles land on the same site, they react instantaneously such that one of them disappears from the system.

Taking the continuum limit by sending the lattice spacing $a \to 0$, and rescaling $Da^2 \mapsto D$ and $\beta a \mapsto \beta$, the equation of motion (4.257), together with the boundary conditions (4.258), for continuum density of empty intervals $E(x,t)$ readily follows and the mean density is given by

$$\bar{\varrho}(t) = -\left.\frac{\partial E(x,t)}{\partial x}\right|_{x=0} . \quad \text{(S.22)}$$

More details on the empty-interval method can be found in a recent book [54], which also treats an extension of the method to the diffusion-annihilation process $2A \to \emptyset$. For a list of $1D$ models with two-particle reactions solvable by the empty-interval method, see [5].

37 The scaling ansatz is only compatible with (4.257) if $z = 2$, when it leads to $4Df''(u) + uf'(u) = 0$ with $u = x/\sqrt{t}$. A first integration gives $f'(u) = f_1 \exp(-\frac{u^2}{8D})$ and further leads to $f(u) = f_0 + f_1\sqrt{2\pi D}\,\mathrm{erf}\left(\frac{x}{\sqrt{8Dt}}\right)$ and where erf is the error function [2]. The boundary conditions (4.258) fix the constants $f_{0,1}$ such that, for $D = \gamma$ and $\beta = 0$ one has

$$E(x,t) = 1 - \mathrm{erf}\left(\frac{x}{\sqrt{8Dt}}\right).$$

Using (S.22), the long-time behaviour of the mean particle-density becomes

$$\bar{\varrho}(t) = -\left.\frac{\partial E(x,t)}{\partial x}\right|_{x=0} = \frac{1}{\sqrt{8Dt}}\frac{2}{\sqrt{\pi}}\exp\left.\left(-\frac{x^2}{8Dt}\right)\right|_{x=0} = \frac{1}{\sqrt{2\pi D}}\cdot t^{-1/2}$$

This is an algebraic decay, where the amplitude is universal in the sense that it is independent of the initial density. While this is qualitatively similar to the results of the mean-field analysis of exercise 35, the decay exponent $y = 1/2$ is different from the mean-field result $y_{\mathrm{MF}} = 1$. The experimental data in Table 4.11 for the exponent y are close to $1/2$ and are clearly distinct from 1. This confirms the importance of fluctuation effects for the correct description of the kinetics of excitons on long confined polymer chains and in particular shows that mean-field equations such as the reaction-diffusion equations (4.255) are not applicable in low dimensions $d < d_c$. The independence of the exponent y of the initial density has also been confirmed for TMMC [401].

A Green's function method gives the solution for all times and in particular reproduces the above long-time behaviour [53, 54].

38 Since the mean density changes only if two particles react, consider the average time between two collisions. To find this, assume [598] that one of the particles is mobile and starts a random walk from the origin, while the second particle sits at a fixed distance $|\boldsymbol{r}|$ from the origin. The normalised Green's function of the diffusion equation $\partial_t \varrho = D\Delta\varrho$ is $G(t,\boldsymbol{r}) = (4\pi Dt)^{-d/2}\exp(-|\boldsymbol{r}|^2/(4Dt))$. The squared distance in a single direction, say x, covered during the time t is then

$$\langle x^2 \rangle = \frac{\displaystyle\int_{\mathbb{R}^d} \mathrm{d}\boldsymbol{r}\; x^2 G(t,\boldsymbol{r})}{\displaystyle\int_{\mathbb{R}^d} \mathrm{d}\boldsymbol{r}\; G(t,\boldsymbol{r})} \sim t$$

Hence the total volume covered is $V \sim t^{d/2}$ and the mean density should behave as $\varrho(t) \sim V^{-1} \sim t^{-d/2}$. For $d=1$ this agrees with the exact calculation of exercise 37. For $d>2$, this is slower than the homogeneous diffusive behaviour, and $\varrho(t) \sim t^{-1}$ for $d>2$. Therefore $d_c = 2$ is the upper critical dimension for diffusion-coagulation.

For a two-species annihilation $A+B \to \emptyset$, one observes first that the particles will rapidly separate into A-rich and B-rich regions. The mean distance covered by the particles is $\ell \sim t^{1/2}$. Consider a region with an excess of, say, A particles, and of linear size ℓ. The remaining particles are those which are spared by a fluctuation, hence there will remain $n \sim n_A - n_B \sim \sqrt{\ell^d} \sim t^{d/4}$ particles at time t. Hence the density of either species is $\varrho_{A,B}(t) \sim t^{d/4}\ell^d \sim t^{-d/4}$. The upper critical dimension for this process is $d_c = 4$ [598].

We mention rigorous bounds which confirm these heuristic arguments [78, 79]. If the initial densities of A's and B's are equal, $\varrho_A(0) = \varrho_B(0)$, then there exist positive constants c_d, C_d such that

$$\begin{aligned} c_d \leq t^{d/4}\varrho_A(t) = t^{d/4}\varrho_B(t) \leq C_d \quad &; \text{ if } d<4 \\ c_d \leq t\varrho_A(t) = t\varrho_B(t) \leq C_d \quad &; \text{ if } d>4 \end{aligned}$$

However, if initially $\varrho_A(0) < \varrho_B(0)$, then

$$c_d \exp\left(-\lambda_d \gamma g_d(t)\right) \leq \varrho_A(t) \leq C_d \exp\left(-\Lambda_d \gamma g_d(t)\right)$$

where $\gamma = \varrho_B(0) - \varrho_A(0)$, λ_d and Λ_d are positive constants and

$$g_d(t) = \begin{cases} \sqrt{t} & ; \text{ if } d=1 \\ t/\ln t & ; \text{ if } d=2 \\ t & ; \text{ if } d>2 \end{cases}$$

Also, $\rho_B(t) \to \gamma$ as $t \to \infty$.

39 The allowed reactions are, between pairs or triplets of nearest neighbour sites

$$A\emptyset \overset{D}{\leftrightarrow} \emptyset A \;\;, \;\; AA \overset{\gamma}{\to} A\emptyset, \emptyset A \;\;, \;\; A\emptyset A \overset{2D\lambda}{\to} AAA$$

The empty-interval method works for $D=\gamma$ and we set $D=1$ from now on. If $E_n(t)$ is the probability that n consecutive sites are empty, the mean particle-density is $\varrho(t) = (1-E_1(t))\, a^{-1}$, where a is the lattice constant. On a periodic chain with L sites, we have the equations of motion

$$\begin{aligned} E_0(t) &= 1, \;\; E_L(t) = 0 \\ \frac{\mathrm{d}E_1}{\mathrm{d}t}(t) &= 2\left[E_0(t) - 2E_1(t) + E_2(t)\right] - 2\lambda\left[E_1(t) - 2E_2(t) + E_3(t)\right] \quad \text{(S.23)} \\ \frac{\mathrm{d}E_n}{\mathrm{d}t}(t) &= 2\left[E_{n-1}(t) - 2E_n(t) + E_{n+1}(t)\right] \;\; ; \qquad 2 \leq n \leq L-1 \end{aligned}$$

where the last boundary condition holds if one has at least one particle on the lattice. The λ-dependent term follows from the relations $P(\bullet\circ\circ) = E_2 - E_3$ and $P(\bullet\circ\bullet) + P(\bullet\circ\circ) = P(\bullet\circ) = E_1 - E_2$, hence $P(\bullet\circ\bullet) = E_1 - 2E_2 + E_3$.

These equations are solved through the ansatz $E_n(t) = \sum_\omega a_n(\omega) e^{-2\omega t}$ which leads to the eigenvalue problem

$$\begin{bmatrix} 0 & 0 & 0 & \cdots & & & & 0 \\ 1 & -2-\lambda & 1+2\lambda & -\lambda & 0 & \cdots & & 0 \\ 0 & 1 & -2 & 1 & 0 & \cdots & & 0 \\ & 0 & 1 & -2 & 1 & 0 & \cdots & 0 \\ \vdots & & \ddots & \ddots & \ddots & \ddots & \ddots & \vdots \\ & & & 0 & 1 & -2 & 1 & 0 \\ & & & & 0 & 1 & -2 & 1 \\ 0 & \cdots & & & & 0 & 0 & 0 \end{bmatrix} \begin{bmatrix} a_0 \\ a_1 \\ a_2 \\ \\ \vdots \\ \\ a_{L-1} \\ a_L \end{bmatrix} = -\omega \begin{bmatrix} a_0 \\ a_1 \\ a_2 \\ \\ \vdots \\ \\ a_{L-1} \\ a_L \end{bmatrix}$$

involving an $(L+1)\times(L+1)$ matrix $\hat{\Omega}$. Since the solution $a_n(0) = 1 - n/L$ with eigenvalue $\omega = 0$ describes the steady state with a single diffusing particle, the model has *two* steady states, one corresponding to the empty lattice and the other one being the translation-invariant superposition of all single-particle states with an average density $\varrho_{\rm av} = 1/L$. The other boundary condition implies $a_L(\omega) = 0$ and we reduce to an eigenvalue problem of a $(L-1)\times(L-1)$ matrix if $\omega \neq 0$

$$\begin{bmatrix} -2-\lambda & 1+2\lambda & -\lambda & 0 & \cdots & & 0 \\ 1 & -2 & 1 & 0 & \cdots & & 0 \\ 0 & 1 & -2 & 1 & 0 & & \\ \vdots & \ddots & \ddots & \ddots & \ddots & \ddots & \vdots \\ & & 0 & 1 & -2 & 1 & 0 \\ & & & 0 & 1 & -2 & 1 \\ 0 & \cdots & & & 0 & 1 & -2 \end{bmatrix} \begin{bmatrix} a_1 \\ a_2 \\ \\ \vdots \\ \\ a_{L-2} \\ a_{L-1} \end{bmatrix} = -\omega \begin{bmatrix} a_1 \\ a_2 \\ \\ \vdots \\ \\ a_{L-2} \\ a_{L-1} \end{bmatrix} . \tag{S.24}$$

Equation (S.24) is solved through the ansatz

$$a_n(\omega) = A e^{ikn} + B e^{-ikn} \quad ; \quad \omega \neq 0$$

which leads to the dispersion relation $\omega = \omega(k) = 2(1 - \cos k)$. The allowed values of k are obtained by inserting the ansatz into the first line of (S.24) and taking the boundary condition $a_L(\omega) = 0$ into account. The resulting system of two linear equations

$$\begin{aligned} A\left(\lambda\left(e^{ik} - 2e^{2ik} + e^{3ik}\right) + 1\right) + B\left(\lambda\left(e^{-ik} - 2e^{-2ik} + e^{-3ik}\right) + 1\right) &= 0 \\ A e^{ikL} + B e^{-ikL} &= 0 \end{aligned}$$

has a non-trivial solution if k is a solution of

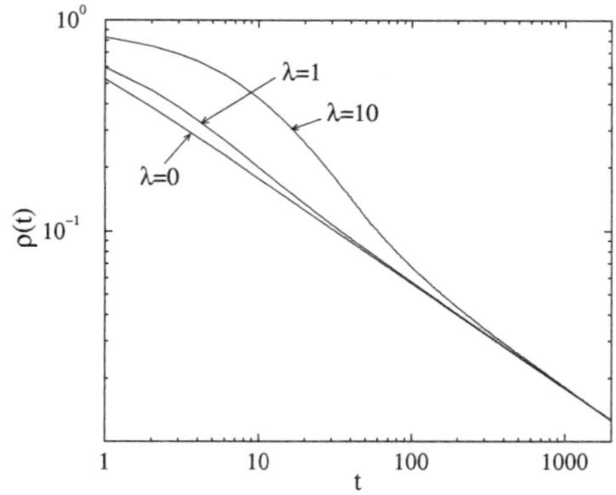

Fig. S.4 Mean particle-density for the $1D$ diffusion-coagulation-production process with $D = \gamma$ and for several values of the production rate λ. Reprinted with permission from [273].

$$\tan kL = \frac{4\lambda \sin(2k) \sin^2(k/2)}{4\lambda \cos(2k) \sin^2(k/2) - 1}. \tag{S.25}$$

We call the solutions of (S.25) k_m, where $m = 1, \ldots, L-1$. In addition, we can include the stationary solution by letting $k_0 = 0$. Then

$$E_n(t) = \left(1 - \frac{n}{L}\right) + \sum_{m=0}^{L-1} C_m \sin\left(k_m(n-L)\right) e^{-2\,\omega_m\, t},$$

where $\omega_m = \omega(k_m) = 2(1 - \cos k_m)$ and the C_m are real constants which must be determined from the initial conditions.

Closed-form solutions of (S.25) exist for $\lambda = 0$, where $k_m = m\pi/L$ and for $\lambda \to \infty$, where $k_m = m\pi/(L-2)$. For general values of λ, we have

$$k_m = \frac{m\pi}{L} - \frac{2\pi^3 \lambda m^3}{L^4} + \ldots \quad ; \quad m = 0, 1, \ldots, L-1$$

Since asymptotically $k_m \sim L^{-1}$ for all values of λ, the production reaction cannot induce a phase transition, at least if $D = \gamma$. This is illustrated in Fig. S.4 which shows that for sufficiently long times, one always returns to the asymptotic form $\varrho(t) \sim t^{-1/2}$.

However, for $D > \gamma$ a phase transition can be found [491]. This may already be seen at the level of the pair-approximation (we leave it to the reader to work out the details for himself). Using the parametrisation $\gamma = p(1-D)$ and $2D\lambda = (1-p)(1-D)$, the critical line is

$$D = \frac{p(p-1)}{p^2 + 3p - 2}$$

Simulational results for the critical point and some exponents in $1D$ are listed in Table S.2.

Table S.2 Estimates for the critical point $p_c(D)$ and of some exponents of the $1D$ diffusion-coagulation-production process. The data are from [491].

D	0.1	0.2	0.4	0.7
p_c	0.1129(1)	0.17975(8)	0.2647(1)	0.3528(2)
α		0.263(9)	0.268(8)	0.275(8)
β		0.57(1)	0.58(1)	0.57(1)

It was pointed out recently [466] that through a stochastic similarity transformation (see exercise 26) the model treated here may be related to a diffusion-annihilation process $2A \to \emptyset$ with the additional back reaction $A \to 3A$, see [464, 465].

40 From the master equation, one has the reaction-diffusion equation

$$\partial_t \varrho(t,\boldsymbol{r}) = D\Delta\varrho(t,\boldsymbol{r}) - \lambda\varrho^3(t,\boldsymbol{r})$$

where Δ is the spatial Laplacian. Hölder's inequality implies

$$\frac{1}{|\Omega|}\int_\Omega \mathrm{d}\boldsymbol{r}\;\varrho \le \left(\frac{1}{|\Omega|}\int_\Omega \mathrm{d}\boldsymbol{r}\;\varrho^3\right)^{1/3}$$

and therefore

$$\begin{aligned}
\frac{\mathrm{d}\bar{\varrho}(t)}{\mathrm{d}t} &= \frac{1}{|\Omega|}\int_\Omega \mathrm{d}\boldsymbol{r}\;\partial_t\varrho(t,\boldsymbol{r}) \\
&= \underbrace{\frac{D}{|\Omega|}\int_{\partial\Omega}\mathrm{d}\boldsymbol{\sigma}\cdot\boldsymbol{\nabla}\varrho(t,\boldsymbol{r})}_{\le 0} - \frac{\lambda}{|\Omega|}\int_\Omega \mathrm{d}\boldsymbol{r}\;\varrho^3(t,\boldsymbol{r}) \\
&\le -\lambda\bar{\varrho}(t)^3
\end{aligned}$$

which implies $\bar{\varrho}(t) \le \varrho_0/\sqrt{1+2\varrho_0^2\lambda t} \le (2\lambda)^{-1/2}t^{-1/2}$. For finite volumes $|\Omega| < \infty$, this can be considerably strengthened to exponentially small bounds. The bound obtained here is already close to the exact behaviour, since the system is at its upper critical dimension.

For the pair mean-field theory, we write $\varrho = p(\bullet)$ and $u = p(\bullet\bullet)$. With the usual abbreviations $v = p(\bullet\circ) = \varrho - u$ and $w = p(\circ\circ) = 1 - 2\varrho + u$, the reaction rates are given in the following table.

event	rate	$\Delta N_\bullet$	$\Delta N_{\bullet\bullet}$	
$\bullet\bullet\bullet\bullet\bullet \to \bullet\circ\circ\circ\bullet$	$\lambda u^4/\varrho^3$	-3	-4	
$\bullet\bullet\bullet\bullet\circ \to \bullet\circ\circ\circ\circ$	$\lambda u^3 v/\varrho^3$	-3	-3	$\times 2$
$\circ\bullet\bullet\bullet\circ \to \circ\circ\circ\circ\circ$	$\lambda u^2 v^2/\varrho^3$	-3	-2	
$\bullet\bullet\circ\circ \to \bullet\circ\bullet\circ$	$Duvw/(\varrho(1-\varrho))$	0	-1	$\times 2$
$\circ\bullet\circ\bullet \to \circ\circ\bullet\bullet$	$Dv^3/(\varrho(1-\varrho))$	0	1	$\times 2$

The equations of motion are

$$\dot{a} = -3\lambda \frac{u^2}{\varrho}$$
$$\dot{u} = -2\lambda \frac{u^2(\varrho + u)}{\varrho^2} + 2D \frac{(\varrho - u)(\varrho^2 - u)}{\varrho(1-\varrho)}$$

such that for $D \to \infty$, one has $u = \varrho^2$ and recovers the site-approximation $\dot{\varrho} = -3\lambda\varrho^3$.

In general, the discussion reduces to the analysis of the dynamical system

$$\frac{\mathrm{d}u}{\mathrm{d}\varrho} = \frac{2}{3}\frac{\varrho + u}{\varrho} - 2D\frac{(\varrho - u)(\varrho^2 - u)}{(1-\varrho)u^2}$$

For $D = 0$, only triplets will remain active. Their density is, in the pair-approximation, $\vartheta = u^2/\varrho$. Then

$$\frac{u - 2\varrho}{u_0 - 2\varrho_0} = \left(\frac{\varrho}{\varrho_0}\right)^{2/3}$$

from which $\vartheta \sim \varrho^{1/3} \lesssim t^{-1/6}$ follows (for large times, when $\varrho \to 0$).

41 Since $\frac{\partial \varrho'}{\partial t'} = \Lambda^{-\beta/\nu_\perp} \frac{\partial \varrho'}{\partial t}\frac{\partial t}{\partial t'} = \Lambda^{-\beta/\nu_\perp - z}\frac{\partial \varrho}{\partial t}$ and similarly $\frac{\partial \varrho'}{\partial \boldsymbol{r}'} = \Lambda^{-\beta/\nu_\perp - 1}\frac{\partial \varrho}{\partial \boldsymbol{r}}$, (4.260), first taken with the primed variables, reduces to

$$\partial_t \varrho = D\Lambda^{z-2}\Delta\varrho + \Lambda^{z-1/\nu_\perp}\tau\varrho - \Lambda^{z-\beta/\nu_\perp}\varrho^2$$

which is identical to (4.260) if one takes $z = \frac{1}{\nu_\perp} = \frac{\beta}{\nu_\perp} = 2$. Then the mean-field exponents $z_{\mathrm{MF}} = 2$, $\nu_{\perp,\mathrm{MF}} = \frac{1}{2}$ and $\beta_{\mathrm{MF}} = 1$ are recovered.

42 The mean-field equation is (λ, μ are reaction rates)

$$\partial_t \varrho = D\Delta\varrho - \lambda\varrho^3 + \mu\varrho(1-\varrho)$$

At the critical point $\mu = 0$, one finds $\varrho(t) \sim t^{-1/2}$, see exercise 40. Under the transformation (4.261), with λ, D assumed to be fixed, the mean-field equation becomes

$$\partial_t \varrho = D\Lambda^{z-2}\Delta\varrho + \Lambda^{z-1/\nu_\perp}\mu\varrho - \underbrace{\Lambda^{z-\beta/\nu_\perp - 1/\nu_\perp}\mu\varrho^2}_{\sim \Lambda^{-1}\mu\varrho^2} - \Lambda^{z-2\beta/\nu_\perp}\lambda\varrho^3$$

This is indeed invariant, if $z = \frac{1}{\nu_\perp} = 2\frac{\beta}{\nu_\perp} = 2$. However, the quadratic term in ϱ is suppressed by a factor Λ^{-1}, which means that this term is *irrelevant* (in the RG sense) for the long-time behaviour. One therefore has some freedom in specifying the mean-field equation without modifying the scaling properties of the theory.

43 The Janssen-de Dominicis functional for DP may be written in the form

$$\mathcal{J}[\widetilde{\varrho},\varrho] = \int \mathrm{d}t\,\mathrm{d}\boldsymbol{r}\ \left[\widetilde{\varrho}\partial_t\varrho - D\widetilde{\varrho}\Delta\varrho - \tau\widetilde{\varrho}\varrho + \mu\widetilde{\varrho}\varrho(\varrho-\widetilde{\varrho})\right]$$

Under a scale-transformation $\boldsymbol{r}\mapsto\Lambda\boldsymbol{r}$. $t\mapsto\Lambda^z t$, $\varrho\mapsto\Lambda^x\varrho$ and $\widetilde{\varrho}\mapsto\Lambda^x\widetilde{\varrho}$ (here the rapidity-reversal-symmetry of DP is used, see exercise 34), one has

$$\mathcal{J} = \int \mathrm{d}t\,\mathrm{d}\boldsymbol{r}\ \left[\Lambda^{d+2x}\widetilde{\varrho}\partial_t\varrho - \Lambda^{z-2}D\widetilde{\varrho}\Delta\varrho - \Lambda^{z+d+2x}\tau\widetilde{\varrho}\varrho + \Lambda^{z+d+3x}\mu\widetilde{\varrho}\varrho(\varrho-\widetilde{\varrho})\right]$$

The requirement of invariance of $\mathcal{J}$ fixes $z=2$ and $x=\widetilde{x}=-d/2$ from the kinetic terms and implies the transformations

$$\tau\mapsto\tau'=\Lambda^2\tau\ ,\ \ \mu\mapsto\mu'=\Lambda^{(4-d)/2}\mu$$

Therefore, a scale-invariance is only possible exactly at the critical point $\tau=0$. For $d>4$, $\mu\to 0$ on increasingly larger scales while the opposite is true for $d<4$. Hence $d_c=4$ is the upper critical dimension of DP and for $d>d_c=4$ the non-linearity does not influence the leading critical behaviour which is therefore described by mean-field theory.

An analogous conclusion can be reached by considering directly the scaling behaviour of the Langevin equation (3.10,3.11). Under a dilation $\boldsymbol{r}\mapsto\Lambda\boldsymbol{r}$. $t\mapsto\Lambda^z t$, $\varrho\mapsto\Lambda^x\varrho$ and $\eta\mapsto\Lambda^\chi\eta$, the Langevin equation becomes

$$\partial_t\varrho = \Lambda^{z-2}D\Delta + \Lambda^z\tau\varrho - \Lambda^{x+z}g\varrho^2 + \Lambda^{\chi-x+z}\eta \qquad \text{(S.26)}$$

with the noise correlator $\langle\eta\eta\rangle = \Lambda^{x-z-d-2\chi}\kappa\varrho\delta(t-t')\delta(\boldsymbol{r}-\boldsymbol{r}')$. Scale-invariance of the noise requires $2\chi=x-z-d$ and of the Langevin equation $z=2$ and $x=-z=-2$. Hence the noise term in (S.26) changes by a factor $\Lambda^{(4-d)/2}$, which means that the noise becomes irrelevant for $d>d_c=4$.

44 The profile follows from a straightforward integration of the stationary case $\dot{a}=0$.

45 Combine equations (4.78) with (4.114) with (D13). From the usual scaling forms of the density and the pair connectedness, one now easily establishes

$$P_{\rm sur}(t;\tau) = \lambda^{-\beta'}\widetilde{P}(\lambda^{-\nu_\|}a_t t;\lambda a_\tau\tau) \sim t^{-\beta'/\nu_\|}\,\widetilde{\Pi}\left(t^{1/\nu_\|}\tau\right)$$

as it should be and where $\widetilde{P}$ and $\widetilde{\Pi}$ are scaling functions. Equations (4.96-4.98) now readily follow.

46 Consider the iterative construction of the Cantor set as indicated in Fig. H1a. At the gth iteration (*'generation'*), the empty intervals of size $\ell = (1/3)^g$ are added, hence $g = -\frac{\ln \ell}{\ln 3}$. By inspection, the number n_g of added empty intervals is $n_1 = 1, n_2 = 2, \ldots, n_g = 2^{g-1}$. Hence

$$n(\ell) = 2^{g-1} = e^{(g-1)\ln 2} = \exp\left[\left(-1 - \frac{\ln \ell}{\ln 3}\right)\ln 2\right] \sim \ell^{-d_{\text{f}}}$$

where $d_{\text{f}} = d_{\text{self}} = \frac{\ln 2}{\ln 3}$ is the fractal dimension of the Cantor set.

Finally, since $P(\ell) \sim \frac{\mathrm{d}n(\ell)}{\mathrm{d}\ell}$, one arrives at $P(\ell) \sim \ell^{-1-d_{\text{f}}}$.

47 The scaling of the density, the survival probability and the pair connectedness is

$$\begin{aligned}
\varrho(t, \boldsymbol{r}; \tau, h) &= \lambda^{-\beta} \widetilde{R}\left(\lambda^{-\nu_{\|}} a_t t, \lambda^{-\nu_{\perp}} a_r \boldsymbol{r}; \lambda a_\tau \tau, \lambda^{\sigma} a_h h\right) \\
P_{\text{sur}}(t, \boldsymbol{r}; \tau, h) &= \lambda^{-\beta'} \widetilde{P}\left(\lambda^{-\nu_{\|}} a_t t, \lambda^{-\nu_{\perp}} a_r \boldsymbol{r}; \lambda a_\tau \tau, \lambda^{\sigma} a_h h\right) \\
\Upsilon(t, \boldsymbol{r}; \tau, h) &= \lambda^{-\beta-\beta'} \widetilde{\Upsilon}\left(\lambda^{-\nu_{\|}} a_t t, \lambda^{-\nu_{\perp}} a_r \boldsymbol{r}; \lambda a_\tau \tau, \lambda^{\sigma} a_h h\right)
\end{aligned}$$

In order to make contact with (4.139), one sets $|a_\tau \tau \lambda| = 1$. Then

$$\varrho(t, \boldsymbol{r}; \tau, h) = (a_\tau |\tau|)^{\beta} \widetilde{R}\left(a_t a_\tau^{\nu_{\|}} |\tau|^{\nu_{\|}} t, a_r a_\tau^{\nu_{\perp}} |\tau|^{\nu_{\perp}} \boldsymbol{r}; \pm 1, \frac{a_h}{a_\tau^{\sigma}} \frac{h}{|\tau|^{\sigma}}\right)$$

where ± 1 refers to $\tau > 0$ and $\tau < 0$, respectively. From this, one may identify the spatial correlation length

$$\xi_{\perp}^{-1} = a_r a_\tau^{\nu_{\perp}} |\tau|^{\nu_{\perp}} =: \xi_0^{-1} |\tau|^{\nu_{\perp}}$$

and the metric factors in (4.139) become

$$\begin{aligned}
\xi_0 &= \left(a_r a_\tau^{\nu_{\perp}}\right)^{-1}, \; D_2 = a_h a_\tau^{-1}, \qquad D_0 = a_t a_r^{-z} \\
D_{1\varrho} &= a_r^{-\beta/\nu_{\perp}}, \; D_{1P} = a_r^{-\beta'/\nu_{\perp}}, \; D_{1\Upsilon} = a_r^{-(\beta+\beta')/\nu_{\perp}}
\end{aligned}$$

48 The derivation given here follows Janssen [330]. Consider the probability $p_N(\tau)$ of a bDP cluster of exactly N particles, created from a single seed at the origin of the space-time lattice. For N sufficiently large, one has

$$p_N(\tau) \approx P_N(\tau) = \left\langle \delta(N - \mathcal{N}) e^{\widetilde{n}(0,\boldsymbol{0})} \right\rangle$$

where $\mathcal{N} = \int_0^\infty \mathrm{d}t \int \mathrm{d}\boldsymbol{r}\, n(t, \boldsymbol{r})$. Define further the generating function $F(\tau, k) := \sum_{N=0}^{\infty} p_N(\tau) e^{-kN}$ and one can show that

$$\lim_{k \to +0} F(\tau, k) = \begin{cases} 1 & ; \text{ if } \tau \leq 0 \\ 1 - P_{\text{perc}} & ; \text{ if } \tau > 0 \end{cases}$$

Taking a formal continuum limit,

$$F(\tau,k) - F(-|\tau|,k) = \int_0^\infty \mathrm{d}N\ (P_N(\tau) - P_N(-|\tau|))\, e^{-kN}$$
$$= \left\langle e^{\widetilde{n}(0,\mathbf{0})-k\mathcal{N}}\right\rangle(\tau) - \left\langle e^{\widetilde{n}(0,\mathbf{0})-k\mathcal{N}}\right\rangle(-|\tau|)$$

In the absorbing phase with $\tau < 0$, one has $\lim_{k\to+0}\left\langle e^{-k\mathcal{N}} e^{\widetilde{n}(0,\mathbf{0})}\right\rangle = 1$, hence

$$P_{\mathrm{perc}}(\tau) = 1 - \lim_{k\to+0}\left\langle e^{\widetilde{n}(0,\mathbf{0})-k\mathcal{N}}\right\rangle = -\lim_{k\to+0}\left\langle \widetilde{n}(0,\mathbf{0})e^{-k\mathcal{N}}\right\rangle + \cdots$$

where the neglected terms merely create corrections to scaling.

For bDP one may now use the duality transformation (4.182), with the result

$$P_{\mathrm{perc}}(\tau) = -\lim_{k\to+0}\left\langle n(0,\mathbf{0})e^{+k\widetilde{\mathcal{N}}}\right\rangle = \varrho_{\mathrm{s}}(\tau)$$

with $\widetilde{\mathcal{N}} = \int \mathrm{d}t \int \mathrm{d}\boldsymbol{r}\, \widetilde{n}(t,\boldsymbol{r})$ and k now corresponds to a constant particle source which is turned off at the end of the calculation.

Comparison of the scaling forms gives $\beta = \beta'$.

We stress that the presentation sketched here glosses over several important conceptual points which are carefully described in [330] but which are beyond the scope of this book. There, the reader may also find an extension of the above argument to the time-dependent survival probability $P_{\mathrm{sur}}(t)$ and density $\varrho(t)$.

Problems of Chapter 5

49 The calculation is analogous to exercise 21.

50 The diffusive pair-contact process is defined by the rates (the two states are denoted by $A = \bullet$, $\emptyset = \circ$)

$$\begin{array}{rl}
\bullet\bullet\circ,\ \circ\bullet\bullet \to \bullet\bullet\bullet & \text{with rate } (1-p)(1-D)/2 \\
\bullet\bullet \to \circ\circ & \text{with rate } p(1-D) \\
\bullet\circ \leftrightarrow \circ\bullet & \text{with rate } D\ ,
\end{array}$$

where $0 \le p \le 1$ and $0 \le D \le 1$ are the control parameters. If we let $\varrho(t) = P_\bullet(t)$ and also assume that it is space-independent, one easily finds, using the techniques of appendix E

$$\frac{\mathrm{d}\varrho(t)}{\mathrm{d}t} = (1-p)\,\varrho(t)^2\Big(1-\varrho(t)\Big) - 2p\,\varrho(t)^2 \tag{S.27}$$

where time was rescaled according to $t \mapsto t(1-D)^{-1}$. The long-time behaviour of the mean density is [126]

$$\varrho(t) \simeq \begin{cases} \frac{1-3p}{1-p} + \mathfrak{a}\exp(-t/\tau_{\rm rel}) & ; \text{ if } p < 1/3 \\ \sqrt{3/4}\cdot t^{-1/2} & ; \text{ if } p = 1/3 \\ (3p-1)^{-1}\cdot t^{-1} & ; \text{ if } p > 1/3 \end{cases}$$

where $\tau_{\rm rel} = (1-p)/(1-3p)^2$ and $\mathfrak{a}$ is a constant which depends on the initial conditions. One distinguishes an active phase, where $\varrho(t) \to \varrho_\infty > 0$ for large times, and an absorbing phase, where $\lim_{t\to\infty}\varrho(t) = 0$. Throughout the absorbing phase, density decays algebraically, although mean-field theory is not capable of reproducing the correct decay $\varrho(t) \sim t^{-1/2}$.

For the derivation of the pair approximation, one considers the particle density $\varrho(t) = P_\bullet(t)$ and the pair density $u(t) = P_{\bullet\bullet}(t)$ which again are assumed to be space-independent. Introducing the short-hands $v = P_{\bullet\circ} = \varrho - u$ and $w = P_{\circ\circ} = 1 - 2\varrho + u$, the reactions are listed in the following table, where the symmetry factors are noted explicitly

reaction	$\Delta N_\bullet$	$\Delta N_{\bullet\bullet}$	rate	
● ● ○○ → ● ○ ●●	0	− 1	$Duvw/\varrho(1-\varrho)$	×2
○ ● ○● → ○ ○ ●●	0	+1	$Dv^3/\varrho(1-\varrho)$	×2
● ● ●● → ● ○ ○●	− 2	− 3	$p(1-D)u^3/\varrho^2$	
● ● ●○ → ● ○ ○○	− 2	− 2	$p(1-D)u^2v/\varrho^2$	×2
○ ● ●○ → ○ ○ ○○	− 2	− 1	$p(1-D)uv^2/\varrho^2$	
● ● ○● → ● ● ●●	+1	+2	$\frac{1}{2}(1-p)(1-D)uv^2/\varrho(1-\varrho)$	×2
● ● ○○ → ● ● ●○	+1	+1	$\frac{1}{2}(1-p)(1-D)uvw/\varrho(1-\varrho)$	×2

Adding the contributions, we have the equations of motion [126]

$$\begin{aligned} \dot{\varrho}(t) &= -2(1-D)p\,u(t) + (1-D)(1-p)\,\left(\varrho(t)-u(t)\right)\frac{u(t)}{\varrho(t)} \\ \dot{u}(t) &= -(1-D)p\,u(t)\frac{2u(t)+\varrho(t)}{\varrho(t)} - 2D\frac{(\varrho(t)-u(t))(u(t)-\varrho(t)^2)}{\varrho(t)(1-\varrho(t))} \\ &\quad +(1-D)(1-p)\frac{(\varrho(t)-u(t))(1-u(t))u(t)}{\varrho(t)(1-\varrho(t))} \end{aligned}$$

In the limit $D \to \infty$, one has $u(t) = \varrho(t)^2$ and the site-approximation (S.27) for $\varrho(t)$ is recovered. The critical line is given by

$$p_c(D) = \begin{cases} \dfrac{1}{3} & ; \text{ if } D \geq \frac{1}{7} \\ \frac{1}{5}\dfrac{1+3D}{1-D} & ; \text{ if } D \leq \frac{1}{7} \end{cases}$$

The scaling of both $\varrho(t)$ and of $u(t)$, within the pair-approximation, depends on the value of D. If $D > \frac{1}{7}$, one has $u \sim \varrho^2$ and if $D < \frac{1}{7}$, then $u \sim \varrho$. The

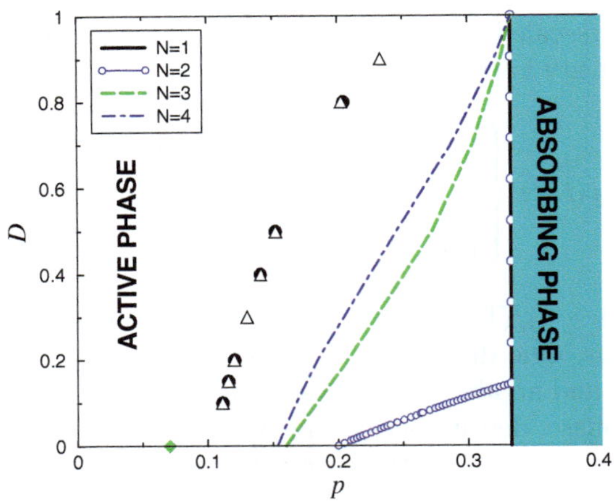

Fig. S.5 Phase diagram of the $1D$ PCPD. The full curves give the critical line according to cluster mean-field theory, from the site- ($N = 1$) to the quartet-approximation ($N = 4$). The filled dots give estimates of $p_c(D)$ from the DMRG and the open triangles from Monte Carlo simulations. The grey diamond gives the critical point $p_c(0)$ of the PCP. Reprinted with permission from [274].

steady-state particle density $\varrho_\infty \sim p_c(D) - p$ for all D, and if $D \neq \frac{1}{7}$ one has $u(t) \sim t^{-1}$ along the critical line. In Fig. S.5 the resulting phase diagram is shown.

However, the expectation of the pair-approximation that there should be a multi-critical point is not confirmed in higher-order approximations when formulated for the $1D$ case. Indeed, because of the three-site interactions, the pair-approximation is still insufficient as is indicated by its prediction $u_\infty = \varrho_\infty(1-3p)/(1-p)$, in contrast with simulational studies, in particular for $D = 0$. This may be corrected in the triplet approximation [461, 444], where the independent variables may be chosen as $P_{\bullet\bullet\bullet}$, $P_{\bullet\bullet\circ}$, $P_{\bullet\circ\circ}$ and $P_{\bullet\circ\bullet}$. Then $p_c(0) \simeq 0.128$ and the steady-state pair density $u_\infty \sim p_c(0) - p$ vanishes linearly, but the critical particle density $\varrho_{\infty,c} \simeq 0.23$ remains finite, as it should be (see [586] for cluster approximations at $D = 0$ up to $N = 12$).

As can be seen from Fig. S.5, the available evidence, either from cluster approximation or numerical simulation (Monte Carlo or DMRG) strongly suggests a single universality class along the transition line, see [492, 274]. However, as we shall see in Vol. 2 of this book, the conclusion of the pair-approximation on the existence of a multi-critical point is confirmed for $d > 2$ in the so-called *bosonic* variant of the PCPD, where an arbitrary number of particles is allowed per site. Furthermore, unpublished simulations of the higher-dimensional PCPD are in qualitative agreement with the pair approximation [178].

51 Intuitively, for $\lambda > \delta$, equal pairs (AA and BB) should disappear more rapidly than unequal pairs (AB). For long times, a typical state should be of the form

$$\cdots A \cdots B \cdots A \cdots B \cdots A \cdots B \cdots B \cdots A \cdots B \cdots A \cdots B \cdots$$

(only the particles A, B were indicated) with many interfaces AB. For $\lambda < \delta$, the opposite should occur, leading to phase-separated typical states

$$\cdots A \cdots A \cdots A \cdots A \cdots A \cdots B \cdots B \cdots B \cdots B \cdots A \cdots A \cdots A \cdots$$

and the minority species should disappear first.

In simple mean-field, one has

$$\dot{a} = -\lambda a^2 - \delta ab \ , \ \dot{b} = -\lambda b^2 - \delta ab$$

hence, with $\rho = b/a$, $\dot{\rho} = (\lambda - \delta)\rho(1-\rho)a$. Introducing the fictitious time $\tau = -\ln a$ such that $\tau \to \infty$ as $a \to 0$ and since $\frac{\mathrm{d}\rho}{\mathrm{d}a} = \frac{\mathrm{d}\rho}{\mathrm{d}\tau}\frac{\mathrm{d}\tau}{\mathrm{d}a} = -\frac{1}{a}\frac{\mathrm{d}\rho}{\mathrm{d}\tau}$, (5.118) follows.

Tracing the vector field $(\rho, \dot{\rho})$, it follows that for $\lambda > \delta$, the fixed point $\rho^* = 1$ is stable, indicating a homogeneous phase with equal numbers of As and Bs. For $\lambda < \delta$, $\rho^* = 1$ is unstable and $\rho \to \infty$ for $\rho_0 > 1$ and $\rho \to 0$ for $\rho_0 < 1$, mimicking the separation of phases with the disappearance of the minority species.

We leave it to the reader to work out the pair-approximation [587].

52 The solution follows the lines of [203]. From (S.1), we have

$$\frac{\partial F}{\partial t} = (1-p)(1-\lambda p)\frac{\partial F}{\partial p} \stackrel{!}{=} H_R(p,q)F$$

hence

$$H_R(p,q) = (1-p)(1-\lambda p)q \ , \ q := \frac{\partial}{\partial p}.$$

The variational equations are

$$\begin{aligned}\partial_t q &= \frac{\partial H_R}{\partial p} &= q(2\lambda p - 1 - \lambda) \\ \partial_t p &= -\frac{\partial H_R}{\partial q} &= -(1-p)(1-\lambda p).\end{aligned}$$

Since $p = 1$ is always a solution, one finds in this case $\dot{q} = (\lambda - 1)q$, which agrees with the equation of motion for $\langle n\rangle(t)$, derived in exercise 2. Hence, for $p = 1$, one may identify $q(t) = \langle n\rangle(t)$.

In order to understand the topology of the phase space of the model, consider the zero-energy curves $H_R(p,q) = 0$. These are the three lines $q = 0$,

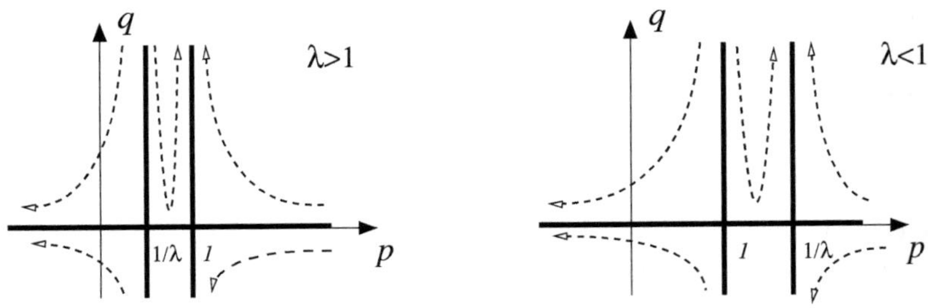

Fig. S.6 Topology of the phase space of the population model and the zero energy curves (thick solid lines) $H_R(p,q)=0$. The dashed lines indicate the evolution according to the variational equations.

$p=1$ and $p=1/\lambda$. In Fig. S.6, we show the resulting diagrams for $\lambda>1$ and for $\lambda<1$ and observe the change of stability at the critical point $\lambda_c=1$.

We leave it to the reader to analyse the contact process or the pair-contact process along the same lines [203].

53 The Volterra-Lotka equation has a fixed point at $(p^*,x^*)=(\beta^{-1},1)$ which is neither stable nor instable. A conserved quantity is

$$C=\ln(xp)-x-\beta p \tag{S.28}$$

From (5.119) is it easily checked that $\dot{C}=0$, but it is more instructive to construct C by considering

$$\frac{\dot{x}}{\dot{p}}=\frac{\mathrm{d}x}{\mathrm{d}p}=\frac{x(\beta p-1)}{p(1-x)}=\frac{x}{1-x}\frac{\beta p-1}{p}$$

Separating the variables and integrating gives (S.28). Hence the system moves along closed curves for a given C, determined from the initial conditions.

The model (5.120) has zero-clines ($\dot{p}=0$ and $\dot{x}=0$, respectively) on the lines $x=p$ and $x=\beta p$. These only coïncide for $\beta=1$ when $x=p$ becomes a fixed line; otherwise only the origin is a fixed point. Equations (5.120) can be decoupled by introducing the ratio $\rho:=x/p$. Then

$$\dot{\rho}=-2\rho+\rho^{\lambda}(\beta^{1-\lambda}+\rho)\ ,\ \dot{p}=p(1-\rho^{\lambda})$$

Tracing the vector field $(\rho,\dot{\rho})$, one easily sees that there is a first-order transition at $\beta_c=2^{1/\lambda(1-\lambda)}\lambda^{1/(\lambda-1)}(1-\lambda)^{1/\lambda}$ and $\rho_c=\left(2(1-\lambda)\right)^{1/\lambda}$ such that real fixed points for ρ only exist for $\beta\leq\beta_c$. At $\rho=\rho_c$ one has $\dot{p}=p(2\lambda-1)$ and the long-time evolution depends only on the sign of $\lambda-\frac{1}{2}$. For $\beta<\beta_c$, one has $p(t)\to 0$ for large times t if the stable fixed point $\rho^*>1$ and $p(t)\to\infty$ if $\rho^*<1$.

List of frequently used Symbols

$\mathrm{O}(x)$	of the order of x
$\simeq$	asymptotically equal
$\sim$	asymptotically proportional
$\mathcal{N}$	total number of sites
$\mathcal{H}$	Hamiltonian of an equilibrium model
$\mathcal{R}$	renormalisation-group transformation
p_c, T_c	critical percolation threshold/temperature
$N(t)$	number of active sites in seed simulations
$\langle \ldots \rangle$	average over many independent realisations
$A, \bullet$	symbol for a site occupied by a particle
$\emptyset, \circ$	symbol for a vacant site
ϱ	order parameter – density of active sites
$\widetilde{\varrho}$	response field associated to order parameter ϱ
τ	generic notation for control parameter
L	lateral system size
z	dynamical exponent $z = \nu_\parallel / \nu_\perp$
d_c	upper critical dimension
$P_t(c)$	probability of finding the system in configuration c
$p_{c \to c'}$	transition probabilities for parallel updates
$\lvert P_t \rangle$	state vector, listing all values of $P_t(c)$
$\mathcal{T}$	transfer matrix for parallel updates
$\langle s \rvert$	sum state over all configurations
$w_{c \to c'}$	transition rates for random sequential updates
$\mathcal{L}$	Liouville operator for random sequential updates
$P_{\mathrm{sur}}(t), P(t)$	survival probability
P_{perc}, P_∞	percolation probability (ultimate survival probability)
$c_\perp(r)$	equal-time correlation function $\langle s_i(t) s_{i+r}(t) \rangle$
$c_\parallel(\Delta t)$	autocorrelation function $\langle s_i(t) s_i(t+\Delta t) \rangle$
$c(t, r; \tau)$	two-point correlation function
$e_\parallel(\Delta t; \tau)$	temporal empty-interval distribution function
$e_\perp(\Delta r; \tau)$	spatial empty-interval distribution function
$d_{f,\parallel}$	temporal fractal dimension
$d_{f,\perp}$	spatial fractal dimension
d_f	spatio-temporal fractal dimension
pbc	subscript indicating periodic boundary conditions
full	subscript denoting a fully occupied initial state
$\Upsilon(t, r; \tau)$	pair connectedness function
$N_a(t)$	average number of active sites in seed simulations
$R_s(t)$	mean square spreading from the origin
Θ	critical initial slip exponent
$N_b(t)$	number of backbone sites
N_b^{red}	total number of red bonds

List of commonly used Abbreviations

BTW	Bak-Tang-Wiesenfeld
CDP	compact directed percolation
CLG	conserved lattice gas
CP	contact process
CTTP	conserved threshold transfer process
DK	Domany-Kinzel model
DMRG	density-matrix renormalisation group
DP	directed percolation
bDP	bond directed percolation
sDP	site directed percolation
DyP	dynamical percolation
FDT	fluctuation-dissipation theorem
FDR	fluctuation-dissipation ratio
FSS	finite-size scaling
GEP	general epidemic process
LCTMRG	light-cone transfer matrix renormalisation group
OJK	Ohta-Jasnow-Kawasaki
PC	parity-conserved model
PCP	pair-contact process
PCPD	pair-contact process with diffusion
PNG	polynuclear growth
QCD	quantum chromodynamics
RG	renormalisation group
TDGL	time-dependent Ginsburg-Landau
TDP	tricritical directed percolation
TMRG	transfer matrix renormalisation group
TTP	threshold transfer process
SOC	self-organised criticality
SOS	solid-on-solid model
ZGB	Ziff-Gulari-Barshad model
$2D$, $3D$	two-dimensional, three-dimensional
$\mathbb{Z}_p$	cyclic group of p elements
$\mathbb{S}_p$	permutation group of p elements

References

1. H.D.I. Abarbanel, J.B. Bronzan, R.L. Sugar, and A.R. White. Reggeon field theory: formulation and use. *Phys. Rep.*, 21:119, 1975.
2. M. Abramowitz and I.A. Stegun. *Handbook of Mathematical Functions.* Dover, New York, 1973.
3. J. Adamek, M. Keller, A. Senftleben, and H. Hinrichsen. Epidemic spreading with long-range infections and incubation times. *J. Stat. Mech.: Theor. Exp.*, :P09002, 2005.
4. I. Affleck. Universal term in the free energy at a critical point and the conformal anomaly. *Phys. Rev. Lett.*, 56:746, 1986.
5. A. Aghamohammadi and M. Khorrami. Models solvable through the empty-interval method. *Eur. Phys. J.*, B47:583, 2005.
6. A. Aharony. Dependence of universal critical behaviour on symmetry and range of interaction. In C. Domb and J.L. Lebowitz, editors, *Phase transitions and critical phenomena*, volume 6, London, 1976. Academic Press.
7. A. Aharony. The ϵ-expansion for exponents and the equation of state in isotropic systems. In C. Domb and M.S. Green, editors, *Phase transitions and critical phenomena*, volume 6, London, 1976. Academic Press.
8. G. Ahlers. Critical phenomena and the superfluid transition in ^{4}He. In J. Ruvalds and T. Regge, editors, *Quantum Liquids*, Amsterdam, 1978. North-Holland.
9. V. Ahlers and A. Pikovsky. Critical Properties of the Synchronization Transition in Space-Time Chaos. *Phys. Rev. Lett.*, 88:254101, 2002.
10. N. Aktekin. The finite-size scaling functions of the four-dimensional Ising model. *J. Stat. Phys.*, 104:1397, 2001.
11. E.V. Albano. Irreversible phase-transitions into non-unique absorbing states in a multicomponent reaction system. *Physica*, A214:426, 1995.
12. R. Albert, A.L. Barabási, N. Carle, and A. Dougherty. Driven interfaces in disordered media: determination of universality classes from experimental data. *Phys. Rev. Lett.*, 81:2926, 1998.
13. F.C. Alcaraz, M. Droz, M. Henkel, and V. Rittenberg. Reaction-diffusion processes, critical dynamics and quantum chains. *Ann. of Phys.*, 230:250, 1994.
14. F.C. Alcaraz, E. Levine, and V. Rittenberg. Conformal invariance and its breaking in a stochastic model of a fluctuating interface. *J. Stat. Mech.: Theor. Exp.*, :P08003, 2006.
15. S. Alexander and T. Holstein. Lattice diffusion and the Heisenberg ferromagnet. *Phys. Rev.*, B18:301, 1978.
16. U. Alon, M. Evans, H. Hinrichsen, and D. Mukamel. Smooth phases, roughening transitions, and novel exponents in one-dimensional growth models. *Phys. Rev.*, E57:4997, 1998.

17. U. Alon, M.R. Evans, H. Hinrichsen, and D. Mukamel. Roughening transition in a one-dimensional growth process. *Phys. Rev. Lett.*, 76:2746, 1996.
18. J. Als-Nielsen. Neutron scattering and spatial correlation near the critical point. In C. Domb and M.S. Green, editors, *Phase transitions and critical phenomena*, volume 5a, London, 1976. Academic Press.
19. J. Als-Nielsen and R.J. Birgeneau. Mean-field theory, the Ginzburg criterion, and marginal dimensionality of phase transitions. *Am. J. Phys.*, 45:554, 1977.
20. P. Alstrøm. Mean-field exponents for self-organized critical phenomena. *Phys. Rev.*, A38:4905, 1988.
21. M.A. Anisimov, S.B. Kiselev, J.V. Sengers, and S. Tang. Crossover approach to global critical phenomena in fluids. *Physica*, A188:487, 1992.
22. T. Antal, M. Droz, and Z. Rácz. Probability distribution of magnetization in the one-dimensional Ising model: effects of boundary conditions. *J. Phys. A Math. Gen.*, 37:1465, 2004.
23. T. Aouaroun and G. Ahlers. Specific heat of ^{4}He confined in cylindrical microchannels and near the superfluid transition. *J. Low Temp. Phys.*, 149:209, 2007.
24. P.F. Arndt. Yang–Lee theory for a nonequilibrium phase transition. *Phys. Rev. Lett.*, 84:814, 2000.
25. P.F. Arndt, S.R. Dahmen, and H. Hinrichsen. Directed percolation, fractal roots, and the Lee-Yang theorem. *Physica*, A295:128, 2001.
26. A.P.F. Atman, R. Dickman, and J.G. Moreira. Phase-diagram of a probabilistic cellular automaton with three-site interactions. *Phys. Rev.*, E67:016107, 2003.
27. F. Bagnoli, N. Boccara, and R. Rechtman. Nature of phase-transitions in a probabilistic cellular automaton with two absorbing states. *Phys. Rev.*, E63:046116, 2001.
28. P. Bak. *How Nature Works*. Springer, Heidelberg, 1996.
29. P. Bak, C. Tang, and K. Wiesenfeld. Self-organized criticality: an explanation of $1/f$ noise. *Phys. Rev. Lett.*, 59:381, 1987.
30. P. Bak, C. Tang, and K. Wiesenfeld. Self-organized criticality. *Phys. Rev.*, A38:364, 1988.
31. D. Balboni, P.-A. Rey, and M. Droz. Universality of a class of annihilation-coagulation models. *Phys. Rev.*, E52:6220, 1995.
32. H. G. Ballesteros, L. A. Fernandez, V. Martín-Mayor, and A. Muñoz-Sudupe. Finite-size scaling and 'perfect' actions: the three-dimensional ising model. *Phys. Lett.*, B441:330, 1998.
33. H. G. Ballesteros, L. A. Fernndez, V. Martín-Mayor, A. MuñozSudupe, G. Parisi, and J. J. Ruiz-Lorenzo. Measures of critical exponents in the four-dimensional site percolation. *Phys. Lett.*, B400:346, 1997.
34. D. Balzarini and K. Ohrn. Coexistence curve of sulfur hexafluoride. *Phys. Rev. Lett.*, 28:840, 1972.
35. J. Banks, V. Dragan, and A. Jones. *Chaos: a mathematical introduction.* Cambridge University Press, Cambridge, 2003.
36. A. C. Barato and H. Hinrichsen. Boundary-induced nonequilibrium phase transition into an absorbing state. *to appear in Phys. Rev. Lett.*, 2008.
37. M.N. Barber. Finite-size scaling. In C. Domb and J. L. Lebowitz, editors, *Phase transitions and critical phenomena*, volume 8, page 146, London, 1983. Academic Press.
38. M.N. Barber and C.J. Hamer. Extrapolation of sequences using a generalised epsilon-algorithm. *J. Austral. Math. Soc.*, B23:229, 1982.
39. V.D. Barger and R.J.N. Philips. *Collider Physics.* Addison-Wesley, Reading, 1987.
40. G.T. Barkema and E. Carlon. Universality in the pair-contact process with diffusion. *Phys. Rev.*, E68:036113, 2003.
41. M. Barmatz, I. Hahn, J.A. Lipa, and R.V. Duncan. Critical phenomena in microgravity: Past, present, and future. *Rev. Mod. Phys.*, 79:1, 2007.
42. L. Baroni, R. Livi, and A. Torcini. Transition to stochastic synchronization in spatially extended systems. *Phys.. Rev.*, E63:036226, 2001.

43. E. Barouch, B.M. McCoy, and T.T. Wu. Zero-field susceptibility of the two-dimensional Ising model near T_c. *Phys. Rev. Lett.*, 31:1409, 1973.
44. A. Barrat. Monte Carlo simulations of the violation of the fluctuation-dissipation theorem in domain-growth processes. *Phys. Rev.*, E57:3629, 1998.
45. F. Baumann and A. Gambassi. Ageing in the contact process: scaling behaviour and universal features. *J. Stat. Mech.: Theor. Exp.*, :P01002, 2007.
46. F. Baumann, M. Henkel, M. Pleimling, and J. Richert. Ageing without detailed balance in the bosonic contact and pair-contact processes: exact results. *J. Phys. A: Math. Gen.*, 38:6623, 2005.
47. F. Baumann and M. Pleimling. Local ageing phenomena close to magnetic surfaces. *Phys. Rev.*, B76:104422, 2007.
48. F. Baumann, S. Stoimenov, and M. Henkel. Local scale-invariances in the bosonic contact and pair-contact processes. *J. Phys. A: Math. Gen.*, 39:4905, 2006.
49. R. Bausch, H.-K. Janssen, and H. Wagner. Renormalized field theory of critical dynamics. *Z. Phys.*, B24:113, 1976.
50. R.J. Baxter. *Exactly solved models in statistical mechanics.* Academic Press, London, 1982.
51. A.A. Belavin, A.M. Polyakov, and A.B. Zamolodchikov. Infinite conformal symmetry in two-dimensional quantum field-theory. *Nucl. Phys.*, B241:333, 1984.
52. M.Y. Belyakov and S.B. Kiselev. Crossover behaviour of the susceptibility of the susceptibility and the specific heat near a second-order phase-transition. *Physica*, A190:75, 1992.
53. D. ben Avraham, M.A. Burschka, and C.R. Doering. Statics and dynamics of a diffusion-limited reaction: anomalous kinetics, non-equilibrium self-ordering and a dynamic transition. *J. Stat. Phys.*, 60:695, 1990.
54. D. ben Avraham and S. Havlin. *Diffusions and Reactions in Fractals and Disordered Systems.* Cambridge University Press, Cambridge, 2000.
55. D. ben Avraham and J. Köhler. Mean-field (n, m)-cluster approximation for lattice models. *Phys. Rev.*, A45:8358, 1992.
56. A. Ben-Hur and O. Biham. Universality in sandpile models. *Phys. Rev.*, E53:R1317, 1996.
57. C.H. Bennett and G. Grinstein. Role of irreversibility in stabilizing complex and nonergodic behaviour in locally interacting discrete systems. *Phys. Rev. Lett.*, 55:657, 1985.
58. B. Berche. Bulk and surface properties in the critical phase of the two-dimensional XY model. *J. Phys. A Math. Gen.*, 36:585, 2003.
59. B.A. Berg. *Markov chain Monte Carlo simulations and their statistical analysis.* World Scientific, Singapour, 2004.
60. A.N. Berker and S. Ostlund. Renormalisation-group calculations of finite systems: order parameter and specific heat for epitaxial ordering. *J. Phys. C: Solid State Phys.*, 12:4961, 1979.
61. T.H. Berlin and M. Kac. The spherical model of a ferromagnet. *Phys. Rev.*, 86:821, 1952.
62. M.J. Berridge. Inositol trisphosphate and calcium signaling. *Nature*, 361:315, 1993.
63. D.D. Betts, A.J. Guttmann, and G.S. Joyce. Lattice-lattice scaling and the generalized law of corresponding states. *J. Phys. C: Solid State Phys.*, 4:1994, 1971.
64. A. Bialas and R. Peschanski. Moments of rapidity distributions as a measure of short-range fluctuations in high-energy collisions. *Nucl. Phys.*, B273:703, 1986.
65. A. Bialas and R. Peschanski. Intermittency in multiparticle production at high energy. *Nucl. Phys.*, B308:857, 1988.
66. O. Biham, E. Milshtein, and O. Maclai. Evidence for universality within the classes of deterministic and stochastic sandpile models. *Phys. Rev.*, E63:061309, 2001.
67. K. Binder. Finite-size scaling analysis of Ising model block distribution functions. *Z. Phys.*, B43:119, 1981.

68. K. Binder. Critical behaviour at surfaces. In C. Domb and J. L. Lebowitz, editors, *Phase transitions and critical phenomena*, volume 8, London, 1983. Academic Press.
69. K. Binder, M. Nauenberg, V. Privman, and A. P. Young. Finite-size tests of hyperscaling. *Phys. Rev.*, B31:1498, 1985.
70. H.W.J. Blöte, J.L. Cardy, and M.P. Nightingale. Conformal invariance, the central charge, and universal finite-size amplitudes at criticality. *Phys. Rev. Lett.*, 56:742, 1986.
71. H.W.J. Blöte and H.J. Hilhorst. Nonuniversality, exponent asymmetry and surface magnetisation in an inhomogeneous square Ising lattice. *Phys. Rev. Lett.*, 51:2015, 1983.
72. R.A. Blythe. An introduction to phase transitions in stochastic dynamical systems. *J. Phys. Conf. Series*, 40:1, 2006.
73. E. Bonabeau and P. Lederer. On $1/f$ power spectra. *J. Phys. A: Math. Gen.*, 27:L243, 1994.
74. J. Bottéro. *Mesopotamia: Writing, Reasoning and the Gods.* The University of Chicago Press, London, 1992.
75. J.P. Bouchaud and M.E. Cates. Triangular and uphill avalanches of a tilted sandpile. *Granular Matter*, 1:101, 1998.
76. M. Bousquet-Mélou. New enumerative results on two-dimensional directed animals. *Discrete Math.*, 180:73, 1998.
77. M. Bär, M. Falcke, H. Levine, and L. Tsimring. Discrete stochastic modeling of calcium channel dynamics. *Phys. Rev. Lett.*, 84:5664, 2000.
78. M. Bramson and J.L. Lebowitz. Asymptotic behaviour of densities in diffusion-dominated annihilation reactions. *Phys. Rev. Lett.*, 61:2397, 1988.
79. M. Bramson and J.L. Lebowitz. Asymptotic behaviour of densities in diffusion-dominated annihilation reactions (erratum). *Phys. Rev. Lett.*, 62:694, 1989.
80. J.G. Brankov, N.S. Tonchev, and D.M. Danchev. *Theory of critical phenomena in finite-size systems.* World Scientific, Singapour, 2000.
81. A. J. Bray, B. Derrida, and C. Godrèche. Nontrivial algebraic decay in a soluble model of coarsening. *Europhys. Lett.*, 27:175, 1994.
82. E. Brézin. An investigation of finite size scaling. *J. Physique*, 43:15, 1982.
83. E. Brézin, Guillou J.C. Le and J. Zinn-Justin. Field theoretical approach to critical phenomena. In C. Domb and M.S. Green, editors, *Phase transitions and critical phenomena*, volume 6, London, 1976. Academic Press.
84. E. Brézin, D.J. Wallace, and K.G. Wilson. Feynman-graph expansion for the equation of state near the critical point. *Phys. Rev.*, B7:232, 1973.
85. C. Brezuidenhout and G. Grimmett. The critical contact process dies out. *Ann. Prob.*, 18:1462, 1990.
86. N.V. Brilliantov. Effective magnetic Hamiltonian and Ginzburg criterion for fluids. *Phys. Rev.*, E58:2628, 1998.
87. S.R. Broadbent and J.M. Hammersley. Percolation processes: I. crystals and mazes. *Proc. Camb. Phil. Soc.*, 53:629, 1957.
88. J.B. Bronzan and J.W. Dash. Erratum: higher-order epsilon terms in Reggeon field-theory. *Phys. Rev.*, D12:1850, 1974.
89. J.B. Bronzan and J.W. Dash. Higher-order epsilon terms in Reggeon field-theory. *Phys. Rev.*, D10:4208, 1974.
90. J.B. Bronzan and J.W. Dash. Higher-order ϵ-terms in the renormalization group approach to Reggeon field-theory. *Phys. Rev.*, D12:1850, 1974.
91. B.J. Brosilow and R.M. Ziff. Effects of A-desorption on the first-order transition in the $A - B2$ reaction model. *Phys. Rev.*, A46:4534, 1992.
92. E. Brunet and D. ben Avraham. On the relation between one-species diffusion-limited coalescence and annihilation in one dimension. *J. Phys. A Math. Gen.*, 38:3247, 2005.
93. P. Brunet and L. Limat. Defects and spatiotemporal disorder in a pattern of falling liquid columns. *Phys. Rev.*, E70:046207, 2004.

94. S.V. Buldyrev, A.L. Barabási, F. Caserta, S. Havlin, H.E. Stanley, and T. Vicsek. Anomalous interface roughening in porous media: Experiment and model. *Phys. Rev.*, A45:R8313, 1992.
95. S.V. Buldyrev, A.L. Barabási, S. Havlin, J. Kertész, H.E. Stanley, and H.S. Xenias. Anomalous interface roughening in porous media: experiment and model. *Physica*, A191:220, 1992.
96. R. Bulirsch and J. Stoer. Fehlerabschätzungen und Extrapolation mit rationalen Funktionen vom Richardsontypus. *Numer. Math.*, 6:413, 1964.
97. T.W. Burkhardt and I. Guim. Universal scaling form of the correlation length in Ising strips with periodic, free, fixed, and mixed boundary conditions. *Phys. Rev.*, B35:1799, 1987.
98. R. Cafiero, A. Gabrielli, and M.A. Muñoz. Disordered one-dimensional contact process. *Phys. Rev.*, E57:5060, 1998.
99. G. Caldarelli. Mean-field theory for ordinary and hot sandpiles. *Physica*, A252:295, 1998.
100. C.G. Callan, S. Coleman, and R. Jackiw. A new improved energy-momentum tensor. *Ann. of Phys.*, 59:42, 1970.
101. M. Campostrini, M. Hasenbusch, A. Pelissetto, and E. Vicari. Theoretical estimates of the critical exponents of the superfluid transition in 4He by lattice methods. *Phys. Rev.*, B74:144506, 2006.
102. L. Canet. Reaction-diffusion processes and non-perturbative renormalisation group. *J. Phys. A Math. Gen.*, 39:7901, 2006.
103. L. Canet and H. Chaté. Non-perturbative Approach to Critical Dynamics. `cond-mat/0610468`, 2006.
104. L. Canet, H. Chaté, and B. Delamotte. Quantitative phase diagrams of branching and annihilating random walks. *Phys. Rev. Lett.*, 92:255703, 2004.
105. L. Canet, H. Chaté, B. Delamotte, I. Dornic, and M. A. Muñoz. Nonperturbative fixed point in a nonequilibrium phase transition. *Phys. Rev. Lett.*, 95:1006001, 2005.
106. L. Canet, H. Chaté, B. Delamotte, I. Dornic, and M. A. Muñoz. What can be learnt from the nonperturbative renormalization group ? *Condensed Matter Phys.*, 8:163, 2005.
107. L. Canet, B. Delamotte, O. Deloubrière, and N. Wschebor. Non-perturbative renormalization-group study of reaction-diffusion processes and directed percolation. *Phys. Rev. Lett.*, 92:105703, 2004.
108. L. Canet and H.J. Hilhorst. Site mean-field theory for reaction-diffusion processes. *J. Stat. Phys.*, 125:513, 2006.
109. M. Caponeri and S. Ciliberto. Thermodynamic aspects of the transition to spatiotemporal chaos. *Physica*, D58:365, 1992.
110. S. Caracciolo, M.S. Causo, A. Pelissetto, P. Rossi, and E. Vicari. Crossover phenomena in spin models with medium-range interactions and self-avoiding walks with medium-range jumps. *Phys. Rev.*, E64:046130, 2001.
111. J. L. Cardy. Field theoretic formulation of an epidemic process with immunisation. *J. Phys. A: Math. Gen.*, 16:L709, 1983.
112. J. L. Cardy. SLE for theoretical physicists. *Ann. of Phys.*, 318:81, 2005.
113. J. L. Cardy and P. Grassberger. Epidemic models and percolation. *J. Phys. A: Math. Gen.*, 18:L267, 1985.
114. J. L. Cardy and U.C. Täuber. Field-theory of branching and annihilating random walks. *J. Stat. Phys.*, 90:1, 1998.
115. J.L. Cardy. Critical behaviour at an edge. *J. Phys. A Math. Gen.*, 16:3617, 1983.
116. J.L. Cardy. Conformal invariance and surface critical behaviour. *Nucl. Phys.*, B240 [FS 12]:514, 1984.
117. J.L. Cardy. Conformal invariance and universality in finite-size scaling. *J. Phys. A Math. Gen.*, 17:L385, 1984.
118. J.L. Cardy. Effect of boundary conditions on the operator content of two-dimensional conformally invariant theories. *Nucl. Phys.*, B275:200, 1986.

119. J.L. Cardy. Central charge and universal combinations of amplitudes in two-dimensional theories away from criticality. *Phys. Rev. Lett.*, 60:2709, 1988.
120. J.L. Cardy. Conformal invariance and statistical mechanics. In É. Brézin and J. Zinn-Justin, editors, *Fields, strings and critical phenomena, Les Houches XLIX*, Amsterdam, 1990. North Holland.
121. J.L. Cardy. Proportion of unaffected sites in a reaction-diffusion process. *J. Phys. A: Math. Gen.*, 28:L19, 1995.
122. J.L. Cardy. *Scaling and Renormalization in Statistical Physics.* Cambridge University Press, Cambridge, 1996.
123. J.L. Cardy and R.L. Sugar. Directed percolation and Reggeon field-theory. *J. Phys. A Math. Gen.*, 13:L423, 1980.
124. J.L. Cardy and U.C. Täuber. Theory of branching and annihilating random walks. *Phys. Rev. Lett.*, 77:4780, 1996.
125. E. Carlon, M. Henkel, and U. Schollwöck. Density-matrix renormalization group and reaction–diffusion processes. *Eur. Phys. J.*, B12:99, 1999.
126. E. Carlon, M. Henkel, and U. Schollwöck. Critical properties of the reaction-diffusion model $2A \to 3A$, $2A \to 0$. *Phys. Rev.*, E63:036101, 2001.
127. P.M. Chaikin and T.C. Lubensky. *Principles of Condensed Matter Physics.* Cambridge University Press, Cambridge, 1995.
128. H. Chaté, 2003 and 2008. `private communication`.
129. C. Chatelain. A far-from-equilibrium fluctuation-dissipation relation for an Ising–Glauber-like model. *J. Phys. A Math. Gen.*, 36:10739, 2003.
130. X.S. Chen and V. Dohm. Nonuniversal finite-size scaling in anisotropic systems. *Phys. Rev.*, E70:056136, 2004.
131. A. Chessa, E. Marinari, A. Vespignani, and S. Zapperi. Mean-field behaviour of the sandpile model below the upper critical dimension. *Phys. Rev.*, E57:6241, 1998.
132. A. Chessa, H.E. Stanley, A. Vespignani, and S. Zapperi. Universality in sandpiles. *Phys. Rev.*, E59:12, 1999.
133. A. Chessa, A. Vespignani, and S. Zapperi. Critical exponents in stochastic sandpile models. *Comp. Phys. Comm.*, 121:299, 1999.
134. M. Collins. *Magnetic critical scattering.* Oxford University Press, Oxford, 1989.
135. J. T. Cox and D. Griffeath. Diffusive clustering in the two-dimensional voter model. *Annals of Probability*, 14:347, 1986.
136. J.K. Cullum and R.A. Willoughby. *Lanczos algorithms for large symmetric eigenvlaue computations: I. Theory.* SIAM, Philadelphia, 2002.
137. J. Kamphorst Leal da Silva and R. Dickman. The pair-contact process in two dimensions. *Phys. Rev.*, E60:5126, 1999.
138. R. da Silva, N.A. Alves, and J.R. Drugowich de Felício. Universality and scaling of the critical behaviour of the two-dimensional Blume-Capel model in short-time dynamics. *Phys. Rev.*, E66:026130, 2002.
139. A. Daerr and S. Douady. Two types of avalanche behaviour in granular media. *Nature*, 399:241, 1999.
140. S. R. Dahmen, L. Sittler, and H. Hinrichsen. Multicritical behavior of the diluted contact process. *J. Stat. Mech.: Theor. Exp.*, :P01011, 2007.
141. S. M. Dammer and H. Hinrichsen. Spreading with immunization in high dimensions. *J. Stat. Mech*, page P07011, 2004.
142. S.M. Dammer, S.R. Dahmen, and H. Hinrichsen. Yang-Lee zeros for a nonequilibrium phase transition. *J. Phys. A Math. Gen.*, 35:4527, 2002.
143. P. K. Das and P. Sen. Probability distribution of persistent spins in an Ising chain. *J. Phys. A: Math. Gen.*, 37:7179, 2004.
144. S. Dattagupta. *Relaxation phenomena in condensed-matter physics.* Academic Press, London, 1987.
145. F. Daviaud, M. Bonetti, and M. Dubois. Transition to turbulence via spatiotemporal intermittency in one-dimensional Rayleigh–Bénard convection. *Phys. Rev.*, A42:3388, 1992.

146. F. Daviaud, J. Lega, P. Bergé, P. Coullet, and M. Dubois. Spatio-temporal intermittency in a $1D$ convective pattern: Theoretical model and experiments. *Physica*, D55:287, 1992.
147. J. de Gier and F.H.L. Essler. Exact spectral gap of the asymmetric exclusion process with open boundaries. *J. Stat. Mech.*, :P12011, 2006.
148. L. de Jongh and A.R. Midema. Experiments on simple magnetic model systems. *Adv. Phys.*, 23:1, 1974.
149. F. de los Santos, M. M. T. da Gama, and M. A. Muñoz. Stochastic theory of non-equilibrium wetting. *Europhys. Lett.*, 57:803, 2002.
150. J.R.G. de Mendonça. Precise critical exponents for the basic contact process. *J. Phys. A Math. Gen.*, 32:L467, 1999.
151. M.M. de Oliveira and R. Dickman. Moment ratios for the pair-contact process with diffusion. *Phys. Rev.*, E74:011124, 2006.
152. J.-M. Debierre and L. Turban. Two-dimensional Ising model with multispin interactions. *J. Phys. A Math. Gen.*, 16:3571, 1982.
153. M.M. Degen, I. Mutzabazi, and C.D. Andereck. Transition to weak turbulence via spatiotemporal intermittency in the Taylor–Dean system. *Phys. Rev.*, E53:3495, 1996.
154. G. Delfino. Universal amplitude ratios in the two-dimensional Ising model. *Phys. Lett.*, B419:291, 1998.
155. E. Demler, W. Hanke, and S.-C. Zhang. $SO(5)$-theory of antiferromagnetism and superconductivity. *Rev. Mod. Phys.*, 76:909, 2004.
156. M.P.M. den Nijs. A relation between the temperature exponents of the eight-vertex and q-state Potts model. *J. Phys. A Math. Gen.*, 12:1857, 1979.
157. B. Derrida. How to extract information from simulations of coarsening at finite temperature. *Phys. Rev.*, E55:3705, 1997.
158. B. Derrida, A. J. Bray, and C. Godrèche. Nontrivial exponents in the zero-temperature dynamics of the 1d Ising and Potts models. *J. Phys. A: Math. Gen.*, 27:L357, 1994.
159. B. Derrida, L. de Seeze, and C. Itzykson. Fractal structure of zeros in hierarchical models. *J. Stat. Phys.*, 33:559, 1983.
160. B. Derrida, V. Hakim, and V. Pasquier. Exact first-passage exponents of 1d domain growth: Relation to a reaction-diffusion model. *Phys. Rev. Lett.*, 75:751, 1995.
161. B. Derrida, V. Hakim, and V. Pasquier. Exact exponent for the number of persistent spins in the zero-temperature dynamics of the one-dimensional potts model. *J. Stat. Phys.*, 85:763, 1996.
162. B. Derrida, V. Hakim, and R. Zeitak. Persistent spins in the linear diffusion approximation of phase ordering and zeros of stationary Gaussian processes. *Phys. Rev. Lett.*, 77:2871, 1996.
163. D. Dhar. The collapse of directed animals. *J. Phys. A Math. Gen.*, 30:L847, 1987.
164. D. Dhar. Self-organized critical state of sandpile automaton models. *Phys. Rev. Lett.*, 64:1613, 1990.
165. D. Dhar. Some results and a conjecture for Manna's stochastic sandpile model. *Physica*, A270:69, 1999.
166. D. Dhar. Theoretical studies of self-organized criticality. *Physica*, A369:29, 2006.
167. D. Dhar and S.N. Majumdar. Abelian sandpile model on the Bethe lattice. *J. Phys. A Math. Gen.*, 23:4333, 1990.
168. P. di Francesco, P. Mathieu, and D. Sénéchal. *Conformal field-theory.* Springer, Heidelberg, 1997.
169. R. Dickman. Mean-field theory of the driven diffusive lattice gas. *Phys. Rev.*, A38:2588, 1988.
170. R. Dickman. Nonuniversality and critical-point shift in systems with infinitely many absorbing configurations. *cond-mat/9909347*, 1999.
171. R. Dickman. Reweighting in non-equilibrium simulations. *Phys. Rev.*, E60:R2441, 1999.

172. R. Dickman. Critical exponents for the restricted sandpile. *Phys. Rev.*, E73:036131, 2006.
173. R. Dickman. Absorbing-state phase transitions: exact solution of small systems. *Phys. Rev.*, E77:030102(R), 2008.
174. R. Dickman, M. Alava, M.A. Muñoz, J. Peltola, A. Vespignani, and S. Zapperi. Critical behaviour of a one-dimensional fixed-energy stochastic sandpile. *Phys. Rev.*, E64:056104, 2001.
175. R. Dickman and J. Kamphorst Leal da Silva. Moment ratios for absorbing-state phase-transitions. *Phys. Rev.*, E58:4266, 1998.
176. R. Dickman and M.A.F. de Menezes. Nonuniversality in the pair-contact process with diffusion. *Phys. Rev.*, E66:045101, 2002.
177. R. Dickman, W. Rabelo, and G. Ódor. Pair-contact process with a particle source. *Phys. Rev.*, E65:016118, 2001.
178. R. Dickman and T. Tomé. First-order phase-transition in a one-dimensional nonequilibrium model. *Phys. Rev.*, A44:4833, 1991.
179. R. Dickman, T. Tomé, and M.J. de Oliveira. Sandpiles with height restrictions. *Phys. Rev.*, E66:016111, 2002.
180. R. Dickman and A.Y. Tretyakov. Hyperscaling in the Domany–Kinzel cellular automaton. *Phys. Rev.*, E52:3218, 1995.
181. R. Dickman, A. Vespignani, and S. Zapperi. Self-organized criticality as an absorbing-state phase-transition. *Phys. Rev.*, E57:5095, 1998.
182. R. Dickman and R.R. Vidigal. Path-integral representation for a stochastic sandpile. *J. Phys. A Math. Gen.*, 35:7269, 2002.
183. H.W. Diehl. Field-theoretical approach to critical behaviour at surfaces. In C. Domb and J. L. Lebowitz, editors, *Phase transitions and critical phenomena*, volume 10, page 76, London, 1987. Academic Press.
184. H.W. Diehl and M. Shpot. Massive field-theory approach to surface critical behaviour in three-dimensional systems. *Nucl. Phys.*, B528:595, 1998.
185. V. Dohm. Anisotropy and restricted universality of critical phenomena. *J. Phys. A Math. Gen.*, 39:L259, 2006.
186. M. Doi. Second quantization representation for classical many-particle system. *J. Phys. A: Math. Gen.*, 9:1465, 1976.
187. M. Doi. Stochastic theory of diffusion-controlled reaction. *J. Phys. A: Math. Gen.*, 9:1479, 1976.
188. E. Domany and W. Kinzel. Equivalence of cellular automata to Ising models and directed percolation. *Phys. Rev. Lett.*, 53:311, 1984.
189. C. Domb and D.L. Hunter. On critical behavior of ferromagnets. *Proc. Phys. Soc. Lond.*, 86:1147, 1965.
190. C.J. De Dominicis. Techniques de renormalisation de la théorie des champs et dynamique des phénomènes critiques. *J. Physique (Colloque)*, 37:247, 1976.
191. I. Dornic, H. Chaté, J. Chave, and H. Hinrichsen. Critical coarsening without surface tension: the voter universality class. *Phys. Rev. Lett.*, 87:5701, 2001.
192. I. Dornic, H. Chaté, and M. A. Muñoz. Integration of Langevin Equations with Multiplicative Noise and the Viability of Field Theories for Absorbing Phase Transitions. *Phys. Rev. Lett.*, 94:100601, 2005.
193. I. Dornic and C. Godrèche. Large deviations and nontrivial exponents in coarsening systems. *J. Phys. A Math. Gen.*, 31:5413, 1998.
194. S. Douady and A. Daerr. Avalanches and inclined chutes. In H.J. Herrmann et al., editor, *Physics of Dry Granular Media*, page 339, New York, 1998. Kluwer Academic Publishers.
195. A. Dougherty and N. Carle. Distribution of avalanches in interfacial motion in a porous medium. *Phys. Rev.*, E58:2889, 1998.
196. B. Drossel, S. Clar, and F. Schwabl. Crossover from percolation to self-organized criticality. *Phys. Rev.*, E50:2399, 1994.

197. M. Droz, A. L. Ferreira, and A. Lipowski. Splitting the voter Potts model critical point. *Phys. Rev.*, E67:056108, 2003.
198. R. Durrett. On the growth of the one-dimensional contact processes. *Ann. Prob.*, 8:890, 1980.
199. R. Durrett and D. Griffeath. Contact processes in several dimensions. *Z. für Wahrsch.*, 59:535, 1982.
200. S. Dutta and S. K. Roy. Persistence exponents and scaling in two-dimensional XY model and a nematic model. *J. Phys. A: Math. Gen.*, 38:5859, 2005.
201. V. Privman (ed). *Finite-size scaling and numerical simulation of statistical systems.* World Scientific, Singapour, 1990.
202. M. Ehsasi, M. Matloch, O. Frank, J.H. Block, K. Christmann, F.S. Rys, and W. Hirschwald. Steady and Nonsteady Rates of Reaction in a Heterogeneously Catalyzed Reaction - Oxidation of CO on Platinum, Experiments and Simulations. *J. Chem. Phys.*, 91:4949, 1989.
203. V. Elgart and A. Kamenev. Classification of phase transitions in reaction-diffusion models. *Phys. Rev.*, E74:041101, 2006.
204. T. Enss, M. Henkel, A. Picone, and U. Schollwöck. Ageing phenomena without detailed balance: the contact process. *J. Phys. A Math. Gen.*, 37:10479, 2004.
205. T. Enss and U. Schollwöck. On the choice of the density-matrix in the stochastic TMRG. *J. Phys. A Math. Gen.*, 34:7769, 2001.
206. R. Erdem. Magnetic relaxation in a spin-1 Ising model near the second-order phase transition point. *J. Mag. Magn. Mat.*, `at press`, 2008.
207. R. Erdem and M. Keskin. Dynamics of a spin-1 Ising system in the neighbourhood of equilibrium states. *Phys. Rev.*, E64:026102, 2001.
208. J.W. Essam. Directed compact percolation: cluster size and hyperscaling. *J. Phys. A Math. Gen.*, 22:4927, 1989.
209. J.W. Essam, K. de'Bell, J. Adler, and F.M. Bhatti. Analysis of extended series for bond percolation on the directed square lattice. *Phys. Rev.*, B33:1982, 1986.
210. J.W. Essam and M.E. Fisher. Padé approximant studies of lattice gas and Ising ferromagnet below critical point. *J. Chem. Phys.*, 38:802, 1963.
211. J.W. Essam, A.J. Guttmann, and K. de'Bell. On two-dimensional directed percolation. *J. Phys. A Math. Gen.*, 21:3815, 1988.
212. J.W. Essam, A.J. Guttmann, I. Jensen, and D. TanlaKishani. Directed percolation near a wall. *J. Phys. A Math. Gen.*, 29:1619, 1996.
213. M.R. Evans, Y. Kafri, H.M. Koduvely, and D. Mukamel. Phase-separation and coarsening in one-dimensional driven diffusive systems: Local dynamics leading to long-range Hamiltonians. *Phys. Rev.*, E58:2764, 1998.
214. K. Falconer. *Fractal geometry: mathematical foundations and applications.* Wiley, New York, 1990.
215. H.A. Fernandes, J.R. Drugowich de Felício, and A.A. Caparica. Short-time behaviour of a classical ferromagnet with double-exchange interaction. *Phys. Rev.*, B72:054434, 2005.
216. A. M. Ferrenberg and R. H. Swendsen. Optimized Monte Carlo data analysis. *Phys. Rev. Lett.*, 63:1195, 1989.
217. M.E. Fisher. Quantum corrections to critical-point behavior. *Phys. Rev. Lett.*, 16:11, 1966.
218. M.E. Fisher. Renormalization group in critical phenomena and quantum field theory. Philadelphia, 1974. Temple University.
219. M.E. Fisher. The renormalization group in the theory of critical behaviour. *Rev. Mod. Phys*, 46:597, 1974.
220. M.E. Fisher and M.N. Barber. Scaling theory for finite-size effects in the critical region. *Phys. Rev. Lett.*, 28:1516, 1972.
221. H.C. Fogedby. Langevin equation for continuous time Lévy flights. *Phys. Rev.*, E50:1657, 1994.

222. G. Forgacs. Ground-state correlations and universality in two-dimensional fully frustrated systems. *Phys. Rev.*, B22:4473, 1980.
223. C.M. Fortuin. On the random cluster model. II. The percolation model. *Physica*, 58:393, 1972.
224. C.M. Fortuin and P.W. Kasteleyn. On the random cluster model. I. Introduction and relation to other models. *Physica*, 57:536, 1972.
225. D.P. Foster and C. Gérard. Critical behaviour of fully frustrated q-state Potts pile-up-domino model. *Phys. Rev.*, B70:014411, 2004.
226. D.P. Foster, C. Gérard, and I. Puha. Critical behaviour of fully frustrated Potts models. *J. Phys. A Math. Gen.*, 34:5183, 2001.
227. L. Frachebourg and P. L. Krapivsky. Exact results for kinetics of catalytic reactions. *Phys. Rev.*, E53:R3009, 1996.
228. P. Fröjdh, M. Howard, and K.B. Lauritsen. Directed percolation with a wall or edge. *J. Phys. A Math. Gen.*, 31:2311, 1998.
229. P. Fröjdh, M. Howard, and K.B. Lauritsen. Directed percolation and other systems with absorbing states: Impact of boundaries. *Int. J. Mod. Phys.*, B15:1761, 2001.
230. J. Fuchs, J. Schelter, F. Ginelli, and H. Hinrichsen. Local Persistence in the Directed Percolation Universality Class. *to be published in J. Stat. Mech.: Theor. Exp.*, 2008.
231. R. Garćia-Pelayo, I. Salazar, and W. C. Schieve. A branching process model for sandpile avalanches. *J. Stat. Phys.*, 72:167, 1993.
232. W. Genovese and M. A. Muñoz. Recent results on multiplicative noise. *Phys. Rev.*, E60:69, 1999.
233. F. Ginelli, V. Ahlers, R. Livi, D. Mukamel, A. Pikovsky, A. Politi, and A. Torcini. From multiplicative noise to directed percolation in wetting transitions. *Phys. Rev.*, E68:065102, 2003.
234. F. Ginelli, R. Livi, A. Politi, and A. Torcini. Relationship between directed percolation and the synchronization transition in spatially extended systems. *Phys. Rev.*, E67:046217, 2003.
235. V.L. Ginzburg. Some remarks on the phase-transitions of the second kind and the microscopic theory of ferroelectric materials. *Sov. Phys. Solid State*, 2:1824, 1960.
236. C. Godrèche, F. Krzakala, and F. Ricci-Tersenghi. Non-equilibrium critical dynamics of the ferromagnetic Ising model with Kawasaki dynamics. *J. Stat. Mech.: Theor. Exp.*, :P04007, 2004.
237. C. Godrèche, J.-M. Luck, M.R. Evans, D. Mukamel, S. Sandow, and E.R. Speer. Spontaneous symmetry-breaking: Exact results for a biased random walk model of an exclusion process. *J. Phys. A Math. Gen.*, 28:6039, 1995.
238. R.L. Graham, D.E. Knuth, and O. Patashnik. *Concrete Mathematics*. Addison-Wesley, New York, 1994.
239. P. Grassberger. On the Hausdorff dimension of fractal attractors. *J. Stat. Phys.*, 26:173, 1981.
240. P. Grassberger. On phase-transitions in Schlögl's second model. *Z. Phys.*, B47:365, 1982.
241. P. Grassberger. Directed percolation in $2 + 1$ dimensions. *J. Phys. A Math. Gen.*, 22:3673, 1989.
242. P. Grassberger. Are damage spreading transitions generically in the universality class of directed percolation? *J. Stat. Phys.*, 79:13, 1995.
243. P. Grassberger. Directed percolation: results and open problems. In S. Pury *et al.*, editor, *Nonlinearities in complex systems*, New Dehli, India, 1997. Narosa Publishing.
244. P. Grassberger. Tricritical directed percolation in $2 + 1$-dimensions. *J. Stat. Mech.: Theor. Exp.*, :P01004, 2006.
245. P. Grassberger, H. Chaté, and G. Rousseau. Spreading in media with long-time memory . *Phys. Rev.*, E55:2488, 1997.
246. P. Grassberger and A. de la Torre. Reggeon field-theory (Schögl's first model) on a lattice: Monte Carlo calculation of critical behaviour. *Ann. of Phys.*, 122:373, 1979.

247. P. Grassberger and S.S. Manna. Some more sandpiles. *J. Physique*, 51:1077, 1990.
248. P. Grassberger and I. Procaccia. Measuring the strangeness of strange attractors. *Physica*, D9:189, 1983.
249. P. Grassberger and Y.-C. Zhang. Self-organized formulation of standard percolation phenomena. *Physica*, A224:169, 1996.
250. R.B. Griffiths. Thermodynamic functions for fluids and ferromagnets near the critical point. *Phys. Rev.*, 158:176, 1967.
251. R.B. Griffiths. Dependence of critical indices on a parameter. *Phys. Rev. Lett.*, 24:1479, 1970.
252. G. Grimvall. *Thermophysical properties of materials*. North Holland, Amsterdam, 1986.
253. G. Grinstein. Generic scale-invariance and self-organized criticality. In A. McKane, M. Droz, J. Vannimenus, and D. Wolf, editors, *Scale Invariance, Interfaces, and Non-Equilibrium Dynamics*, London, 1995. Plenum Press.
254. G. Grinstein, Z.W. Lai, and R.J. Browne. Critical phenomena in a nonequilibrium model of heterogeneous catalysis. *Phys. Rev.*, A40:4820, 1989.
255. D. Grüneberg and A. Hucht. Universal finite-size scaling analysis of Ising models with long-range interactions at the upper critical dimensionality: Isotropic case. *Phys. Rev.*, E69:036104, 2004.
256. I. Guim, T.W. Burkhardt, and T. Xue. Surface behaviour of the fully frustrated Ising model. *Phys. Rev.*, B42:10298, 1990.
257. B. Gutenberg and C.F. Richter. *Seismicity of the earth and associated phenomena*. Princeton University Press, Princeton, New Jersey, 1954.
258. H.A. Gutowitz, J.D. Victor, and B.W. Knight. Local structure-theory for cellular automata. *Physica*, D28:18, 1987.
259. A.J. Guttmann. Asymptotic analysis of power-series expansions. In C. Domb and J.L. Lebowitz, editors, *Phase transitions and critical phenomena*, volume 13, London, 1989. Academic Press.
260. L.-H. Gwa and H. Spohn. Bethe solution for the dynamical-scaling exponent of the noisy Burgers equation. *Phys. Rev.*, A46:844, 1992.
261. H. Hadwiger. *Vorlesungen über Inhalt, Oberfläche und Isoperimetrie*. Springer, Heidelberg, 1957.
262. B.I. Halperin and T.C. Lubensky. On the analogy between smectic A liquid crystals and superconductors. *Solid State Comm.*, 14:997, 1974.
263. B.I. Halperin, T.C. Lubensky, and S.-K. Ma. First-order phase transitions in superconductors and smectic-A liquid crystals. *Phys. Rev. Lett.*, 32:292, 1974.
264. C.J. Hamer and M.N. Barber. Finite-lattice methods in quantum Hamiltonian field theory: I. The Ising model. *J. Phys. A Math. Gen.*, 14:241, 1981.
265. O. Al Hammal, H. Chaté, I. Dornic, and M. A. Muñoz. Langevin description of critical phenomena with two symmetric absorbing states. *Phys. Rev. Lett.*, 94:230601, 2005.
266. A. Hankey and H.E. Stanley. Systematic application of generalized homogeneous functions to static scaling, dynamical scaling, and universality. *Phys. Rev.*, B6:3515, 1972.
267. T.E. Harris. Contact interactions on a lattice. *Ann. Prob.*, 2:969, 1974.
268. M. Hasenbusch, A. Pelissetto, and E. Vicari. Multicritical behaviour in the fully frustrated XY model and related systems. *J. Stat. Mech.: Theor. Exp.*, :P12002, 2005.
269. B. Hede, J. Kertész, and T. Vicsek. Self-affine fractal clusters : conceptual questions and numerical results for directed percolation. *J. Stat. Phys.*, 64:829, 1991.
270. M. Henkel. *Phase Transitions and Conformal Invariance*. Springer, Heidelberg, 1999.
271. M. Henkel. Reaction–diffusion processes and their connection with integrable quantum spin chains. In A. Kundu, editor, *Classical and quantum nonlinear integrable systems*, page 256, Bristol, 2003. IOP Press.
272. M. Henkel and H.J. Herrmann. The Hamiltonian spectrum of directed percolation. *J. Phys. A Math. Gen.*, 23:4369, 1990.

273. M. Henkel and H. Hinrichsen. Exact solution of a reaction-diffusion process with three-site interactions. *J. Phys. A Math. Gen.*, 34:1561, 2001.
274. M. Henkel and H. Hinrichsen. The non-equilibrium phase-transition of the pair-contact process with diffusion. *J. Phys. A Math. Gen.*, 37:R117, 2004.
275. M. Henkel, E. Orlandini, and J.E. Santos. Reaction-diffusion processes from equivalent integrable quantum chains. *Ann. of Phys.*, 259:163, 1997.
276. M. Henkel, E. Orlandini, and G.M. Schütz. Equivalences between stochastic systems. *J. Phys. A Math. Gen.*, 28:6335, 1995.
277. M. Henkel and R. Peschanski. Intermittency studies in directed bond percolation. *Nucl. Phys.*, B390:637, 1993.
278. M. Henkel and V. Privman. New mechanism for mass gap scaling and transfer matrix study for $(1+1)D$ directed percolation. *Phys. Rev. Lett.*, 65:1777, 1990.
279. M. Henkel and U. Schollwöck. Universal finite-size scaling amplitudes in anisotropic scaling. *J. Phys. A Math. Gen.*, 34:3333, 2001.
280. M. Henkel and G. Schütz. Finite-lattice extrapolation algorithms. *J. Phys. A: Math. Gen.*, 21:2617, 1988.
281. M. Henkel and G. Schütz. Boundary-induced phase transitions in equilibrium and non-equilibrium systems. *Physica*, A206:187, 1994.
282. M. Henkel and F. Seno. Phase diagram of branched polymer collapse. *Phys. Rev.*, E53:3662, 1996.
283. R. Hilfer, T. Rage, and B. Virgin. Local percolation probabilities for a natural sandstone. *Physica*, A241:105, 1997.
284. H.J. Hilhorst and J.M. van Leeuwen. Nonuniversal and anomalous decay of boundary spin correlations in inhomogeneous Ising systems. *Phys. Rev. Lett.*, 47:1188, 1983.
285. H. Hinrichsen. Stochastic lattice models with several absorbing states. *Phys. Rev.*, E55:219, 1997.
286. H. Hinrichsen. First-order transitions in fluctuating 1+1-dimensional nonequilibrium systems. *unpublished*, 2000.
287. H. Hinrichsen. Nonequilibrium critical phenomena and phase-transitions into absorbing states. *Adv. Phys.*, 49:815, 2000.
288. H. Hinrichsen. Pair-contact process with diffusion: a new type of non-equilibrium behaviour? *Phys. Rev.*, E63:036102, 2001.
289. H. Hinrichsen. Is local scale invariance a generic property of ageing phenomena ? *J. Stat. Mech.: Theor. Exp.*, page L06001, 2006.
290. H. Hinrichsen. Non-equilibrium Phase Transitions with Long-Range Interactions. *J. Stat. Mech.: Theor. Exp.*, page P07066, 2007.
291. H. Hinrichsen. Dynamical response function of the disordered kinetic Ising model. *submitted to J. Stat. Mech: Theor. Exp.*, 2008.
292. H. Hinrichsen and M. Howard. A model for anomalous directed percolation. *Eur. Phys. J.*, B7:635, 1999.
293. H. Hinrichsen, A. Jiménez-Dalmaroni, Y. Rozov, and E. Domany. Flowing sand – a physical realization of directed percolation. *Phys. Rev. Lett.*, 83:4999, 1999.
294. H. Hinrichsen, A. Jiménez-Dalmaroni, Y. Rozov, and E. Domany. Flowing sand – a possible physical realization of directed percolation. *J. Stat. Phys.*, 98:1149, 2000.
295. H. Hinrichsen and H.M. Koduvely. Numerical study of local and global persistence in directed percolation. *Eur. Phys. J.*, B5:257, 1998.
296. H. Hinrichsen, R. Livi, D. Mukamel, and A. Politi. A model for nonequilibrium transitions in two dimensions. *Phys. Rev. Lett.*, 79:2710, 1997.
297. H. Hinrichsen, R. Livi, D. Mukamel, and A. Politi. First-order phase transition in a (1+1)-dimensional nonequilibrium wetting process. *Phys. Rev.*, E61:R1032, 2000.
298. H. Hinrichsen, R. Livi, D. Mukamel, and A. Politi. Wetting under non-equilibrium conditions. *Phys. Rev.*, E68:041606, 2003.
299. H. Hinrichsen and G. Ódor. Correlated initial conditions in directed percolation. *Phys. Rev.*, E58:311, 1998.

300. H. Hinrichsen, O. Stenull, and H. K. Janssen. Multifractal current distribution in random diode networks. *Phys. Rev.*, E65:045104, 2002.
301. H. Hinrichsen, J. S. Weitz, and E. Domany. An algorithm-independent definition of damage spreading - application to directed percolation. *J. Stat. Phys.*, 88:617, 1997.
302. R. Hocken and M.R. Moldover. Ising critical exponents in real fluids: An experiment. *Phys. Rev. Lett.*, 37:29, 1976.
303. J. Hooyberghs, F. Iglói, and C. Vanderzande. Strong Disorder Fixed Point in Absorbing-State Phase Transitions. *Phys. Rev. Lett.*, 90:100601, 2003.
304. J. Hooyberghs, F. Iglói, and C. Vanderzande. Absorbing-state phase-transitions with quenched disorder. *Phys. Rev.*, E69:066140, 2004.
305. B. Houchmandzadeh. Clustering of diffusing organisms. *Phys. Rev.*, E 66:052902, 2002.
306. R.M.F. Houtappel. Order-disorder in hexagonal lattices. *Physica*, 16:425, 1950.
307. M. Howard, P. Fröjdh, and K.B. Lauritsen. Surface critical behavior in systems with non-equilibrium phase transitions. *Phys. Rev.*, E61:167, 2000.
308. M. Howard and C. Godrèche. Persistence in the Voter model: continuum reaction-diffusion approach. *J. Phys. A: Math. Gen.*, 31:L209, 1998.
309. M. Howard and U.C. Täuber. 'Real' vs 'imaginary' noise in diffusion-limited reactions. *J. Phys. A: Math. Gen.*, 30:7721, 1997.
310. M. Hoyuelos, E.V. Albano, and H.O. Mártin. A multilayer contact process. *J. Phys. A Math. Gen.*, 30:431, 1997.
311. C.-K. Hu. Boundary conditions and scaling functions of percolation models. *J. Phys. A Math. Gen.*, 27:L813, 1994.
312. C.-K. Hu, C.Y. Lin, and J.A. Chen. Universal scaling functions in critical phenomena. *Phys. Rev. Lett.*, 75:193, 1995.
313. X. Hu. Bicritical and tetracritical phenomena and scaling properties of the SO(5) theory. *Phys. Rev. Lett.*, 87:057004, 2001.
314. J. Hubbard and P. Schofield. Wilson theory of a liquid–vapour critical point. *Phys. Lett.*, 40A:245, 1972.
315. C. Hyver. Impossibility of existence of undamped oscillations in linear chemical systems. *J. Theor. Biol.*, 36:133, 1972.
316. F. Iglói, I. Peschel, and L. Turban. Inhomogeneous systems with unusual critical behaviour. *Adv. Phys.*, 42:683, 1993.
317. F. Iglói and L. Turban. First-order line transitions: a general scaling theory with irrelevant variables. *Phys. Rev*, B47:3404, 1993.
318. E.V. Ivashkevich. Boundary height correlations in two-dimensional Abelian sandpile. *J. Phys. A Math. Gen.*, 27:3643, 1994.
319. E.V. Ivashkevich. Critical behaviour of the sandpile model as a self-organized branching process. *Phys. Rev. Lett.*, 76:3368, 1996.
320. E.V. Ivashkevich, D.V. Ktitarev, and V.B. Priezzhev. Critical exponents for boundary avalanches in two-dimensional Abelian sandpile. *J. Phys. A Math. Gen.*, 27:L585, 1994.
321. E.V. Ivashkevich, D.V. Ktitarev, and V.B. Priezzhev. Waves of topplings in an Abelian sandpile. *Physica*, A209:347, 1994.
322. L.F. Jaffe. Classes and mechanisms of calcium waves. *Cell Calcium*, 14:736, 1993.
323. W. Janke. Introduction to simulation techniques. In M. Henkel, M. Pleimling, and R. Sanctuary, editors, *Ageing and the glass transition*, Heidelberg, 2007. Springer.
324. S.A. Janowsky and C.A. Laberge. Exact solutions for a mean-field Abelian sandpile. *J. Phys. A Math. Gen.*, 26:L973, 1993.
325. H.-K. Janssen. On a Lagrangean for classical field dynamics and renormalization group calculations of dynamical critical properties. *Z. Phys.*, B23:377, 1976.
326. H.-K. Janssen. On the nonequilibrium phase transition in reaction-diffusion systems with an absorbing stationary state. *Z. Phys.*, B42:151, 1981.
327. H.-K. Janssen. Renormalized field-theory of dynamical percolation. *Z. Phys.*, B58:311, 1985.

328. H.-K. Janssen. On the renormalized field-theory of nonlinear critical relaxation. In G. Győrgi et *al.*, editor, *From phase-transitions to chaos*, Singapore, 1992. World Scientific.
329. H.-K. Janssen. Renormalization field-theory of the Gribov process with quenched disorder. *Phys. Rev.*, E55:6253, 1997.
330. H.-K. Janssen. Directed percolation with colors and flavors. *J. Stat. Phys.*, 103:801, 2001.
331. H.-K. Janssen. Survival and percolation probabilities in the field-theory of growth models. *J. Phys.: Cond. Mat.*, 17:S1973, 2005.
332. H.-K. Janssen, Ü. Kutbay, and K. Oerding. Equation of state for directed percolation. *J. Phys. A Math. Gen.*, 32:1809, 1999.
333. H.-K. Janssen, S. Lübeck, and O. Stenull. Finite-size scaling in non-equilibrium systems at the instance of directed percolation. *Phys. Rev.*, E76:041126, 2007.
334. H.-K. Janssen, M. Müller, and O. Stenull. A generalized epidemic process and tricritical dynamical percolation. *Phys. Rev.*, E70:026114, 2004.
335. H.-K. Janssen, K. Oerding, Wijland F. van and H.J. Hilhorst. Lévy-flight spreading of epidemic processes leading to percolation clusters. *Eur. Phys. J.*, B7:137, 1999.
336. H.-K. Janssen, B. Schaub, and B. Schmittmann. Finite-size scaling for directed percolation and related stochastic evolution processes. *Z. Phys.*, B71:377, 1988.
337. H.-K. Janssen, B. Schaub, and B. Schmittmann. The effects of surfaces on directed percolation and related stochastic evolution processes. *Z. Phys.*, B72:111, 1988.
338. H.-K. Janssen, B. Schaub, and B. Schmittmann. New universal short-time scaling behaviour of critical relaxation processes. *Z. Phys.*, B73:539, 1989.
339. H. K. Janssen and O. Stenull. Random Resistor-Diode Networks and the Crossover from Isotropic to Directed Percolation. *Phys. Rev.*, E62:3173, 2000.
340. H.-K. Janssen and O. Stenull. Logarithmic corrections in dynamical isotropic percolation. *Phys. Rev.*, E68:036131, 2003.
341. H.-K. Janssen and O. Stenull. Logarithmic corrections in directed percolation. *Phys. Rev.*, E69:016125, 2004.
342. H.-K. Janssen and U.C. Täuber. The field-theory approach to percolation processes. *Ann. of Phys.*, 315:147, 2005.
343. H.K. Janssen, B. Schaub, and B. Schmittmann. New universal short-time scaling behaviour of critical relaxation processes. *Z. Physik*, B73:539, 1989.
344. F. Jasch and H. Kleinert. Fast-convergent resummation algorithm and critical exponents of ϕ^4-theory in three dimensions. *J. Math. Phys.*, 42:52, 2001.
345. D. Jasnow and M. Wortis. High-temperature critical indices for the classical anisotropic Heisenberg model. *Phys. Rev.*, 176:739, 1968.
346. H. J. Jensen. Lattice gas as a model of $1/f$ noise. *Phys. Rev. Lett.*, 64:3103, 1990.
347. H. J. Jensen, K. Christensen, and H. C. Fogedby. Dynamical and spatial aspects of sandpile cellular automata. *J. Stat. Phys.*, 63:653, 1991.
348. H.J. Jensen, K. Christensen, and H. C. Fogedby. $1/f$ noise, distribution of lifetimes, and a pile of sand. *Phys. Rev.*, B40:7425, 1989.
349. I. Jensen. Critical behaviour of the three-dimensional contact process. *Phys. Rev.*, A45:R563, 1992.
350. I. Jensen. Critical behaviour of the pair-contact process. *Phys. Rev. Lett.*, 70:1465, 1993.
351. I. Jensen. Low-density series expansions for directed percolation on square and triangular lattices. *J. Phys. A Math. Gen.*, 29:7013, 1996.
352. I. Jensen. Temporally disordered bond percolation on the directed square lattice. *Phys. Rev. Lett.*, 77:4988, 1996.
353. I. Jensen. Low-density series expansions for directed percolation, I. A new efficient algorithm with applications to the square lattice. *J. Phys. A Math. Gen.*, 32:5233, 1999.
354. I. Jensen. Low-density series expansions for directed percolation II: The square lattice with a wall. *J. Phys. A Math. Gen.*, 32:6055, 1999.

355. I. Jensen. Low-density series expansions for directed percolation: III. Some two-dimensional lattices. *J. Phys. A Math. Gen.*, 37:6899, 2004.
356. I. Jensen. Low-density series expansions for directed percolation IV. Temporal disorder. *J. Phys. A Math. Gen.*, 38:1441, 2005.
357. I. Jensen and R. Dickman. Nonequilibrium phase-transitions in systems with infinitely many absorbing states. *Phys. Rev.*, E48:1710, 1993.
358. I. Jensen and R. Dickman. Time-dependent perturbation theory for diffusive non-equilibrium lattice models. *J. Phys. A Math. Gen.*, 26:L151, 1993.
359. I. Jensen and R. Dickman. Time-dependent perturbation theory for non-equilibrium lattice models. *J. Stat. Phys.*, 71:89, 1993.
360. I. Jensen, H. C. Fogedby, and R. Dickman. Critical exponents for an irreversible surface reaction model. *Phys. Rev.*, A41:3411, 1990.
361. I. Jensen and A.J. Guttmann. Series expansions of the percolation probability for directed square and honeycomb lattices. *J. Phys. A Math. Gen.*, 28:4813, 1995.
362. I. Jensen and A.J. Guttmann. Series expansions for two-dimensional directed percolation. *Nucl. Phys. B (Proc. Suppl.)*, 47:835, 1996.
363. A. Jiménez-Dalmaroni and H. Hinrichsen. Epidemic processes with immunization. *Phys. Rev.*, E68:036103, 2003.
364. D. Jones, A. Love, and M. Moore. Phase transitions in superfluid ^{3}He. *J. Phys. C: Solid State Phys.*, 9:743, 1976.
365. B. D. Josephson. Equation of state near the critical point. *J. Phys. C: Solid State Phys.*, 2:1113, 1969.
366. G.S. Joyce. Critical properties of the spherical model. In C. Domb and M.S. Green, editors, *Phase transitions and critical phenomena*, volume 2, London, 1972. Academic Press.
367. G.A. Baker Jr. *Quantitative theory of critical phenomena*. Academic Press, New York, 1990.
368. L.P. Kadanoff. The introduction of the idea that exponents could be derived from real-space scaling arguments. *Physics (NY)*, 2:263, 1966.
369. L.P. Kadanoff. Critical behaviour, universality and scaling. In M.S. Green, editor, *Proceedings of the 1970 Varenna summer school on critical phenomena*, New York, 1971. Academic Press.
370. L.P. Kadanoff. Notes on Migdal's recursion formula. *Ann. of Phys.*, 100:359, 1976.
371. C. Kaiser and L. Turban. Fractal dimensions of confined clusters in two-dimensional directed percolation. *J. Phys. A Math. Gen.*, 27:L579, 1994.
372. C. Kaiser and L. Turban. Surface shape and local critical behaviour in two-dimensional directed percolation. *J. Phys. A Math. Gen.*, 28:351, 1995.
373. E. Kamke. *Differentialgleichungen: Lösungsmethoden und Lösungen, vol.2, 4th edition.* Akademische Verlagsgesellschaft, Leipzig, 1959.
374. K. Kaneda and Y. Okabe. Finite-size scaling for the Ising model on the Möbius strip and the Klein bottle. *Phys. Rev. Lett.*, 86:2134, 2001.
375. K. Kaneda, Y. Okabe, and M. Kikuchi. Shape effects of finite-size scaling functions for anistropic three-dimensional Ising models. *J. Phys. A Math. Gen.*, 32:7263, 1999.
376. M. Kardar, G. Parisi, and Y.-C. Zhang. Dynamic scaling of growing interfaces. *Phys. Rev. Lett.*, 56:889, 1986.
377. K. Katajantie, M. Laine, R. Rummukainen, and M. Shaposhnikov. Is there a hot electroweak phase transition at $m_H \geq m_W$. *Phys. Rev. Lett.*, 77:2887, 1996.
378. M. Katori and H. Kobayashi. Mean-field theory of avalanches in self-organized critical states. *Physica*, A229:461, 1996.
379. M. Kaufmann and R.B. Griffiths. Exactly soluble Ising models on hierarchical lattices. *Phys. Rev.*, B24:496, 1981.
380. M. Kaulke and I. Peschel. A DMRG study of the q-symmetric Heisenberg chain. *Eur. Phys. J.*, B5:727, 1998.
381. J. Keizer. On the solutions and the steady-states of a master equation. *J. Stat. Phys.*, 6:67, 1972.

382. J. Keizer, Y.-X. Li, J. Rinzel, and S. Stojilkovic. InsP3-induced Ca2+ excitability of the endoplasmic reticulum. *Mol. Biol. Cell*, 6:945, 1995.
383. J. Keizer and G. D. Smith. Spark-to-wave transition: saltatory transmission of Ca2+ waves in cardiac myocytes. *Bioph. Chem.*, 72:87, 1998.
384. A. Kemper, A. Gendiar, T. Nishino, A. Schadschneider, and J. Zittartz. Stochastic light-cone CTMRG: a new DMRG approach to stochastic models. *J. Phys. A Math. Gen.*, 36:29, 2003.
385. A. Kemper, A. Schadschneider, and J. Zittartz. Transfer-matrix DMRG for stochastic models: The Domany-Kinzel cellular automaton. *J. Phys. A Math. Gen.*, 34:L279, 2001.
386. R. Kenna. Homotopy in statistical physics. *Condensed Matter Phys.*, 9:283, 2006.
387. J. Kertész and L.B. Kiss. The noise spectrum in the model of self-organised criticality. *J. Phys. A Math. Gen.*, 23:L433, 1990.
388. J. Kertész and D.E. Wolf. Anomalous roughening in growth processes. *Phys. Rev. Lett.*, 62:2571, 89.
389. Y. C. Kim, M. A. Anisimov, J. V. Sengers, and E. Luijten. Crossover critical behaviour in the three-dimensional Ising model. *J. Stat. Phys.*, 110:591, 2003.
390. W. Kinzel. Percolation structures and processes. In G. Deutscher, R. Zallen, and J. Adler, editors, *Ann. Isr. Phys. Soc.*, volume 5, page 425, Bristol, 1983. Adam Hilger.
391. W. Kinzel. Phase transition of cellular automata. *Z. Phys.*, B58:229, 1985.
392. H. Kleinert. Critical exponents from seven-loop strong-coupling ϕ^4-theory in three dimensions. *Phys. Rev.*, D60:085001, 1999.
393. H. Kleinert and V.I. Yukalov. Self-similar variational perturbation theory for critical exponents. *Phys. Rev.*, E71:026131, 2005.
394. M. Knezević and J. Vannimenus. On directed interacting animals and directed percolation. *J. Phys. A Math. Gen.*, 35:2725, 2002.
395. J. Kockelkoren and H. Chaté. Absorbing phase transitions of branching-annihilating random Walks. *Phys. Rev. Lett.*, 90:125701, 2003.
396. P.-M. König, R. Roth, and K. Mecke. Morphological thermodynamics of fluids: shape-dependence of free energies. *Phys. Rev. Lett.*, 93:160601, 2004.
397. R. Kopelman, C.S. Li, and Z.-Y. Shi. One-dimensional exciton fusion kinetics indilute polymer blends. *J. Luminescence*, 45:40, 1990.
398. L. Környei, M. Pleimling, and F. Iglói. Non-equilibrium critical dynamics of the two-dimensional Ising model quneched from a correlated initial state. `arXiv:0710.2829`, 2007.
399. K. Krebs, M.P. Pfannmüller, B. Wehefritz, and H. Hinrichsen. Finite-size scaling studies of one-dimensional reaction-diffusion systems I. *J. Stat. Phys.*, 78:1429, 1995.
400. R. Kree, B. Schaub, and B. Schmitmann. Effects of pollution on critical population dynamics. *Phys. Rev.*, A39:2214, 1989.
401. R. Kroon, H. Fleurent, and R. Sprik. Diffusion-limited exciton fusion reactions inone-dimensional tetramethylammonium maganese trichloride (TMMC). *Phys. Rev.*, E47:2462, 1993.
402. J. Krug. Boundary-induced phase transitions in driven diffusive systems. *Phys. Rev. Lett.*, 67:1882, 1991.
403. D. V. Ktitarev, S. Lübeck, P. Grassberger, and V.B. Priezzhev. Scaling of waves in the Bak-Tang-Wiesenfeld sandpile model. *Phys. Rev.*, E61:81, 2000.
404. S. Kwon and Y. Kim. Double domain structure of the pair-contact process with diffusion. *Phys. Rev.*, E75:042103, 2007.
405. D. Landau and K. Binder. *Monte Carlo simulations in statistical physics.* Cambridge University Press, Cambridge, 2000.
406. E. Landau. *Handbuch über die Lehre von der Verteilung der Primzahlen.* Teubner, Leizig, 1909.
407. L.D. Landau and E.M. Lifschitz. *Lehrbuch der theoretischen Physik,* Band V. Akademie Verlag, Berlin, 1966.

408. K. B. Lauritsen, P. Fröjdh, and M. Howard. Surface critical behavior in systems with absorbing states. *Phys. Rev. Lett.*, 81:2104, 1998.
409. K. B. Lauritsen, K. Sneppen, M. Markosová, and M. H. Jensen. Directed percolation with an absorbing boundary. *Physica*, A247:1, 1997.
410. L. Laurson, M. J. Alava, and S. Zapperi. Power spectra of self-organized critical sandpiles. *J. Stat. Mech.: Theor. Exp.*, :L11001, 2005.
411. I.D. Lawrie and S. Sarbach. Theory of tricritical points. In C. Domb and J. L. Lebowitz, editors, *Phase transitions and critical phenomena*, volume 9, London, 1984. Academic Press.
412. J. Lechleiter and D. Clapham. Molecular mechanisms of intracellular Ca2+ excitability in Xenopus laevis oocytes. *Cell*, 69:283, 1992.
413. B. P. Lee and A. D. Rutenberg. Persistence, poisoning, and autocorrelations in dilute coarsening. *Phys. Rev. Lett.*, 79:4843, 1997.
414. B.P. Lee. Renormalisation-group calculation for the reaction $kA \to \emptyset$. *J. Phys. A Math. Gen.*, 27:2633, 1994.
415. B.P. Lee and J.L. Cardy. Renormalization-group study of the $A+B$ diffusion-limited reaction. *J. Stat. Phys.*, 80:971, 1995.
416. B.P. Lee and J.L. Cardy. Erratum: Renormalization-group study of the $A + B$ diffusion-limited reaction. *J. Stat. Phys.*, 87:951, 1997.
417. T.D. Lee and C.N. Yang. Statistical theory of equation of state and phase transition: II. Lattice gas and Ising model. *Phys. Rev.*, 87:410, 1952.
418. C. Lehner, N. Rajewsky, D.E. Wolf, and J. Kertész. Anomalous roughening in growth processes. *Physica*, A164:81, 1990.
419. H.W. Lewis and G.H. Wannier. Spherical model of a ferromagnet. *Phys. Rev.*, 88:682, 1952.
420. H.W. Lewis and G.H. Wannier. Spherical model of a ferromagnet. *Phys. Rev.*, 90:1131, 1953.
421. R. Liebman. *Statistical mechanics of periodic frustrated Ising systems, Lecture Notes in Physics* **251**. Springer, Heidelberg, 1986.
422. T.M. Liggett. *Interacting particle systems.* Springer, Heidelberg, 1985.
423. F. Linder, J. Tran-Gia, S. R. Dahmen, and H. Hinrichsen. Long-range epidemic spreading with immunization. *submitted to Phys. Rev. E*, 2008.
424. J.A. Lipa, J.A. Nissen, D.A. Stricker, D.R. Swanson, and T.C.P. Chui. Specific heat of liquid helium in zero gravity very near the lambda point. *Phys. Rev.*, B68:174518, 2003.
425. D.-J. Liu, N. Pavlenko, and J.W. Evans. Crossover between mean-field and Ising critical behaviour in a lattice-gas reaction-diffusion model. *J. Stat. Phys.*, 114:101, 2004.
426. E. Lorenz and W. Janke. Ageing phenomena in phase-ordering kinetics in Potts models. *Diplomarbeit, Uni Leipzig July 2005, see* `http://www.physik.uni-leipzig.de/~lorenz/`, 2005.
427. E. Lorenz and W. Janke. Numerical tests of local scale-invariance in ageing q-state Potts models. *Europhys. Lett.*, 77:10003, 2007.
428. S. Lübeck. Crossover phenomenon in self-organized critical sandpile models. *Phys. Rev.*, E62:6149, 2000.
429. S. Lübeck. Moment analysis of the probability distributions of different sandpile models. *Phys. Rev.*, E61:204, 2000.
430. S. Lübeck. Scaling behaviour of the conserved threshold transfer process. *Phys. Rev.*, E66:046114, 2002.
431. S. Lübeck. Scaling behaviour of the order parameter and its conjugated field in an absorbing phase transition around the upper critical dimension. *Phys. Rev.*, E65:046150, 2002.
432. S. Lübeck. Universal behaviour of crossover scaling functions for continuous phase-transitions. *Phys. Rev. Lett.*, 90:210601, 2003.

433. S. Lübeck. Universal scaling behaviour of non-equilibrium phase-transitions. *Int. J. Mod. Phys.*, B18:3977, 2004.
434. S. Lübeck. Violation of the Widom scaling law for effective crossover exponents. *Phys. Rev.*, E69:066101, 2004.
435. S. Lübeck. Crossover scaling in the Domany-Kinzel cellular automaton. *J. Stat. Mech.: Theor. Exp.*, :P09009, 2006.
436. S. Lübeck. Tricritical directed percolation. *J. Stat. Phys.*, 123:193, 2006.
437. S. Lübeck and P.C. Heger. Universal finite-size scaling behaviour and universal dynamical scaling behaviour of absorbing phase-transitions with a conserved field. *Phys. Rev.*, E68:056102, 2003.
438. S. Lübeck and P.C. Heger. Universal scaling behaviour at the upper critical dimension of non-equilibrium continuous phase transitions. *Phys. Rev. Lett.*, 90:230601, 2003.
439. S. Lübeck and A. Hucht. Mean-field scaling function of the universality class of absorbing phase-transitions with a conserved field. *J. Phys. A Math. Gen.*, 35:4853, 2002.
440. S. Lübeck and H.-K. Janssen. Finite-size scaling of directed percolation above the upper critical dimension. *Phys. Rev.*, E72:016119, 2005.
441. S. Lübeck and A. Misra. Persistence distributions in a conserved lattice gas with absorbing states. *Eur. Phys. J.*, B26:75, 2002.
442. S. Lübeck and K.D. Usadel. Numerical determination of the avalanche exponents of the Bak-Tang-Wiesenfeld model. *Phys. Rev.*, E55:4095, 1997.
443. S. Lübeck and K.D. Usadel. The Bak-Tang-Wiesenfeld sandpile model around the upper critical dimension. *Phys. Rev.*, E56:5138, 1997.
444. S. Lübeck and R.D. Willmann. Universal scaling behaviour of directed percolation and the pair-contact process in an external field. *J. Phys. A Math. Gen.*, 35:10205, 2002.
445. S. Lübeck and R.D. Willmann. Universal scaling behaviour of directed percolation around the upper critical dimension. *J. Stat. Phys.*, 115:1231, 2004.
446. S. Lübeck and R.D. Willmann. Scaling behaviour of the directed percolation universality class. *Nucl. Phys.*, B718:341, 2005.
447. J.-M. Luck. Corrections to finite-size-scaling laws and convergence of transfer-matrix methods. *Phys. Rev.*, B31:3069, 1985.
448. E. Luijten and K. Binder. Nature of crossover from classical to Ising-like critical behaviour. *Phys. Rev.*, E58:4060, 1998.
449. E. Luijten and H.W.J. Blöte. Boundary between long-range and short-range critical behavior in systems with algebraic interactions. *Phys. Rev. Lett.*, 89:025703, 2002.
450. E. Luijten, H.W.J. Blöte, and K. Binder. Medium-range interactions and crossover to classical critical behaviour. *Phys. Rev.*, E54:4626, 1996.
451. E. Luijten, H.W.J. Blöte, and K. Binder. Nonmomotonic crossover of the effective susceptibility exponent. *Phys. Rev. Lett.*, 79:561, 1997.
452. S.-K. Ma, C. Dasgupta, and C.-K. Hu. Random Antiferromagnetic Chain. *Phys. Rev. Lett.*, 43:1434, 1979.
453. C. Maes, K. Netocny, and B. Shergelashvili. A selection of non-equilibrium issues. `math-ph/0701047`, 2007.
454. S. Majumdar, A.J. Bray, S.J. Cornell, and C. Sire. Global persistence exponent for critical dynamics. *Phys. Rev. Lett.*, 77:3704, 1996.
455. S. N. Majumdar and A. J. Bray. Persistence of manifolds in nonequilibrium critical dynamics. *Phys. Rev. Lett.*, 91:030602, 2003.
456. S.N. Majumdar and D. Dhar. Height correlations in the Abelian sandpile model. *J. Phys. A Math. Gen.*, 24:L357, 1991.
457. S.N. Majumdar and D. Dhar. Equivalence between the Abelian sandpile model and the $q \to 0$ limit of the Potts model. *Physica*, A185:129, 1992.
458. H.A. Makse, S. Buldyrev, H. Leschhorn, and H.E. Stanley. The pinning paths of an elastic interface. *Europhys. Lett.*, 41:251, 1998.

459. S.S. Manna. Large-Scale simulation of avalanche cluster distribution in sandpile model. *J. Stat. Phys.*, 59:509, 1990.
460. S.S. Manna. Two-state model of self-organized criticality. *J. Phys. A Math. Gen.*, 24:L363, 1991.
461. M.C. Marques, M.A. Santos, and J.F.F. Mendes. Static critical behaviour in the inactive phase of the pair-contact process. *Phys. Rev.*, E65:066110, 2001.
462. J. Marro and R. Dickman. *Nonequilibrium phase-transitions in lattice models.* Cambridge University Press, Cambridge, 1999.
463. P.C. Martin, E.D. Siggia, and H.A. Rose. Statistical dynamics of classical systems. *Phys. Rev.*, A8:423, 1973.
464. T.M. Masser and D. ben Avraham. A method of intervals for the study of diffusion-limited annihilation, $A + A \to \emptyset$. *Phys. Rev.*, E63:066108, 2001.
465. T.O. Masser and D. ben Avraham. Correlation functions for diffusion-limited annihilation, $A + A \to \emptyset$. *Phys. Rev.*, E64:062101, 2001.
466. P. Mayer and P. Sollich. Ageing in one-dimensional coagulation-diffusion processes and the Fredrickson-Andersen model. *J. Phys. A Math. Theor.*, 40:5823, 2007.
467. B. M. McCoy and T. T. Wu. *The two-dimensional Ising model.* Havard University Press, Cambridge, Mass., 1973.
468. P. Meakin and D.J. Scalapino. Simple models for heterogeneous catalysis: phase transition-like behaviour in nonequilibrium systems. *J. Chem. Phys.*, 87:731, 1987.
469. K. Mecke. *Integralgeometrie in der statistischen Physik.* Verlag Harri Deutsch, Frankfurt/Main, 1994.
470. J.F.F. Mendes, R. Dickman, M. Henkel, and M.C. Marques. Generalized scaling for models with multiple absorbing states. *J. Phys. A Math. Gen.*, 27:3019, 1994.
471. M. De Menech. Comment on Universality in sandpiles. *Phys. Rev.*, E70:028101, 2004.
472. M. De Menech, A.L. Stella, and C. Tebaldi. Rare events and breakdown of simple scaling in the Abelian sandpile model. *Phys. Rev.*, E58:2677, 1998.
473. N. Menyhárd and G. Ódor. Non-equilibrium phase transitions in one-dimensional kinetic Ising models. *J. Phys. A: Math. Gen.*, 29:7739, 1996.
474. N. Menyhárd and G. Ódor. Non-Markovian persistence at the parity conserving point of a one-dimensional nonequilibrium kinetic Ising model. *J. Phys. A Math. Gen.*, 30:8515, 1997.
475. S. Michalland, M. Rabaud, and Y. Couder. Transition to chaos by spatio-temporal intermittency in directional viscous fingering. *Europhys. Lett.*, 22:17, 1993.
476. A.A. Migdal. Phase transitions in gauge and spin-lattice systems. *Sov. Phys. JETP*, 42:743, 1976.
477. K.K. Mon and K. Binder. Finite-size scaling and the crossover to mean-field critical behaviour in the two-dimensional Ising model with medium-ranged interactions. *Phys. Rev.*, E48:2498, 1993.
478. L. Mondaini and E.C. Marino. Sine-Gordon/Coulomb-gas soliton correlation functions and an exact evaluation of the Kosterlitz–Thouless critical exponent. *J. Stat. Phys.*, 118:767, 1995.
479. A.G. Moreira. Critical dynamics of the contact process with quenched disorder. *Phys. Rev.*, E54:3090, 1996.
480. H. Mori and K.J. McNeil. Critical dimensionality for normal fluctuations of macrovariables in nonequilibrium states. *Prog. Theor. Phys.*, 57:770, 1977.
481. M.A. Muñoz, R. Dickman, A. Vespignani, and S. Zapperi. Avalanche and spreading exponents in systems with absorbing states. *Phys. Rev.*, E59:6175, 1999.
482. M.A. Muñoz, G. Grinstein, and R. Dickman. Phase structure of systems with infinite numbers of absorbing states. *J. Stat. Phys.*, 91:541, 1998.
483. M.A. Muñoz, G. Grinstein, R. Dickman, and R. Livi. Critical behaviour of systems with many absorbing states. *Phys. Rev. Lett.*, 76:451, 1996.
484. O. Narayan and A.P. Young. Convergence of Monte Carlo simulations to equilibrium . *Phys. Rev.*, E64:021104, 2001.

485. M.E.J. Newman and G.T. Barkema. *Monte Carlo methods in statistical physics.* Clarendon Press, Oxford, 1999.
486. J.F. Nicoll and P.C. Albright. Crossover functions by renormalization-group matching: Three-loop results. *Phys. Rev.*, B31:4576, 1985.
487. B. Nienhuis, E.K. Riedel, and M. Schick. Magnetic exponents of the two-dimensional q-state Potts model. *J. Phys. A Math. Gen.*, 13:L189, 1980.
488. J.D. Noh and H. Park. Universality class of absorbing transitions with continuously varying critical exponents. *Phys. Rev.*, E69, 2004.
489. S.P. Obukhov. The problem of directed percolation. *Physica*, A101:145, 1980.
490. G. Ódor. Critical behaviour of the one-dimensional annihilation-fission process $2A \to \emptyset$, $2A \to 3A$. *Phys. Rev.*, E62:R3027, 2000.
491. G. Ódor. Phase transition of the one-dimensional coagulation-production process. *Phys. Rev.*, E63:067104, 2001.
492. G. Ódor. Critical behaviour of the one-dimensional diffusive pair-contact process. *Phys. Rev.*, E67:016111, 2003.
493. G. Ódor. Critical behaviour of the two-dimensional $2A \to 3A$, $4A \to \emptyset$ binary system. *Phys. Rev.*, E70:026119, 2004.
494. G. Ódor. Phase transitions of the binary production $2A \to 3A$, $4A \to \emptyset$ model. *Phys. Rev.*, E69:026119, 2004.
495. G. Ódor. The role of diffusion in branching and annihilation random walk models. *Phys. Rev.*, E70:066122, 2004.
496. G. Ódor. Universality classes in nonequilibrium lattice systems. *Rev. Mod. Phys.*, 76:663, 2004.
497. G. Ódor. Phase transition of triplet reaction-diffusion models. *Phys. Rev.*, E73:047103, 2006.
498. G. Ódor, M.C. Marques, and M.A. Santos. Phase transition of a two-dimensional binary spreading model. *Phys. Rev.*, E65:056113, 2002.
499. K. Oerding and F. van Wijland. Global persistence in directed percolation. *J. Phys. A Math. Gen.*, 31:7011, 1998.
500. K. Oerding, F. van Wijland, J.P. Leroy, and H.J. Hilhorst. Fluctuation-induced first-order transition in a nonequilibrium steady-state. *J. Stat. Phys.*, 99:1365, 2000.
501. T. Ohtsuki and T. Keyes. Crossover in nonequilibrium multicritical phenomena of reaction-diffusion systems. *Phys. Rev.*, A36:4434, 1987.
502. T. Ohtsuki and T. Keyes. Nonequilibrium critical phenomena in one-component reaction-diffusion systems. *Phys. Rev.*, A35:2697, 1987.
503. L. Onsager. Crystal statistics. I. A two-dimensional model with an order-disorder transition. *Phys. Rev.*, 65:117, 1944.
504. M. Paessens and G.M. Schütz. Phase-transitions and correlations in the bosonic pair-contact process with diffusion: exact results. *J. Phys. A: Math. Gen.*, 37:4709, 2004.
505. H. Park, J. Köhler, I.M. Kim, D. ben Avraham, and S. Redner. Excluded volume effects in heterogeneous catalysis – reactions between dollars and dimes. *J. Phys. A Math. Gen.*, 26:2071, 1993.
506. K. Park, H. Hinrichsen, and I.M. Kim. Binary spreading process with parity-conservation. *Phys. Rev.*, E63:065103(R), 2001.
507. K. Park, H. Hinrichsen, and I.M. Kim. Phase transition in a triplet process. *Phys. Rev.*, E66:025101(R), 2002.
508. K. Park and I.M. Kim. Well-defined set of exponents for a pair-contact process with diffusion . *Phys. Rev.*, E66:027106, 2002.
509. S.-C. Park and H. Park. Driven pair-contact process with diffusion. *Phys. Rev. Lett.*, 94:065701, 2005.
510. R. Pastor-Satorras and A. Vespignani. Field-theory of absorbing phase-transitions with a nondiffusive conserved field. *Phys. Rev.*, E62:5875, 2000.
511. R. Pastor-Satorras and A. Vespignani. Reaction-diffusion system with self-organized critical behaviour. *Eur. Phys. J.*, B19:583, 2001.

512. A.Z. Patashinskii and V.L. Pokrovskii. Behaviour of ordered systems near transition point. *Sov. Phys. JETP*, 23:292, 1966.
513. A. Pelissetto, P. Rossi, and E. Vicari. Crossover scaling from classical to nonclassical critical behaviour. *Phys. Rev.*, E58:7146, 1998.
514. A. Pelissetto, P. Rossi, and E. Vicari. Mean-field expansion for spin models with medium-range interactions. *Nucl. Phys.*, B554:552, 1999.
515. A. Pelissetto and E. Vicari. Critical phenomena and renormalisation-group theory. *Physics Reports*, 368:549, 2002.
516. L. Peliti. Renormalisation of fluctuation effects in the $A+A$ to A reaction. *J. Phys. A Math. Gen.*, 19:L365, 1986.
517. K.A. Penson, R. Jullien, and P. Pfeuty. Phase transitions in systems with multispin interactions. *Phys. Rev.*, B26:6334, 1982.
518. I. Peschel, K. Hallberg, and X. Wang (eds). *Proceedings of the first international DMRG workshop.* Springer, Heidelberg, 1999.
519. I. Peschel, V. Rittenberg, and U. Schultze. Spectra of quantum chains without the Yang-Baxter equation. *Nucl. Phys.*, B430:633, 1994.
520. P. Pfeuty and G. Toulouse. *Introduction to the renormalization group and critical phenomena.* John Wiley & Sons, Chichester, 1994.
521. Gennes P.G. de˙ *Scaling concepts on polymer physics.* Cornell University, Ithaca, 1979.
522. C. Pirat, A. Naso, J.-L. Meunier, P. Maïssa, and C. Mathis. Transition to spatiotemporal chaos in a two-dimensional hydrodynamic system. *Phys. Rev. Lett.*, 94:134502, 2005.
523. M. Pleimling. Critical phenomena at perfect and non-perfect surfaces. *J. Phys. A Math. Gen.*, 37:R79, 2004.
524. M. Pleimling and F. Iglói. Nonequilibrium critical relaxation at a first-order phase transition point. *Europhys. Lett.*, 79:56002, 2007.
525. M. Plischke and B. Bergersen. *Equilibrium statistical physics.* World Scientific, Singapour, 1994.
526. J. Polchinski. Scale and conformal invariance in quantum field theory. *Nucl. Phys.*, B303:226, 1988.
527. A.M. Polyakov. Conformal symmetry of critical fluctuations. *Sov. Phys. JETP Lett.*, 12:381, 1970.
528. Y. Pomeau. Front motion, metastability and subcritical bifurcations in hydrodynamics. *Physica*, D23:3, 1986.
529. R.B. Potts. Some generalized order-disorder transformations. *Proc. Camb. Phil. Soc.*, 48:106, 1952.
530. J. Prasad and R. Kopelman. Molecular reaction kinetics inside channel pores: delayed fluorescence of naphtalene in methanol. *Chem. Phys. Lett.*, 157:535, 1989.
531. W.H. Press, S.A. Teukolsky, W.T. Vetterling, and B.P. Flannery. *Numerical recipes (2nd edition).* Cambridge University Press, Cambridge, 1992.
532. V.B. Priezzhev. Structure of the two-dimensional sandpile. *J. Stat. Phys.*, 74:955, 1994.
533. V.B. Priezzhev, D.V. Ktitarev, and E.V. Ivashkevich. Formation of avalanches and critical exponents in an Abelian sandpile model. *Phys. Rev. Lett.*, 76:2093, 1994.
534. V. Privman and M.E. Fisher. Finite-size effects at first-order transitions. *J. Stat. Phys.*, 33:385, 1983.
535. V. Privman and M.E. Fisher. Universal critical amplitudes in finite-size scaling. *Phys. Rev.*, B30:322, 1984.
536. V. Privman, P.C. Hohenberg, and A. Aharony. Universal critical-point amplitude relations. In C. Domb and J.L. Lebowitz, editors, *Phase transitions and critical phenomena*, volume 14, London, 1991. Academic Press.
537. A. Jiménez Dalmaroni. Directed percolation with incubation times. *Phys. Rev.*, E74:011123, 2006.

538. A.P. Prudnikov, Yu.A. Brychkov, and O.I. Marichev. *Integrals and series, vol. 1.* Gordon and Breach, New York, 1986.
539. A.P. Prudnikov, Yu.A. Brychkov, and O.I. Marichev. *Integrals and series, vol. 2.* Gordon and Breach, New York, 1986.
540. A.P. Prudnikov, Yu.A. Brychkov, and O.I. Marichev. *Integrals and series, vol. 3.* Gordon and Breach, New York, 1990.
541. J.J. Ramasco, M.A. Muñoz, and C.A. da Silva Santos. Numerical study of the Langevin theory for fixed-energy sandpiles. *Phys. Rev.*, E69:045105, 2004.
542. M.A. Furman R.C. Brower and M. Moshe. Critical exponents for the Reggeon quantum spin model. *Phys. Lett.*, 76B:213, 1978.
543. S. Redner. *A Guide to First Passage Processes.* Cambridge University Press, Cambridge, U.K., 2001.
544. P.-A. Rey and M. Droz. A renormalization group study of a class of reaction-diffusion models, with particles input. *J. Phys. A Math. Gen.*, 30:1101, 1997.
545. F. Ricci-Tersenghi. Measuring the fluctuation-dissipation ratio in glassy systems with no perturbing field. *Phys. Rev.*, E68:065104(R), 2003.
546. E.K. Riedel and F.J. Wegner. Scaling approach to anisotropic magnetic systems statics. *Z. Phys.*, 225:195, 1969.
547. E.K. Riedel and F.J. Wegner. Tricritical exponents and scaling fields. *Phys. Rev. Lett.*, 29:349, 1972.
548. E.K. Riedel and F.J. Wegner. Effective critical and tricritical exponents. *Phys. Rev.*, B9:294, 1974.
549. U. Ritschel and P. Czerner. Near-surface long-range order at the ordinary transition. *Phys. Rev. Lett.*, 77:3645, 1996.
550. U. Ritschel and H.W. Diehl. Dynamical relaxation and universal short-time behaviour in finite systems: the renormalization-group approach. *Nucl. Phys.*, B464:512, 1996.
551. V. Riva and J.L. Cardy. Scale and conformal invariance in field theory: a physical counterexample. *Phys. Lett.*, B622:339, 2005.
552. M. Rossi, R. Pastor-Satorras, and A. Vespignani. The universality class of absorbing phase transition with a conserved field. *Phys. Rev. Lett.*, 85:1803, 2000.
553. L. Roters and K.D. Usadel. Dimensional crossover and driven interfaces in disordered ferromagnets. *Phys. Rev.*, E65:027101, 2002.
554. P. Rupp, R. Richter, and I. Rehberg. Critical exponents of directed percolation measured in spatio-temparal intermittency. *Phys. Rev.*, E67:036209, 2003.
555. M. Saharay and P. Sen. Persistence in an antiferromagnetic Ising system with conserved magnetisation. *Physica*, A318:243, 2003.
556. M. Scheucher and H. Spohn. A soluble kinetic model for spinodal decomposition. *J. Stat. Phys.*, 53:279, 1988.
557. B. Schmittmann and R.K.P. Zia. Statistical mechanics of driven diffusive systems. In C. Domb and J. L. Lebowitz, editors, *Phase transitions and critical phenomena*, volume 17, page 1, London, 2000. Academic Press.
558. J. Schnakenberg. Network theory of microscopic and macroscopic behaviour of master equation systems. *Rev. Mod. Phys.*, 48:571, 1976.
559. P. Schofield. Parametric representation of the equation of state near a critical point. *Phys. Rev. Lett.*, 22:606, 1969.
560. U. Schollwöck. The density-matrix renormalization-group. *Rev. Mod. Phys.*, 77:259, 2005.
561. U. Schollwöck. Time-dependent density-matrix renormalization-group methods. *J. Phys. Soc. Japan (Suppl.)*, 74:246, 2005.
562. L. Schülke and B. Zheng. Monte Carlo measurement of the global persistence exponent. *Phys. Lett.*, A233:93, 1997.
563. G.M. Schütz. Generalized Bethe ansatz solution to an asymmetric exclusion process on a ring with a defect. *J. Stat. Phys.*, 71:471, 1993.
564. G.M. Schütz. Reaction-diffusion processes of hard-core particles. *J. Stat. Phys.*, 79:243, 1995.

565. G.M. Schütz. Exactly solvable models for many-body systems far from equilibrium. In C. Domb and J. L. Lebowitz, editors, *Phase transitions and critical phenomena*, volume 19, page 1, London, 2000. Academic Press.
566. G.M. Schütz and E. Domany. Phase transitions in an exactly soluble one-dimensional exclusion process. *J. Stat. Phys.*, 72:277, 1993.
567. P. Sen and S. Dasgupta. Persistence and dynamics in the ANNNI chain. *J. Phys. A Math. Gen.*, 37:11949, 2004.
568. A. Levelt Sengers, R. Hocken, and J.V. Sengers. Critical-point universality and fluids. *Physics Today*, 28:42 (December), 1977.
569. J.M.H. Levelt Sengers. From van der Waals' equation to the scaling law. *Physica*, A73:73, 1974.
570. J.V. Sengers and J.M.H. Levelt Sengers. Critical phenomena in classical fluids. In C.A. Croxton, editor, *Progress in liquid physics*, New York, 1978. John Wiley&Sons.
571. T. Shimada, Y. Murase, S. Yukawa, N. Ito, and K. Aihara. A simple model of evolving ecosystems. *Artificial Life and Robotics*, 11:153, 2008.
572. C.S. Simões and J.R. Drugowich de Felício. Dynamic critical exponents of the Ising model with multispin interactions. *Mod. Phys. Lett.*, B15:487, 2001.
573. H. Simon. Concentration for one- and two-species one-dimensional reaction-diffusion systems. *J. Phys. A Math. Gen*, 28:6585, 1995.
574. F. Smallenburg and G.T. Barkema. Universality class of the pair-contact process with diffusion. `arXiv:0710.3692`, 2007.
575. J. Smoller. *Shock waves and reaction-diffusion equations (2nd edition).* Springer, Heidelberg, 1994.
576. H. Spohn. *Large-scale dynamics of interacting particles.* Springer, Heidelberg, 1991.
577. H.E. Stanley. Spherical model as the limit of infinite spin dimensionality. *Phys. Rev.*, 176:718, 1968.
578. D. Stauffer and A. Aharony. *Introduction to Percolation Theory.* Taylor & Francis, London, 1992.
579. W.-H. Steeb. *Chaos und Quantenchaos in dynamischen Systemen.* B.I. Wissenschaftsverlag, Mannheim, 1994.
580. O. Stenull and H. K. Janssen. Nonlinear random resistor diode networks and fractal dimensions of directed percolation clusters. *Phys. Rev.*, E64:016135, 2001.
581. O. Stenull and H. K. Janssen. Renormalized field theory of resistor diode percolation. *J. Stat. Phys.*, 104:1273, 2001.
582. J. Stephenson. Ising-model spin correlations on the triangular lattice. III. Isotropic antiferromagnetic lattice. *J. Math. Phys.*, 11:413, 1970.
583. J.F. Stilck, R. Dickman, and R.R. Vidigal. Series expansion for a stochastic sandpile. *J. Phys. A Math. Gen.*, 37:1145, 2004.
584. S.H. Strogatz. *Nonlinear dynamics and chaos.* Addison-Wesley, Reading, 1994.
585. L.C.E. Struik. *Physical ageing in amorphous polymers and other materials.* Elsevier, Amsterdam, 1978.
586. A. Szolnoki. Dynamical mean-field approximation for a pair-contact process with a particle source. *Phys. Rev.*, 66:057102, 2002.
587. F. Tabatabaee and A. Aghamohammadi. Cluster approximation solution of a two-species annihilation model. *Phys. Rev.*, E66:066136, 2003.
588. B. Tadić and D. Dhar. Emergent spatial structures in critical sandpiles. *Phys. Rev. Lett.*, 79:1519, 1997.
589. K.A. Takeuchi, M. Kuroda, H. Chaté, and M. Sano. Directed percolation criticality in turbulent liquid crystals. *Phys. Rev. Lett.*, 99:234503, 2007.
590. C. Tang and P. Bak. Mean-field theory of self-organized critical phenomena. *J. Stat. Phys.*, 51:797, 1988.
591. L.H. Tang and H. Leschhorn. Pinning by directed percolation. *Phys. Rev.*, A45:R8309, 1992.

592. U.C. Täuber. Field-theory approaches to nonequilibrium dynamics. In M. Henkel, M. Pleimling, and R. Sanctuary, editors, *Ageing and the glass transition*, Heidelberg, 2006. Springer.
593. U.C. Täuber, M. Howard, and B.P. Vollmayr-Lee. Applications of field-theoretic renormalization-group methods to reaction-diffusion problems. *J. Phys. A Math. Gen.*, 38:R79, 2005.
594. C. Tebaldi, M. De Menech, and A.L. Stella. Multifractal scaling in the Bak-Tang-Wiesenfeld Sandpile and edge events. *Phys. Rev. Lett.*, 83:3952, 1999.
595. N.S. Tonchev. Finite-size scaling in anisotropic systems. *Phys. Rev.*, E75:031110, 2007.
596. A. Toom. On critical phenomena in interacting growth systems. I. General. *J. Stat. Phys.*, 74:91, 1994.
597. A.L. Toom. Stable and attractive trajectories in multicomponent systems. In R.L. Dobrushin and Y.G. Sinai, editors, *Multicomponent Random Systems*, volume 6, page 549, New York, 1980. Marcel Dekker Ltd.
598. D. Toussaint and F. Wilczek. Particle-antiparticle annihilation in diffusive motion. *J. Chem. Phys.*, 78:2642, 1983.
599. L. Turban. Self-dual anisotropic two-dimensional Ising models with multispin interactions. *J. Physique Lettres*, 43:L259, 1982.
600. L. Turban. Self-dual Ising chain in a transverse field with multispin interactions. *J. Phys. C Solid State Phys.*, 15:L65, 1982.
601. D.L. Turcotte. Self-organized criticality. *Rep. Prog. Phys.*, 62:1377, 1999.
602. F. van Wijland. Universality class of nonequilibrium phase-transitions with infinitely many absorbing states. *Phys. Rev. Lett.*, 89:190602, 2002.
603. F. van Wijland, K. Oerding, and H.J. Hilhorst. Wilson renormalization of a reaction–diffusion process. *Physica*, A251:179, 1998.
604. C. Vanderzande. *Lattice models of polymers.* Cambridge University Press, Cambridge, 1998.
605. M. Vergeles, A. Maritan, and J. Banavar. Mean-field theory of sandpiles. *Phys. Rev.*, E55:1998, 1997.
606. D. Vernon and M. Howard. Branching and annihilating Levy flights. *Phys. Rev.*, E63:041116, 2001.
607. A. Vespignani, R. Dickman, M.A. Muñoz, and S. Zapperi. Driving, conservation, and absorbing states in sandpiles. *Phys. Rev. Lett.*, 81:5676, 1998.
608. A. Vespignani, R. Dickman, M.A. Muñoz, and S. Zapperi. Absorbing-state phase-transition in fixed-energy sandpiles. *Phys. Rev.*, E62:4564, 2000.
609. A. Vespignani and S. Zapperi. Order parameter and scaling fields in self-organized criticality. *Phys. Rev. Lett.*, 78:4793, 1997.
610. A. Vespignani and S. Zapperi. How self-organized criticality works: A unified mean-field picture. *Phys. Rev.*, E57:6345, 1998.
611. J. Villain. Spin glass with non-random interactions. *J. Phys. C: Solid State Phys.*, 10:1717, 1977.
612. C.A. Voigt and R.M. Ziff. Epidemic analysis of the second-order transition in the Ziff-Gulari-Barshad surface-reaction model. *Phys. Rev.*, E56:R6241, 1997.
613. T. Vojta. Broadening of a nonequilibrium phase transition by extended structural defects. *Phys. Rev.*, E70:026108, 2004.
614. T. Vojta and M. Dickison. Critical behavior and Griffiths effects in the disordered contact process. *Phys. Rev.*, E72:036126, 2005.
615. T. Vojta and M. Y. Lee. Nonequilibrium phase transition on a randomly diluted lattice. *Phys. Rev. Lett.*, 96:035701, 2006.
616. G.H. Wannier. Antiferromagnetism: The triangular Ising net. *Phys. Rev.*, 79:357, 1950.
617. G.H. Wannier. Antiferromagnetism: The triangular Ising net. *Phys. Rev.*, B7:5017, 1973.

618. P.G. Watson. Formation of invariants from critical amplitudes of ferromagnets. *J. Phys. C: Solid State Phys.*, 2:1883, 1969.
619. I. Webman, D. ben Avraham, A. Cohen, and S. Havlin. Dynamical phase transitions in a random environment. *Phil. Mag.*, B77:1401, 1998.
620. F.J. Wegner. Corrections to scaling laws. *Phys. Rev.*, B5:4529, 1972.
621. F.J. Wegner. The critical state, general aspects. In C. Domb and M.S. Green, editors, *Phase transitions and critical phenomena*, volume 6, London, 1976. Academic Press.
622. F.J. Wegner and E.K. Riedel. Logarithmic corrections to the molecular-field behavior of critical and tricritical systems. *Phys. Rev.*, B7:248, 1973.
623. M. Weigel and W. Janke. Monte Carlo study of the scaling of universal correlation lengths in three-dimensional $O(n)$ spin models. *Phys. Rev.*, B62:6343, 2000.
624. E.J. Weniger. Nonlinear sequence-transformations for the accelarationof convergence and the summation of divergent series. *Comp. Phys. Rep.*, 10:194, 1989.
625. S.R. White. Density-matrix formulation for quantum renormalization-groups. *Phys. Rev. Lett.*, 69:2863, 1992.
626. S.R. White. Density-matrix algorithms for quantum renormalization-groups. *Phys. Rev.*, B48:10345, 1993.
627. B. Widom. Equation of state in the neighborhood of the critical point. *J. Chem. Phys.*, 43:3898, 1965.
628. H. Willaime, O. Cardoso, and P. Tabeling. Spatiotemporal intermittency in lines of vortices. *Phys. Rev.*, E48:288, 1993.
629. K.G. Wilson. Renormalization group and critical phenomena. I. Renormalization group and the Kadanoff scaling picture. *Phys. Rev.*, B4:3174, 1971.
630. K.G. Wilson. Renormalization group and critical phenomena. II. Phase-space cell analysis of critical behavior. *Phys. Rev.*, B4:3184, 1971.
631. K.G. Wilson and M.E. Fisher. Critical exponents in 3.99 dimensions. *Phys. Rev. Lett.*, 28:240, 1972.
632. K.G. Wilson and J. Kogut. The renormalization group and the ϵ expansion. *Phys. Rep.*, 12C:75, 1974.
633. J. Wintterlin, S. Völkening, T.V.W. Janssens, T. Zambelli, and G. Ertl. Identification of the "Active Sites" of a Surface Catalyzed Reaction. *Science*, 273:1688, 1996.
634. S. Wolfram. Statistical mechanics of cellular automata. *Rev. Mod. Phys.*, 55:601, 1983.
635. F.Y. Wu. The Potts model. *Rev. Mod. Phys.*, 54:235, 1982.
636. F.Y. Wu. erratum. *Rev. Mod. Phys.*, 55:315, 1983.
637. J. Wu and D.C. Mattis. Partially frustrated Ising models in two dimensions. *Phys. Rev.*, B67:224414, 2003.
638. C.N. Yang and T.D. Lee. Statistical theory of equations of state and phase transitions: 2. Lattice gas and Ising model. *Phys. Rev.*, 87:420, 1952.
639. C.N. Yang and T.D. Lee. Statistical theory of equations of state and phase transitions: I. Theory of condensation. *Phys. Rev.*, 87:404, 1952.
640. J.M. Yeomans. *Statistical mechanics of phase transitions.* Oxford University Press, Oxford, 1992.
641. S. Y. Yoon, S. Kwon, and Y. Kim. Non-local pair-contact process with two offsprings. *J. Kor. Phys. Soc.*, 46:1392, 2005.
642. S. Zapperi, K.B. Lauritsen, and H.E. Stanley. Self-organized branching process: mean-field theory for avalanches. *Phys. Rev. Lett.*, 75:4071, 1995.
643. S.-C. Zhang. A unified theory based on SO(5) symmetry of superconductivity and antiferromagnetism. *Science*, 275:1089, 1997.
644. D. Zhong and D. ben Avraham. Universality class of two-offspring branching annihilating random walks. *Phys. Lett.*, A209:333, 1995.
645. J. Zhuo, S. Redner, and H. Park. Critical behaviour of an interacting surface-reaction model. *J. Phys. A Math. Gen.*, 26:4197, 1993.

646. R.M. Ziff and B.J. Brosilow. Investigation of the first-order phase-transition in the $A - B2$ reaction model using a constant-coverage kinetic ensemble. *Phys. Rev.*, A46:4630, 1992.
647. R.M. Ziff, E. Gulari, and Y. Barshad. Kinetic phase-transitions in an irreversible surface-reaction model. *Phys. Rev. Lett.*, 56:2553, 1986.
648. J. Zinn-Justin. *Quantum field-theory and critical phenomena.* Oxford University Press, Oxford, 1989.

List of Tables

List of Figures

Index

Theoretical and Mathematical Physics

Non-Equilibrium Phase Transitions
Volume I: Absorbing Phase Transitions
By M. Henkel, H. Hinrichsen and S. Lübeck

Quantum Probability and Spectral Analysis of Graphs
By A. Hora and N. Obata

From Nucleons to Nucleus
Concepts of Microscopic Nuclear Theory
By J. Suhonen

Concepts and Results in Chaotic Dynamics: A Short Course
By P. Collet and J.-P. Eckmann

The Theory of Quark and Gluon Interactions
4th Edition
By F. J. Ynduráin

From Microphysics to Macrophysics
Methods and Applications of Statistical Physics
Volume I
By R. Balian

From Microphysics to Macrophysics
Methods and Applications of Statistical Physics
Volume II
By R. Balian

Statistical Methods in Quantum Optics 2
Non-Classical Fields
By H. J. Carmichael

Titles published before 2006 in *Texts and Monographs in Physics*

The Statistical Mechanics of Financial Markets
3rd Edition
By J. Voit

Magnetic Monopoles
By Y. Shnir

Coherent Dynamics of Complex Quantum Systems
By V.M. Akulin

Geometric Optics on Phase Space
By K. B. Wolf

General Relativity
By N. Straumann

Quantum Entropy and Its Use
By M. Ohya and D. Petz

Statistical Methods in Quantum Optics 1
By H. J. Carmichael

Operator Algebras and Quantum Statistical Mechanics 1
By O. Bratteli and D.W. Robinson

Operator Algebras and Quantum Statistical Mechanics 2
By O. Bratteli and D.W. Robinson

Aspects of Ergodic, Qualitative and Statistical Theory of Motion
By G. Gallavotti, F. Bonetto and G. Gentile

The Frenkel-Kontorova Model
Concepts, Methods, and Applications
By O. M. Braun and Y. S. Kivshar

The Atomic Nucleus as a Relativistic System
By L. N. Savushkin and H. Toki

The Geometric Phase in Quantum Systems
Foundations, Mathematical Concepts, and Applications in Molecular and Condensed Matter Physics
By A. Bohm, A. Mostafazadeh, H. Koizumi, Q. Niu and J. Zwanziger

Relativistic Quantum Mechanics
2nd Edition
By H. M. Pilkuhn

Physics of Neutrinos and Applications to Astrophysics
By M. Fukugita and T. Yanagida

High-Energy Particle Diffraction
By E. Barone and V. Predazzi

Foundations of Fluid Dynamics
By G. Gallavotti

Many-Body Problems and Quantum Field Theory An Introduction
2nd Edition
By Ph.A. Martin, F. Rothen, S. Goldfarb and S. Leach

Statistical Physics of Fluids
Basic Concepts and Applications
By V. I. Kalikmanov

Statistical Mechanics A Short Treatise
By G. Gallavotti

Quantum Non-linear Sigma Models
From Quantum Field Theory
to Supersymmetry, Conformal Field Theory,
Black Holes and Strings
By S.V. Ketov

Perturbative Quantum Electrodynamics and Axiomatic Field Theory
By O. Steinmann

The Nuclear Many-Body Problem
By P. Ring and P. Schuck

Magnetism and Superconductivity
By L.-P. Lévy

Information Theory and Quantum Physics
Physical Foundations for Understanding
the Conscious Process
By H. S. Green

Quantum Field Theory in Strongly Correlated Electronic Systems
By N. Nagaosa

Quantum Field Theory in Condensed Matter Physics
By N. Nagaosa

Conformal Invariance and Critical Phenomena
By M. Henkel

Statistical Mechanics of Lattice Systems
Volume 1: Closed-Form and Exact Solutions
2nd Edition
By D. A. Lavis and G. M. Bell

Statistical Mechanics of Lattice Systems
Volume 2: Exact, Series
and Renormalization Group Methods
By D. A. Lavis and G. M. Bell

Fields, Symmetries, and Quarks
2nd Edition
By U. Mosel

Renormalization An Introduction
By M. Salmhofer

Multi-Hamiltonian Theory of Dynamical Systems
By M. Błaszak

Quantum Groups and Their Representations
By A. Klimyk and K. Schmüdgen

Quantum The Quantum Theory of Particles, Fields, and Cosmology
By E. Elbaz

Effective Lagrangians for the Standard Model
By A. Dobado, A. Gómez-Nicola,
A. L. Maroto and J.R. Peláez

Scattering Theory of Classical and Quantum *N*-Particle Systems
By. J. Derezinski and C. Gérard

Quantum Relativity A Synthesis
of the Ideas of Einstein and Heisenberg
By D. R. Finkelstein

The Mechanics and Thermodynamics of Continuous Media
By M. Šilhavý

Local Quantum Physics Fields, Particles, Algebras
2nd Edition
By R. Haag

Relativistic Quantum Mechanics and Introduction to Field Theory
By F. J. Ynduráin

Supersymmetric Methods in Quantum and Statistical Physics
By G. Junker

Path Integral Approach to Quantum Physics An Introduction
2nd printing
By G. Roepstorff

Finite Quantum Electrodynamics
The Causal Approach
2nd edition
By G. Scharf

From Electrostatics to Optics
A Concise Electrodynamics Course
By G. Scharf

Geometry of the Standard Model of Elementary Particles
By A. Derdzinski

Quantum Mechanics II
By A. Galindo and P. Pascual

MIX
Papier aus verantwortungsvollen Quellen
Paper from responsible sources
FSC® C105338

FSC
www.fsc.org

If you have any concerns about our products,
you can contact us on
ProductSafety@springernature.com

In case Publisher is established outside the EU,
the EU authorized representative is:
Springer Nature Customer Service Center GmbH
Europaplatz 3, 69115 Heidelberg, Germany

Printed by Libri Plureos GmbH
in Hamburg, Germany